Sedimentology (2000) **47** (Suppl. 1), 1

EDITORIAL

Millennium Reviews

Millennia come and millennia go. Perhaps it is in the Earth sciences that millennial markers find their true insignificance, where they pale in comparison with the vastness of geologic time. Indeed, even in human terms, millennial markers are arbitrary: what about year zero, and should we not celebrate the start of the new millennium at the beginning of 2001? To what calendar should we refer? The year 2000 is 6236 according to the first Egyptian calendar and 1420 according to the Moslem calendar. Even the current calendar is probably wrong with respect to some estimated dates of the birth of Christ – the third millennium perhaps should have started in 1997. All good reasons then, not to mark the start of 2000 with a special Millennium Supplement to *Sedimentology*. So why this issue?

Sedimentology has seen many significant advances and changes over the past 40 years, ranging from facies modelling to sequence stratigraphy; chemostratigraphy to basin analysis; and the integration of studies of physical, chemical and, increasingly, biological processes in the interpretation and prediction of sedimentary environments and products. The subject is becoming ever more interdisciplinary and applied, and now has far more links to other physical sciences. Research and debate are continuing afresh as we move into this new interdisciplinary phase, and promise many developments and increased uses of our subject. Now, with the switch to a new century and millennium on the front cover of the journal, seemed a good time to publish a series of review papers concerning some key current areas of research. We hope that these will provide comprehensive starting points for those wishing to become acquainted with an area, act as stimuli for debate and, hopefully, provide awareness and ideas for future research avenues. No issue of this sort can, of course, ever be truly comprehensive in its coverage: these reviews concern only selected snippets from the wide scope of sedimentology, and each has, of necessity, been selective in its own area. We are grateful to all our colleagues who contributed to discussions concerning areas 'ripe for review' and also suggested possible authors.

Additionally, the International Association of Sedimentologists (IAS) has the central remit to 'promote the study of sedimentology and the interchange of research, particularly where international cooperation is desirable'. The IAS seeks to achieve this through the publication of *Sedimentology*, the organization and sponsorship of conferences, meetings and workshops, and the sponsorship of young scientists and of those from developing countries. Therefore, the IAS strongly feels that this supplement, provided free to all subscribers, further contributes to its key aims and the development of the subject as a whole. It marks the continuing commitment of the IAS to encouraging the best research in our science.

In assembling this supplement, we are indebted to many people. We sincerely thank the reviewers of these papers for ensuring that the rigour and quality of the review process were the same as for any other paper submitted to *Sedimentology*. In addition, we are grateful to the authors for the time and great effort that they devoted to completing these papers on schedule so that this supplement could be mailed with the first regular issue of 2000. We are very much aware of the considerable time and sacrifices writing these papers demanded, and hope that this will be repaid by both their use and the debate they may stimulate.

We hope this supplement will be enjoyable, stimulating, useful and sometimes controversial. This is what good reviews should do.

Jim Best, Chris Fielding, Ian Jarvis and Peter Mozley

Sedimentology (2000), **47** (Suppl. 1), 2–48

Fluvial responses to climate and sea-level change: a review and look forward

MICHAEL D. BLUM* and TORBJÖRN E. TÖRNQVIST†
**Department of Geosciences, University of Nebraska – Lincoln, 214 Bessey Hall, Lincoln, NE 68588-0340, USA (E-mail: mblum1@unl.edu)*
†Department of Earth and Environmental Sciences, University of Illinois at Chicago, 845 West Taylor Street, Chicago, IL 60607-7059, USA (E-mail: tor@uic.edu)

ABSTRACT

Fluvial landforms and deposits provide one of the most readily studied Quaternary continental records, and alluvial strata represent an important component in most ancient continental interior and continental margin successions. Moreover, studies of the long-term dynamics of fluvial systems and their responses to external or 'allogenic' controls, can play important roles in research concerning both global change and sequence-stratigraphy, as well as in studies of the dynamic interactions between tectonic activity and surface processes. These themes were energized in the final decades of the twentieth century, and may become increasingly important in the first decades of this millennium.

This review paper provides a historical perspective on the development of ideas in the fields of geomorphology/Quaternary geology vs. sedimentary geology, and then summarizes key processes that operate to produce alluvial stratigraphic records over time-scales of 10^3–10^6 years. Of particular interest are changes in discharge regimes, sediment supply and sediment storage en route from source terrains to sedimentary basins, as well as changes in sea-level and the concept of accommodation. Late Quaternary stratigraphic records from the Loire (France), Mississippi (USA), Colorado (Texas, USA) and Rhine–Meuse (The Netherlands) Rivers are used to illustrate the influences of climate change on continental interior rivers, as well as the influence of interacting climate and sea-level change on continental margin systems.

The paper concludes with a look forward to a bright future for studies of fluvial response to climate and sea-level change. At present, empirical field-based research on fluvial response to climate and sea-level change lags behind: (a) the global change community's understanding of the magnitude and frequency of climate and sea-level change; (b) the sequence-stratigraphic community's desire to interpret climate and, especially, sea-level change as forcing mechanisms; and (c) the modelling community's ability to generate numerical and physical models of surface processes and their stratigraphic results. A major challenge for the future is to catch up, which will require the development of more detailed and sophisticated Quaternary stratigraphic, sedimentological and geochronological frameworks in a variety of continental interior and continental margin settings. There is a particular need for studies that seek to document fluvial responses to allogenic forcing over both shorter (10^2–10^3 years) and longer (10^4–10^6 years) time-scales than has commonly been the case to date, as well as in larger river systems, from source to sink. Studies of Quaternary systems in depositional basin settings are especially critical because they can provide realistic analogues for interpretation of the pre-Quaternary rock record.

Keywords Climate change, fluvial sedimentology, Quaternary geology, sea-level change, sequence stratigraphy.

INTRODUCTION

Studies of continental stratigraphic records, Quaternary and ancient, were energized in the final decades of the twentieth century by two somewhat distinct trends that emerged within the geosciences. First, there was increasingly widespread recognition that human activities were altering climatic and environmental conditions over very short time-scales. Since the stratigraphic record is the only empirical record of global change (Burke *et al.*, 1990), studies of the magnitude and frequency of past global changes, Quaternary and ancient, were seen as a means to provide a context for observed historical trends and predicted near-future conditions. Second, development of sequence-stratigraphic concepts and methods (e.g. Vail *et al.*, 1977; Jervey, 1988; Posamentier & Vail, 1988; Posamentier *et al.*, 1988) provided a potential unifying framework for much of sedimentary geology, especially coastal and marine strata, but their application to continental deposits has been controversial.

At a basic level, global change and sequence-stratigraphic research have in common a focus on changes in process through time in response to external forcing, and it is hard to envisage any scenario for the first decades of this millennium in which these issues will be less important. In addition, an important third trend has begun to emerge, as there is now a deeper appreciation of the magnitude and frequency of tectonic activity in plate boundary, continental interior and passive margin settings, and relationships between tectonic activity and surface processes (e.g. Molnar & England, 1990; Molnar *et al.*, 1993; Merritts & Ellis, 1994; Pinter & Brandon, 1997; Burbank & Pinter, 1999). This third trend will further energize studies of continental records, since integrative models of this kind must be tested against, and refined by, studies of Quaternary and ancient strata.

Fluvial landforms and deposits provide one of the most readily studied Quaternary continental records, and alluvial strata represent an important component in most ancient continental interior and continental margin successions. Moreover, modern fluvial environments are critical to human activities, and ancient fluvial deposits are important repositories for hydrocarbons, groundwater and other resources. Studies of the long-term dynamics of fluvial systems, and their responses to external or 'allogenic' controls, therefore have a broad range of applications. This paper provides an overview of fluvial responses to climate and sea-level change, and attempts to: (a) develop a historical perspective; (b) outline a basic process framework, with illustrations of fluvial responses to climate and sea-level change from Late Quaternary contexts; and (c) suggest key areas of emphasis for the first decades of this millennium.

HISTORICAL STARTING POINTS

Studies of fluvial responses to climate and sea-level changes have evolved somewhat differently within the fields of geomorphology and Quaternary geology vs. sedimentary geology.

Geomorphology and Quaternary geology

Several concepts from the late 1800s and early 1900s have played key roles in studies of fluvial response to climate and sea-level change. One of these was base level, defined by Powell (1875) as the lower limit to which rivers can erode their valleys, with his 'grand' or ultimate base level corresponding to sea-level. A second concept was that of the graded stream, which Gilbert (1877) suggested was adjusted in terms of channel slope, or longitudinal profile, so that the available discharge could transport the amount of sediment supplied by the drainage network, and the channel bed was neither aggrading nor degrading. Mackin's (1948) discussion of the graded stream is better known and more detailed, and also focused on equilibrium long profiles that are adjusted to prevailing conditions of discharge and sediment load.

One of the first empirical studies of fluvial response to climate change was by Penck & Brückner (1909), whose examination of terraces along tributaries to the Danube in southern Germany gave birth to the concept of four Pleistocene glacial–interglacial cycles. This model for Quaternary climate change remained dominant until the development of oxygen-isotope records from ocean basins in the 1960s and 1970s, and confirmation of the Milankovitch theory of glaciation (Hays *et al.*, 1976). Critical to the present discussion, Penck & Brückner (1909) linked aggradation to glacial-period sediment supply, and incision with floodplain abandonment and terrace formation to interglacial periods (Fig. 1A). By contrast, Lamothe (1918) in the Somme Valley of France, and Fisk (1944) in the Lower Mississippi Valley of the USA, focused on the role of glacio-eustasy. Fisk's (1944) well-known model suggested valley incision, floodplain abandonment and terrace formation during sea-level fall and lowstand, followed by valley aggradation during transgression and highstand (Fig. 1B,C).

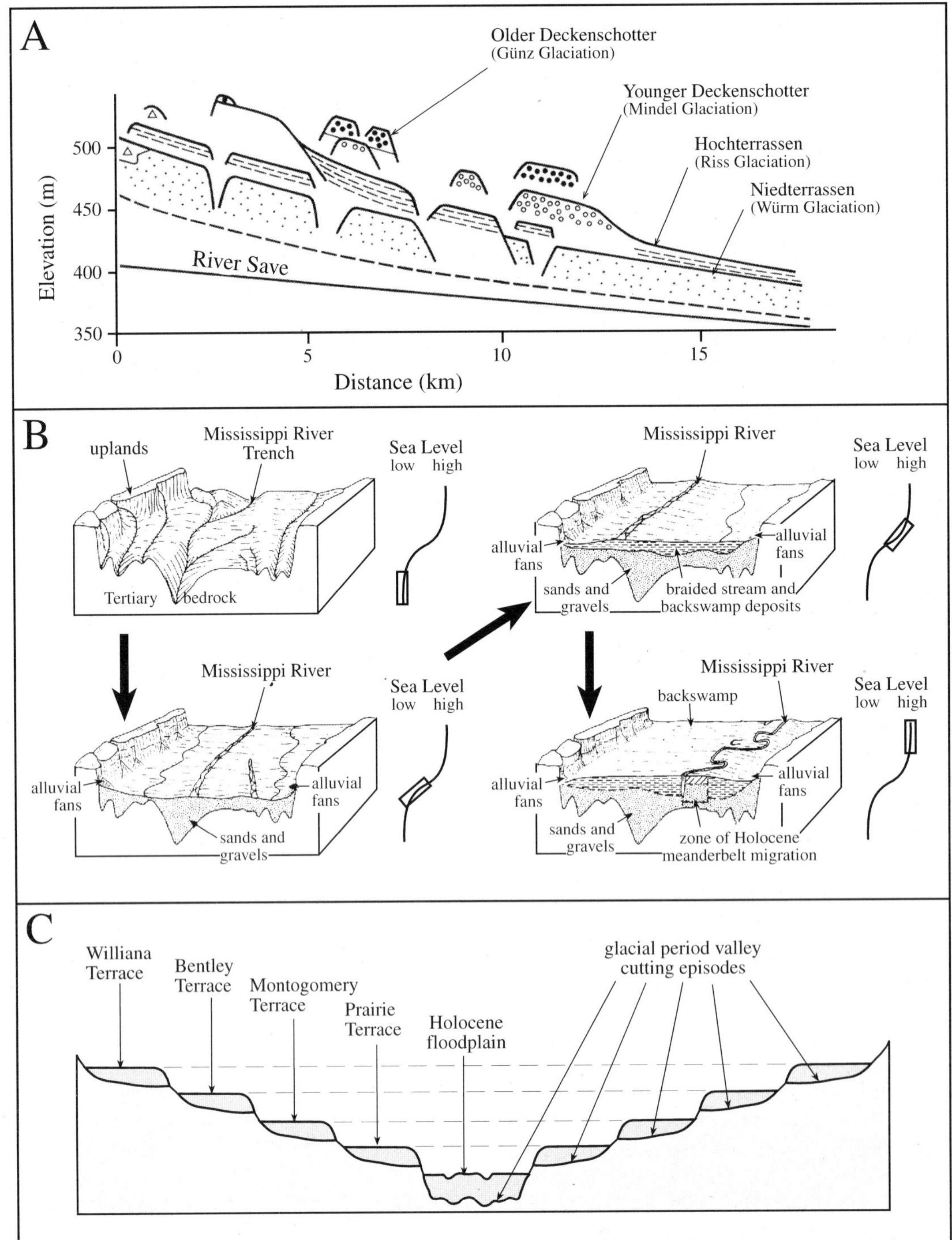

Fig. 1. (A) Penck & Brückner's (1909) model for terraces in southern Germany, linking aggradation to glacial cycles, and incision to interglacials (adapted from Bowen, 1978; original from Penck & Brückner, 1909). (B) Fisk's (1944) model for valley development during an individual eustatic cycle, illustrating cutting during sea-level fall, and the different stages of filling during sea-level rise and highstand. (C) Fisk's model for terraces of the Gulf Coastal Plain in Louisiana (USA), linking incision to glacial cycles, and aggradation to interglacials (B and C modified from Fisk, 1944).

This 'continental interior' vs. 'continental margin' dichotomy, or a focus on upstream vs. downstream controls depending on study area location, was still prevalent in the last decades of the twentieth century.

Upstream controls in continental interiors

As described by Butzer (1980), a 'glacial' vs. 'interglacial' dichotomy, following Penck & Brückner (1909), also persisted for some time, and it was commonly assumed that present floodplains represented the Holocene, the first terrace was from the last Pleistocene glacial, the second terrace was from the penultimate glacial, and so on. Clear turning points can be traced to the development of radiocarbon dating (Libby, 1952) and to the hydraulic geometry studies of Leopold & Maddock (1953). Radiocarbon dating techniques provided an opportunity to develop chronological frameworks for climate change from palaeobiological and other indicators, and, independently, for alluvial successions. Moreover, the effective limits of radiocarbon eventually served to focus attention on the last full-glacial and the Holocene (last 20 kyr). Leopold & Maddock (1953), by contrast, initiated a generation of efforts that developed quantitative relations between discharge regimes, sediment load and measures of channel geometry (e.g. width, depth, sinuosity, meander wavelength, slope).

Several papers from the 1960s illustrate the immediate impact of hydraulic geometry. Among these are Schumm's (1965) general model for fluvial response to climatic change, Dury's (1965) studies of 'underfit' streams and Schumm's (1968a) discussion of the concept of channel metamorphosis, based on studies of the Murrumbidgee River, south-eastern Australia (Fig. 2). A subsequent series of influential papers by Schumm and colleagues in the 1970s and early 1980s served to caution against developing general models for fluvial response to climate change. In concise summaries of this work, Schumm (1977, 1991) notes the importance of: (a) spatial and temporal scale; (b) convergence of responses given different external forcing mechanisms; (c) divergence of response following similar external inputs; (d) differential sensitivity of fluvial systems to external controls due to differing thresholds for change; and (e) complex response due to internal system dynamics. These issues suggest that fluvial responses to climate change may be geographically circumscribed, nondeterministic and nonlinear.

Many of these ideas are reflected in review papers by Knox (1983, 1996) and Starkel (1987, 1991), and Bull's (1991) monograph. Knox (1983), for example, summarized the Holocene record from the USA, and argued that climate change produces time-parallel discontinuities in alluvial successions over broad areas, even though the direction of change may differ between regions. He suggested this could be due to the way changes in global atmospheric circulation are manifested regionally, and/or the intrinsic characteristics of each system. Starkel (1987, 1991) also recognized the variability of fluvial response to climate change over the last 15 kyr, and differentiated classes of European rivers on the basis of relationships between tectonic highlands, Pleistocene ice extent, the former periglacial region, isostatic rebound and sea-level change. Starkel's papers emanated from international projects that demonstrate increased interest in fluvial systems and climate change during the last two decades (Gregory, 1983; Gregory *et al.*, 1987,

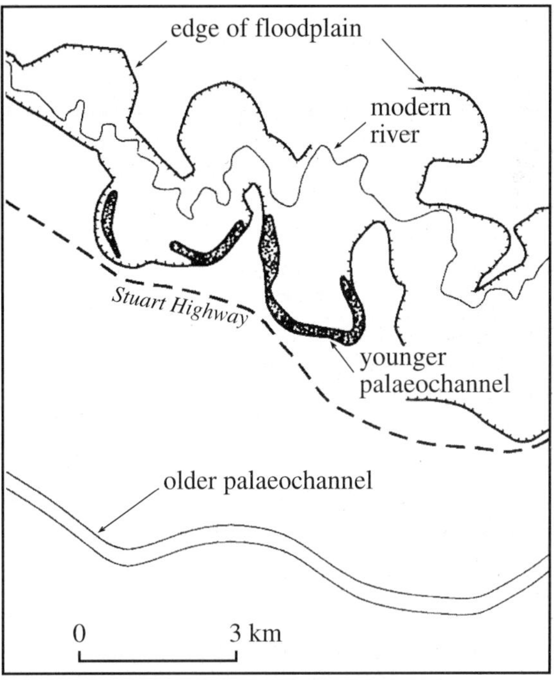

Fig. 2. Concept of channel metamorphosis, introduced by Schumm (1968a) based on studies of the Murrumbidgee River, Australia. Map shows modern highly sinuous Murrumbidgee River, vs. older channel with larger meanders, and an even older low-sinuosity channel (adapted from Schumm & Brakenridge, 1987; original from Schumm, 1968a).

1996; Starkel *et al.*, 1991; Branson *et al.*, 1996; Benito *et al.*, 1998).

Downstream controls in continental margins

Fisk's (1944) investigations of the Lower Mississippi Valley coupled valley evolution to glacio-eustasy and changes in stream gradients, and played a fundamental role in development of models for fluvial response to sea-level change. According to Fisk, the valley was deeply incised and swept clean of sediments to a point some 1000 km upstream from the present Gulf of Mexico shoreline due to increased channel gradients during the Last Glacial Maximum sea-level lowstand. Braided-stream surfaces formed when gradients decreased during latest Pleistocene to middle Holocene sea-level rise, with a transition to a meandering regime during the late Holocene sea-level highstand. Fisk's model was so influential that many workers assigned successively older terraces in the lower reaches of river valleys to successively older interglacials. Extreme examples of this type of thinking can be found in studies along the Gulf Coast of the USA that inferred short-lived periods of sea-level rise for terraces that did not fit into the standard model of four glacial–interglacial cycles (e.g. Bernard & Leblanc, 1965).

Further empirical studies of continental margin systems have been less visible, perhaps because of the perceived power of Fisk's model. It is therefore appropriate that the Mississippi has played a role in reassessment of fluvial responses to sea-level change. Saucier's (1994, 1996) work, for example, shows that the Mississippi was never completely excavated, and did not feel the effects of sea-level change as far upstream as Fisk had envisaged. Saucier also assigned braided-stream surfaces to the last glacial period, and suggested they record fluvial responses to glacially induced changes in discharge and sediment load, whereas the transition to a meandering regime occurred in the earliest Holocene with the loss of a glacial source for sediment and water.

A variety of conceptual, experimental and theoretical models of river response to base-level change appeared in the 1970s and 1980s. Leopold & Bull (1979), for example, used observations from modern rivers to suggest the upstream effects of base-level rise were limited, whereas Begin *et al.* (1981) and Begin (1988) used flume studies coupled with a diffusion-based model to show that vertical lowering of base-level (as along a fault block) results in long-profile lowering by nickpoint migration at rates that diminish in the upstream direction. Slingerland & Snow (1988) developed a numerical sediment-transport model to simulate long-profile evolution in response to base-level fall, and suggested a series of complex autogenic oscillations of cutting and filling might result from a single allogenic perturbation. At a broader conceptual level, Summerfield (1985) suggested the influence of sea-level fall will vary with geometric relationships between the slope of the coastal plain and the slope of the shelf as it becomes emergent (Fig. 3). For example, Brown & Wilson (1988) and Leckie (1994) discuss the Canterbury Plains of New Zealand, where aggradation occurred during glacial periods because of newly exposed shelf gradients that were less

A Slope of Coastal Plain < Slope of Shelf

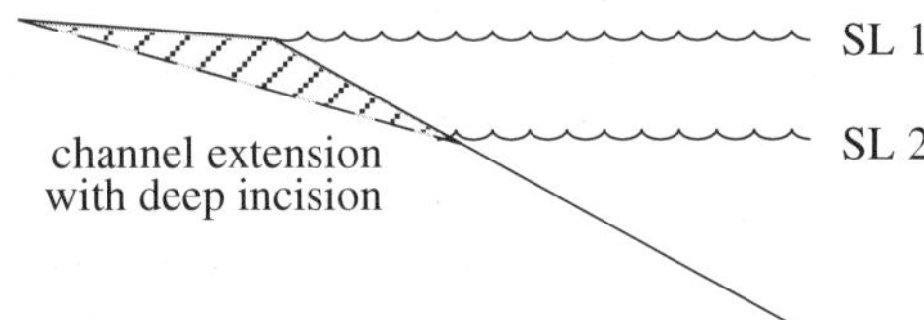

B Slope of Coastal Plain > Slope of Shelf

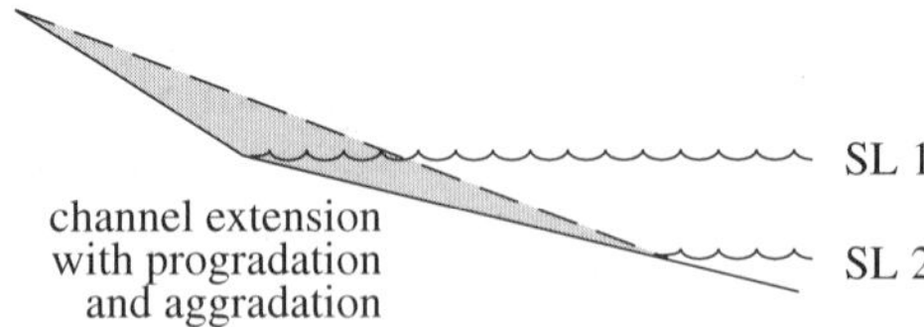

C Slope of Coastal Plain = Slope of Shelf

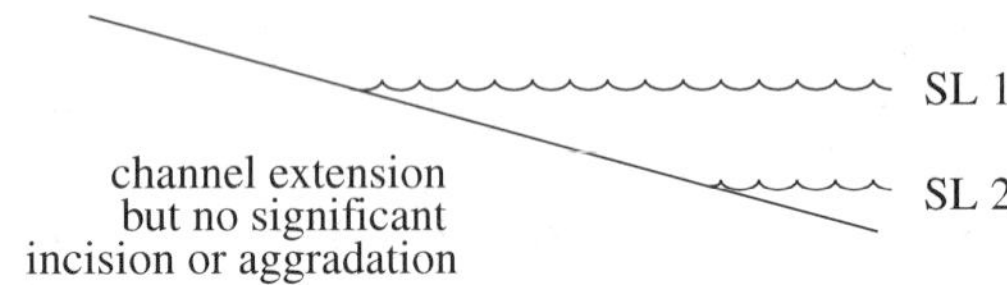

Fig. 3. Model for fluvial response to sea-level fall as a function of coastal plain and shelf gradients. (A) Incision through coastal prism (depositional shoreline break) due to steeper gradients exposed on the inner shelf as sea-level falls; (B) aggradation and progradation due to sea-level fall across a shelf that has a gradient shallower than the coastal plain; (C) channel extension with little incision, other than that necessary to contain the channel itself, due to no gradient differences between coastal plain and exposed shelf during sea-level fall. Adapted from Summerfield (1985).

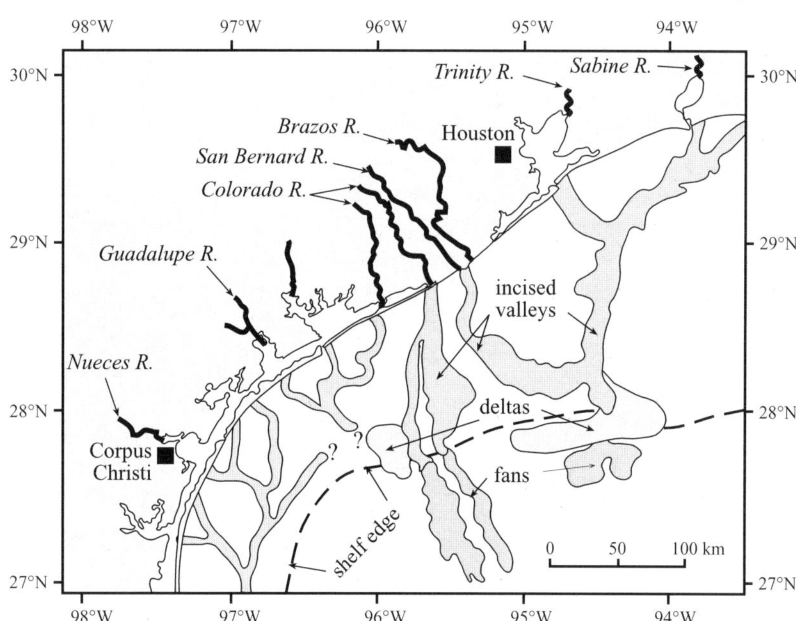

Fig. 4. Valleys and deltas of the north-west Gulf of Mexico shelf during the last full-glacial period, as identified in high-resolution seismic data (adapted from Suter & Berryhill, 1985; Anderson *et al.*, 1996).

steep than the permanently subaerial alluvial plains.

Most recently, workers have examined interactions between upstream climatic and downstream sea-level controls in the lowermost reaches of coastal plain rivers. As discussed later, these efforts have concentrated on rivers of the Texas Gulf Coastal Plain (USA) and on the Rhine–Meuse of The Netherlands. The 1980s and 1990s also witnessed the emergence of high-resolution seismic-stratigraphic studies on continental shelves. Suter & Berryhill (1985), for example, demonstrated channel extension and delta progradation across the Gulf of Mexico shelf during the last sea-level fall and lowstand. Subsequent work by others continued to document the extent and variability of fluvial channels and deltaic strata on the shelves and at shelf edges (Fig. 4). Important examples include work by Morton & Suter (1996) and Anderson *et al.* (1996) for the north-western Gulf of Mexico (USA), and Tesson *et al.* (1993; also Tesson & Gensous, 1998) for the Rhone shelf of the Meditteranean.

Sedimentary geology

Eustatic sea-level change played an important role in explanations of cyclicity in pre-Quaternary sedimentary rocks during the first two-thirds of the twentieth century (e.g. Wanless & Weller, 1932; Sloss, 1963; Wheeler, 1964). However, a period of waning enthusiasm for sea-level change accompanied development of plate tectonics on one hand, and facies models on the other (Dott, 1992), such that studies in the 1960s and 1970s commonly focused on tectonic activity coupled with autogenic processes to explain stratal patterns. Sedimentary geologists have also long recognized the importance of climate in the production of sediments, as well as the range of depositional systems, environments and facies, but attempts to explain changes in alluvial successions through time with reference to climate change are a recent innovation.

There is little question that plate tectonics plays the overriding role in organization of landmasses into sediment source regions, transport routes and sedimentary basins, and is thus primarily responsible for the accumulation of thick successions of alluvial strata. The following discussion refers readers to Miall (1996) for a comprehensive discussion of the role of tectonic activity in the development of alluvial successions, and instead highlights alternative trends from the last three decades. Of greatest significance is the emergence of sequence-stratigraphic models and their emphasis on the interwoven concepts of sea-level, base level and accommodation. Most recently, workers have begun to examine links between alluvial successions and climate change.

Sea-level, base level and accommodation

Development of sequence-stratigraphic concepts, beginning with publications of Vail and colleagues from Exxon Production Research in *AAPG Memoir 26* (Payton, 1977) initiated a wave of research into the responses of depositional systems to sea-level change. Publication of the first Exxon cycle charts (Vail *et al.*, 1977) gave rise to new views on the pervasiveness of sea-level changes through geological time.

The origins of this wave of research, as it pertains to 'fluvial sequence stratigraphy', lagged behind development of the general model, and can be traced to *SEPM Special Publication 42* (Wilgus *et al.*, 1988), which contained two papers that catalysed discussion over the next decade. The first was by Jervey (1988) which introduced the term 'accommodation' as the space available below base level for sediments to accumulate, a function of eustatic sea-level change and subsidence (i.e. relative sea-level change). The second was by Posamentier & Vail (1988), in which fluvial response to sea-level change played a prominent role in the development of conceptual models for 'systems tracts', the basic building blocks in the Exxon model. Their process framework rests on the premise that rivers adjust longitudinal profiles in response to sea-level change, and they argued that: (a) widespread fluvial deposition characterizes the early stages of relative sea-level fall; (b) valley incision and sediment bypass characterizes the late stages of sea-level fall and lowstand; and (c) filling of valleys occurs during transgression and highstand.

The Exxon views on sequence stratigraphy have been carefully scrutinized, accepted by some, and criticized by others (e.g. Galloway, 1989; Miall, 1991, 1996; Helland-Hansen & Gjelberg, 1994; Shanley & McCabe, 1994). Some criticisms are grounded on scientific shortcomings, but others reflect an unfortunate use of genetic terminology for Exxon systems tracts (see Van Wagoner, 1995a, for a historical perspective), as well as contradictions within and between publications from the Exxon group as a whole. For example, numerous figures in Posamentier & Vail (1988) show widespread fluvial deposition during sea-level fall, but the text commonly infers fluvial incision and sediment bypass during that time. Moreover, Posamentier & Vail (1988) illustrate the development of systems tracts with respect to eustasy, and/or various combinations of eustasy and subsidence. By contrast, Van Wagoner *et al.* (1988) provide an operational view that follows the original definition by Mitchum *et al.* (1977), and systems tracts are 'defined objectively' on the basis of key bounding surfaces, position within a sequence, stratal geometry and

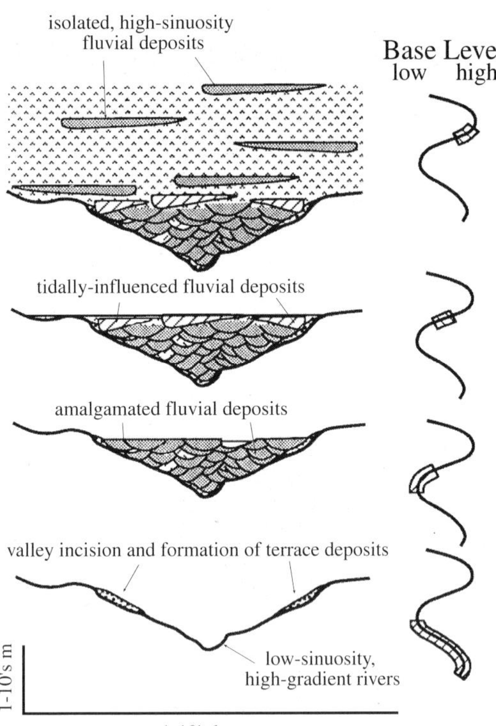

Fig. 5. Shanley & McCabe's (1993, 1994) model for development of depositional sequences in mixed fluvial–coastal–marine strata of the Cretaceous Western Interior basin of the USA, in response to base-level change. They illustrate the concept of incision and sediment bypass during base-level fall that is common to many sequence stratigraphic models, followed by valley filling, first with amalgamated fluvial deposits, then tidally influenced strata and finally low net-to-gross fluvial deposits.

facies associations, and the 'terms lowstand and highstand are not meant to imply a unique period of time or position on a cycle of eustatic or relative change of sea-level'.

Nevertheless, *SEPM Special Publication 42* resulted in a flurry of interest in the concepts of sea-level, base level and accommodation, and in how fluvial and other continental systems fit within the mostly marine-derived sequence-stratigraphic model (see the review by Shanley & McCabe, 1994). 'Incised valleys' emerged as a focus of attention; according to Van Wagoner *et al.* (1990), 'incised valleys' contain significant erosional relief, truncate older strata, juxtapose fluvial or estuarine sandstone on marine deposits, and define a significant basinward shift of facies due to relative sea-level fall. Dalrymple *et al.* (1994) defined two types of 'incised valleys', those

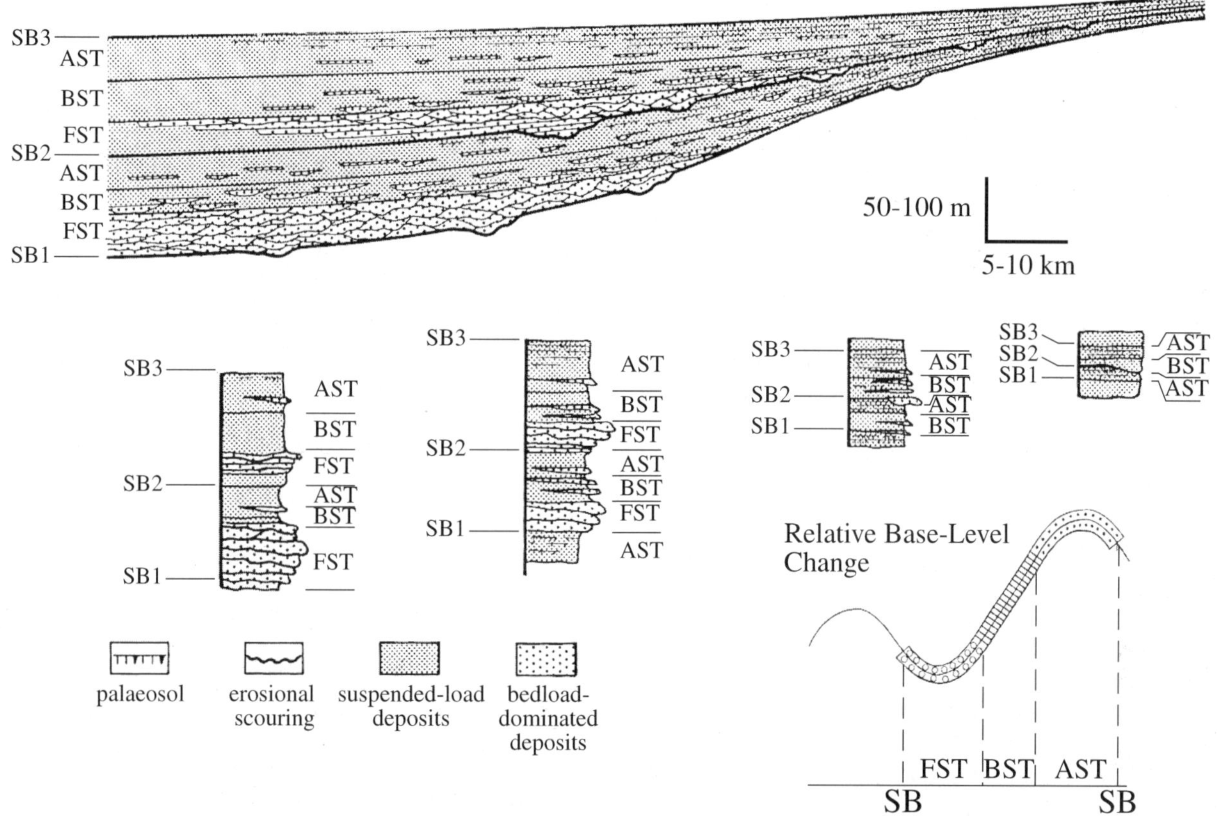

Fig. 6. Hinterland depositional sequence model of Legarreta and colleagues, based on Cretaceous nonmarine strata in Argentina. FST = forestepping systems tract, BST = backstepping systems tract, AST = aggradational systems tract, SB = sequence boundary. Downdip is to the left. Modified from Legarreta & Uliana (1998).

formed by relative sea-level fall, and those formed in response to some other mechanism. They argued the first type should have sequence boundaries at their base and will be filled by predictable successions of fluvial and estuarine facies due to relative sea-level rise (Zaitlin *et al.*, 1994), whereas the second type will be difficult to recognize since they may occur within a succession of fluvial strata, and have no significance in the Exxon sequence-stratigraphic sense.

A variety of explanatory models for fluvial sequence stratigraphy have been published that focus on linking fluvial and nearshore marine strata. Shanley & McCabe (1991, 1993, 1994) recognized depositional sequences based on interbedded fluvial and marine strata from the Cretaceous of the Western Interior foreland basin of Utah (USA). Their interpretations are grounded on correlations between alluvial strata and coeval shorelines that step basinward or landward, as well as recognition of systematic vertical changes in sandbody amalgamation and/or isolation (Fig. 5). Posamentier & Allen (1993) proposed a variant on the standard sequence model that applies to foreland basins, whereas Howell & Flint (1996) discussed active extensional settings. Van Wagoner (1995b) developed a detailed model for Cretaceous foreland basin strata of the Book Cliffs in Utah (USA) that, among other things, has served as a catalyst for debate in the literature (see Yoshida *et al.*, 1998; Van Wagoner, 1998) and, more informally, among the sedimentology community as a whole.

Fully nonmarine models have been proposed as well. Legarreta & Uliana (1998) propose a model for Mesozoic continental basins in Argentina that includes: (a) a coarsening-upwards 'forestepping' systems tract that is laterally confined and dominated by amalgamated channel sandstones; (b) a fining-upwards 'backstepping' systems tract; and (c) an 'aggradational' systems tract (Fig. 6). Wright & Marriott (1993) propose a generic model that is not based on actual examples, but incorporates systems tracts terminology and emphasizes relationships between alluvial architecture, palaeosols and changes in accommodation that relate specifically to floodplain sedimentation. Gibling & Bird (1994) and Aitken & Flint (1995), among others, use Carboniferous examples from Canada and the USA to discuss overall sequence architecture, the role of palaeosols, and the recognition, variability and importance of 'interfluve' sequence boundaries. Most recently, Martinsen *et al.* (1999) use the accommodation to sediment supply ratio (A/S ratio) described in Muto & Steel (1997) to interpret changes in alluvial architecture.

Climate change

Studies that link fluvial deposition to climate change in the pre-Quaternary record are relatively sparse when compared with the volume of literature on fluvial responses to sea-level or base-level change that has been generated following introduction of sequence-stratigraphic concepts. By far the most common theme is the interpretation of ancient climate from individual features (e.g. palaeosols) within alluvial successions, as opposed to interpretations of fluvial response to climate change. Much of this work is summarized by Parrish (1998) and not repeated here. Instead the discussion below focuses on examples where workers have specifically interpreted depositional responses to climate change.

A number of workers have inferred climate change as a control based on transitions between fluvial and other strata. Over time-scales of 10^6–10^7 years, Olsen & Larsen (1993) suggest the Late Devonian of East Greenland can be subdivided into depositional complexes similar in concept to Exxon's systems tracts. Interpreted changes from braidplain environments to terminal fluvial systems that pass laterally to aeolian environments, then to a meandering stream-dominated setting were attributed to shifts towards more arid and then more humid conditions. A number of studies also interpret transitions between fluvial and other strata due to climate changes over time-scales of 10^4–10^5 years (Milankovitch scales). These include Cojan's (1993) work on Upper Cretaceous to Palaeocene fluvial and lacustrine strata in the Provence Basin of southern France, Yang & Nio's (1993) studies of Permian Rotliegendes strata of the Dutch North Sea, Clemmensen *et al.*'s (1994) summary of interbedded fluvial and aeolian rocks of Permian to Triassic age in the UK, and Dubiel *et al.*'s (1996) discussion of interfingering fluvial and aeolian strata within the Permian Cutler Group of the western USA.

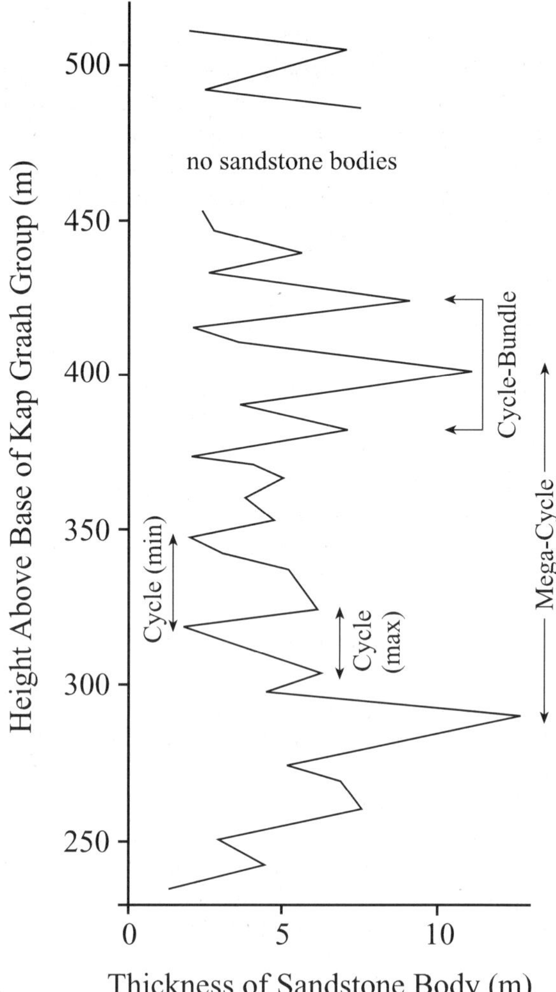

Fig. 7. Interpreted cyclical changes in sandbody thicknesses deposited by meandering rivers from the Andersson Land Formation, Devonian Kap Graah Group of East Greenland (after Olsen, 1994). Cycle (min) indicates cycles defined by minimum thicknesses, whereas Cycle (max) indicates the opposite definition. Mega-cycles are interpreted to reflect the Milankovitch eccentricity period, calculated at 95 kyr for the Devonian, whereas individual cycles are interpreted to represent precessional periods of 17 or 20 kyr.

Changes within fully alluvial successions have been ascribed to the influence of Milankovitch-scale climate changes as well. Olsen (1990, 1994), for example, examined the Late Devonian Andersson Land Formation of east Greenland, the axial meandering stream-dominated complex described above. He identified two scales of cyclicity in the thickness of channel sand bodies, which were interpreted to represent climatically controlled changes in discharge regimes at Milankovitch-scale periodicities of 100 kyr and ~20 kyr (Fig. 7). Smith (1994) suggested that changes in channel sandbody geometry, sand-to-mud ratios, sediment accumulation rates and types of palaeosols in Plio-Pleistocene strata from Arizona (USA) occurred during a tectonically quiescent period and might instead reflect climate changes.

Modelling fluvial responses to climate change in the pre-Quaternary record has not been common practice, but a number of important efforts can be identified. Perlmutter & Matthews (1989) propose a cyclostratigraphic model that emphasizes Milankovitch-scale climate change superimposed on tectonically driven basin evolution. Key model components include weathering and sediment production, discharge regimes, and transport efficiency as they relate to variations in climate, as well as how these factors control the spatial patterns of clastic environments and temporal evolution of clastic successions. Paola *et al.* (1992) use a diffusion-based model to examine the roles of sediment flux, rates of subsidence, changes in sediment grain-size (gravel vs. sand) and water supply on patterns of deposition in foreland basins at time steps of 10^7 and 10^5 years. This study is not specifically concerned with causal mechanisms, but the authors note that sediment flux, changes in grain-size and variations in water supply can have several causes, including climate change. Variations in sediment flux and grain size are important in both time steps, with both resulting in progradation or retrogradation of gravel fronts. By contrast, cyclical changes in water supply play a significant role in shorter time steps only, resulting in progradational cycles during time periods when water supply is increasing. Paola *et al.* (1992) suggest that over time-scales of 10^5 years (Milankovitch time-scales), patterns of deposition for foreland basin settings most closely reflect surface processes, whereas over longer time-scales basin subsidence and other tectonically controlled factors become dominant.

GENERAL PROCESS FRAMEWORK

Promising avenues of research lie at the interface between Quaternary and ancient fluvial systems, especially studies of fluvial responses to climate and sea-level changes at time-scales of 10^6 years or less. The following section focuses on how rivers produce stratigraphic records over these time-scales; the section concentrates on alluvial channels because of their mobile beds and banks, the rapidity with which such systems can respond to allogenic controls, and the fact that such rivers provide the common link between Quaternary and ancient records.

Channel geometry and depositional style

Process-based interpretations of fluvial responses to climate and sea-level change should be linked in some way to controls on the mutually dependent variables of channel geometry and depositional style. Numerous discussions of alluvial channel process and form, as well as fluvial depositional processes and systems, are readily available elsewhere and not repeated here (e.g. Knox, 1976; Leopold & Bull, 1979; Richards, 1982; Miall, 1985, 1995, 1996; Schumm, 1985; Ferguson, 1987; Nanson & Croke, 1992; Bridge, 1993a,b, 1995; Orton & Reading, 1993; Van den Berg, 1995; Brierley, 1996; Nanson & Knighton, 1996; Knighton, 1998; Fielding, 1999). Instead, four fundamental relationships can be distilled from this body of literature: (a) cross-sectional and plan view geometry of a 'graded stream' is adjusted to prevailing discharge regimes and sediment loads, and maintains a statistically constant size and shape over some interval of time (10^2–10^3 years); (b) fluvial facies and depositional systems are linked to channel geometry through the spatial distribution of depositional landforms and environments; (c) changes in the relationship between discharge and sediment load result in changes in one or more geometric variables, for example width, depth, width-to-depth ratio, sinuosity or slope; and (d) changes in channel geometry result in changes in the lateral and vertical distribution of depositional environments and facies.

Aggradation and degradation: changes in sediment storage

Aggradation or degradation often accompanies changes in channel geometry and depositional style, and results in changes in sediment storage per unit valley length due to changes in channel bed and floodplain elevation. Width-averaged bedload transport rates (q_s) reflect stream power per unit channel length, $\Omega = \gamma QS$ (where Ω = stream power, γ = specific weight, Q = discharge and S = channel bed slope;

Bagnold, 1973). Channel beds aggrade when sediment supply exceeds maximum bedload transport rates, degrade when the reverse is true and maintain stable long profiles when there is a balance between stream power and sediment supply (Fig. 8).

Theoretical models to describe aggradation or degradation are widely published in the engineering, geomorphological and sedimentological literature (e.g. Torres & Jain, 1984; Wyroll, 1988; Paola *et al.*, 1992; Leeder & Stewart, 1996). Such models are ultimately based on the sediment continuity equation, which can be written in its simplest one-dimensional form as:

$$dz/dt + dq_s/dx = 0 \qquad (1)$$

where z = bed elevation, t = time, q_s = width-averaged sediment transport rate and x = distance along the channel. For a channel bed that is in equilibrium ('graded') and neither aggrading nor degrading, $dz/dt = 0$, hence $dq_s/dx = 0$. From this simple relationship, three end-member situations for aggradation can be derived: (a) sediment overloading (holding discharge constant) from an upstream point source (typical of an alluvial fan) will result in a downstream-tapering wedge of channel-bed aggradation (i.e. the 'downfilling' of Schumm, 1993); (b) sediment overloading from upstream and lateral sources (drainage network feeding a main valley axis) will result in a relatively constant thickness of channel-bed aggradation; and (c) base-level rise under conditions of constant sediment supply will result in downstream decreases in sediment transport rates, and downstream increasing rates of channel-bed aggradation (the 'backfilling' of Schumm, 1993). Degradation can be triggered in a variety of ways, and from both upstream and downstream directions, but always results in channel incision and floodplain abandonment, net sediment removal, and production of an unconformity.

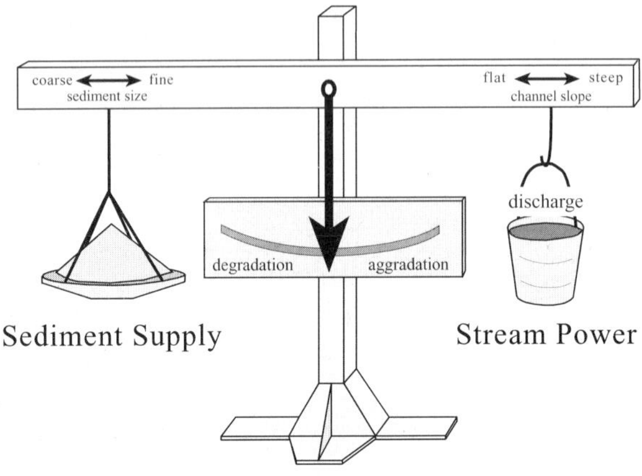

Fig. 8. Balance model for aggradation and degradation of alluvial channels, emphasizing changes in the relationship between discharge and sediment supply. Channels aggrade when sediment supply exceeds transport capacity of the discharge regime, and degrade when the reverse is true. Redrawn from a widely circulated diagram that originated as an unpublished drawing by W. Borland of the USA Bureau of Reclamation, and was based on an equation by Lane (1955).

A simple two-dimensional view of channel-bed aggradation and degradation is shown in Fig. 9A, but may be complicated by a number of factors. For example, under equilibrium conditions channels could migrate laterally but undergo no significant aggradation or degradation, which translates into no net change in sediment storage. This does not mean that a stable stream cannot leave behind a deposit, since lateral migration of a channel belt results in cannibalization of older sediments and emplacement of new deposits. In this case, interpretation of true aggradation and development of multistoried channel belts vs. simple lateral migration is not a simple issue, and should be grounded on observations of deposit thicknesses that exceed the vertical distance between the maximum depth of autogenic scour (Salter, 1993) and the upper limit to floodplain accretion as a channel belt migrates laterally. For example, recent studies demonstrate that depths of autogenic scour at flow confluences and at channel bends can be up to five times deeper than the mean channel depth (Best & Ashworth, 1997), and depths of scour can be significantly below contemporary sea-level positions.

Perhaps equally important are cases where there is little, if any, change in channel-bed elevation, but floodplain height increases or decreases (e.g. Blum & Valastro, 1994; Fig. 9B). For example, floodplain abandonment can take place without significant channel incision due to a reduction in peak discharges and overbank flooding. Lateral migration of a channel belt then produces a basal erosional unconformity that traces laterally up to soils that developed on the abandoned floodplain surface. By contrast, formerly stable terrace surfaces can be buried by renewed overbank flooding if flood magnitudes increase. Finally, changes in channel-bed and floodplain elevation may occur at different rates, and in some instances in opposite directions (cf.

Tebbens *et al.*, 1999), due to changes in channel geometry. This variety of possible geometric responses suggests that inferences of fluvial aggradation and/or incision are less than straightforward, and should be based on evidence that bears on changes in both channel-bed and floodplain elevations, wherever possible.

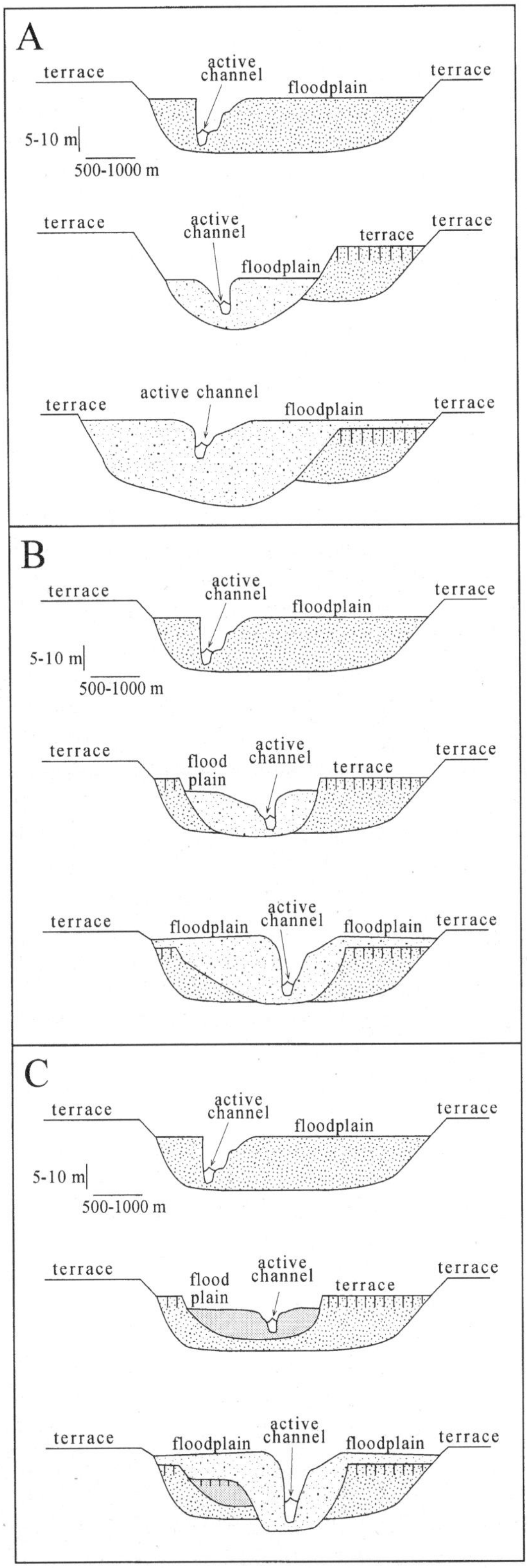

Controls on changes in discharge, sediment supply and sediment storage

Changes in channel geometry, depositional style and net sediment storage are dependent variables over time-scales of interest here, and reflect prevailing conditions of discharge and sediment supply. What controls changes in discharge and sediment supply over these time-scales?

Discharge regimes

Alluvial channel size reflects the volume of water that must be transmitted on a relatively frequent basis, with recurrence intervals for channel-forming discharges that range from 1 to 30 years (Williams, 1978; Knox, 1983; Lewin, 1989). Overviews of flood characteristics in various climates can be found in Baker *et al.* (1988), McMahon *et al.* (1987, 1992) and Waylen (1996), whereas Pazzaglia *et al.* (1998) discuss possible relationships between rates of uplift, rates of erosion, river long profiles and flood hydrographs. Critical to the present discussion is the potential impact of climate change on flood magnitude and frequency, and how this might vary to the extent that global climate change impacts different river systems in the same or different ways.

In this context, Knox (1983) summarizes a variety of historical period data that illustrate

Fig. 9. Changes in sediment storage within valley cross-sections, and possible stratigraphic results. Each scenario begins with lateral migration of a single-story channel belt. (A) Channel-bed degradation and floodplain abandonment with terrace formation, then channel-bed aggradation to produce a multistory channel belt, with floodplain deposition over previously stable terrace surfaces characterized by palaeosols (hachures). (B) Reduction in channel size, floodplain abandonment and terrace formation without significant channel-bed degradation, then channel enlargement with floodplain deposition over previously stable terrace surfaces characterized by palaeosols. (C) Significant reduction in channel size accompanied by channel-bed aggradation, floodplain abandonment and terrace formation, followed by channel enlargement accompanied by channel-bed degradation and floodplain deposition over previously stable terrace surfaces characterized by palaeosols.

how climate change can impact flood regimes of individual rivers. For small drainages in the mid-latitude USA, he shows that flood discharge per unit area increases (holding recurrence intervals constant) with decreasing precipitation due to decreases in vegetation cover and corresponding increases in surface runoff. Moreover, he uses a long time series of annual floods from the upper Mississippi River to illustrate relationships between flood magnitudes and independently identified changes in global atmospheric circulation. These data show that circulation changes have significant impacts on the magnitude of rare events (recurrence intervals > 10 years), but lesser impact on more frequent floods (Fig. 10). More recently, Knox (1987, 1993) documents changes in middle to late Holocene channel-forming and extreme floods, and concludes that minor changes in climate produced changes in bankfull discharges ranging from 70 to 130% of present-day values.

Another important approach is illustrated by Probst (1989), who examines the spatial structure of runoff fluctuations from larger European rivers during the historical period, and shows periods with positive and negative deviations from long-term averages. Most commonly all rivers varied in-phase, but at other times runoff regimes of rivers within present-day oceanic and continental regimes (Rhone, Loire, Garonne, Seine and Rhine) were out-of-phase with rivers in the core Mediterranean region of southern Europe (Ebro, Guadalquivir and Po). From these relationships, it might be envisaged that discharge regimes in the two areas should respond simultaneously to global climate change, but in different directions due to the regional manifestations of changes in global atmospheric circulation. Similar in-phase and out-of-phase relationships between regional precipitation changes and changes in atmospheric circulation are well documented (e.g. Lau, 1988; Hurrell, 1995; Rogers, 1997; Dai *et al.*, 1997).

In aggregate, these studies serve to illustrate four points that are of particular relevance. First, changes in atmospheric circulation through time have a direct impact on the magnitude and frequency of floods. Second, these changes are not directly proportional to increases or decreases in precipitation, but are instead buffered by other variables that influence runoff rates, especially vegetation. Third, flood regimes are especially sensitive to changes in climatic inputs, such that small changes in climate can significantly increase or decrease flood magnitudes. Finally, responses of flood regimes to climate change should be geographically circumscribed since global changes in circulation may result in simultaneous changes in discharge regimes everywhere, but some regions will be out-of-phase with others in terms of the direction of change.

Sediment supply

Routing of sediments in river systems is a topic that resides at the interface between studies of modern processes, Quaternary landscape evolution and development of the pre-Quaternary stratigraphic record. Hovius & Leeder (1998) note that sedimentary basins record erosion of the continents, whereas drainage networks are the

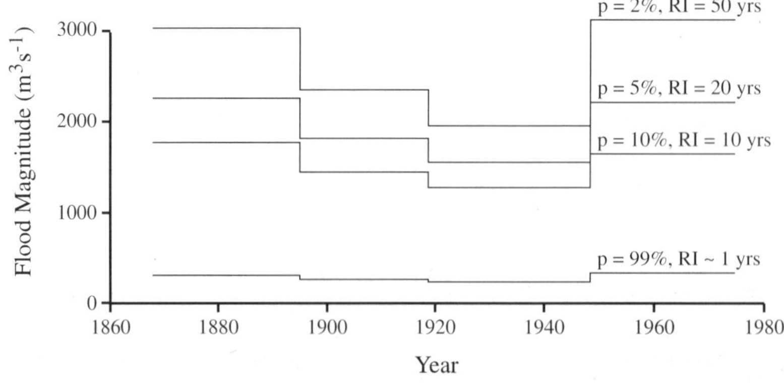

Fig. 10. Changes in flood magnitudes due to changes in atmospheric circulation, Mississippi River at St. Paul, Minnesota, USA (after Knox, 1983). Data represent a time series of all floods that exceeded $368\,m^3\,s^{-1}$, and was disaggregated into time periods defined by dominant types of atmospheric circulation (meridional vs. zonal). Flood probabilities were then calculated for each period, with p = the probability that a flood with a given magnitude will be equalled or exceeded in any given year, and RI = average recurrence interval.

'negative imprint' of this same record, and their existence must be inferred from stratigraphic data. Especially pertinent, then, are the allogenic controls on erosion from upland source terrains and delivery to depositional basins.

There is a long history to discussions of climatic controls on sediment supply, with studies by Langbein & Schumm (1958) and Fournier (1960) among the first to develop quantitative relationships between climate, erosion and sediment yield. Langbein & Schumm (1958), in particular, indicate that sediment yield increases with increasingly effective precipitation through the arid to semi-arid transition, then decreases towards more humid climates due to the increasing importance of vegetation. Compilations by Walling & Webb (1983, 1996) suggest the Langbein & Schumm (1958) curve may be applicable for the midcontinent USA where it was derived, but when other climate regimes, especially subtropical and tropical, are taken into account then sediment yield increases with precipitation at values greater than $\sim 1000\,\mathrm{mm\,yr^{-1}}$. Because of the mitigating effect of vegetation in each of these relationships, they are most applicable for that part of Earth history when land plants have been common. In the absence of land plants, sediment supply would most likely increase as precipitation increases (e.g. Schumm, 1968b).

There is an equally long history to discussions of relationships between sediment yield and measures of relief or rates of uplift. Indeed, Walling & Webb's (1983, 1996) compilations show the highest rates of sediment yield in orogenic belts that surround the Pacific Ocean. Moreover, specific sediment yields (yields per unit area) decrease as drainage area increases, which emphasizes the importance of small and steep catchments for sediment production. Milliman & Syvitski (1992), for example, show that up to 80% of sediments delivered to the oceans is derived from small catchments within active orogens, and suggest that sediment yield is a log-linear function of drainage area and maximum elevation.

Hovius (1998) presents the most comprehensive study of sediment supply to date, based on examination of 97 intermediate and large drainage basins around the world. He indicates that specific sediment yield can be estimated using the following empirical relationship:

$$\ln E = -0{\cdot}416 \ln A + 4{\cdot}26 \times 10^{-4} H + 0{\cdot}15 T + 0{\cdot}095 T_{\mathrm{R}} + 0{\cdot}0015 R + 3{\cdot}58 \qquad (2)$$

where E is specific sediment yield ($\mathrm{t\,km^{-2}\,yr^{-1}}$), A is drainage basin area ($\mathrm{km^2}$), H is mean height (relief) of the drainage basin (m), T and T_{R} are mean annual temperature and annual temperature range (°C), and R is specific runoff ($\mathrm{mm\,km^{-2}\,yr^{-1}}$). This equation explains only 49% of the observed variance and, although vegetation was not considered, Hovius suggests that inclusion of other variables does not substantially improve the explanatory power. Hovius (1998) also uses Milliman & Syvitski's (1992) larger data set to show that specific sediment yield varies with tectonic setting and rates of rock uplift, with contractional orogens far outpacing other settings, and low-relief cratons yielding very low values.

To the extent that drainage area, tectonic setting, average rates of uplift and total relief remain relatively constant for individual river systems over time-scales of 10^3–10^5 years, Hovius' (1998) equation suggests significant variations in specific sediment yields with changes in climatic parameters such as temperature and runoff. However, Leeder *et al.* (1998) model erosion rates and discharge regimes during the late Pleistocene full-glacial and the Holocene, contrasting the USA Basin and Range province with the Mediterranean region, and in doing so illustrate a key point that deserves emphasis. In much the same way as changes in discharge regimes can vary regionally, Leeder *et al.* (1998) show the same global climate change can produce out-of-phase responses for the Basin and Range and Mediterranean regions, when measured in terms of runoff and sediment yield, sediment flux to the depositional basin, relative changes in lake or sea-level and development of a stratigraphic record (Fig. 11). In addition to the above, an important factor that cannot be gleaned readily from the present interglacial world is the high rates of erosion and sediment production that accompanies widespread glaciation (e.g. Hallet *et al.*, 1996).

Sediment storage en-route to depositional basins

At any point in time, there is a disparity between the liberation of sediments from uplands and sediment yields past some point in the drainage network. The difference is related to sediments that are temporarily stored in the drainage network as colluvial, alluvial fan and fluvial deposits. This amount can be considered in the context of the sediment delivery ratio (Walling, 1983),

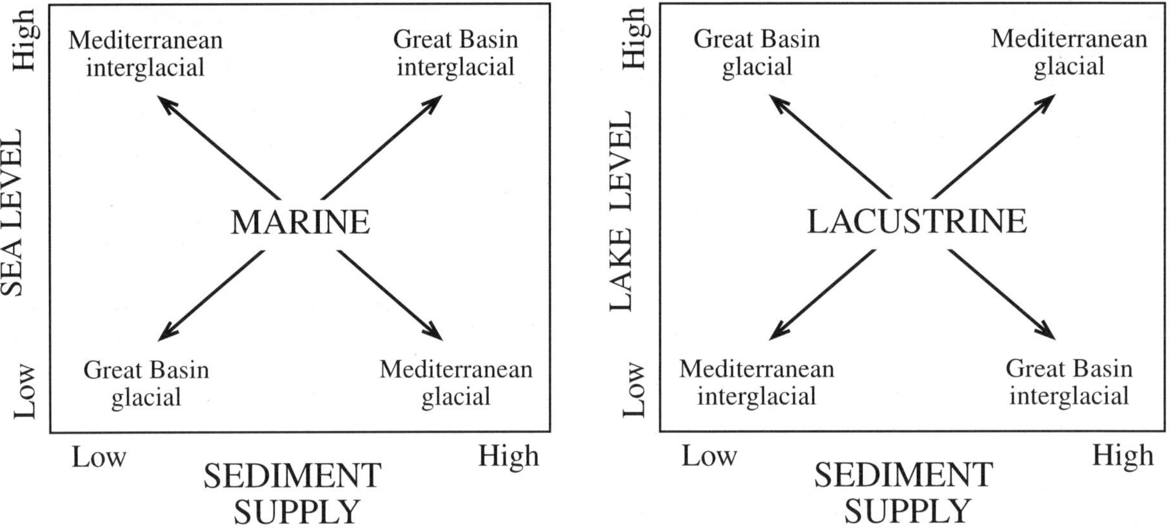

Fig. 11. Model for changes in sediment supply in Mediterranean vs. USA Great Basin climates during glacial and interglacial periods. Diagram on left illustrates conditions that might apply to a marine basin, whereas diagram on the right illustrates conditions for a lacustrine basin. Both diagrams illustrate a succession of possible out-of-phase responses. After Leeder *et al.* (1998).

defined as $S_d = S_y/ER$, where S_y is sediment yield past a given point in the drainage network, and ER equals total erosion from the drainage basin above that point.

Compilation of sediment delivery ratios from a variety of settings indicates that they typically decrease as drainage area increases and stream gradients decrease, with very high values for very small drainage basins, and values as low as 0·1–0·2 for drainage areas > 1000 km^2 (Walling, 1983). Data from modern settings are strongly biased by human activities that have increased rates of erosion, increased sediment storage within small catchments and trapped sediments within artificial reservoirs. For example, Meade *et al.* (1990) note that roughly 10% of all sediment eroded from the conterminous USA presently reaches the ocean basins, but up to 20% of all sediment is trapped by reservoirs constructed on a few large rivers. As a result, continent-wide sediment delivery of 20–30% (S_d = 0·2–0·3) or more may be more representative.

Changes in the sediment delivery ratio can significantly imprint the stratigraphic record. Hovius' (1998) view that drainage networks are the negative imprint of basin filling would correctly imply that basin margin sediment delivery over geological time must approach 100%. It follows that over shorter periods of time, individual segments, or perhaps entire drainage networks, must have delivery ratios greater than average values, and must in some cases be greater than 1 during periods when previously stored sediments are flushed basinward.

Fig. 12. (A) Contrasts between 'vacuum cleaner' vs. 'conveyor belt' models for sediment supply to basin margins. The vacuum cleaner derives all sediment from more distal parts of the basin by means of excavation of a valley at the basin margin, whereas the conveyor belt relies on sediment delivery from a large hinterland drainage network, and does not require deep incision with complete sediment bypass during sea-level fall. (B) Calculation of relative contribution of vacuum cleaner vs. conveyor belt model, using the palaeovalley fill from the last glacial cycle of the Colorado River, Texas, as an example. Dip and strike dimensions and thickness of palaeovalley fill based on data presented in Blum (1993) and Aslan & Blum (1999), with data on sediment supply taken from Hovius (1998). For vacuum cleaner calculation, V = volume of palaeovalley fill, ρ = average density of earth materials, D = duration of glacial cycle falling stage and lowstand (kyr), and VC = mass of sediments produced by complete excavation of the valley over duration D. For conveyor belt calculation, sy = specific sediment yield (tons km^{-2} yr^{-1}), A = drainage area (km^2), SY = total sediment yield from drainage network (t yr^{-1}), D = as above. For both calculations, CB = mass of sediments delivered from the conveyor belt during time interval D.

Role of sea-level change

The following discussion asks a simple question – exactly how does sea-level change affect a fluvial system? To begin, it is difficult to envisage how sea-level or base-level change can have a significant impact on discharge regimes or liberation of sediments, since these are mostly upstream-controlled. Nevertheless, since much of the literature implies relative sea-level

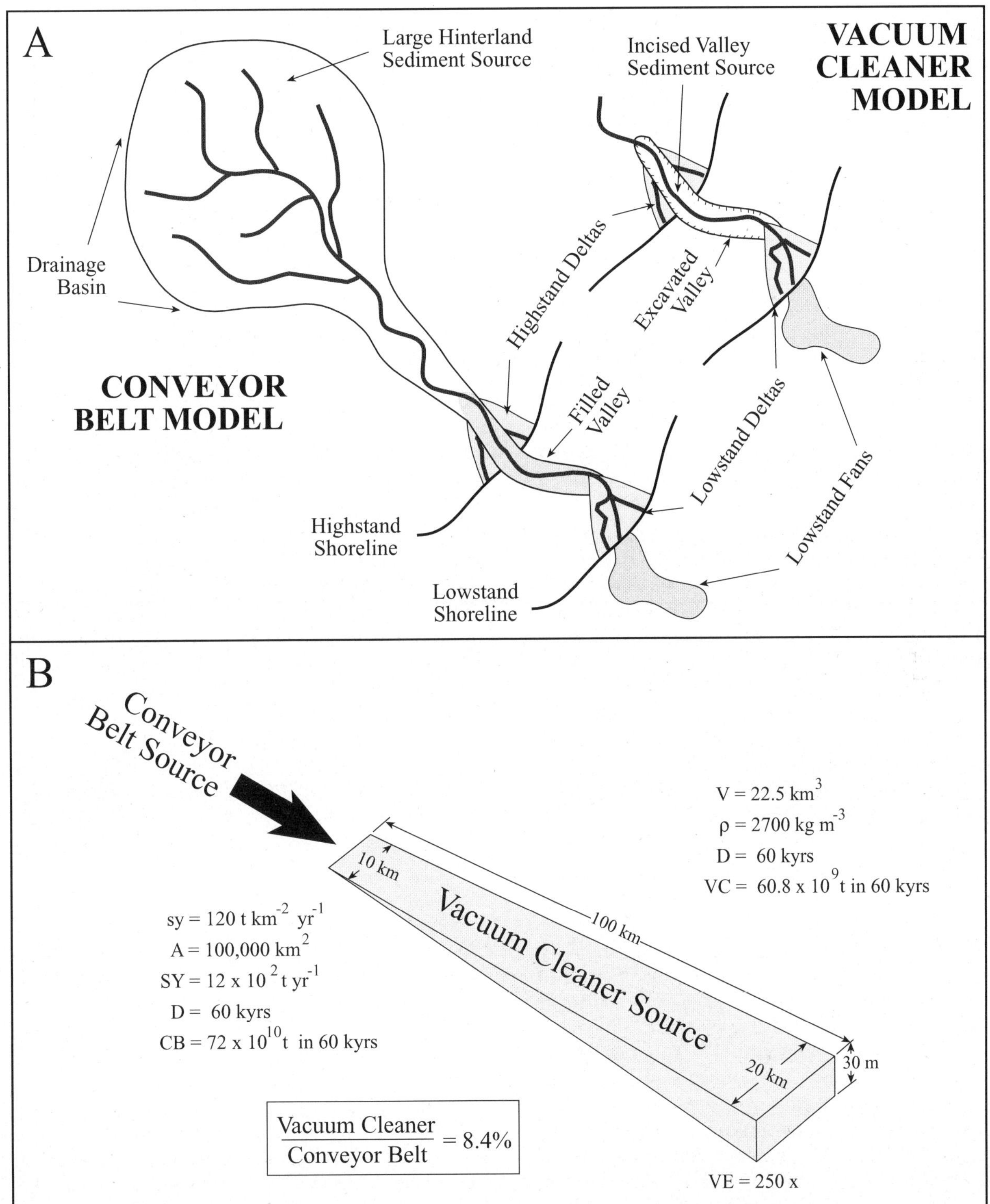

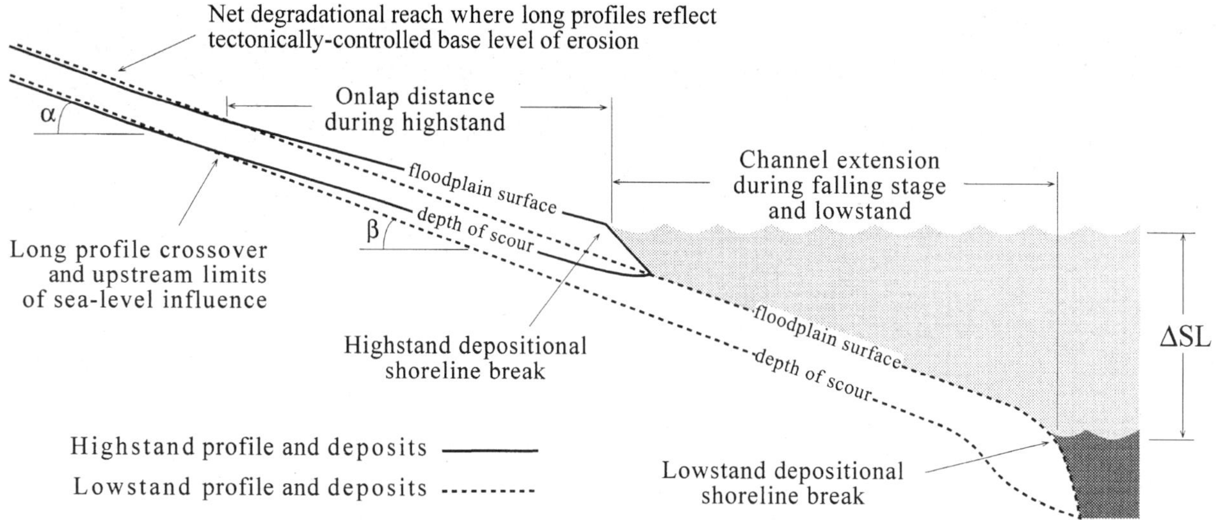

Fig. 13. Definition sketch for fluvial response to sea-level change along a continental margin with a distinct highstand depositional shoreline break. Diagram illustrates concepts of channel extension during sea-level fall and lowstand, vs. upstream limits of onlap during sea-level rise and highstand. Also, average slope of the channel above the limits of sea-level influence is shown as changes, and reflect a long-term tectonically controlled base-level of erosion.

controls over sediment supply, it is important to first look at what sea-level change does not do.

Nowhere is confusion more apparent than in sequence-stratigraphic models for 'incised valleys' (hereafter valleys or palaeovalleys, since all valleys are incised, by definition) that assume relative sea-level fall produces incision, and an upstream-propagating wave of stream rejuvenation, which produces sediments that entirely bypass the coastal plain and newly emergent shelf to provide a critical volume of sediment for systems tracts further basinward. This is referred to here as a 'vacuum cleaner' model for sediment supply, as contrasted with a 'conveyor belt' model, where sediments are continuously delivered to the basin margin from a large inland drainage (Fig. 12A). Figure 12(B) illustrates these models with reference to the Colorado River, Texas Gulf Coast (USA), and shows that the volume of sediments produced by complete 'vacuum cleaner' excavation of a valley during the last glacio-eustatic falling stage and lowstand would be an order of magnitude less than the total volume delivered by the 'conveyor belt' during that same interval of time.

The 'vacuum cleaner' model has intellectual roots in Fisk (1944), yet subsequent work in the Lower Mississippi Valley clearly shows that complete sediment bypass was not the case (e.g. Saucier, 1994, 1996; Blum *et al.*, 2000). The same can be said for other large fluvial systems of the Gulf Coastal Plain of the USA (Blum *et al.*, 1995; Blum & Price, 1998), for large rivers of southern Europe such as the Rhone and Po (Mandier, 1988; Amorosi *et al.*, 1999), the Rhine–Meuse system in north-west Europe (e.g. Pons, 1957; Berendsen *et al.*, 1995; Kasse *et al.*, 1995; Törnqvist, 1998) and the Nile (Stanley & Warne, 1993). In fact, it is difficult to find an example where complete sediment bypass in response to base-level fall has actually been documented on Quaternary river systems that have significant hinterland drainage areas and discharge to a depositional basin. Hence, to the extent that Quaternary systems provide ground-truth for sequence-stratigraphic models, the body of evidence actually points in a different direction. These conclusions are supported by modelling results of Leeder & Stewart (1996), which demonstrate that high rates of sediment supply can overcome base-level fall, and result in progradation without significant incision.

Given a continental margin setting with a pronounced depositional shoreline break (Fig. 13), and a shelf that is as steep or steeper than the coastal plain, why should rivers not completely bypass sediments during sea-level fall? Simply put, given the influx of sediments from the 'conveyor belt', complete sediment bypass of a coastal plain and shelf during sea-level fall would require significant downstream increases in slope and stream power ($\beta > \alpha$ in Fig. 13), and corresponding downstream increases in rates of sediment transport. In most cases, some incision through a highstand depositional shoreline break takes place as the channel extends across the newly exposed shelf (Posamentier *et al.*, 1992; Blum, 1993; Talling, 1998; Törnqvist *et al.*, 2000), with corresponding net removal of previously stored sediment, but it seems unlikely that the depth of incision will produce slopes that exceed those further upstream from the limits of sea-level influence (i.e. $\beta > \alpha$ is not likely). Hence if the valley upstream from the limits of long profile adjustment aggrades due to sediment overloading, or laterally migrates due to equilibrium conditions, this mode of behaviour is simply translated further basinward. Periods of complete sediment bypass should only occur when sediment supply from upstream sources decreases relative to maximum sediment transport rates, such that the valley further upstream is also incising.

On the other hand, sea-level rise along a continental margin with the configuration described above (Fig. 13) results in channel shortening, decreases in the distance over which sediments can be stored and, in most cases, flattening of channel slope. Discharge is conserved, but reductions in slope will result in corresponding downstream decreases in stream power and sediment transport rates. These conditions result in net valley aggradation that in turn may play a significant role in adjustments of channel geometry and depositional style (discussed more fully below). However, even within long periods of overall net valley aggradation driven by sea-level rise (time-scales of 10^3–10^4 years), significant intervals of incision and/or sediment bypass can occur if sediment supply decreases relative to stream power, as shown by Blum (1990, 1993) for the Colorado River of the Texas Gulf Coast (USA), Törnqvist (1998) for the Rhine–Meuse and Blum *et al.* (2000) for the Lower Mississippi Valley.

Turning back to the original question, the fundamental responses to sea-level change appear to be channel extension or shortening, coupled with changes in the elevation of channel bases and floodplain surfaces, in order to keep pace with a shoreline that is advancing or retreating and changing its elevation. All other adjustments take place within this context, and should be regarded as nondeterministic, since they depend on alluvial valley, coastal plain, shoreface and shelf gradients (see Fig. 3), discharge and sediment supply from the drainage network and local physiographical factors. With respect to the latter, Leckie (1994) provides an example from the Canterbury Plains of New Zealand, where incision has occurred during Holocene sea-level rise because of steep alluvial plain gradients in combination with high-energy wave attack. Moreover, Schumm (1993), Wescott (1993) and Ethridge *et al.* (1998) note that fluvial systems possess several degrees of freedom by which they can adjust channel parameters to changes in base level, a discussion not repeated here.

Many recent papers (e.g. Butcher, 1990; Nummedal *et al.*, 1993; Posamentier & Weimer, 1993; Schumm, 1993; Koss *et al.*, 1994; Merritts *et al.*, 1994; Shanley & McCabe, 1994; Leeder & Stewart, 1996) have emphasized the issue of how far inland sea-level actually controls fluvial incision and aggradation. Again, Fisk's (1944) work in the Lower Mississippi Valley has been extremely influential, postulating eustatic control more than 1000 km upstream from the modern shoreline. This view has been challenged by more recent studies in the same area (e.g. Saucier, 1994, 1996) that, in turn, are now cited as critical evidence in numerous recent reviews (Shanley & McCabe, 1994; Emery & Myers, 1996; Miall, 1996, 1997; North, 1996). Shanley & McCabe (1993) propose strong generalizations, suggesting that sea-level control typically extends 100–150 km updip from the coeval shoreline. However, in most of these studies it is not clear how the landward limit of sea-level control is actually defined.

For this paper, and for the purposes of discussion and comparison, the landward limit is defined as the upstream extent of coastal onlap due to sea-level rise (see Fig. 13). In late Quaternary contexts, this consists of the intersection between the modern floodplain and the floodplain surface from the Last Glacial Maximum (~ 20 ka), and is not necessarily the same as the landward extent of incision due to sea-level fall (cf. Ethridge *et al.*, 1998). By contrast, the basinward limit of sea-level influence is defined as the maximum distance of channel extension and delta progradation during sea-level fall. Table 1 summarizes a variety of data from fluvial systems from Europe and the Gulf Coastal Plain of the USA. These data suggest the landward limit of onlap due to sea-level rise correlates to hinterland sediment supply, and is inversely related to the gradient of the onlapped floodplain surface. Accordingly, this distance is highly variable and ranges from at least 300–400 km

Table 1. Relationship between drainage area, onlapped lowstand floodplain gradient and aggradation distance for six late Quaternary fluvial systems. Channel extension and onlap distances are measured with respect to the average highstand shoreline. Data from the Rhone taken from Mandier (1988) and Tesson *et al.* (1993).

River system	Drainage area (km^2)	LGM floodplain gradient ($cm\ km^{-1}$)	LGM channel extension (km)	Holocene onlap distance (km)
Nueces (Texas, USA)	40 000	50	> 200	40
Trinity (Texas, USA)	50 000	20	150	90
Rhone (France)	99 000	130	50–60	100
Colorado (Texas, USA)	100 000	80	100–120	90
Rhine–Meuse (The Netherlands)	254 000	25	800	150
Mississippi (Louisiana, USA)	3 344 000	15	150 – 200*	> 300 – 400

* The present birdfoot delta is also extended close to this amount.

for the low-gradient, high-sediment-supply Mississippi River, to ~40 km for steep-gradient, low-sediment-supply systems like the Nueces River of the Texas Gulf Coastal Plain. Channel extension during lowstand reflects shelf physiography, and ranges from 800 km or so for the Rhine, 100–200 km for rivers of the Texas Gulf Coastal Plain (Colorado, Trinity, Nueces), to very short distances for the Mississippi, since it now discharges to the shelf edge. The Rhine is a special case, since during the last lowstand it flowed across the very low-gradient North Sea, then was diverted south-westward through the Strait of Dover and discharged to the Atlantic (Gibbard, 1995; Bridgland & D'Olier, 1995).

Role of accommodation

Few would argue that the concept of accommodation plays a critical role in sedimentology in the new millennium. This section again asks a basic question: what exactly is accommodation?

To begin, many workers recognize that Jervey's (1988) definition of accommodation uses a 'stratigraphic' (as opposed to 'geomorphic') definition of base level that can be traced to Barrell (1917), Sloss (1962) and Wheeler (1964) among others. These differing definitions for base level are not trivial. The geomorphic view defines the lower limits to erosion, and extends upstream from sea-level into the subaerial landscape as a 'base level of erosion' that defines the graded longitudinal profile (see Bull, 1991), but has no significance offshore beyond the fluvial system. By contrast, the stratigraphic definition defines the upper limits to deposition, and was intended to be used in marine systems as well. However, as Wheeler (1964) noted, stratigraphic base level does not control anything – it is not a cause but an effect. In the fluvial context, stratigraphic base-level changes in response to changes in geomorphic base level (lake or sea-level), climate-related and geomorphic parameters (flood magnitudes, sediment supply, channel geometries, etc.) and rates of uplift or subsidence. It follows from the above that accommodation is not exactly a control either. Instead, increases or decreases in accommodation are at least in part a response to changes in geomorphic base level, which defines the lower limits to accommodation, and in fluvial contexts the upper limits are defined by changes in discharge regimes and sediment supply as they control equilibrium floodplain heights.

It also seems that accommodation, as it is commonly used, somewhat imprecisely mixes processes that operate over a range of rates and temporal scales; it is difficult to reconcile the time-scales over which sediments are deposited in the first place, whether or not those deposits will be preserved in the stratigraphic record, and the manner in which ancient alluvial successions are interpreted in terms of changes in accommodation or an accommodation/sediment supply ratio. It may be more precise to consider these issues in much the same manner as Kocurek & Havholm (1993) and Kocurek (1998), who differentiate 'accumulation' vs. 'preservation' space in the context of aeolian sequences. In a fluvial context, accumulation space ('process-scale' or real-time accommodation) would be the volume of space that can be filled within present process regimes, and is fundamentally governed by the relationship between stream power and sediment load, and how this changes in response to geomorphic base level. However, preservation in the stratigraphic record occurs when subsidence lowers these deposits below possible depths of incision and removal. As accumulations subside towards this depth, the probability of preservation increases, but it is only assured when this depth is reached. Hence preservation space (net accommodation?) lies below the maximum depths of incision. For a continental margin, like that shown in Fig. 13, preservation space would lie below the lowstand longitudinal profile.

Accommodation as a term in the vocabulary of sedimentary geology is certainly here to stay. However, it might be most useful to use accommodation explicitly in the same manner as Kocurek & Havholm (1993) and Kocurek (1998) use accumulation space, and future discussions should be explicit about deposition- vs. preservation-scale processes, since they operate at different rates, as well as over different time-scales, but ultimately result in what is preserved in the stratigraphic record. While accommodation remains a useful concept, the stratigraphic concept of base level is the same as the upper limits of accommodation – it is not a control on anything, has outlived its usefulness, and it should simply be abandoned.

MAGNITUDE AND FREQUENCY OF CLIMATE AND SEA-LEVEL CHANGE

Fluvial systems respond to climate changes within their catchments and/or relative sea-level changes that occur at their mouths. Over time-scales of interest here, climate change occurs because of global-scale changes in radiation balance and atmospheric or ocean circulation, and sea-level changes have a globally coherent 'eustatic' component. However, global-scale changes are manifested in specific regions in a variety of ways. The following section outlines key elements of climate and sea-level change, as viewed from the Quaternary vs. the Phanerozoic.

The Quaternary

The Quaternary provides a unique perspective on the dynamism of the Earth's climate and oceanographic systems, and has been the subject of intense study over the past three decades. Recent publications that summarize records and mechanisms of Quaternary climate and/or sea-level change include Crowley & North (1991), Broecker (1995), Pirazzoli (1996), Lowe & Walker (1997) and Bradley (1999).

The most significant development in Quaternary palaeoclimatology over the last 3–4 decades has been the new picture of glacial–interglacial cycles that emerged from studies of oxygen-isotope records in deep-sea sediments (e.g. Shackleton & Opdyke, 1973; Imbrie *et al.*, 1984; Chappell & Shackleton, 1986; Shackleton, 1987). Isotope records are responsible for the widespread acceptance of the Milankovitch theory of climate change (Hays *et al.*, 1976; Imbrie & Imbrie, 1979), and provide a model for major changes in global climate and ice volume, as well as the corresponding eustatic component of sea-level over the last several millions of years (see the recent review by Pillans *et al.*, 1998). There has been a high-amplitude 100-kyr periodicity to full glacial–interglacial cycles during the middle to late Pleistocene, but lower amplitude 40-kyr cycles dominated prior to that. One important implication pertains to the concept of 'average conditions', since only a small percentage of the last 10^6 years resembled full interglacial (present-day) or full glacial (20 ka) conditions (with eustatic sea-level at −120 to −130 m). As noted by Porter (1989), up to 80% of any middle to late Pleistocene 100-kyr glacial cycle witnessed global temperatures that were cooler than full interglacial conditions, but not as cold as a full glacial, and eustatic sea-level was at −15 to −85 m (Fig. 14A).

More recently, work on ice cores and the marine record shows that significant climate changes at scales of 10^3 and 10^4 years are superimposed on the 100-kyr glacial–interglacial cycles discussed above (e.g. Johnsen *et al.*, 1992; Bond *et al.*, 1993, 1997; Grootes *et al.*, 1993; Bond & Lotti, 1995; Behl & Kennett, 1996), and provides an exciting picture of the complexity of the last glacial, as well as the Pleistocene–Holocene transition. 'Dansgaard–Oeschger cycles' (D-O cycles), for example, consist of abrupt warming over a period of decades to centuries, followed by cooling trends that last 1–2 kyr. D-O cycles are themselves bundled together into cooling trends that last 10–15 kyr ('Bond cycles'), culminate with massive iceberg discharges and deposition of large quantities of ice-rafted debris into the North Atlantic ('Heinrich events'), and are followed by abrupt warming (Fig. 14B). The causal mechanisms for these cycles are not completely understood, but they must in some way reflect the internal dynamics, and feedback within, the coupled atmosphere–cryosphere–ocean system (Bradley, 1999), and their effects are propagated around the ocean basins via

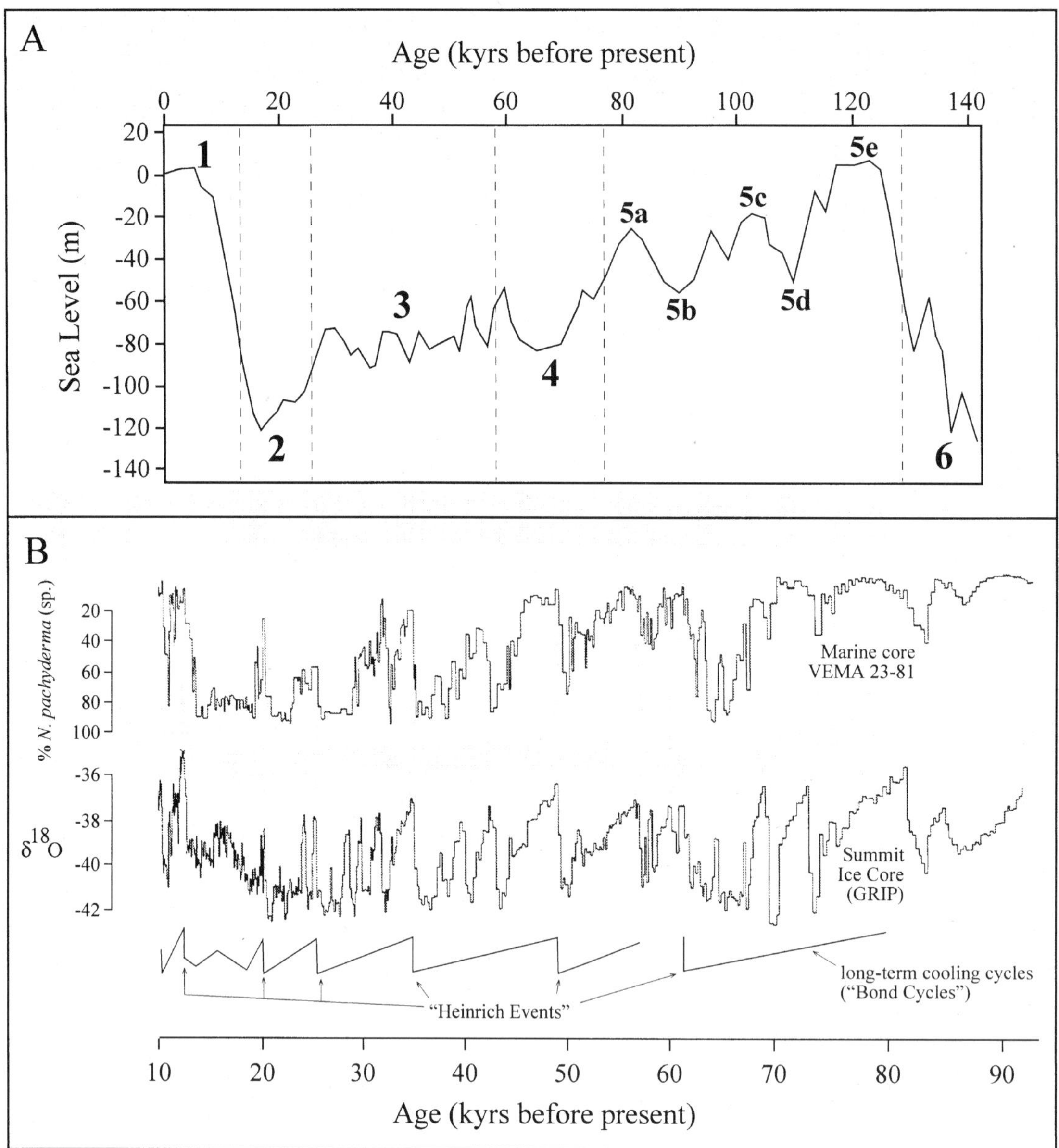

Fig. 14. (A) Eustatic sea-level curve based on precision U–Th dating of corals from Huon Peninsula, Papua New Guinea (after Chappell *et al.*, 1996), with oxygen isotope stages (boundaries shown by dashed lines). (B) Comparison of the proportion of the planktonic coldwater foraminifera *Neogloboquadrina pachyderma* (sinistral) in marine core VM 23–81 (after Bond *et al.*, 1993).

impacts on the large-scale thermohaline circulation and formation of North Atlantic deep water (Broecker, 1997).

Regional climate changes are more complex than the global picture due to the way global changes in atmospheric circulation are manifested in different regions; for the last 20 kyr, many regions show significant departures in magnitude and/or direction from global climatic trends (COHMAP Project Members, 1988; Wright *et al.*, 1993; Kutzbach *et al.*, 1997). Local sea-level changes also most commonly depart from eustasy in both rate and magnitude, and sometimes in direction, due to local to regional tectonics, mantle response to accumulation and melting of ice sheets, glacio- and hydro-isostatic deformation of coastlines, and migration of ocean water mass to geoidal lows (Tushingham & Peltier,

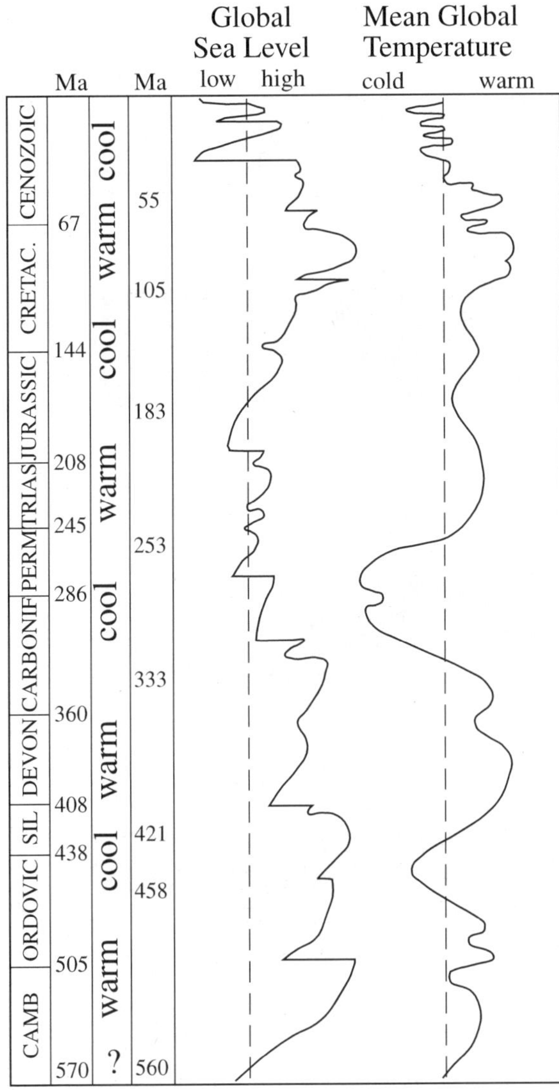

Fig. 15. Estimated eustatic sea-level and global temperature changes through the Phanerozoic, illustrating warm and cool modes. Adapted from Frakes *et al.* (1992).

1992; Lambeck, 1993; Fleming *et al.*, 1998; Peltier, 1998). Strong eustatic signatures may be present at some localities, but all sea-level change is essentially relative.

Pre-Quaternary

A broad range of climate and sea-level change is evident from the Phanerozoic record (Frakes, 1979; Crowley & North, 1991; Frakes *et al.*, 1992; Parrish, 1998), and only small parts of this record may resemble the Quaternary 'ice-house'. Perhaps the most important distinction within the Phanerozoic is between what Frakes *et al.* (1992) refer to as alternating 'warm' and 'cool' modes with average durations of 50 Myr, and the long-wavelength cycles of relative sea-level change evident in Haq *et al.* (1987, 1988; Fig. 15). For the purposes of this paper, such climate changes are initiated by the agglomeration and break-up of supercontinents, and/or by migrations of landmasses into different latitudinal belts, and are beyond the time-scales of interest here.

Of great interest, however, are the externally forced cyclical climate and sea-level changes that might go on within these longer-term modes. For example, there is little reason to believe that Milankovitch-scale radiation-forced changes have not been significant throughout the Phanerozoic (Barron & Moore, 1994), even though their periodicity, amplitudes and phase relations may have changed. Indeed, calculations by Berger *et al.* (1989) and Berger & Loutre (1994) show that wavelengths of the present 100 kyr eccentricity cycle have varied little since 500 Ma, whereas the obliquity (currently 41 kyr) and precessional (currently 23 and 19 kyr) components were shorter at 500 Ma (~ 30 kyr, 19 and 16 kyr, respectively) due to changes in Earth–Moon distance and day length. Even though Milankovitch-scale changes were present throughout the Phanerozoic, they operated within a context defined by plate tectonics and long-term evolution of the biosphere. Accordingly, this type of external forcing should have different manifestations in cold modes, when continental landmasses were located in more polar latitudes and glaciation was widespread, than in warm modes when evidence for significant glaciation is lacking (Frakes *et al.*, 1992).

Two caveats should be kept in mind. First, Broecker (1997) stresses the possible amplification of relatively limited orbital forcing by changes in the ocean thermohaline circulation during considerable parts of Phanerozoic history. Second, and equally important, is the amplification or suppression of flood regimes that occurs even with minor changes in climate, and the manner by which such changes are manifested in channel geometry, fluvial depositional styles, sediment transport rates and changes through time in net sediment storage. The key question is how to recognize pre-Quaternary fluvial responses to climate change, sea-level change, or some combination of the two, and disentangle them from tectonic effects or autogenic processes.

FLUVIAL RESPONSE TO CLIMATE AND SEA-LEVEL CHANGE: RECOGNITION

Potential fluvial responses to climate and sea-level change vary by position within the system, and the ability to store sediment over various lengths of time. At one extreme are the high stream powers and high sediment delivery ratios common in the upper reaches of drainage networks. To the extent that these settings often possess bedrock channels with discontinuous floodplains, and short residence times for sediment as a whole, there is little potential for records of fluvial response to climate change that span significant lengths of time ($>10^4$ years). By contrast, lower-gradient mixed bedrock/alluvial valleys with significant drainage areas ($>10^3$ km^2) commonly have continuous floodplains and low sediment delivery ratios. These settings typically show long-term rates of incision that may outpace rates of rock uplift (Hovius, 1999), and serve as the 'conveyor belts' for sediment delivery to basin margins, with sediment residence times measured in multiples of 10^4–10^6 years. As a result, mixed bedrock/alluvial valleys commonly contain records of fluvial landscape evolution over late Cenozoic time-scales, but have little to no actual preservation space.

Alluvial valleys within subsiding basins represent the opposite extreme, and provide the critical link between Quaternary and ancient systems. Present-day continental interior basins occur within tectonically active drylands, where drainage integration has not proceeded to the extent that fluvial systems reach marine basins, or in subsiding reaches of fluvial systems that are en-route to a marine basin. Rift and pull-apart basins of the arid western USA provide examples of the former, whereas the Himalayan foreland provides an example of the latter. Continental margin systems are numerous, but much of the present-day landscape is drained by a relatively few large, long-lived rivers (e.g. Potter, 1978) that discharge to passive margins or flooded forelands, whereas smaller fluvial systems typically drain to rifted or sheared margins. Sediments delivered to basins enter preservation space when they subside below maximum depths of incision, but until that time they can be remobilized by a variety of processes, and transported out of the basin (continental interior) or elsewhere within the basin (both continental interior and continental margin settings).

Types of response

Fluvial responses to climate and sea-level change can be disaggregated into stratigraphic, morphological and sedimentological components. As defined here, stratigraphic responses (stratigraphic architecture) are produced by aggradation due to increasing accumulation space, degradation as accumulation space decreases and lateral migration when there is no change in accumulation space. These types of responses serve to differentiate the larger-scale aspects of alluvial successions that mostly reflect allogenic controls (climate change, sea-level change, tectonic activity). Lower-gradient mixed bedrock/alluvial valleys typically contain flights of downward-stepping terraces and underlying fills that demarcate long-term valley incision punctuated by relatively short periods (10^3–10^4 years) of lateral migration and perhaps minor aggradation. Fluvial systems in subsiding basins are characterized by just the opposite, long-term net aggradation, punctuated by relatively short periods (10^3–10^4 years) of incision and/or lateral migration. Stratigraphic responses are perhaps best defined in terms of unconformity-bounded stratigraphic units of varying scale (i.e. the 'allostratigraphic units' of the North American Commission on Stratigraphic Nomenclature, 1983; see also Walker, 1990, 1992; Autin, 1992; Blum & Valastro, 1994); definition of allostratigraphic units is independent of lithological characteristics, but should include: (a) basal erosional unconformities that trace laterally and upwards, eventually truncating older landscape or palaeolandscape elements that are commonly defined by palaeosols; and (b) upper boundaries that represent intact or truncated former floodplain surfaces that are/were defined by palaeosols or coals (Fig. 16).

For lower-gradient mixed bedrock/alluvial valleys, the vertical separation between allostratigraphic units reflects the frequency of responses to allogenic forcing, but also must correlate to long-term rates of uplift; with increasing rates of uplift, the preservation of older units becomes increasingly fragmentary. By contrast, for continental margin systems, vertical separation between allostratigraphic units reflects the frequency of responses, sediment supply and relative sea-level change (eustasy vs. subsidence), and with increasing rates of subsidence, preservation in the stratigraphic record becomes increasingly likely. Two scenarios highlight this relationship. First, contrast slowly subsiding vs.

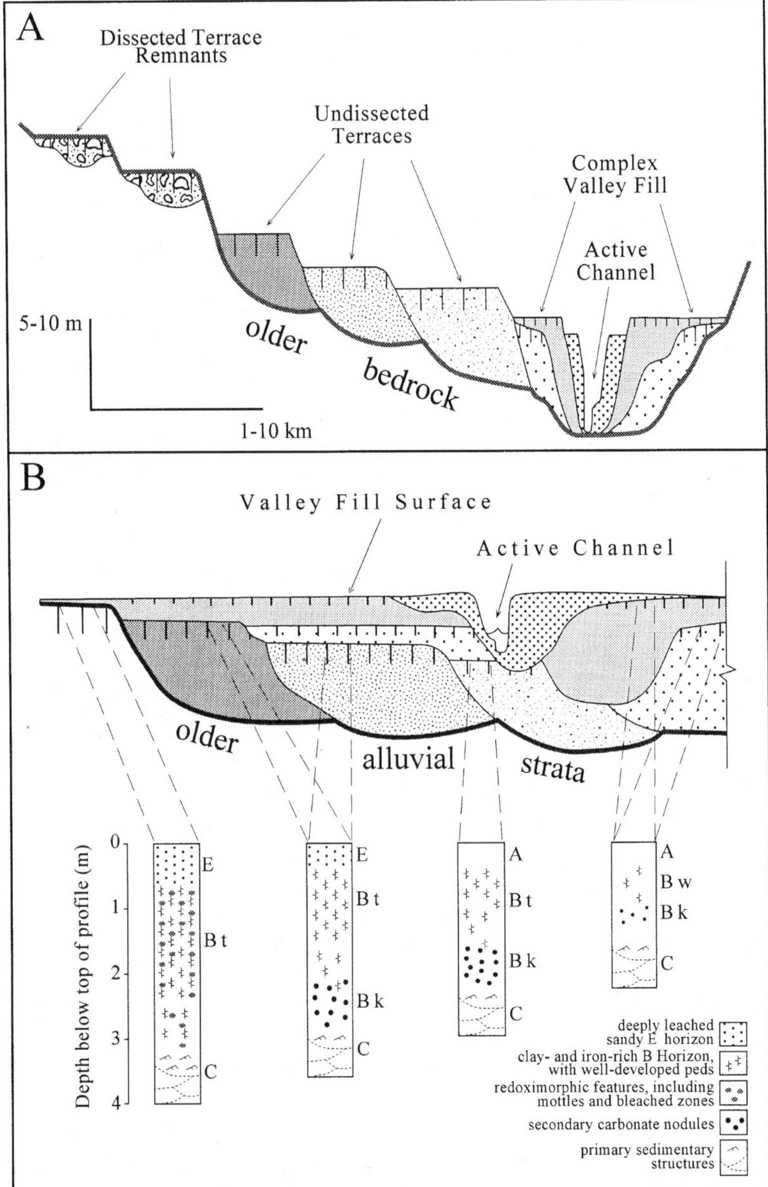

Fig. 16. Concept of allostratigraphic units of varying scale (scale bar applies to both diagrams, and vertical hatchures represent soil profiles with the relative degree of soil development represented by length of hatchure). (A) Mixed bedrock/alluvial valley setting, with flights of downward-stepping terrace surfaces and underlying fills, and a complex valley fill with multiple allostratigraphic units. Older and higher terraces are commonly discontinuous and dissected, with significant thicknesses of strata removed; they also may represent long periods of time, and multiple lateral migration, aggradation and incision events within a limited elevation range that are not readily differentiated due to degree of preservation. Younger mostly intact terraces and the complex valley fill may represent the last glacial cycle. (B) Continental margin situation, where sea-level has fallen then risen, resulting in downward-stepping then aggradational successions of allostratigraphic units bounded by palaeosols. Vertical hatchures represent palaeosols and, in each diagram, the solid grey line encloses a single 100-kyr glacial–interglacial cycle. In (B), the relative degree of palaeosol development is shown schematically by the length of the hatchure (sandy facies only, with muddy facies slightly different), and is a proxy for the duration of subaerial exposure before burial. As defined here, unconformities that bound allostratigraphic units are equivalent in concept to the higher-order bounding surfaces of Miall (1996), the basal valley fill unconformity in (B) would be a strongly time-transgressive equivalent to the Exxon sequence boundary as classically defined, and the well-developed alluvial plain palaeosol would be equivalent to the 'interfluve sequence boundaries' discussed by various workers. After the work of Blum & Price (1998) on the Colorado River, Texas (USA).

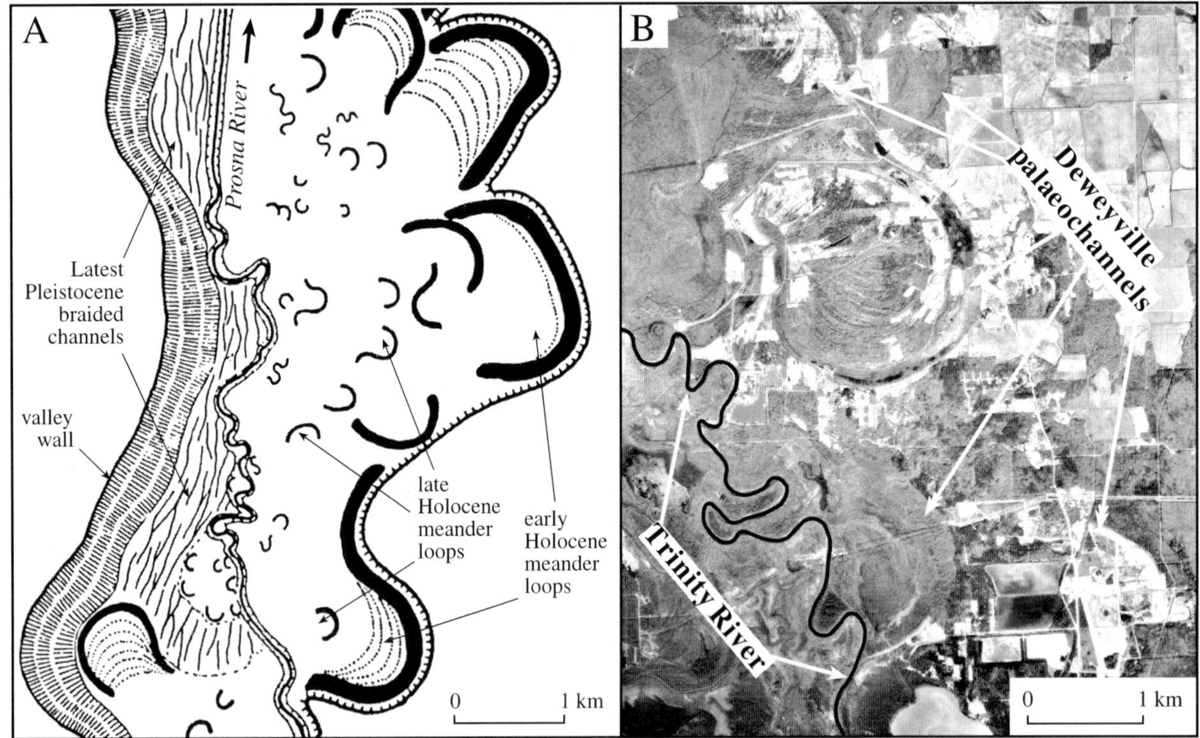

Fig. 17. (A) Channel changes in plan-view interpreted for the Prosna River in Poland (after Schumm & Brakenridge, 1987; original from Kozarski & Rotnicki, 1977), with latest Pleistocene braided channels, followed by large early Holocene meandering channels, and small late Holocene meandering channels. (B) Contrast between palaeochannels (Deweyville allostratigraphic units) from the last glacial cycle (OIS 2–4) and the modern channel of Trinity River, Texas Gulf Coastal Plain (see Blum *et al.*, 1995).

rapidly subsiding continental margins during eustatic sea-level fall; relative sea-level fall will be greater along the slowly subsiding margin, hence vertical separation of allostratigraphic units will be significantly greater. Second, during eustatic sea-level rise, relative rates of rise will be significantly greater along the rapidly subsiding margin, as will vertical separation of allostratigraphic units. Indeed, in rapidly subsiding settings, key stratigraphic markers like palaeosols may not form during eustatic sea-level fall due to lack of separation between units, or during eustatic sea-level rise due to continuous deposition and burial.

Morphological responses reflect changes in channel geometry, and occur within the context provided by differentiation of allostratigraphic units. Clear and unambiguous recognition of plan-view changes is limited to Late Quaternary examples, where channel patterns are preserved on terrace surfaces. Examples include braided to sinuous single-channel transitions that occurred along many European rivers (Fig. 17A; Starkel, 1991), and changes in meander dimensions within single-channel systems of the Texas Gulf Coast (Fig. 17B; Blum *et al.*, 1995). In theory, changes in cross-sectional dimensions (width, depth, width-to-depth ratio) can be recognized in both Quaternary and ancient settings, although in practice a number of complicating factors arise (see Ethridge & Schumm, 1978; Williams, 1987; Rotnicki, 1991; Wohl & Enzel, 1996). A well-studied Holocene example is provided by Knox (1987, 1993), who cored a large number of palaeochannel fills from small streams in Wisconsin (USA) to establish changes in channel cross-sectional area through time. A promising technique for the future may be the estimates of palaeochannel depth from detailed descriptions of the heights of cross-strata and strata sets. This method is based on establishment of relationships between thicknesses of preserved cross-strata and the dimensions of parent bedforms (Best & Bridge, 1992; Bridge, 1997; Leclair *et al.*, 1997), coupled with empirical relations that relate dune height to palaeoflow depth.

As defined here, sedimentological responses can be viewed within the context of alluvial

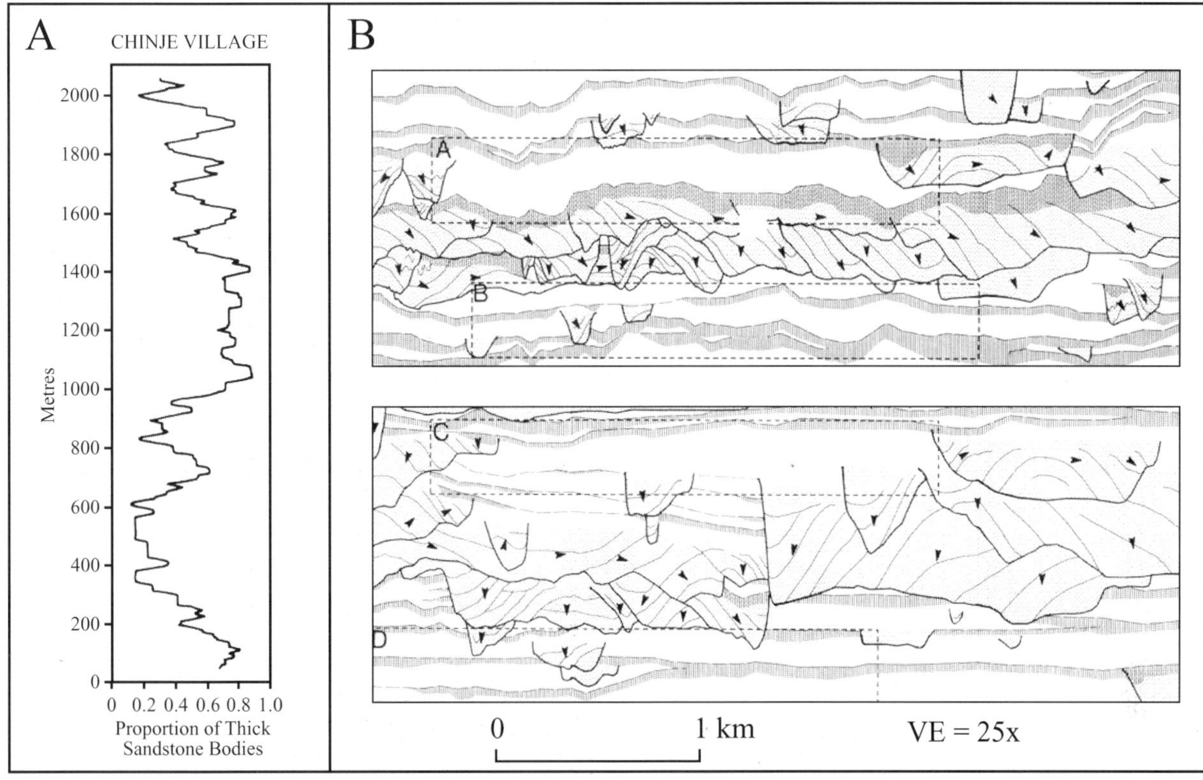

Fig. 18. (A) Proportion of thick sandstone bodies in the Siwalik succession at Chinje Village, Pakistan, relative to mudstone-dominated intervals. The diagram illustrates changes in sandstone proportions through time at several possible levels of cyclicity (after Willis, 1993b). (B) Line drawings of palaeosol-bounded stratigraphic units in the Siwalik succession. Stippled pattern illustrates sandbodies, white indicates mudstones, vertical hatchures indicate palaeosols and arrows indicate palaeocurrent directions (after Willis & Behrensmeyer, 1994).

architecture, initially described as the geometry, proportion and spatial distribution of different types of fluvial deposits (Leeder, 1978; Allen, 1978). Sedimentological responses reflect changes in channel geometry and the spatial distribution of depositional environments, and also occur within the context provided by differentiation of allostratigraphic units and bounding unconformities, or stratigraphic architecture as defined here. Sedimentological responses can be defined in a number of ways, depending on the types of data available, and range from simple descriptions of textural changes to detailed description and quantification of changes in the proportions of different lithofacies, facies associations and bounding surfaces. Outstanding detailed descriptions of alluvial strata can be found in the series of papers that discuss the Neogene Siwalik succession of northern Pakistan (Willis, 1993a,b; Willis & Behrensmeyer, 1994; Khan *et al.*, 1997; Zaleha, 1997). Among the many characteristics highlighted are successions of palaeosol-bounded floodplain facies, and quasi-cyclical vertical changes in mean grain-size and the proportions of thick sandstone bodies relative to muddy floodplain facies (Fig. 18).

Palaeosols deserve special mention (Wright, 1986; Retallack, 1986, 1990; Mack *et al.*, 1993; Birkeland, 1999), and a critical issue is the degree to which palaeosols represent allogenic or autogenic controls. There is, for example, a long history to the use of palaeosols to subdivide and correlate flights of Quaternary terraces and underlying fills; a key underlying premise in studies of this kind is the soil chronosequence and how soil profiles are increasingly well developed with increasing durations of subaerial exposure (see Birkeland, 1999, for a recent summary). By contrast, studies of Holocene soils as analogues for pre-Quaternary strata (e.g. Aslan & Autin, 1998), or pre-Quaternary palaeosols themselves (e.g. Allen, 1974; Leeder, 1975; Bown & Kraus, 1987; Kraus, 1987; Willis & Behrensmeyer, 1994; Behrensmeyer *et al.*, 1995), have more commonly focused on how palaeosol characteristics are related to aggradation rates, distance to active channel belts or floodplain hydrology. However, the advent of sequence stratigraphy has drawn attention to the use of palaeosols as key stratigraphic markers that trace from interfluves into adjacent palaeovalleys, and therefore serve as sequence-bounding unconformities (e.g. Wright & Marriott, 1993; Gibling & Wightman, 1994; Aitken & Flint, 1996; Blum & Price, 1998; McCarthy & Plint, 1998). Moreover, it is important to realize that palaeosols with lesser degrees of development, or perhaps different characteristics, may serve as boundaries to smaller-scale allostratigraphic units (e.g. Blum & Price, 1998; see Fig. 16B). Indeed, most important from a stratigraphic point of view may be differences in the duration of periods of nondeposition and the degree of development of pedogenic characteristics, which may provide an index of the time-scales over which different allostratigraphic units form. Clearly, palaeosols deserve close scrutiny in the future.

FLUVIAL RESPONSE TO CLIMATE CHANGE: EXAMPLES

Recognition of fluvial response to climate change in continental interiors often becomes a question of spatial and temporal scale, the degree of temporal resolution possible given existing dating methods and predilection of the investigator for autogenic (everything is 'complex response') vs. allogenic (everything is climate change) controls. There are few 'smoking guns' for any particular model, and within the Quaternary community, standards for correlation of changes in alluvial successions increase with decreasing age, such that debate can ensue regarding discrepencies of 10^1–10^2 years as the age of deposits approach the period of historical monitoring. On the one hand it seems unlikely that autogenic mechanisms can result in mappable stratigraphic units that can be traced over significant distances within, or correlated between, river systems, whereas on the other hand there is every reason to assume that fluvial systems are sensitive to climatically controlled changes in discharge and sediment load. However, there is also every reason to assume that: (a) fluvial response to global climate change will vary spatially such that different regions may be out of phase; and (b) the frequency of adjustments to global climate change will vary between river systems due to different thresholds for change, i.e. differential sensitivity.

Because of the numerous uncertainties, the most convincing studies interpret chronologically controlled alluvial stratigraphic frameworks within the context of independent records of climate change. A number of reviews or regional syntheses have been published, with most focusing on the last 20 kyr and often concerning small systems. These include Knox's (1983) summary of Holocene records from the USA, Bull's (1991) documentation of Quaternary histories for several fluvial systems in the western USA & New Zealand, Starkel's (1991) summary of late Pleistocene and Holocene changes in European rivers, Bravard's (1992) summary of late Pleistocene and Holocene data from France (especially the Rhone), Macklin & Lewin's (1993) discussion of Holocene records in England, Schirmer's (1995) discussion of central European (mostly German) rivers during the last 20 kyr and Knox's (1996) overview of global records from the last 20 kyr. There is no reason to provide an additional review here; instead, two case studies are used to demonstrate contrasting styles of fluvial responses to climate change.

Loire River, France

The Loire is one of the larger rivers in Europe, with a drainage area of ~ 120 000 km^2. The Loire is typical of many mid-latitude continental interior fluvial systems in that it has a flight of downward-stepping terraces and underlying fills that demarcate long-term valley incision punctuated by relatively short periods of lateral migration and minor aggradation.

Recent work by Straffin *et al.* (2000) in the upper Loire recognized five allostratigraphic units from the last glacial period (oxygen-isotope stages 2–4; hereafter OIS 2–4) through the Holocene (OIS-1; Fig. 19). Terrace surfaces with braided channel traces occur at elevations of 17 and 12 m above the modern low-water channel and define the upper boundaries to two glacial-age allostratigraphic units (T6 and T5). Underlying fills are 6–10 m in thickness, and interpreted to represent high concentrations of coarse sand and gravel along valley axes under periglacial climatic conditions, and moderate-magnitude floods produced by snowmelt runoff. Inter-vening periods of valley incision represent diminished sediment supply due to increased slope stability, and/or perhaps increased flood magnitudes and sediment transport capacity associated with moderately warm periods. The last significant period of valley incision corresponds to the latest

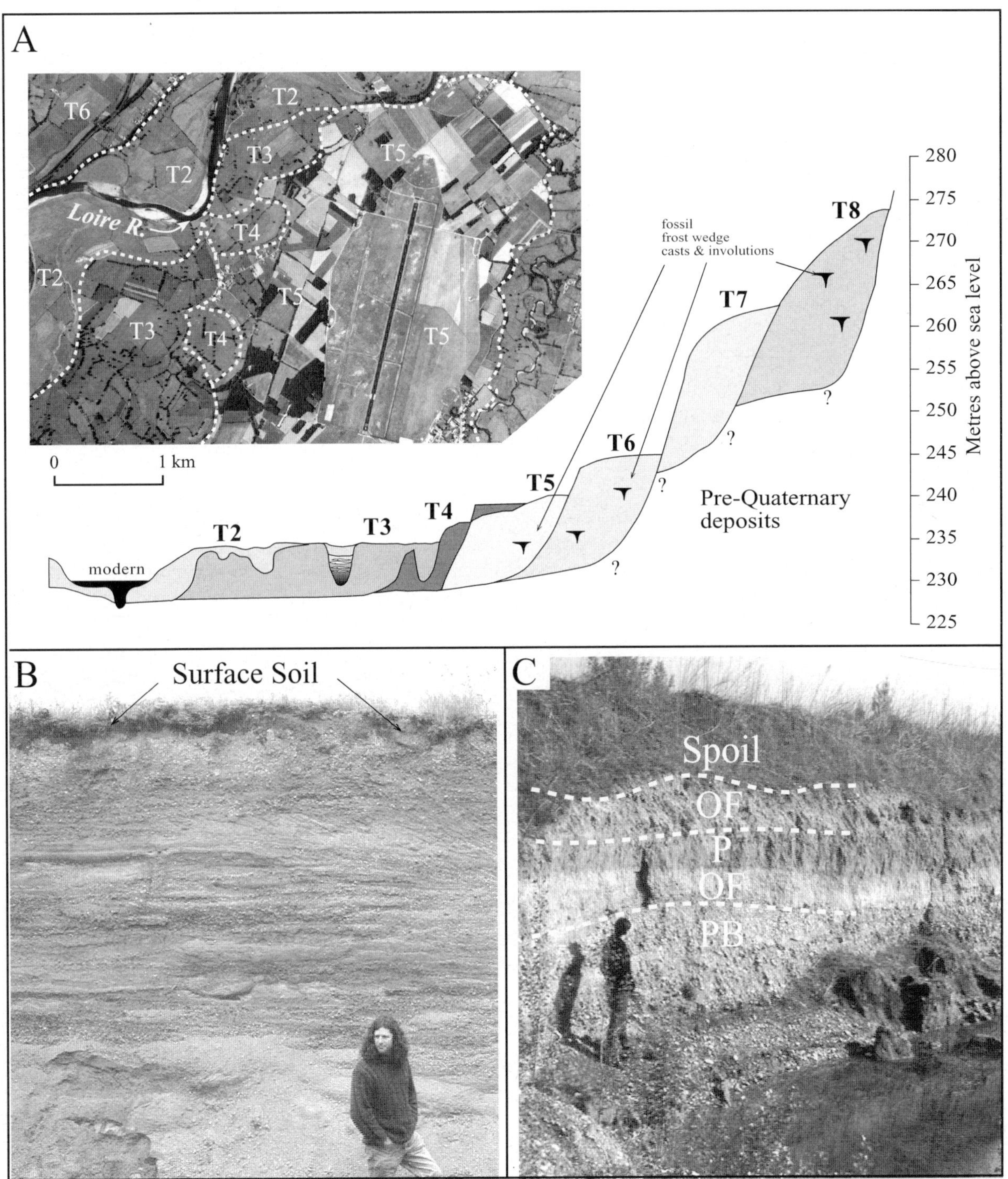

Fig. 19. (A) Aerial photo and schematic geomorphic and stratigraphic cross-section from a portion of the Loire River, France, illustrating large late Pleistocene braided stream surfaces (T5) and Holocene surfaces with clear high-sinuosity single-channel traces (modified from Straffin *et al.*, 2000). (B) Typical glacial period facies, with sets of trough cross-strata, but no overbank facies (unit T5 in (A)). (C) Typical late Holocene facies, illustrating overbank facies with palaeosols resting on point bar gravel (unit T2 in (A)). PB = point bar gravel, OF = overbank fine sand, P = weakly developed palaeosol.

Pleistocene–Holocene transition, and Holocene allostratigraphic units are part of a complex valley fill, some 5–7 m in thickness, that was deposited by a succession of single-channel systems of varying sinuosity. Early to middle Holocene units display high-sinuosity meandering planforms, whereas late Holocene units have lower sinuosity channels (Fig. 19A).

Stratigraphic, morphological and sedimentological responses of the Loire to climate change during the last glacial cycle can be discussed at two contrasting scales. First, as is the case for many larger European rivers, transition from a glacial to interglacial climate produced fundamental transformations in channel geometry and depositional style. Glacial-age units record deposition by braided channels, and systematically lack fine sandy and muddy overbank strata (Fig. 19B), whereas Holocene units were deposited by sinuous single channels, and contain appreciable thicknesses of fine-grained overbank facies (Fig. 19C). Second, high-amplitude, low-frequency adjustments characterized the relatively long glacial period, with only two extended periods of sediment storage punctuated by episodes of valley incision over a period of 50–60 kyr. By contrast, low-amplitude, high-frequency adjustments have been typical of the relatively short Holocene, with little if any net valley incision or net changes in sediment storage, and only minor adjustments in channel geometry. However, relatively frequent changes in climatically controlled flood regimes have resulted in alternating periods of floodplain stabilization with soil development vs. high-magnitude flooding and burial of palaeosols by overbank facies.

Over longer time periods, records of fluvial activity in the Loire are strongly biased in favour of the long glacial periods, whereas short interglacial periods like the Holocene would seem to have a low potential to be recorded in the landscape. This type of situation may be typical of many continental interior river systems in the unglaciated mid-latitudes and subtropics.

Mississippi River, USA

The Mississippi River served as the principal conduit for meltwater and sediment discharging from the southern margins of Pleistocene ice sheets, and the Lower Mississippi Valley (LMV) provides an example of the response of a large, long-lived, low-gradient fluvial system to the growth and decay of ice sheets. Indeed, landforms and deposits of the LMV (Fig. 20A) testify to the scale, complexity and dynamism of this continental-scale fluvial system, as do the changing interpretations through time.

Blum *et al.* (2000) used stratigraphic relations between chronologically controlled loess units and subjacent fluvial deposits to provide a model for LMV history (Fig. 20B) that differs from both Fisk (1944) and Saucier (1994). In this model: (a) major braided stream surfaces are either late Illinoian (OIS 6) or late Wisconsin age (OIS 2); (b) there is little evidence for preserved fluvial deposits from the last interglacial period (OIS 5) through much of the valley; and (c) deposits from the early Wisconsin substage (OIS 4) are preserved in protected settings only. At a finer level of detail, a difference in elevation of up to 5 m between three successive late Wisconsin terraces indicates that large braided stream surfaces were rapidly constructed, incised and abandoned in response to high-frequency fluctuations in the delivery of meltwaters and outwash from the former ice margin and the drainage of large proglacial lakes. The tremendous volumes of meltwaters during this time also resulted in the simultaneous occupation of two large-scale channel courses (Western and Eastern Lowlands), as well as breaching of a new bedrock gap (Thebes Gap) through which the modern river flows. Subsurface data are not sufficient to distinguish allostratigraphic units; however, from surface mapping, three major units might be inferred to represent the last deglaciation (a period of 10–12 kyr) within the main valley and in overflow courses (Fig. 20C).

Unlike the Loire, where flights of terraces record long-term valley incision punctuated by relatively short periods (10^4 years) of lateral migration and minor aggradation, the stratigraphic, morphological and sedimentological record of the low-gradient LMV at any point in time will always be biased in favour of the most recent period of deglaciation. Maximum floodplain elevations of only 100 m can be found at distances of 1000 km upstream from the present highstand shoreline, so there is little opportunity for long-term progressive downcutting and development of terrace flights, and strata from the penultimate glacial period (OIS 6) are only preserved due to the wholesale abandonment of that channel course in favour of the present combined Mississippi–Ohio valley. Within the context of these longer-term controls, large fluctuations in discharge and sediment loads delivered from the ice margin during deglaciation serve to ensure that relatively short periods of time (~ 10 kyr) at the end of successive 100-kyr glacial cycles (late OIS 6 and late OIS 2) dominate the landscape, and deposits from earlier parts of a glacial cycle or previous interglacial periods are not widely preserved.

Pre-Quaternary records

As discussed above, several examples of pre-Quaternary fluvial response to climate change have been described in the literature. It is clear that records of this kind must be more widespread than presently interpreted, but reasonable confidence in such interpretations is hampered by a number of factors. Among these, the temporal resolution provided by existing dating techniques is the most obvious and most important limitation. High-resolution palaeomagnetic chronologies, for example those of the Neogene Siwalik succession (Khan *et al.*, 1997), permit calculation of average time periods represented

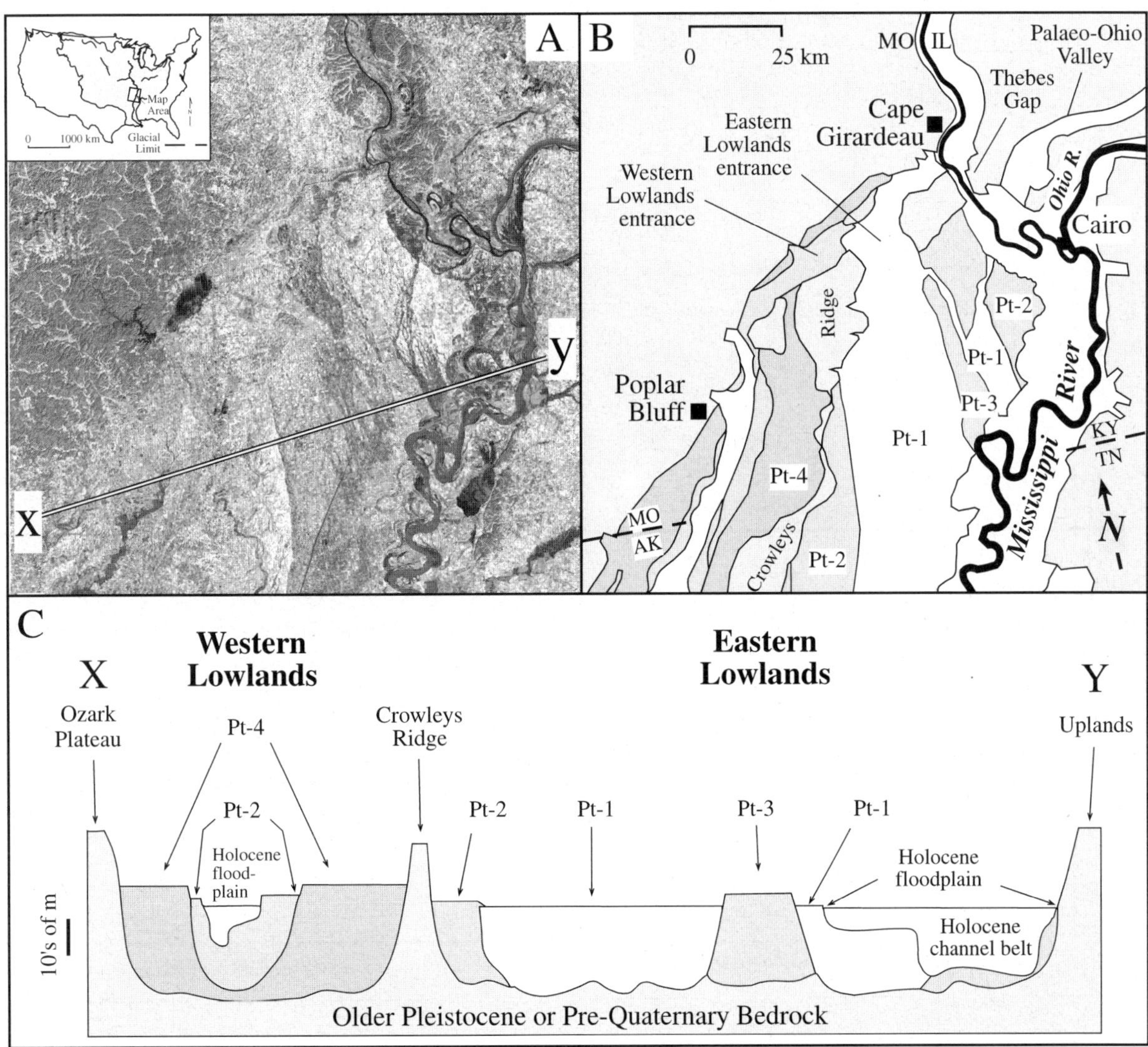

Fig. 20. (A) Landsat image of northern part of the Lower Mississippi alluvial valley (USA), with braided late Pleistocene vs. sinuous meandering Holocene channel belts. (B) Surficial geological map of the area shown in (A). (C) Schematic cross-section of inferred stratigraphic relationships along x–y in (A). Pt-4 = loess-covered terrace and underlying glacio-fluvial deposits from the OIS 6 glacial period, Pt-3 = terrace and underlying glacio-fluvial deposits from the OIS 2 full-glacial period, Pt-2 and Pt-1 = braided stream surfaces and underlying deposits from the OIS 2 late-glacial period. Thicknesses of deposits are schematic only, but data in Saucier indicates 50–60 m is not uncommon. Holocene floodplain in Western Lowlands represents deposits of smaller streams that head on the Ozark Plateau to the west. After Blum *et al.* (2000).

by specific unconformity-bounded intervals of deposition that are well within the Milankovitch scale, and would seem to offer the best chance to evaluate fluvial responses to climate change. Within relatively well-constrained successions such as these, periods of incision and floodplain abandonment with formation of palaeosols, as well as changes in channel morphology and depositional style that occur across unconformities (especially when defined in part by palaeosols), are excellent candidates for interpreting fluvial responses to climate change.

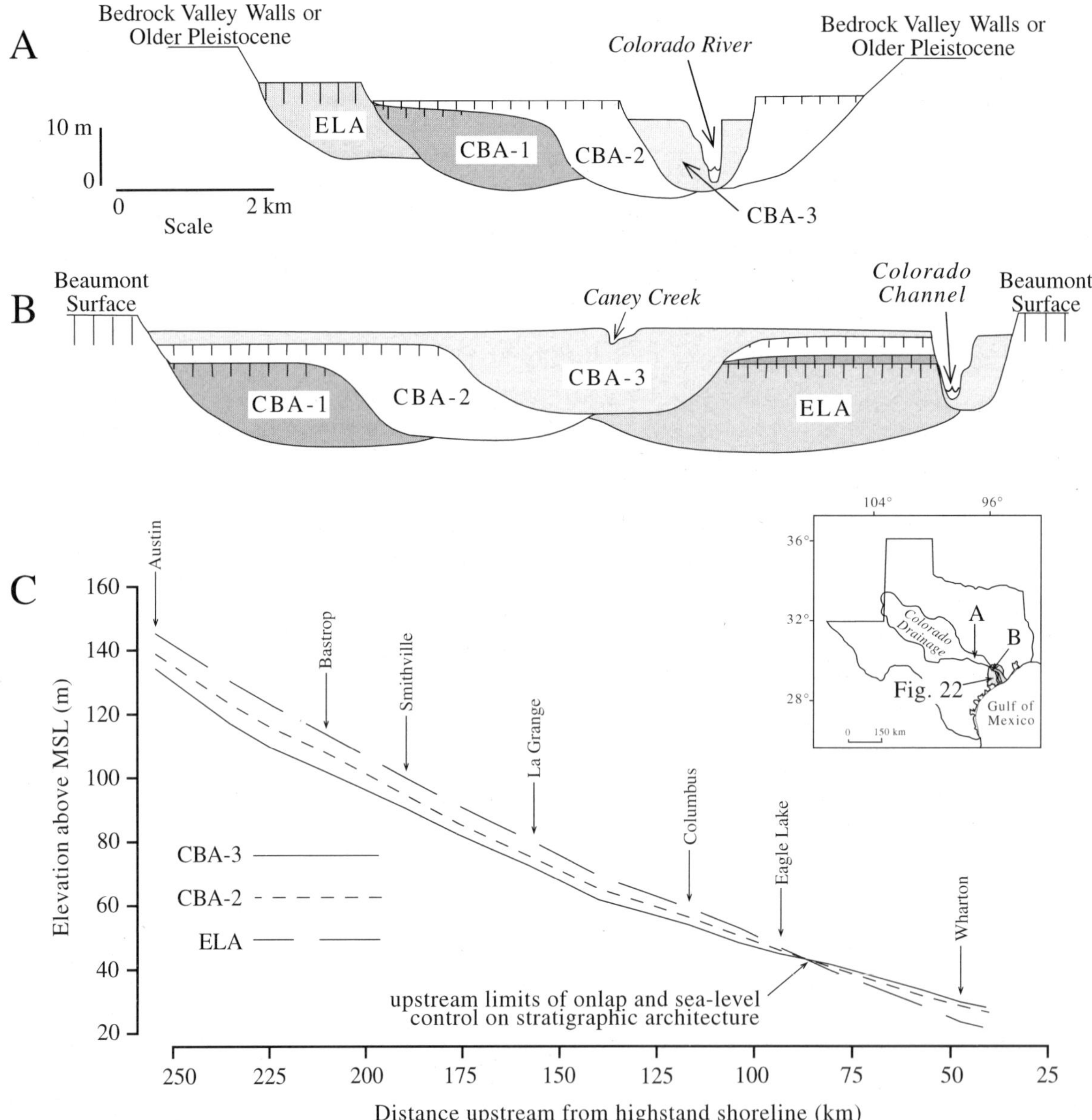

Fig. 21. Schematic cross-sections of the late Pleistocene and Holocene valley fill of the Colorado River, Texas Coastal Plain, illustrating downstream persistence of allostratigraphic units with downstream changes in stacking patterns. (A) The bedrock-confined degradational valley near La Grange, Texas, and (B) the lower valley near Wharton. Beaumont surface is the last interglacial highstand surface (see Fig. 21A), ELA = Eagle Lake Alloformation of late Pleistocene full-glacial age, whereas CBA-1–3 = Columbus Bend Alloformation Members 1–3, of latest Pleistocene and Holocene age. Caney Creek was the active channel until 200–300 years ago, when the lower 90 km avulsed to its present position. (C) Long profiles for ELA, CBA-2 and CBA-3, illustrating crossover point, and upstream limits of onlap. Inset map shows locations for (A) and (B), with long profile in (C) extending between these two locations. Modified from Blum & Price (1998).

FLUVIAL RESPONSES TO CLIMATE AND SEA-LEVEL CHANGE: EXAMPLES

At some point upstream, fluvial systems become completely independent of high-frequency ($<10^6$ years) sea-level change. However, the reverse is not true, and there is no point moving in the downstream direction where the influence of climate and climatically controlled changes in discharge and sediment supply can go to zero – if it does not rain, nothing happens. Moreover, since glacio-eustasy is a function of climate change, it seems likely that some sort of autocorrelation will arise between the effects of climate vs. sea-level change. It follows from the above that continental margin fluvial systems must in some way reflect the influence of both controlling mechanisms, although in some cases the influences of sea-level change and subsidence may overwhelm the recognizable signature of other factors. Nevertheless, Quaternary systems in continental margin settings provide unique opportunities to isolate the nature of these interactions, as discussed in the two case studies below.

Colorado River, Texas

From studies of late Pleistocene and Holocene strata of the Colorado River, Gulf Coastal Plain of Texas (USA), Blum (1993) defined interactions between climate and sea-level change in terms of: (a) the downstream continuity of allostratigraphic units and component facies as a reflection of climatic controls; and (b) the corresponding downstream changes in stratigraphic architecture as a reflection of the increasing influence of sea-level change (Fig. 21). Subsequent work confirms these relationships for palaeovalleys that span the last 300–400 kyr (Blum & Price, 1998). Individual palaeovalleys correspond to 100-kyr glacial–interglacial cycles, and are partitioned by lowering of sea-level below highstand depositional shoreline breaks, when channels incise and valley axes become fixed in place as they extend across the subaerially exposed shelf. During falling stage and lowstand, multiple episodes of lateral migration, minor aggradation and degradation occur within the extended valleys in response to climatic controls on discharge and sediment supply. Transgression and highstand then results in channel shortening, and valleys begin to fill; as valley filling nears completion, overbank facies spread laterally and bury soils that developed on the downdip margins of the previous highstand alluvial plain. Complete valley filling promotes avulsion, with relocation of valley axes before the next sea-level fall, such that successive 100-kyr palaeovalley fills have a distributary pattern (Fig. 22A). Individual palaeovalleys range from 15 to 20 m in thickness at the updip limits of sea-level influence, to >35 m at the present highstand shoreline, whereas they extend 15–40 km along strike.

Within individual unconformity-bounded palaeovalley sequences, alluvial architecture reflects interactions between changes in climate and sea-level (Fig. 22B). During the glacial period, regional climate was more temperate than today and the shoreline was in a midshelf position or further basinward. Allostratigraphic units from this time period consist of amalgamated channel-belt sands with few overbank facies, and are bounded by well-drained palaeosols; these units can be 10–15 m in thickness, and extend 2–5 km along strike. Transition to the interglacial period resulted in increases in sea surface size and temperature, 'subtropical' land surface temperatures, and deeper inland penetration of tropical moisture, that in turn produced increases in the depth and flashiness of floods. Superimposed on this overall trend were higher-frequency climate changes as well (Toomey *et al.*, 1993). As a result, multiple episodes of aggradation, lateral migration and some incision with sediment bypass occurred contemporaneously with the late stages of transgression and sea-level highstand, but in response to climatic controls on discharge regimes and sediment supply. Even in upstream reaches, beyond any influence of sea-level, Holocene units contain an abundance of fine-grained overbank facies (as was the case for the Loire as well), and sand bodies are less amalgamated (Blum & Valastro, 1994; Blum *et al.*, 1994). These characteristics are enhanced within the lower valley by a forced shortening of channels and flattening of gradients.

Aslan & Blum (1999) also show that in far downstream reaches, rates of valley filling and style of avulsion during sea-level rise and highstand play a strong role in alluvial architecture. During early transgression, when the shoreline is still some distance basinward, avulsion occurs by reoccupation of abandoned falling stage and lowstand channels, with erosion and reworking of older channel-belt sands (Fig. 23A). By contrast, accumulation space is rapidly generated as the shoreline moves closer, rates of valley filling increase and avulsion occurs by repeated diversion into floodplain depressions. This produces successions of massive to laminated floodbasin muds that encase thin (< 5 m) ribbon-like crevasse channels and thin (< 2–3 m) sheet-like splay sands (Fig. 23B), and comprise up to 50% of the total valley fill (comparable to the 'avulsion deposits' of Smith *et al.*, 1989). Deposits of individual avulsions are separated by massive to slickensided muds or buried soil horizons that represent periods of slow sediment accumulation or floodplain stability between episodes of avulsive deposition. As filling of accumulation space nears completion, avulsion by channel reoccupation again becomes the dominant process, and again results in amalgamated channel belts

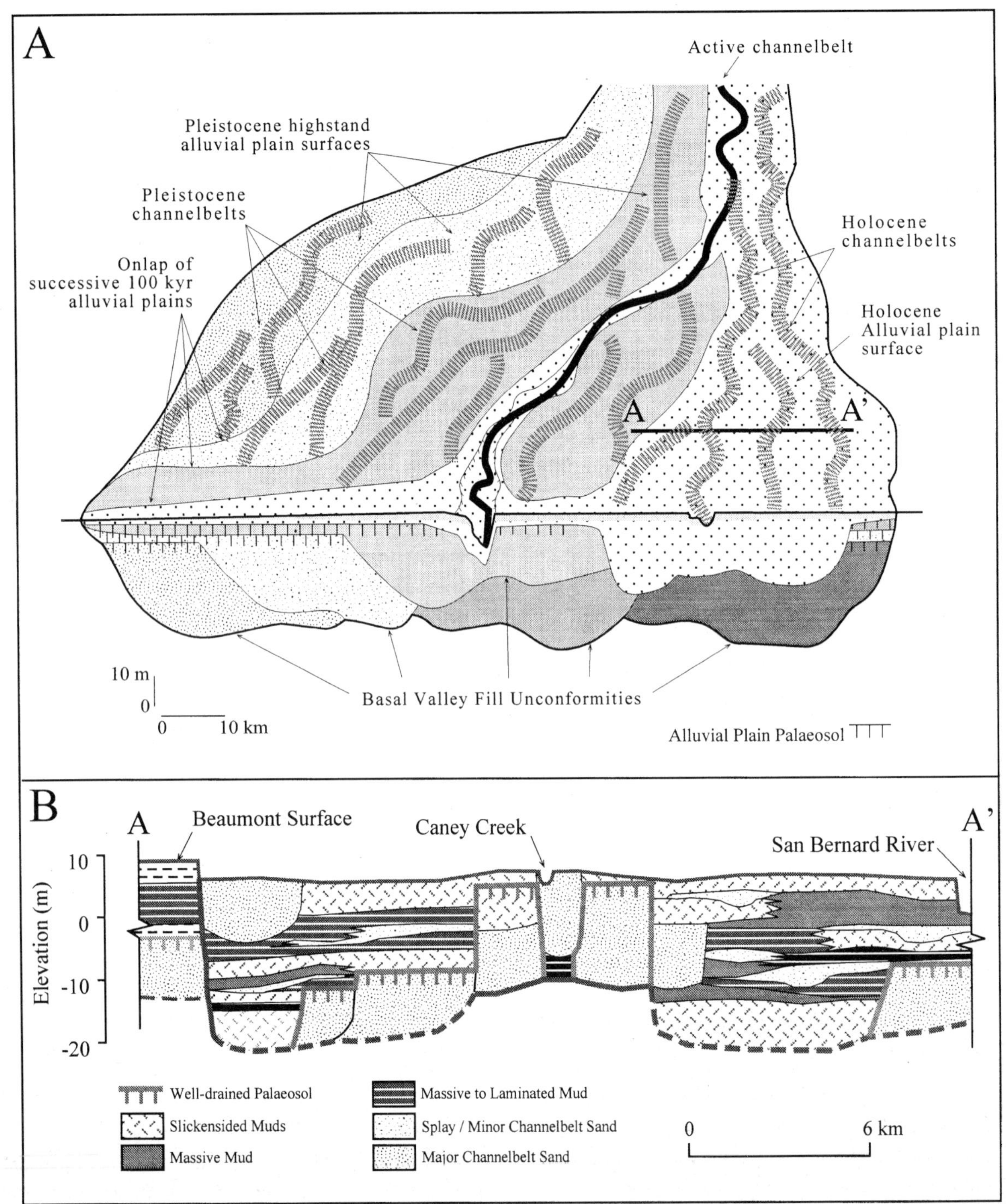

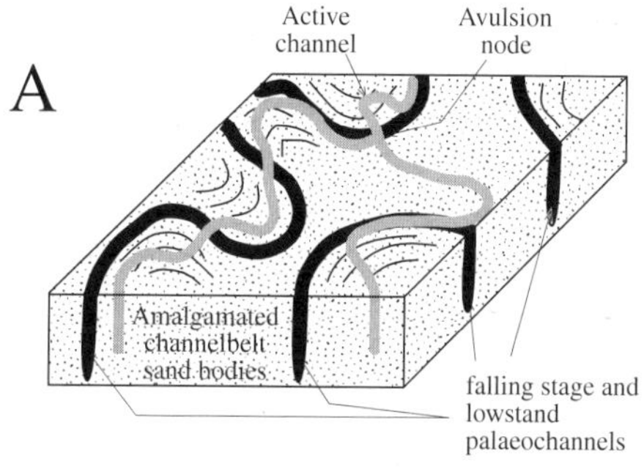

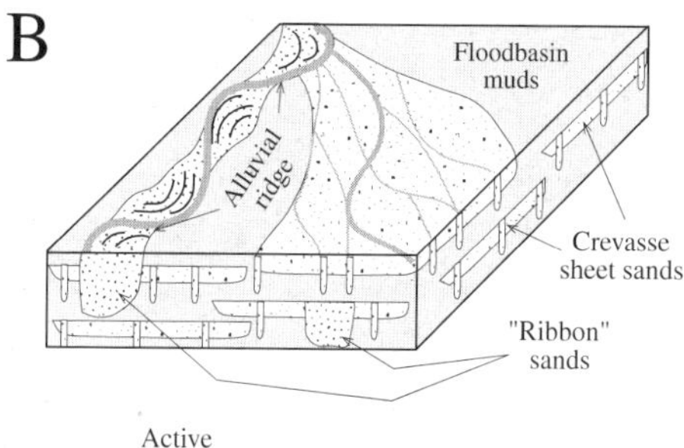

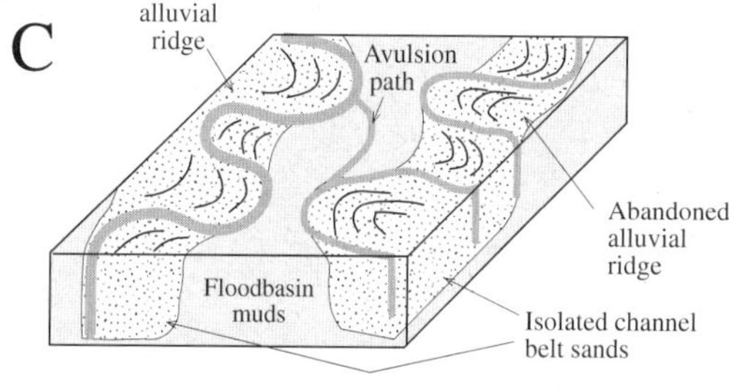

Fig. 23. Model for changes in avulsion styles during valley filling in response to sea-level rise. (A) Avulsion by reoccupation of falling stage and lowstand channels during early stages of valley filling, when rates of floodplain aggradation are very low. (B) Avulsion by frequent crevassing into floodplain depression in response to rapid increases in accumulation space that correspond to high rates of sea-level rise. (C) Avulsion by reoccupation of older channel belts, when sea-level highstand has been reached, and accumulation space is nearly filled (after Aslan & Blum, 1999).

(Fig. 23C). During the most recent avulsion, the Colorado River abandoned its valley from the last glacial cycle, and reoccupied a Pleistocene channel belt from the previous interglacial highstand (see Fig. 22A).

Rhine–Meuse Rivers, The Netherlands

Studies of the Rhine–Meuse Rivers during the last glacial cycle also clearly document interactions between upstream and downstream controls. During much of the last glacial (OIS 2–4) the Rhine mouth was located beyond the Strait of Dover, up to 800 km downstream from the present shoreline (Gibbard, 1995; Bridgland & D'olier, 1995), and direct sea-level influence was limited to reaches that are now submerged in the English Channel. Throughout the glacial cycle as a whole, periglacial conditions with sparse vegetation and continuous permafrost alternated with a temperate climate comparable to present-day, and it is generally assumed that sediment supply to the Rhine and Meuse was considerably higher when periglacial conditions prevailed (e.g. Vandenberghe, 1995). Recent work by Törnqvist *et al.* (2000) suggests that during the sea-level fall associated with OIS 4, substantial deposition occurred near the present Dutch shoreline despite net incision, similar to the Colorado River, as described above. The Rhine and Meuse were also predominantly aggrading braided systems during the Last Glacial Maximum sea-level lowstand (OIS 2, 25–13 ka; Kasse *et al.*, 1995), and climate change continued to play a paramount role during deglaciation as sea-level rose rapidly. As examples, during the relatively warm Bølling–Allerød interstadial (the last D-O Cycle), which corresponds to extreme rates of eustatic sea-level rise (meltwater pulse 1 of

---

Fig. 22. (A) Perspective map view and schematic cross-section of Pleistocene (Beaumont Formation) and Holocene alluvial plains of the Colorado River, Texas, illustrating large-scale stratigraphic architecture dominated by multiple cross-cutting incised valley fills. Individual valley fills are interpreted to represent 100 kyr glacial–interglacial cycles based on physical stratigraphic relations, mostly downdip onlap and burial of successively older alluvial plain palaeosols, especially the onlap of floodplain facies and palaeosols from one highstand by floodplain facies and palaeosols from successively younger highstands. A series of thermoluminescence ages support these age relations. The Holocene alluvial plain surface was recently abandoned, when the channel avulsed to reoccupy an older Pleistocene meanderbelt. Location shown in inset map of Fig. 21. After Blum & Price (1998). (B) Interpreted post-Beaumont valley fill cross-section along line A–A′, based on a series of continuous cores with 2 km spacing. Lowstand channelbelt sandbody equivalent to ELA in Fig. 20, whereas strata that overlie well-drained palaeosols are equivalent to CBA (individual members not differentiated). After Aslan & Blum (1999).

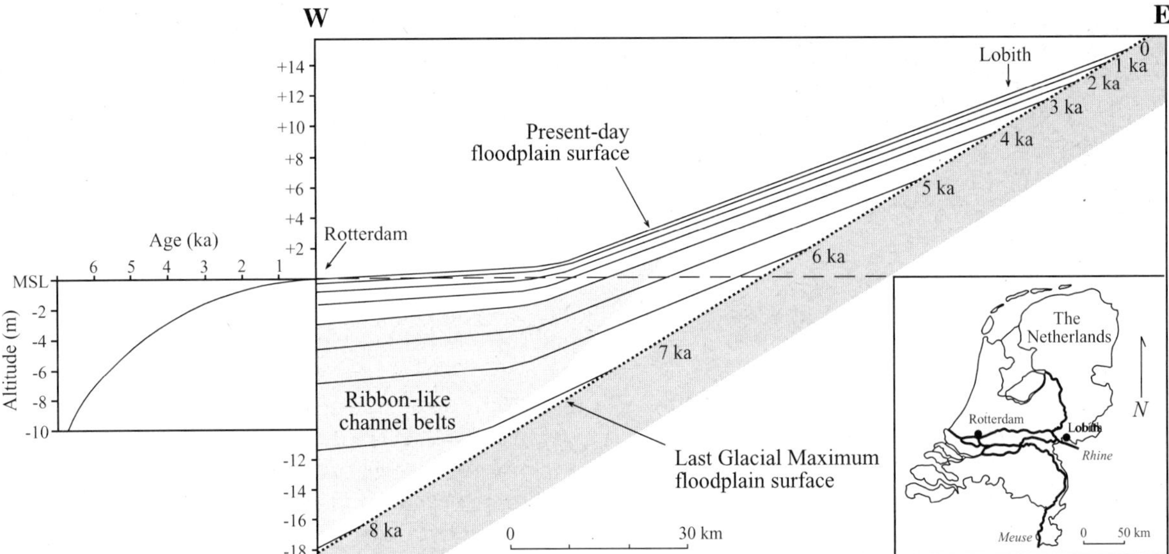

Fig. 24. Holocene rise of the (ground)water table in the Rhine–Meuse Delta as a function of relative sea-level rise (relative sea-level curve indicated to the left; mainly after Van de Plassche, 1982), primarily based on ~100 ^{14}C ages of basal peats (Van Dijk *et al.*, 1991). Spacing of the (ground)water gradient lines (isochrons with ages) define rates of creation of accumulation space. Ribbon-like channel belts encased in thick, organic-rich overbank successions dominate downdip areas, whereas sheet-like channel belts are typical of conditions with lower rates of increase in accumulation space. Note the intersection point between the Last Glacial Maximum and modern longitudinal profiles to the far east, representing the landward limit of onlap.

Fairbanks, 1989), as well as during the earliest part of the Holocene (meltwater pulse 2), the Rhine–Meuse incised in response to reductions of sediment supply due to re-establishment of dense vegetation (Pons, 1957; Kasse *et al.*, 1995; Berendsen *et al.*, 1995; Törnqvist, 1998).

Well-documented rates of rise in (ground)-water tables in the present Rhine–Meuse delta during the middle to late Holocene are interpreted as a proxy for sea-level rise (Van Dijk *et al.*, 1991; Fig. 24), which is in turn interpreted to have played a prominent role in the development of alluvial architecture. The database for interpretation of the Holocene record includes >200 000 boreholes (an average of 50–100 boreholes km^{-2}) plus some 1000 radiocarbon ages (e.g. Berendsen, 1982; Törnqvist, 1993; Berendsen *et al.*, 1994), and permits unparalleled documentation of changes in alluvial architecture through time. For the middle Holocene, prior to 4 ka, rates of sea-level controlled (ground)water-table rise were very high (typically >1·5 mm yr^{-1}). The record from this time consists of a succession of anastomosed (Törnqvist, 1993) or straight (Makaske, 1998) distributaries with ribbon-like channel belts (width-to-thickness ratios <15) encased in lacustrine muds and organics (Van der Woude, 1984), and avulsion frequency was very high (Törnqvist, 1994). By contrast, meandering single-channel distributaries with sheet-like channel belts (width-to-thickness ratios >50) and more homogeneous overbank deposits have dominated the late Holocene sea-level highstand, and avulsion frequency was significantly less. There are striking similarities between the middle to late Holocene Rhine–Meuse, the Texas Gulf Coast examples described above and recent re-evaluations of the middle to late Holocene of the Lower Mississippi Valley (Aslan & Autin, 1999).

Pre-Quaternary records

How can interactions between climate and sea-level change be recognized in the pre-Quaternary record? The bad news is that interactions defined by the downstream continuity of allostratigraphic units and component facies, with corresponding downstream changes in stratigraphic architecture, may be difficult to identify. For example, the intersection between lowstand and highstand floodplain surfaces defined along both the Texas Coast and in

The Netherlands will not be preserved, which may be typical of most passive margin settings where subsidence rates decrease landward. Indeed, as a general rule, stratigraphic units deposited during falling stage to lowstand should occur deeper within palaeovalley fills, and stand the best chance of preservation, whereas strata deposited during highstand will more commonly be removed during subsequent sea-level falls (cf. Törnqvist, 1995; Blum & Price, 1998; Törnqvist *et al.*, 2000). However, the good news is there is no reason to assume that interactions between climate and sea-level change will not occur, and downstream continuity of strata should be the rule, not the exception. Moreover, future interpretations of these interactions will at least in part rely on the well-tested method of developing modern analogues, only at 'sequence' time-scales (see discussion below).

In this context, it is worthy of note that models developed from the pre-Quaternary record (e.g. Shanley & McCabe, 1991, 1993, 1994; Wright & Marriott, 1993; Olsen *et al.*, 1995; Currie, 1997; Zhang *et al.*, 1997) infer relationships between alluvial architecture and sea-level change that are very similar to what is observed in Late Quaternary successions. Although details differ (compare Shanley & McCabe, 1993; Wright & Marriott, 1993), a common pattern would be amalgamated high net-to-gross channel belts that rest on what is interpreted to be a sequence boundary, and overlain by isolated ribbon-like sand bodies that are encased in laterally extensive mudstone-dominated successions. Many workers have interpreted the sequence boundary to reflect incision and sediment bypass during sea-level fall, and the amalgamated character of overlying sand bodies reflects low accommodation, whereas the more isolated sand bodies upsection reflect increasing accommodation due to relative sea-level rise and/or highstand. For the Late Quaternary examples discussed above, the basal palaeovalley unconformity would be interpreted as a composite erosion surface, cut during multiple periods of incision alternating with lateral migration and minor aggradation during falling stage and lowstand. Moreover, the contrast between laterally amalgamated and more isolated sand bodies that interfinger with mud-rich successions reflect fundamentally different glacial vs. interglacial climate regimes, respectively, upon which are superimposed sea-level controls on overall palaeovalley-fill architecture.

SYNTHESIS AND A LOOK FORWARD

The discussion above has addressed the history of research on fluvial responses to climate and sea-level change, outlined a basic process framework for investigations of this kind, and provided examples of fluvial response to climate and sea-level change from work on Quaternary systems. The Mississippi River, which has long played an important role in the fluvial community, can be used to illustrate a key concept that may play a role in the future. In a previous summary of Mississippi River response to glaciation and deglaciation, Wright (1987) noted that an idealized model would predict some crossover point where aggradation due to glacial sediment loading gives way downstream to incision due to glacio-eustatic lowstand. Wright recognized that this crossover is more conceptual than real; the synthesis here takes this a step further and suggests this crossover is best viewed as a metaphor for how workers in different settings, Quaternary and ancient, have focused on causal mechanisms that reside in their own backyard. A bright future for studies of fluvial response to climate and sea-level change can be envisaged for the new millennium, one that will become increasingly integrative with respect to both upstream and downstream controls, as well as how they operate in tandem to produce the landscape and stratigraphic record. This review concludes by suggesting some key issues for the future.

Quaternary systems

The first 2–3 decades of this millennium should be an exciting time to study Quaternary fluvial systems, with challenges that await at both ends of the temporal scale. On the one hand, it will be important to keep pace with research in high-frequency climate and sea-level change. For example, how do fluvial systems respond to the Dansgaard–Oeschger events that are so prominent in records from Greenland and the North Atlantic? Holocene records clearly show that river systems respond in dramatic fashion to relatively subtle climate changes, but are there upper limits to the frequency of responses that can be interpreted from the late Pleistocene and Holocene stratigraphic record, and were river systems during the glacial period as sensitive to rapid climate changes as their interglacial counterparts? To date, much of the high-profile, well-funded work has focused on documentation of climate changes *per se*. However, humans live on, and interact with, the Earth's surface, which has received considerably less attention and funding. Understanding the spatial dimensions and significance of high-frequency climate and sea-level changes, and their effects on surface processes, will require more precise, systematically defined, whole-system (from source to sink) and/or regional stratigraphic and geochronological frameworks than are now widely available.

On the other hand, it is equally important to study longer-wavelength cycles of fluvial landscape evolution. It is fair to say that most detailed research on fluvial response to climate change during the last 3–4 decades has focused on the last 20 kyr. While important, this relatively short period encompasses only the tail end of a single 100-kyr glacial to interglacial cycle. Very little is actually known about fluvial responses to complete glacial–interglacial cycles, or how the first 70–80% of the last cycle have preconditioned fluvial processes during the last 20–30% of the cycle. A number of studies have begun to address fluvial responses to climate change at Milankovitch scales. Antoine (1994), for example, uses stratigraphic relationships between loess sheets and fluvial deposits in the Somme Valley of France to develop a relative-age geochronological framework, and then infers cycles of aggradation, degradation and terrace formation at the 100-kyr scale. Bridgland (1994) also infers 100-kyr cyclicity for terraces of the Thames in England. Moreover, terraces of the Meuse in the southern Netherlands span the entire Quaternary, and have been used to develop numerical models of long-term fluvial response to climate change and tectonic uplift, suggesting 100-kyr cycles of aggradation and incision (Veldkamp & Van den Berg, 1993). Investigations such as these also provide an opportunity to link studies of fluvial response to climate and sea-level change over longer time-scales with new ideas on relationships between tectonics and surface processes (e.g. Burbank & Pinter, 1999).

Two issues emerge as fundamental to future advancements in understanding the responses of Quaternary fluvial systems to climate and sea-level change over both shorter and longer time-scales. First, there is a clear need to develop and apply more sophisticated geophysical techniques to explore three-dimensional relationships at both the stratigraphic and alluvial architecture scales. Applications of ground-penetrating radar (GPR) and shallow high-resolution seismic will become increasingly important in this regard. GPR has, for example, been applied in a number of fluvial contexts, but mostly at the alluvial architecture scale, and more specifically with reference to internal structure at the channel bar scale (e.g. Huggenberger, 1993; Bridge *et al.*, 1995, 1998). Future applications of high-resolution geophysical techniques in Late Quaternary continental strata are not without problems, especially at the stratigraphic architecture scale, but may prove to be essential for developing linkages between continental strata and the high-resolution records that have been, and will continue to be, documented for alluvial–deltaic strata on the now submerged shelves. This, in turn, must play a critical role in developing an understanding of the relative importance of upstream vs. downstream controls at both the stratigraphic and alluvial architecture scales, and the evolution of fluvial systems from source to sink.

Second, and equally important, is a corresponding need to develop, refine and apply new dating techniques. Radiocarbon has served as the clear reference standard for almost five decades, and has had a profound impact on understanding the dynamics of terrestrial systems over the last 20–30 kyr. However, many records remain poorly constrained because they lack organic matter in key stratigraphic contexts, and the time period covered by radiocarbon represents only 30–40% of a single glacial–interglacial cycle. On the near horizon, application and refinement of the various luminescence techniques seem to offer the most promise. In theory, luminescence techniques measure the time elapsed since sand- and silt-sized grains were exposed to solar radiation during transport, and can provide numerical ages for deposits within the middle to late Pleistocene and Holocene. For fluvial deposits, optically stimulated luminescence (OSL) has considerable advantages over the older technique of thermoluminescence (TL), since only minutes of exposure are required during transport (Duller, 1996; Aitken, 1998). Both TL and OSL have been used in fluvial contexts; examples include TL dating of fluvial deposits in the Lake Eyre basin of Australia (Nanson *et al.*, 1988, 1992), Murrumbidgee River palaeochannels in south-eastern Australia (Page *et al.*, 1996), Saharan wadi systems (Blum *et al.*, 1998) and Texas Gulf Coast fluvial deposits (Blum & Price, 1998), as well as OSL dating of fluvial deposits of the Ebro drainage in Spain (Fuller *et al.*, 1998), the Loire in France (Straffin *et al.*, 2000) and the Rhine–Meuse in The Netherlands (Törnqvist *et al.*, 2000).

Pre-Quaternary systems

Two fundamental issues can be identified that should play a role in interpretations of fluvial responses to high-frequency (time-scales of 10^4–10^6 years) climate and sea-level change in the pre-Quaternary record. First, future interpretations must at least in part rely on the development of 'sequence-scale' Quaternary analogues. There is, at present, a paucity of good analogues that: (a) extend beyond the relatively short latest Pleistocene and Holocene period into geologically resolvable time-scales (multiples of 10^4–10^6 years); (b) include depositional basin settings where preservational processes operate; and (c) describe fluvial responses to allogenic controls in sedimentological terms that facilitate recognition in the ancient record. Further detailed study of Quaternary analogues may also provide a mechanism by which cross-fertilization of methods and perspectives from geomorphology and sedimentology can occur, and traditional disciplinary and other barriers can be dismantled.

Second, and perhaps more difficult, will be changes in the philosophy of interpretation. It seems clear that the development of sequence stratigraphy resulted in a paradigm shift with respect to how sedimentologists view the responses of fluvial systems to sea-level change. However, the same intellectual hurdle has yet to be crossed with respect to the significance of climate change, which both causes and covaries with sea-level, and is one of the dominant controls on changes in sediment delivery to depositional basins through time. Clear recognition and interpretation of fluvial responses to climate change certainly remains difficult, and for the time being may require well-constrained geochronological frameworks that demonstrate rates of change comparable to known climate forcing functions, coupled with an approach where possible alternatives are eliminated and the only one that remains must be the most viable. However, part of the hurdle will involve a willingness to explicitly consider climate change as a forcing mechanism; for example, many features described in ancient successions are remarkably similar in scale and kind to that observed in Late Quaternary records, but they are commonly interpreted in terms of autogenic variation superimposed on tectonic controls. No analogues clearly demonstrate, as opposed to assume, this particular model, and the assumption of autogenic variation may not be as justified today as it might have been in the past given present views on the magnitude, frequency and significance of climate change.

Importance of modelling

Although modelling efforts are discussed elsewhere in this issue (Paola, 2000), it seems clear that integration of different 'fluvial' models will be a major challenge for the near future (cf. Nystuen, 1998). As appropriately stated by Paola in this issue, presently available numerical models should be considered temporary building blocks that will ultimately merge together. Four broad classes of models can be distinguished: (a) the predominantly autogenic and exclusively aggradational numerical models of alluvial architecture (e.g. Bridge & Leeder, 1979; Mackey & Bridge, 1995; Heller & Paola, 1996); (b) sequence-stratigraphic models that focus on incision and/or aggradation in downdip parts of the fluvial system (e.g. Nummedal *et al.*, 1993; Leeder & Stewart, 1996); (c) numerical models that focus primarily on long-term fluvial incision and/or sediment production in the hinterland (Veldkamp & Van den Berg, 1993; Leeder *et al.*, 1998); and (d) the unscaled and scaled physical (analogue) models that have provided important qualitative insight into fluvial responses to varying rates and directions of sea-level change (Wood *et al.*, 1993, 1994; Koss *et al.*, 1994; Paola, 2000), or how alluvial architecture varies with aggradation rate (e.g. Ashworth *et al.*, 1999).

In the future, linkage of models that focus on both continental interiors and continental margins (e.g. Veldkamp & Van Dijke, 1998), as well as integrating model results with real-world data (e.g. Leeder *et al.*, 1996), must become increasingly important so as to more clearly understand relations between sediment production, transport and deposition at time-scales that reside at the interface between the traditional interests of geomorphologists/Quaternary geologists and sedimentary geologists (10^3–10^6 years). As noted by Paola in this issue, model development can proceed at higher rates than field studies; hence the first decades of this millennium offer significant opportunities for new and creative field-based investigations that are designed to constrain and validate quantitative models.

ACKNOWLEDGEMENTS

We appreciate the invitation from the *Sedimentology* editorial board to write this paper.

Comments by Peter Friend, Mike Leeder and editor Jim Best were very helpful. We also appreciate numerous discussions with our various collaborators, students and colleagues, but stress that this is a selective review that reflects our own perspectives on the subject matter.

REFERENCES

Aitken, J.F. and **Flint, S.S.** (1995) The application of high-resolution sequence stratigraphy to fluvial systems: a case study from the Upper Carboniferous Breathitt Group, eastern Kentucky, USA. *Sedimentology*, **42**, 3–30.

Aitken, J.F. and **Flint, S.S.** (1996) Variable expressions of interfluvial sequence boundaries in the Breathitt Group (Pennsylvanian), eastern Kentucky, USA. In: *High Resolution Sequence Stratigraphy: Innovations and Applications* (Ed. by J. A. Howell and J. F. Aitken), *Spec. Publ. Geol. Soc.*, **104**, 193–206.

Aitken, M.J. (1998) *An Introduction to Optical Dating*. Oxford University Press, Oxford.

Allen, J.R.L. (1974) Studies in fluviatile sedimentation: implications of pedogenic carbonate units, Lower Old Red Sandstone, Anglo-Welsh outcrop. *Geol. J.*, **9**, 181–208.

Allen, J.R.L. (1978) Studies in fluviatile sedimentation: an exploratory quantitative model for the architecture of avulsion-controlled suites. *Sedim. Geol.*, **21**, 129–147.

Amorosi, A., **Colalongo, M.L.**, **Fusco, F.**, **Pasini, G.** and **Fiorini, F.** (1999) Glacio-eustatic control of continental-shallow marine cyclicity from Late Quaternary deposits of the southeastern Po Plain, northern Italy. *Quat. Res.*, **52**, 1–13.

Anderson, J.B., **Abdulah, K.**, **Sarzalejo, S.**, **Siringin, F.** and **Thomas, M.A.** (1996) Late Quaternary sedimentation and high-resolution sequence stratigraphy of the East Texas Shelf. In: *Geology of Siliciclastic Shelf Seas* (Ed. by M. De Batist and P. Jacobs), *Spec. Publ. Geol. Soc.*, **117**, 95–124.

Antoine, P. (1994) The Somme valley terrace system (northern France); a model of river response to Quaternary climatic variations since 800,000 BP. *Terra Nova*, **6**, 453–464.

Ashworth, P.J., **Best, J.L.**, **Peakall, J.** and **Lorsong, J.A.** (1999) The influence of aggradation rate on braided alluvial architecture: field study and physical scale-modelling of the Ashburton River gravels, Canterbury Plains, New Zealand. In: *Current Research in Fluvial Sedimentology: Proceedings of 6th International Conference on Fluvial Sedimentology* (Ed. by N. D. Smith and J. J. Rogers), *Spec. Publ. Int. Assoc. Sediment.*, **28**, 333–346.

Aslan, A. and **Autin, W.J.** (1998) Holocene flood-plain soil formation in the southern lower Mississippi Valley: Implications for interpreting alluvial paleosols. *Geol. Soc. Am. Bull.*, **110**, 433–449.

Aslan, A. and **Autin, W.J.** (1999) Evolution of the Holocene Mississippi River floodplain, Ferriday, Louisiana: insights on the origin of fine-grained floodplains. *J. Sedim. Res.*, **69**, 800–815.

Aslan, A. and **Blum, M.D.** (1999) Contrasting styles of Holocene avulsion, Texas Gulf Coastal Plain. In: *Current Research in Fluvial Sedimentology: Proceedings of 6th International Conference on Fluvial Sedimentology* (Ed. by N. D. Smith and J. J. Rogers), *Spec. Publ. Int. Assoc. Sediment.*, **28**, 193–209.

Autin, W.J. (1992) Use of alloformations for definition of Holocene meanderbelts in the middle Amite River, southeastern Louisiana. *Geol. Soc. Am. Bull.*, **104**, 233–241.

Bagnold, R.A. (1973) The nature of saltation and bedload transport in water. *Proc. Royal Society, London*, **322A**, 473–504.

Baker, V.R., **Kochel, R.C.** and **Patton, P.C.** (Eds). (1988) *Flood Geomorphology*. John Wiley, New York.

Barrell, J. (1917) Rhythms and the measurement of geological time. *Geol. Soc. Am. Bull.*, **28**, 745–904.

Barron, E.J. and **Moore, G.T.** (1994) *Climate Model Application in Palaeoenvironmental Analysis*. SEPM Short Course no. 33.

Begin, Z.B. (1988) Application of diffusion-erosion model to alluvial channels which degrade due to base-level lowering. *Earth Surf. Proc. Landf.*, **13**, 487–500.

Begin, Z.B., **Meyer, D.F.** and **Schumm, S.A.** (1981) Development of longitudinal profiles of alluvial channels in response to base-level lowering. *Earth Surf. Proc. Landf.*, **6**, 49–68.

Behl, R.J. and **Kennett, J.P.** (1996) Brief interstadial events in the Santa Barbara basin, NE Pacific, during the past 60 kyr. *Nature*, **379**, 243–246.

Behrensmeyer, A.K., **Willis, B.J.** and **Quade, J.** (1995) Floodplains and palaeosols of Pakistan Neogene and Wyoming Palaeogene deposits: a comparative study. *Palaeogeogr. Palaeoclimatol. Palaeoecol.*, **115**, 37–60.

Benito, G., **Baker, V.R.** and **Gregory, K.J.** (Eds). (1998) *Palaeohydrology and Environmental Change*. John Wiley, Chichester.

Berendsen, H., **Hoek, W.** and **Schorn, E.** (1995) Late Weichselian and Holocene river channel changes of the rivers Rhine and Meuse in the Netherlands (Land van Maas en Waal). *Palâoklimaforschung*, **14**, 151–171.

Berendsen, H.J.A. (1982) De genese van het landschap in het zuiden van de provincie Utrecht, een fysisch-geografische studie. *Utr. Geogr. Studies*, **25**, 1–255.

Berendsen, H.J.A., **Faessen, E.L.J.H.** and **Kempen, H.F.J.** (1994) *Zand in Banen. Zanddiepte-Attentiekaarten van het Gelders Rivierengebied*. Provincie Gelderland, Arnhem.

Berger, A., **Loutre, M.F.** and **Dehant, V.** (1989) Astronomical frequencies for pre-Quaternary palaeoclimate studies. *Terra Nova*, **1**, 474–479.

Berger, A. and **Loutre, M.F.** (1994) Astronomical forcing through geologic time. In: *Orbital Forcing and Cyclic Sequences* (Ed. by P. L. de Boer and D. G. Smith), *Spec. Publ. Int. Assoc. Sediment.*, **19**, 15–24.

Bernard, H.A. and **Leblanc, R.J.** (1965) Resume of the Quaternary Geology of the Northwestern Gulf of Mexico Province. In: *The Quaternary of the United States* (Ed. by H. E. Wright and D. G. Frey), pp. 137–185. Princeton University Press, Princeton.

Best, J.L. and **Ashworth, P.J.** (1997) Scour in large braided rivers and the recognition of sequence stratigraphic boundaries. *Nature*, **387**, 275–277.

Best, J.L. and **Bridge, J.S.** (1992) The morphology and dynamics of low amplitude bedwaves upon upper stage plane beds and the preservation of planar laminae. *Sedimentology*, **39**, 737–752.

Birkeland, P.W. (1999) *Soils and Geomorphology*. Oxford University Press, New York.

Blum, M.D. (1990) Climatic and eustatic controls on Gulf Coastal Plain fluvial sedimentation: an example from the Late Quaternary of the Colorado River, Texas. In: *Sequence Stratigraphy as an Exploration Tool: Concepts and Practices* *in the Gulf Coast* (Ed. by J. M. Armentrout and B. F. Perkins) Gulf Coast Section of the Society of Economic Palaeontologists and Mineralogists Foundation, Proceedings of the 11th Annual Research Conference, pp. 71–83.

Blum, M.D. (1993) Genesis and architecture of incised valley fill sequences: a Late Quaternary example from the Colorado River, Gulf Coastal Plain of Texas. In: *Siliciclastic Sequence Stratigraphy: Recent Developments and Applications* (Ed. by P. Weimer and H. W. Posamentier), *Mem. Am. Assoc. Petrol. Geol.*, **58**, 259–283.

Blum, M.D. and **Price, D.M.** (1998) Quaternary alluvial plain construction in response to interacting glacio-eustatic and climatic controls, Texas Gulf Coastal Plain. In: *Relative Role of Eustasy, Climate, and Tectonism in Continental Rocks* (Ed. by K. W. Shanley and P. J. McCabe), *Spec. Publ. Soc. Econ. Paleont. Miner.*, **59**, 31–48.

Blum, M.D. and **Valastro, S. Jr** (1994) Late Quaternary sedimentation, Lower Colorado River, Gulf Coastal Plain of Texas. *Geol. Soc. Am. Bull.*, **106**, 1002–1016.

Blum, M.D., Toomey III, R.S. and **Valastro, S. Jr** (1994) Fluvial response to Late Quaternary climatic and environmental change, Edwards Plateau of Texas. *Palaeogeogr. Palaeoclimatol. Palaeoecol.*, **108**, 1–21.

Blum, M.D., Morton, R.A. and **Durbin, J.M.** (1995) 'Deweyville' terraces and deposits of the Texas Gulf Coastal Plain. *Trans. Gulf Coast Assoc. Geol. Soc.*, **45**, 53–60.

Blum, M.D., Kocurek, G., Deynoux, M., Swezey, C., Lancaster, N., Price, D. and **Pion, J.-C.** (1998) Quaternary wadi-lacustrine-aeolian depositional cycles and sequences, Chott Rharsa Basin, Tunisia. In: *Quaternary Deserts and Climatic Change* (Ed. by A. S. Alsharhan, K. Glennie, G. L. Whittle and C. G. St. C. Kendall), pp. 539–552. Balkema Press, Rotterdam.

Blum, M.D., Guccione, M.J., Wysocki, D.A. and **Robnett, P.C.** (2000) Late Pleistocene evolution of the Central Mississippi Valley, Southen Missouri to Arkansas. *Geol. Soc. Am. Bull.*, **112**, 221–235.

Bond, G., Broecker, W., Johnsen, S., McManus, J., Labeyrie, L., Jouzel, J. and **Bonani, G.** (1993) Correlations between climate records from North Atlantic sediments and Greenland ice. *Nature*, **365**, 143–147.

Bond, G., Showers, W., Cheseby, M., Lotti, R., Almasi, P., Demenocal, P., Priore, P., Cullen, H., Hajdas, I. and **Bonani, G.** (1997) A pervasive millennial-scale cycle in North Atlantic Holocene and glacial climates. *Science*, **278**, 1257–1266.

Bond, G.C. and **Lotti, R.** (1995) Iceberg discharges into the North Atlantic on millennial time scales during the last glaciation. *Science*, **267**, 1005–1010.

Bowen, D.Q. (1978) *Quaternary Geology*. Pergamon Press, Oxford.

Bown, T.M. and **Kraus, M.J.** (1987) Integration of channel and floodplain suites, I. Developmental sequence and lateral relations of alluvial paleosols. *J. Sedim. Petrol.*, **57**, 587–601.

Bradley, R.S. (1999) *Paleoclimatology*. Academic Press, San Diego.

Branson, J., Brown, A.G. and **Gregory, K.J.** (Eds). (1996) Global Continental Changes: The Context of Palaeohydrology. *Spec. Publ. Geol. Soc. London,* **115**.

Bravard, J.-P. (1992) Rapport no. 2: Les rythmes d'evolution morphologique des vallèes Francaises au Tardiglaciaire et l'Holocene. *Bull. Assoc. Geographes Francais.*

Bridge, J.S. (1993a) Description and interpretation of fluvial deposits: a critical perspective. *Sedimentology*, **40**, 801–810.

Bridge, J.S. (1993b) The interaction between channel geometry, water flow, sediment transport and deposition in braided rivers. In: *Braided Rivers* (Ed. by J. L. Best and C. S. Bristow), *Spec. Publ. Geol. Soc. Lond.*, **75**, 13–71.

Bridge, J.S. (1995) Description and interpretation of fluvial deposits: a critical perspective. *Sedimentology*, **42**, 382–389.

Bridge, J.S. (1997) Thickness of sets of cross-strata and planar strata as a function of formative bedwave geometry and migration, and aggradation rate. *Geology*, **25**, 971–974.

Bridge, J.S. and **Leeder, M.R.** (1979) A simulation model of alluvial stratigraphy. *Sedimentology*, **26**, 617–644.

Bridge, J.S., Alexander, J., Collier, R.E.L., Gawthorpe, R.L. and **Jarvis, J.** (1995) Ground-penetrating radar and coring used to document the large-scale structure of point-bar deposits. *Sedimentology*, **42**, 839–852.

Bridge, J.S., Collier, R.E.L. and **Alexander, J.** (1998) Large-scale structure of Calamus River deposits (Nebraska, USA) revealed using ground-penetrating radar. *Sedimentology*, **45**, 977–986.

Bridgland, D.R. (1994) *Quaternary of the Thames. Geological Conservation Review Series 7.* Chapman & Hall, London.

Bridgland, D.R. and **D'olier, B.** (1995) The Pleistocene evolution of the Thames and Rhine drainage systems in the southern North Sea Basin. In: *Island Britain: a Quaternary Perspective* (Ed. by R. C. Preece), *Spec. Publ. Geol. Soc.*, **96**, 27–45.

Brierley, G.J. (1996) Channel morphology and element assemblages: A constructivist approach to facies modelling. In: *Advances in Fluvial Dynamics and Stratigraphy* (Ed. by P. A. Carling and M. R. Dawson), pp. 263–298. John Wiley, Chichester.

Broecker, W.S. (1995) *The Glacial World According to Wally.* Eldigio Press, Palisades.

Broecker, W.S. (1997) Thermohaline circulation, the achilles heel of our climate system: Will man-made CO2 upset the current balance? *Science*, **278**, 1582–1588.

Brown, L.J. and **Wilson, D.D.** (1988) Stratigraphy of Late Quaternary deposits of the northern Canterbury Plains. *NZ J. Geol. Geoph.*, **31**, 305–335.

Bull, W.B. (1991) *Geomorphic Responses to Climate Change.* Oxford University Press, Oxford.

Burbank, D.W. and **Pinter, N.** (1999) Landscape evolution: the interactions of tectonics and surface processes. *Basin Res.*, **11**, 1–6.

Burke, K., Francis, P. and **Wells, G.** (1990) Importance of the geological record in understanding global change. *Palaeogeogr. Palaeoclimatol. Palaeoecol. (Glob. Plan. Change Sect.)*, **89**, 193–204.

Butcher, S.W. (1990) The nickpoint concept and its implications regarding onlap to the stratigraphic record. In: *Quantitative Dynamic Stratigraphy* (Ed. by T. A. Cross), pp. 375–385. Prentice Hall, Englewood Cliffs.

Butzer, K.W. (1980) Holocene alluvial sequences: problems of dating and correlation. In: *Timescales in Geomorphology* (Ed. by R. A. Cullingford, D. A. Davidson and J. Lewin), pp. 131–142. John Wiley, New York.

Chappell, J. and **Shackleton, N.J.** (1986) Oxygen isotopes and sea-level. *Nature*, **324**, 137–140.

Chappell, J., Omura, A., Ezat, T., McCulloch, M., Pandolfi, J., Ota, Y. and **Pillans, B.** (1996) Reconciliation of late Quaternary sea-levels derived from coral terraces at Huon Peninsula with deep sea oxygen isotope records. *Earth Planet. Sci. Lett.*, **141**, 227–236.

Clemmensen, L.B., Øxnevad, I.E.I. and **De Boer, P.L.** (1994) Climatic controls on ancient sedimentation: some late Palaeozoic and Mezosoic examples from NW Europe and the Western Interior of the USA. In: *Orbital Forcing and Cyclic Sequences* (Ed. by P. L. de Boer and D. G. Smith), *Spec. Publ. Int. Assoc. Sediment.*, **19,** 439–457.

COHMAP Project Members. (1988) Climatic changes of the last 18,000 years: Observations and model simulations. *Science*, **241,** 1043–1052.

Cojan, I. (1993) Alternating fluvial and lacustrine sedimentation: tectonic and climatic controls (Provence Basin, S. France, Upper Cretaceous/Palaeocene). In: *Alluvial Sedimentation* (Ed. by M. Marzo and C. Puigdefabregas), *Spec. Publ. Int. Assoc. Sediment.*, **17,** 425–438.

Crowley, T. and **North, G.R.** (1991) *Palaeoclimatology.* Oxford University Press, New York.

Currie, B.S. (1997) Sequence stratigraphy of nonmarine Jurassic–Cretaceous rocks, central Cordilleran foreland-basin system. *Geol. Soc. Am. Bull.*, **109,** 1206–1222.

Dai, A., Fung, I.Y. and **Del Genio, A.D.** (1997) Surface observed global land precipitation variations during 1900–88. *J. Clim.*, **10,** 2943–2962.

Dalrymple, R.W., Boyd, R. and **Zaitlin, B.A.** (Eds) (1994) History of research, types and internal organisation of incised-valley systems: Introduction to the Volume. In: *Incised-Valley Systems: Origin and Sedimentary Sequences* (Ed. by R. W. Dalrymple, R. Boyd and B. A. Zaitlin), *Spec. Publ. Soc. Econ. Paleont. Miner.*, **51,** 3–10.

Dott, R.H. Jr (1992) An introduction to the ups and downs of eustasy. In: *Eustasy: the Historical Ups and Downs of a Major Geological Concept* (Ed. by R. H. Dott, Jr), *Geol. Soc. Am. Mem.*, **180,** 1–16.

Dubiel, R.F., Huntoon, J.E., Condon, S.M. and **Stanesco, J.D.** (1996) Permian deposystems, palaeogeography, and palaeoclimate of the Paradox Basin and vicinity. In: *Palaeozoic Systems of the Rocky Mountain Region* (Ed. by M. W. Longman and M. D. Sonnenfeld), pp. 427–444. Rocky Mountain Section, SEPM.

Duller, G.A.T. (1996) Recent developments in luminescence dating of Quaternary sediments. *Progr. Phys. Geogr.*, **20,** 127–145.

Dury, G.H. (1965) Principles of Underfit Streams. *United States Geol. Surv. Prof. Pap.*, **452-A.**

Emery, D. and **Myers, K.J.** (Eds). (1996) *Sequence Stratigraphy.* Blackwell Science, Oxford.

Ethridge, F.G. and **Schumm, S.A.** (1978) Reconstructing palaeochannel morphologic and flow characteristics: methodology, limitations and assessment. In: *Fluvial Sedimentology* (Ed. by A. D. Miall), *Can. Soc. Petrol. Geol. Mem.*, **5,** 703–721.

Ethridge, F.G., Wood, L.J. and **Schumm, S.A.** (1998) Cyclic variables controlling fluvial sequence development: Problems and perspectives. In: *Relative Role of Eustasy, Climate, and Tectonism in Continental Rocks* (Ed. by K. W. Shanley and P. J. McCabe), *Spec. Publ. Soc. Econ. Paleont. Miner.*, **59,** 17–29.

Fairbanks, R.G. (1989) A 17,000-year glacio-eustatic sea-level record: influence of glacial melting rates on the Younger Dryas event and deep-ocean circulation. *Nature*, **342,** 637–642.

Ferguson, R. (1987) Hydraulic and sedimentary controls of channel pattern. In: *River Channels: Environment and Process* (Ed. by K. S. Richards), pp. 17–29. Blackwell Science, Oxford.

Fielding, C.R. (1999) Varieties of fluvial form: the relevance to geologists of an expanded reality. In: *Varieties of Fluvial Form* (Ed. by A. J. Miller and A. Gupta), pp. 497–505. John Wiley, Chichester.

Fisk, H.N. (1944) *Geological Investigation of the Alluvial Valley of the Lower Mississippi River.* Mississippi River Commission, Vicksburg.

Fleming, K., Johnston, P., Zwartz, D., Yokoyama, Y., Lambeck, K. and **Chappell, J.** (1998) Refining the eustatic sea-level curve since the Last Glacial Maximum using far- and intermediate-field sites. *Earth Planet. Sci. Lett.*, **163,** 327–342.

Fournier, F. (1960) *Climat et Erosion: la relation entre lı́erosion du sol par lı́eau et les precipitations atmospheriques.* Presse Universitaire de France.

Frakes, L.A. (1979) *Climates Throughout Geologic Time.* Elsevier, Amsterdam.

Frakes, L.A., Francis, J.E. and **Syktus, J.I.** (1992) *Climate Modes of the Phanerozoic.* Cambridge University Press, Cambridge.

Fuller, I.C., Macklin, M.G., Lewin, J., Passmore, D.G. and **Wintle, A.G.** (1998) River response to high-frequency climate oscillations in southern Europe over the past 200 k.y. *Geology*, **26,** 275–278.

Galloway, W.E. (1989) Genetic stratigraphic sequences in basin analysis I: architecture and genesis of flooding-surface bounded depositional units. *Am. Assoc. Petrol. Geol. Bull.*, **73,** 125–142.

Gibbard, P.L. (1995) The formation of the Strait of Dover. In: *Island Britain: a Quaternary Perspective* (Ed. by R. C. Preece), *Spec. Publ. Geol. Soc.*, **96,** 15–26.

Gibling, M.R. and **Bird, D.J.** (1994) Late Carboniferous cyclothems and alluvial paleovalleys in the Sydney Basin, Nova Scotia. *Geol. Soc. Am. Bull.*, **106,** 105–117.

Gibling, M.R. and **Wightman, W.G.** (1994) Palaeovalleys and protozoan assemblages in a Late Carboniferous cyclothem, Sydney Basin, Nova Scotia. *Sedimentology*, **41,** 699–719.

Gilbert, G.K. (1877) *Report on the Geology of the Henry Mountains.* Government Printing Office, Washington, DC.

Gregory, K.J. (Ed). (1983) *Background to Palaeohydrology.* John Wiley, Chichester.

Gregory, K.J., Lewin, J. and **Thornes, J.B.** (Eds). (1987) *Palaeohydrology in Practice.* John Wiley.

Gregory, K.J., Starkel, L. and **Baker, V.R.** (Eds). (1996) *Global Continental Palaeohydrology.* John Wiley.

Grootes, P.M., Stuiver, M., White, J.W.C., Johnsen, S. and **Jouzel, J.** (1993) Comparison of oxygen isotope records from the GISP2 and GRIP Greenland ice cores. *Nature*, **366,** 552–554.

Hallet, B., Hunter, L. and **Bogen, J.** (1996) Rates of erosion and sediment evacuation by glaciers: a review of field data and their implications. *Global Planet. Change*, **12,** 213–235.

Haq, B.U., Hardenbol, J. and **Vail, P.R.** (1987) Chronology of fluctuating sea-levels since the Triassic. *Science*, **235,** 1156–1167.

Haq, B.U., Hardenbol, J. and **Vail, P.R.** (1988) Mesozoic and Cenozoic chronostratigraphy and cycles of sea-level change. In: *Sea-Level Changes: an Integrated Approach* (Ed. by C. K. Wilgus, B. S. Hastings, C. G. S. C. Kendall, H. W. Posamentier, C. A. Ross and J. C. Van Wagoner), *Spec. Publ. Soc. Econ. Paleont. Miner.*, **42,** 71–108.

Hays, J.D., Imbrie, J. and **Shackleton, N.J.** (1976) Variations in the earth's orbit: Pacemaker of the ice ages. *Science*, **194,** 1121–1132.

Helland-Hansen, W. and **Gjelberg, J.G.** (1994) Conceptual basis and variability in sequence stratigraphy: a different perspective. *Sedim. Geol.*, **92**, 31–52.

Heller, P.L. and **Paola, C.** (1996) Downstream changes in alluvial architecture: an exploration of controls on channel-stacking patterns. *J. Sedim. Res.*, **66**, 297–306.

Hovius, N. (1998) Controls on sediment supply by large rivers. In: *Relative Role of Eustasy, Climate, and Tectonism in Continental Rocks* (Ed. by K. W. Shanley and P. J. McCabe), *Spec. Publ. Soc. Econ. Paleont. Miner.*, **59**, 3–16.

Hovius, N. (1999) Macro scale process-systems of mountain belt erosion. In: *Geomorphology and Global Tectonics* (Ed. by M. A. Summerfield), pp. 77–105. John Wiley, Chichester.

Hovius, N. and **Leeder, M.** (1998) Clastic sediment supply to basins. *Basin Res.*, **10**, 1–5.

Howell, J.A. and **Flint, S.S.** (1996) A model for high resolution sequence stratigraphy within extensional basins. In: *High Resolution Sequence Stratigraphy: Innovations and Applications* (Ed. by J. A. Howell and J. F. Aitken). *Spec. Publ. Geol. Soc.*, **104**, 129–138.

Huggenberger, P. (1993) Radar facies: recognition of facies patterns and heterogeneities within Pleistocene Rhine gravels, NE Switzerland. In: *Braided Rivers* (Ed. by J. L. Best and C. S. Bristow), *Spec. Publ. Geol. Soc.*, **75**, 163–176.

Hurrell, J.W. (1995) Decadal trends in the North Atlantic Oscillation regional temperatures and precipitation. *Science*, **269**, 676–679.

Imbrie, J. and **Imbrie, K.** (1979) *Ice Ages.* Harvard University Press, Cambridge, MA.

Imbrie, J., Hays, J.D., Martinson, D.G., McIntyre, A., Mix, A.C., Moreley, J.J., Pisias, N.G., Prell, W.L. and **Shackleton, N.J.** (1984) The orbital theory of Pleistocene climate: support from a revised chronology of the marine $\delta^{18}O$ record. In: *Milankovitch and Climate* (Ed. by A. L. Berger, J. Imbrie, J. Hays, G. Kukla and B. Saltzman), pp. 269–306. Reidel.

Jervey, M.T. (1988) Quantitative geological modeling of siliciclastic rock sequences and their seismic expression. In: *Sea-Level Changes: an Integrated Approach* (Ed. by C. K. Wilgus, B. S. Hastings, C. G. S. C. Kendall, H. W. Posamentier, C. A. Ross and J. C. Van Wagoner), *Spec. Publ. Soc. Econ. Paleont. Miner.*, **42**, 47–69.

Johnsen, S.J., Clausen, H.B., Dansgaard, W., Fuhrer, K., Gundestrup, N., Hammer, C.U., Iversen, P., Jouzel, J., Stauffer, B. and **Steffensen, J.P.** (1992) Irregular glacial interstadials recorded in a new Greenland ice core. *Nature*, **359**, 311–313.

Kasse, K., Vandenberghe, J. and **Bohncke, S.** (1995) Climatic change and fluvial dynamics of the Maas during the late Weichselian and early Holocene. *Paläoklimaforschung*, **14**, 123–150.

Khan, I.A., Bridge, J.S., Kappelman, J. and **Wilson, R.** (1997) Evolution of Miocene fluvial environments, eastern Potwar plateau, northern Pakistan. *Sedimentology*, **44**, 221–251.

Knighton, D. (1998) *Fluvial Forms and Processes.* Edward Arnold, London.

Knox, J.C. (1976) Concept of the Graded Stream. In: *Theories of Landform Development* (Ed. by W. N. Melhorn and R. C. Flemel), pp. 169–198. Publications in Geomorphology, State University of New York at Binghamton, Binghamton, NY.

Knox, J.C. (1983) Responses of river systems to Holocene climates. In: *Late Quaternary Environments of the United States*, Vol. 2, *the Holocene* (Ed. by H. E. Wright and S. C. Porter), pp. 26–41. University of Minnesota Press.

Knox, J.C. (1987) Climatic influence on Upper Mississippi Valley floods. In: *Flood Geomorphology* (Ed. by V. R. Baker, R. C. Kochel and P. C.Patton), pp. 279–300. John Wiley and Sons, New York.

Knox, J.C. (1993) Large increases in flood magnitude in response to modest changes in climate. *Nature*, **361**, 430–432.

Knox, J.C. (1996) Fluvial systems since 20,000 yrs BP. In: *Global Continental Palaeohydrology* (Ed. by K. J. Gregory, L. Starkel and V. R. Baker), pp. 87–108. John Wiley and Sons, New York.

Kocurek, G. (1998) Aeolian system response to external forcing factors – a sequence stratigraphic view of the Saharan region. In: *Quaternary Deserts and Climatic Change* (Ed. by A. S. Alsharhan, K. Glennie, G. L. Whittle and C. G. S. C. Kendall), pp. 327–337. Balkema Press.

Kocurek, G. and **Havholm, K.G.** (1993) Eolian sequence stratigraphy: A conceptual framework. In: *Siliciclastic Sequence Stratigraphy: Recent Developments and Applications* (Ed. by P. Weimer and H. W. Posamentier), *Am. Assoc. Petrol. Geol. Mem.*, **58**, 393–410.

Koss, J.E., Ethridge, F.G. and **Schumm, S.A.** (1994) An experimental study of the effects of base-level change on fluvial, coastal plain and shelf systems. *J. Sedim. Res.*, **B64**, 90–98.

Kozarski, S. and **Rotnicki, K.** (1977) Valley floors and changes of river channel patterns in the north Polish Plain during the late Würm and Holocene. *Quaestiones Geographicae*, **4**, 51–93.

Kraus, M.J. (1987) Integration of channel and floodplain suites, II. Vertical relations of alluvial paleosols. *J. Sedim. Petrol.*, **57**, 602–612.

Kutzbach, J.E., Gallimore, R., Harrison, S.P., Behling, P., Selin, R. and **Laarif, F.** (1997) Climate and biome simulations for the last 21,000 years. *Quat. Sci. Rev.*, **17**, 473–506.

Lambeck, K. (1993) Glacial rebound and sea-level change: an example of a relationship between mantle and surface processes. In: *Relationships Between Mantle Processes and Geological Processes at or Near the Earth's Surface* (Ed. by R. Wortel, U. Hansen and R. Sabadini), *Tectonophysics*, **223**, 15–37.

Lamothe, L. (1918) Les anciennes nappes alluviales et lignes de Rivage du bassin de la Somme et leurs rapports avec celles de la Meditarranee occidentale. *Bull. Soc. Geol., Fr.*, **18**, 3–58.

Langbein, W.B. and **Schumm, S.A.** (1958) Yield of sediment in relation to mean annual sediment. *Trans. Am. Geophys Union.*, **39**, 1076–1084.

Lane, E.W. (1955) The importance of fluvial morphology in hydraulic engineering. *Am. Soc. Civil Eng. Proc.*, **81** (745), 1–17.

Lau, N.-C. (1988) Variability of the observed midlatitude storm tracks in relation to low-frequency changes in the circulation pattern. *J. Atmos. Sci.*, **45**, 2718–2743.

Leckie, D.A. (1994) Canterbury Plains, New Zealand – Implications for sequence stratigraphic models. *Am. Assoc. Petrol. Geol. Bull.*, **78**, 1240–1252.

Leclair, S.F., Bridge, J.S. and **Wang, F.** (1997) Preservation of cross-strata due to migration of subaqueous dunes over aggrading and non-aggrading beds: comparison of experimental data with theory. *Geosci. Canada*, **24**, 55–66.

Leeder, M.R. (1975) Pedogenic carbonates and flood sediment accretion rates: a quantitative model for alluvial arid-zone lithofacies. *Geol. Mag.*, **112**, 257–270.

Leeder, M.R. (1978) A quantitative stratigraphic model for alluvium, with special reference to channel deposit density and interconnectedness. In: *Fluvial Sedimentology* (Ed. by A. D. Miall), *Can. Soc. Petrol. Geol. Mem.*, **5**, 587–596.

Leeder, M.R. and **Stewart, M.D.** (1996) Fluvial incision and sequence stratigraphy: alluvial responses to relative sea-level fall and their detection in the geologic record. In: *Sequence Stratigraphy in British Geology* (Ed. by S. P. Hesselbo and D. N. Parkinson), *Spec. Publ. Geol. Soc.*, **103**, 25–39.

Leeder, M.R., Mack, G.H., Peakall, J. and **Salyards, S.L.** (1996) First quantitative test of alluvial stratigraphic models: Southern Rio Grande rift, New Mexico. *Geology*, **24**, 87–90.

Leeder, M.R., Harris, T. and **Kirkby, M.J.** (1998) Sediment supply and climate change: implications for basin stratigraphy. *Basin Res.*, **10**, 7–18.

Legarreta, L. and **Uliana, M.A.** (1998) Anatomy of hinterland depositional sequences: Upper Cretaceous fluvial strata, Neuquen Basin, West-Central Argentina. In: *Relative Role of Eustasy, Climate, and Tectonism in Continental Rocks* (Ed. by K. W. Shanley and P. J. McCabe), *Spec. Publ. Soc. Econ. Paleont. Miner.*, **59**, 83–92.

Leopold, L.B. and **Bull., W.B.** (1979) Base level, aggradation, and grade. *Proc. Am. Phil. Soc.*, **123**, 168–202.

Leopold, L.B. and **Maddock, T.** (1953) The Hydraulic Geometry of Stream Channels and Some Physiographic Implications. *Prof. Pap. United States Geol. Surv.*, **252**.

Lewin, J. (1989) Floods in fluvial geomorphology. In: *Floods: Hydrological, Sedimentological, and Geomorphological Implications* (Ed. by K. Beven and P. Carling), pp. 265–284. John Wiley, Chichester.

Libby, W.F. (1952) *Radiocarbon Dating.* University of Chicago Press, Chicago.

Lowe, J.J. and **Walker, M.J.C.** (1997) *Reconstructing Quaternary Environments.* Addison-Wesley Longman, Harlow.

Mack, G.H., James, W.C. and **Monger, H.C.** (1993) Classification of paleosols. *Geol. Soc. Am. Bull.*, **105**, 129–136.

Mackey, S.D. and **Bridge, J.S.** (1995) Three-dimensional model of alluvial stratigraphy: theory and application. *J. Sedim. Res.*, **B65**, 7–31.

Mackin, J.H. (1948) Concept of the graded river. *Geol. Soc. Am. Bull.*, **59**, 463–512.

Macklin, M.G. and **Lewin, J.** (1993) Holocene river alluviation in Britain. *Z. Geomorph. N. F. (Suppl.)-Bd.*, **88**, 109–122.

Makaske, B. (1998) Anastomosing rivers. *Nederl. Geogr. Studies*, **249**, 1–287.

Mandier, P. (1988) *Le Relief de la Moyenne VallÈe du Rhone au Tertiaire et au Quaternaire.* Doc. du BGRM, No 151, Paris.

Martinsen, O.J., Ryseth, A., Helland-Hansen, W., Flesche, H., Torkildsen, G. and **Idil, S.** (1999) Stratigraphic base level and fluvial architecture. Ericson Sandstone (Campanian), Rock Springs Uplift, W. Wyoming, USA. *Sedimentology*, **46**, 235–260.

McCarthy, P.J. and **Plint, A.G.** (1998) Recognition of interfluve seqeunce boundaries: Integrating paleopedology and sequence stratigraphy. *Geology*, **26**, 387–390.

McMahon, T.A., Finlayson, B.L., Haines, A. and **Srikanthan, R.** (1987) Runoff variability: a global perspective. In: *The Influence of Climate Change and Climate Variability on the Hydrologic Regime and Water Resources* (Proceedings of the Vancouver Symposium, August 1987). *IAHS Publ.*, **168**, 3–11.

McMahon, T.A., Finlayson, B.L., Haines, A.T. and **Srikanthan, R.** (1992) Global Runoff. *Continental Comparisons of Annual Flows and Peak Discharges.* Catena Paperback, Catena-Verlag.

Meade, R.H., Yuzyk, T.R. and **Day, T.J.** (1990) Movement and storage of sediment in rivers of the United States and Canada. In: *Surface Water Hydrology* (Ed. by M. G. Wolman and H. C. Riggs) *Geological Society of America, Geology of North America Series*, **Vol. 0-1**, pp. 255–280.

Merritts, D. and **Ellis, M.** (1994) Introduction to special section on tectonics and topography. *J. Geophys. Res.*, **99**, 12135–12141.

Merritts, D.J., Vincent, K.R. and **Wohl, E.E.** (1994) Long river profiles, tectonism, and eustasy: a guide to interpreting fluvial terraces. *J. Geophys. Res.*, **99**, 14031–14050.

Miall, A.D. (1985) Architectural element analysis: a new method of facies analysis applied to fluvial deposits. *Earth-Sci. Rev.*, **22**, 261–308.

Miall, A.D. (1991) Stratigraphic sequences and their chronostratigraphic correlation. *J. Sedim. Petrol.*, **61**, 497–505.

Miall, A.D. (1995) Description and interpretation of fluvial deposits: a critical perspective. *Sedimentology*, **42**, 370–384.

Miall, A.D. (1996) *The Geology of Fluvial Deposits.* Springer, Berlin.

Miall, A.D. (1997) *The Geology of Stratigraphic Sequences.* Springer, Berlin.

Milliman, J.D. and **Syvitski, J.P.M.** (1992) Geomorphic / tectonic control on sediment discharge to the ocean: the importance of small mountainous streams. *J. Geol.*, **100**, 525–544.

Mitchum, R.M. Jr, Vail, P.R. and **Thompson III, S.** (1977) Seismic stratigraphy and global changes of sea-level, part 2: the depositional sequence as a basic unit for stratigraphic analysis. In: *Seismic Stratigraphy – Applications to Hydrocarbon Exploration* (Ed. by C. E. Payton), *Am. Assoc. Petrol. Geol. Mem.*, **27**, 53–62.

Molnar, P. and **England, P.** (1990) Late Cenozoic uplift of mountain ranges and global climate change: chicken or egg? *Nature*, **346**, 29–34.

Molnar, P., England, P. and **Martinod, J.** (1993) Mantle dynamics, uplift of the Tibetan Plateau, and the Indian monsoon. *Rev. Geophys.*, **31**, 357–396.

Morton, R.A. and **Suter, J.R.** (1996) Sequence stratigraphy and composition of Late Quaternary shelf margin deltas, northern Gulf of Mexico. *Am. Assoc. Petrol. Geol. Bull.*, **80**, 505–530.

Muto, T. and **Steel, R.J.** (1997) Principles of regression and transgression: the nature of the interplay between accommodation and sediment supply. *J. Sedim. Res.*, **67**, 994–1000.

Nanson, G.C. and **Croke, J.C.** (1992) A genetic classification of floodplains. *Geomorphology*, **4**, 459–486.

Nanson, G.C. and **Knighton, A.D.** (1996) Anabranching rivers; their cause, character and classification. *Earth Surf. Proc. Landf.*, **21**, 217–239.

Nanson, G.C., Young, R.Y., Prices, D.M. and **Rust, B.R.** (1988) Stratigraphy, sedimentology, and Late Quaternary chronology of the channel country of western Queensland. In: *Fluvial Geomorphology of Australia* (Ed. by R. F. Warner), pp. 151–175. Academic Press, Sydney.

Nanson, G.C., Price, D.M. and **Short, S.A.** (1992) Wetting and drying of Australia over the past 300 ka. *Geology*, **20**, 791–794.

North, C.P. (1996) The prediction and modelling of subsurface fluvial stratigraphy. In: *Advances in Fluvial Dynamics and*

Stratigraphy (Ed. by P. A. Carling and M. R. Dawson), pp. 395–508. John Wiley, Chichester.

North American Commission on Stratigraphic Nomenclature. (1983) North American Stratigraphic Code. *Am. Assoc. Petrol. Geol. Bull.*, **67**, 841–875.

Nummedal, D., Riley, G.W. and **Templet, P.L.** (1993) High-resolution sequence architecture: a chronostratigraphic model based on equilibrium profile studies. In: *Sequence Stratigraphy and Facies Associations* (Ed. by H. W. Posamentier, C. P. Summerhayes, B. U. Haq and G. P. Allen), *Spec. Publ. Int. Assoc. Sediment.*, **18**, 55–68.

Nystuen, J.P. (1998) History and development of sequence stratigraphy. In: *Sequence Stratigraphy – Concepts and Applications* (Ed. by F. M. Gradstein, K. O. Sandvik and N. J. Milton), pp. 31–116. Elsevier, Amsterdam.

Olsen, H. (1990) Astronomical forcing of meandering river behavior: Milankovitch cycles in Devonian of East Greenland. *Palaeogeogr. Palaeoclimatol. Palaeoecol.*, **79**, 99–115.

Olsen, H. (1994) Orbital forcing on continental depositional systems – lacustrine and fluvial cyclicity in the Devonian of East Greenland. In: *Orbital Forcing and Cyclic Sequences* (Ed. by P. L. de Boer and D. G. Smith), *Spec. Publ. Int. Assoc. Sediment.*, **19**, 429–438.

Olsen, H. and **Larsen, P.H.** (1993) Structural and climatic controls on fluvial depositional systems: Devonian of East Greenland. In: *Alluvial Sedimentation* (Ed. by M. Marzo and C. Puigdefabregas), *Spec. Publ. Int. Assoc. Sediment.*, **17**, 401–423.

Olsen, T., Steel, R.H., Høgseth, K., Skar, T. and **Røe, S.-L.** (1995) Sequential architecture in a fluvial succession: sequence stratigraphy in the Upper Cretaceous Mesaverde Group, Price Canyon. *Utah. J. Sedim. Res.*, **B65**, 265–280.

Orton, G.J. and **Reading, H.G.** (1993) Variability of delatic processes in terms of sediment supply, with particular emphasis on grain size. *Sedimentology*, **40**, 475–512.

Page, K., Nanson, G. and **Price, D.** (1996) Chronology of Murrumbidgee River palaeochannels on the Riverine Plain, southeastern Australia. *J. Quat. Sci.*, **11**, 311–326.

Paola, C. (2000) Quantitative models of sedimentary basin filling. *Sedimentology*, **47** (Suppl. 1), 121–178.

Paola, C., Heller, P.L. and **Angevine, C.L.** (1992) The large-scale dynamics of grain size variation in alluvial basins, 1. Theory. *Basin Res.*, **4**, 73–90.

Parrish, J.T. (1998) *Interpreting Pre-Quaternary Climate from the Geologic Record.* Columbia University Press, New York.

Payton, C.E. (Ed). (1977) Seismic Stratigraphy – Applications to Hydrocarbon Exploration. *AAPG Mem.*, **27**.

Pazzaglia, F.J., Gardner, T.W. and **Merritts, D.J.** (1998) Bedrock fluvial incision and longitudinal profile development over geologic time scales determined by fluvial terraces. In: *Rivers Over Rock: Fluvial Processes in Bedrock Channels* (Ed. by K. J. Tinkler and E. E. Wohl), *AGU Monogr.*, **107**, 207–235.

Peltier, W.R. (1998) Post-glacial variations in the level of the sea: implications for climate dynamics and solid-earth geophysics. *Rev. Geophys.*, **36**, 603–689.

Penck, A. and **Brückner, E.** (1909) *Die Alpen im Eiszeitalter.* Tauchnitz, Leipzig.

Perlmutter, M.A. and **Matthews, M.D.** (1989) Global cyclostratigraphy: a model. In: *Quantitative Dynamic Straitgraphy* (Ed. by T. A. Cross), pp. 233–260. Prentice Hall, Englewood Cliffs.

Pillans, B., Chappell, J. and **Naish, T.R.** (1998) A review of the Milankovitch climatic beat: template for Plio-Pleistocene sea-level changes and sequence stratigraphy. *Sedim. Geol.*, **122**, 5–21.

Pinter, N. and **Brandon, M.T.** (1997) How erosion builds mountains. *Sci. Am.*, **276**, 74–79.

Pirazzoli, P.A. (1996) *Sea-Level Changes.* John Wiley, Chichester.

Pons, L.J. (1957) De geologie, de bodemvorming en de waterstaatkundige ontwikkeling van het Land van Maas en Waal en een gedeelte van het Rijk van Nijmegen. *Versl. Landbouwk. Onderz.*, **63-11**, 1–156.

Porter, S.C. (1989) Some geological implications of average Quaternary glacial conditions. *Quat. Res.*, **32**, 245–262.

Posamentier, H.W. and **Allen, G.P.** (1993) Siliclastic sequence stratigraphic patterns in foreland ramp-type basins. *Geology*, **21**, 455–458.

Posamentier, H.W. and **Vail, P.R.** (1988) Eustatic controls on clastic deposition II – Sequence and systems tract models. In: *Sea-Level Changes: an Integrated Approach* (Ed. by C. K. Wilgus, B. S. Hastings, C. G. S. C. Kendall, H. W. Posamentier, C. A. Ross and J. C. Van Wagoner), *Spec. Publ. Soc. Econ. Paleont. Miner.*, **42**, 125–154.

Posamentier, H.W. and **Weimer, P.** (1993) Siliciclastic sequence stratigraphy and petroleum geology ñ Where to from here? *Am. Assoc. Petrol. Geol. Bull.*, **77**, 731–742.

Posamentier, H.W., Jervey, M.T. and **Vail, P.R.** (1988) Eustatic controls on clastic deposition I – Conceptual frameworks. In: *Sea-Level Changes: an Integrated Approach* (Ed. by C. K. Wilgus, B. S. Hastings, C. G. S. C. Kendall, H. W. Posamentier, C. A. Ross and J. C. Van Wagoner), *Spec. Publ. Soc. Econ. Paleont. Miner.*, **42**, 109–124.

Posamentier, H.W., Allen, G.P., James, D.P. and **Tesson, M.** (1992) Forced regressions in a sequence stratigraphic framework: Concepts, examples, and exploration significance. *Am. Assoc. Petrol. Geol. Bull.*, **76**, 1687–1709.

Potter, P.E. (1978) Significance and origin of big rivers. *J. Geol.*, **86**, 13–33.

Powell, J.W. (1875) *Exploration of the Colorado River of the West and its Tributaries.* Government Printing Office. Washington, DC.

Probst, J.-L. (1989) Hydroclimatic fluctuations of some European Rivers since 1800. In: *Historical Changes of Large Alluvial Rivers, Western Europe* (Ed. by G. E. Petts), pp. 41–55. John Wiley and Sons, New York.

Retallack, G.J. (1986) Fossil soils as grounds for interpreting long-term controls on ancient rivers. *J. Sedim. Petrol.*, **56**, 1–18.

Retallack, G.J. (1990) *Soils of the Past – An Introduction to Palaeopedology.* Unwin Hyman, Boston.

Richards, K. (1982) *Rivers.* Methuen, London.

Rogers, J.C. (1997) North Atlantic storm track variability and its association to the North Atlantic Oscillation and climate variability of Northern Europe. *J. Clim.*, **10**, 1635–1647.

Rotnicki, K. (1991) Retrodiction of palaeodischarges of meandering and sinuous alluvial rivers and its palaeohydroclimatic implications. In: *Temperate Palaeohydrology* (Ed. by L. Starkel, K. J. Gregory and J. B. Thornes), pp. 431–471. John Wiley, Chichester.

Salter, T. (1993) Fluvial scour and incision: models for their influence on the development of realistic reservoir geometries. In: *Characterization of Fluvial and Aeolian Reservoirs* (Ed. by C. P. North and D. J. Prosser), *Spec. Publ. Geol. Soc.*, **73**, 33–51.

Saucier, R.T. (1994) *Geomorphology and Quaternary Geologic History of the Lower Mississippi Valley.* Mississippi River Commission, Vicksburg.

Saucier, R.T. (1996) A contemporary appraisal of some key Fiskian concepts with emphasis on Holocene meander belt formation and morphology. *Engineering Geol.*, **45**, 67–86.

Schirmer, W. (1995) Valley bottoms in the late Quaternary. *Z. Geomorph. N. F. (Suppl.)-Bd.*, **100**, 27–51.

Schumm, S.A. (1965) Quaternary palaeohydrology. In: *The Quaternary of the United States* (Ed. by H. E. Wright and D. G. Frey), pp. 783–794. Princeton University Press.

Schumm, S.A. (1968a) River Adjustment to Altered Hydrologic Regimen: the Murrumbidgee River and Palaeochannels, Australia. *Prof. Pap. United States Geol. Surv.*, **598**.

Schumm, S.A. (1968b) Speculations concerning palaeohydrologic controls of terrestrial sedimentation. *Geol. Soc. Am. Bull.*, **79**, 1573–1588.

Schumm, S.A. (1985) Explanation and extrapolation in geomorphology: seven reasons for geologic uncertainty. *Trans. Jap Geomorphol. Union.*, **6**, 1–18.

Schumm, S.A. (1977) *The Fluvial System.* John Wiley, New York.

Schumm, S.A. (1991) *To Interpret the Earth. Ten Ways to Be Wrong.* Cambridge University Press. Cambridge.

Schumm, S.A. (1993) River response to baselevel change: Implications for sequence stratigraphy. *J. Geol.*, **101**, 279–294.

Schumm, S.A. and **Brakenridge, G.R.** (1987) River responses. In: *North America and Adjacent Oceans During the Last Deglaciation* (Ed. by W. F. Ruddiman and H. E. Wright), The Geology of North America. K-3, pp. 221–240. Geological Society of America.

Shackleton, N.J. (1987) Oxygen isotopes, ice volume and sea-level. *Quat. Sci. Rev.*, **6**, 183–190.

Shackleton, N.J. and **Opdyke, N.D.** (1973) Oxygen isotope and palaeomagnetic stratigraphy of equatorial Pacific core V28–238: oxygen isotope temperatures and ice volumes on a 10^5 and 10^6 year scale. *Quat. Res.*, **3**, 39–55.

Shanley, K.W. and **McCabe, P.J.** (1991) Predicting facies architecture through sequence stratigraphy: an example from the Kaiparowits Plateau. *Utah. Geology*, **19**, 742–745.

Shanley, K.W. and **McCabe, P.J.** (1993) Alluvial architecture in a sequence stratigraphic framework: a case study from the Upper Cretaceous of southern Utah, U.S.A. In: *Quantitative Modelling of Clastic Hydrocarbon Reservoirs and Outcrop Analogs* (Ed. by S. S. Flint and I. D. Bryant), *Spec. Publ. Int. Assoc. Sediment.*, **15**, 21–55.

Shanley, K.W. and **McCabe, P.J.** (1994) Perspectives on the sequence stratigraphy of continental strata: report of a working group at the 1991 NUNA Conference on High Resolution Sequence Stratigraphy. *Am. Assoc. Petrol. Geol. Bull.*, **74**, 544–568.

Slingerland, R.L. and **Snow, R.S.** (1988) Stability analysis of a rejuvenated fluvial system. *Z. Geomorph. N. F. (Suppl.)-Bd.*, **67**, 93–102.

Sloss, L.L. (1962) Stratigraphic models in exploration. *Am. Assoc. Petrol. Geol. Bull.*, **46**, 1050–1057.

Sloss, L.L. (1963) Sequences in the cratonic interior of North America. *Geol. Soc. Am. Bull.*, **74**, 93–113.

Smith, G.A. (1994) Climatic influences on continental deposition during late-stage filling of an extensional basin, southeastern Arizona. *Geol. Soc. Am. Bull.*, **106**, 1212–1228.

Smith, N.D., Cross, T.A., Dufficy, J.P. and **Clough, S.R.** (1989) Anatomy of an avulsion. *Sedimentology*, **36**, 1–23.

Stanley, D.J. Warne, A.G. (1993) Nile Delta: recent geological evolution and human impact. *Science*, **260**, 628–634.

Starkel, L. (1987) The evolution of European rivers – a complex response. In: *Palaeohydrology in Practice* (Ed. by K. J. Gregory, J. Lewin and J. B. Thornes), pp. 333–339. John Wiley, Chichester.

Starkel, L. (1991) Long-distance correlation of fluvial events in the temperate zone. In: *Temperate Palaeohydrology* (Ed. by L. Starkel, K. J. Gregory and J. B. Thornes), pp. 473–495. John Wiley, Chichester.

Starkel, L., Gregory, K.J. and **Thornes, J.B.** (Eds). (1991) *Temperate Palaeohydrology.* John Wiley, Chichester.

Straffin, E.C., Blum, M.D., Colls, A. and **Stokes, S.** (2000) Alluvial stratigraphy of the Loire and Arroux Rivers, Burgundy, France. *Quaternaire*, **10**, 271–282.

Summerfield, M.A. (1985) Plate tectonics and landscape development on the African continent. In: *Tectonic Geomorphology* (Ed. by M. Morisawa and J. Hack), pp. 27–51. Allen and Unwin, Boston.

Suter, J.R. and **Berryhill, H.L. Jr** (1985) Late Quaternary shelf-margin deltas, northwest Gulf of Mexico. *Am. Assoc. Petrol. Geol. Bull.*, **69**, 77–91.

Talling, P.J. (1998) How and where do incised valleys form if sea-level remains above the shelf edge? *Geology*, **26**, 87–90.

Tebbens, L.A., Veldkamp, A., Westerhoff, W. and **Kroonenberg, S.B.** (1999) Fluvial incision and channel downcutting as a response to Lateglacial and Early Holocene climate change: the River Meuse lower reach. *J. Quat. Sci.*, **14**, 59–75.

Tesson, M., Allen, G.P. and **Ravenne, C.** (1993) Late Pleistocene shelf-perched lowstand wedges on the Rhone continental shelf. In: *Sequence Stratigraphy and Facies Associations* (Ed. by H. W.Posamentier, C. P. Summerhayes, B. U. Haq and G. P. Allen), *Spec. Publ. Int. Assoc. Sediment.*, **18**, 183–196.

Tesson, M. and **Gensous, B.** (1998) Líenregistrement des cycles climatiques et eustatiques quaternaires de marges recentes du bassin Nord-Mediterraneen. *C. R. Academie des Sciences, Paris, Sciences de la Terre Des Planetes*, **326**, 121–127.

Toomey, R.S., Blum, M.D. and **Valastro, S. Jr.** (1993) Late Quaternary climates and environments of the Edwards Plateau, Texas. *Glob. Plan. Change*, **7**, 299–320.

Törnqvist, T.E. (1993) Holocene alternation of meandering and anastomosing fluvial systems in the Rhine-Meuse Delta (central Netherlands) controlled by sea-level rise and subsoil erodibility. *J. Sedim. Petrol.*, **63**, 683–693.

Törnqvist, T.E. (1994) Middle and late Holocene avulsion history of the River Rhine (Rhine-Meuse delta, Netherlands). *Geology*, **22**, 711–714.

Törnqvist, T.E. (1995) Discussion: alluvial architecture of the Quaternary Rhine–Meuse river system in the Netherlands. *Geol. Mijnb.*, **74**, 183–186.

Törnqvist, T.E. (1998) Longitudinal profile evolution of the Rhine–Meuse system during the last deglaciation: interplay of climate change and glacio-eustasy? *Terra Nova*, **10**, 11–15.

Törnqvist, T.E., Wallinga, J., Murray, A.S., De Wolf, H., Cleveringa, P. and **De Gans, W.** (2000) Response of the Rhine-Meuse system (west-central Netherlands) to the last Quaternary glacio-eustatic cycles: a first assessment. *Global Planet. Change*, in press.

Torres, W.F.J. and **Jain, S.C.** (1984) *Aggradation and Degradation of Alluvial Channel Beds.* Iowa Institute for Hydraulic Research Report 274, University of Iowa, Iowa City.

Tushingham, A.M. and **Peltier, W.R.** (1992) Validation of the ICE-3G model of Würm-Wisconsin deglaciation using a global database of relative sea-level histories. *J. Geophys. Res.*, **97**, 3285–3304.

Vail, P.R., **Mitchum, R.M.** and **Thompson III, S.** (1977) Seismic stratigraphy and global changes of sea-level, part 4: global cycles of relative changes of sea-level. In: *Seismic Stratigraphy – Applications to Hydrocarbon Exploration* (Ed. by C. E. Payton), *Am. Assoc. Petrol. Geol. Mem.*, **26**, 83–97.

Van den Berg, J.H. (1995) Prediction of alluvial channel pattern of perennial rivers. *Geomorphology*, **12**, 259–279.

Vandenberghe, J. (1995) Postglacial river activity and climate: state of the art and future prospects. *Palâoklimaforschung*, **14**, 1–9.

Van de Plassche, O. (1982) Sea-level change and water-level movements in the Netherlands during the Holocene. *Meded. Rijks Geol. Dienst*, **36**, 1–93.

Van der Woude, J.D. (1984) The fluviolagoonal palaeoenvironment in the Rhine/Meuse deltaic plain. *Sedimentology*, **31**, 395–400.

Van Dijk, G.J., **Berendsen, H.J.A.** and **Roeleveld, W.** (1991) Holocene water level development in The Netherlands' river area; implications for sea-level reconstruction. *Geol. Mijnb.*, **70**, 311–326.

Van Wagoner, J.C. (1995a) Overview of sequence stratigraphy of foreland basin deposits: Terminology, summary of papers, and glossary of sequence stratigraphy. In: *Sequence Stratigraphy of Foreland Basin Deposits* (Ed. by J. C. Van Wagoner and G. T. Bertram), *Am. Assoc. Petrol. Geol. Mem.*, **64**, ix–xxi.

Van Wagoner, J.C. (1995b) Sequence stratigraphy and marine to nonmarine facies architecture of foreland basin strata, Book Cliffs, Utah, U.S.A. In: *Sequence Stratigraphy of Foreland Basin Deposits* (Ed. by J. C. Van Wagoner and G. T. Bertram), *Am. Assoc. Petrol. Geol. Mem.*, **64**, 137–223.

Van Wagoner, J.C. (1998) Sequence stratigraphy and marine to nonmarine facies architecture of foreland basin strata, Book Cliffs, Utah, U.S.A.: Reply. *Am. Assoc. Petrol. Geol. Bull.*, **82**, 1607–1618.

Van Wagoner, J.C., **Posamentier, H.W.**, **Mitchum, R.M.**, **Vail, P.R.**, **Sarg, J.F.**, **Loutit, T.S.** and **Hardenbol, J.** (1988) An overview of the fundamentals of sequence stratigraphy and some key definitions. In: *Sea-Level Changes: an Integrated Approach* (Ed. by C. K. Wilgus, B. S. Hastings, C. G. S. C. Kendall, H. W. Posamentier, C. A. Ross and J. C. Van Wagoner), *Spec. Publ. Soc. Econ. Paleont. Miner.*, **42**, 39–45.

Van Wagoner, J.C., **Mitchum, R.M.**, **Campion, K.M.** and **Rahmanian, V.D.** (1990) *Siliciclastic Sequence Stratigraphy in Well Logs, Cores, and Outcrops: Concepts for High-Resolution Correlation of Time and Facies: AAPG Methods in Exploration 7*. AAPG, Tulsa.

Veldkamp, A. and **Van den Berg, M.W.** (1993) Three-dimensional modelling of Quaternary fluvial dynamics in a climo-tectonic dependent system. A case study of the Maas record (Maastricht, The Netherlands). *Global Planet. Change*, **8**, 203–218.

Veldkamp, A. and **Van Dijke, J.J.** (1998) Modelling long-term erosion and sedimentation processes in fluvial systems: A case study for the Allier/Loire system. In: *Palaeohydrology and Environmental Change* (Ed. by G. Benito, V. R. Baker and K. J. Gregory), pp. 53–66. John Wiley, Chichester.

Walker, R.G. (1990) Facies modeling and sequence stratigraphy. *J. Sedim. Petrol.*, **60**, 777–786.

Walker, R.G. (1992) Facies, facies models and modern stratigraphic concepts. In: *Facies Models* (Ed. by R. G. Walker and N. P. James), pp. 1–14. Geological Association of Canada, St. John's.

Walling, D.E. (1983) The sediment delivery problem. *J. Hydrol.*, **65**, 209–237.

Walling, D.E. and **Webb, B.W.** (1983) Patterns of sediment yield. In: *Background to Palaeohydrology* (Ed. by K. J. Gregory), pp. 69–100. John Wiley, Chichester.

Walling, D.E. and **Webb, B.W.** (1996) Erosion and sediment yield: a global overview. In: *Erosion and Sediment Yield: Global and Regional Perspectives* (Ed. by D. E. Walling and B. W. Webb), *Int. Assoc. Hydrol. Sci. Publ.*, **236**, 3–20.

Wanless, H.R. and **Weller, J.M.** (1932) Correlation and extent of Pennsylvanian cyclothems. *Geol. Soc. Am. Bull.*, **43**, 1003–1016.

Waylen, P.R. (1996) Global hydrology in relation to palaeohydrological change. In: *Global Continental Palaeohydrology* (Ed. by K. J. Gregory, L. Starkel and V. R. Baker), pp. 61–86. John Wiley, Chichester.

Wescott, W.A. (1993) Geomorphic thresholds and complex response of fluvial systems – some implications for sequence stratigraphy. *Am. Assoc. Petrol. Geol. Bull.*, **77**, 1208–1218.

Wheeler, H.E. (1964) Baselevel, lithostratigraphic surface, and time stratigraphy. *Geol. Soc. Am. Bull.*, **75**, 599–610.

Wilgus, C.K., **Hastings, B.S.**, **Kendall, C.G.S.C.**, **Posamentier, H.W.**, **Ross, C.A.** and **Van Wagoner, J.C.** (Eds) (1988) Sea-Level Changes. An Integrated Approach. *Spec. Publ. Soc. Econ. Paleont. Miner.*, **42**, 1–407.

Williams, G.P. (1978) Bankfull discharge of rivers. *Water Res. Res.*, **14**, 1141–1154.

Williams, G.P. (1987) Palaeofluvial estimates from dimensions of former channels and meanders. In. *Flood Geomorphology* (Ed. by V. R. Baker, R. C. Kochel, and P. C. Patton), pp. 321–334. John Wiley and Sons.

Willis, B. (1993a) Ancient river systems in the Himalayan foredeep, Chinji village area, northern Pakistan. *Sedim. Geol.*, **88**, 1–76.

Willis, B. (1993b) Evolution of Miocene fluvial systems in the Himalayan foredeep through a two kilometer thick succession in northern Pakistan. *Sedim. Geol.*, **88**, 77–122.

Willis, B.J. and **Behrensmeyer, A.K.** (1994) Architecture of Miocene overbank deposits in northern Pakistan. *J. Sedim. Res.*, **B64**, 60–67.

Wohl, E.E. and **Enzel, Y.** (1996) Data for palaeohydrology. In: *Global Continental Palaeohydrology* (Ed. by K. J. Gregory, L. Starkel and V. R. Baker), pp. 23–60. John Wiley, Chichester.

Wood, L.J., **Ethridge, F.G.** and **Schumm, S.A.** (1993) The effects of rate of base-level fluctuation on coastal-plain, shelf and slope depositional systems: an experimental approach. In: *Sequence Stratigraphy and Facies Associations* (Ed. by H. W. Posamentier, C. P. Summerhayes, B. U. Haq and G. P. Allen), *Spec. Publ. Int. Assoc. Sediment.*, **18**, 43–53.

Wood, L.J., **Ethridge, F.G.** and **Schumm, S.A.** (1994) An experimental study of the influence of subaqueous shelf angles on coastal plain and shelf deposits. In: *Siliciclastic Sequence Stratigraphy. Recent Developments and Applications* (Ed. by P. Weimer and H. W. Posamentier), *Am. Ass. Petrol. Geol. Mem.*, **58**, 381–391.

Wright, H.E. Jr (1987) Synthesis: Land south of the ice sheets. In: *North America and Adjacent Oceans During the Last Deglaciation* (Ed. by W. F.Ruddiman and H. E. Wright), GSA Geology of North America, Vol. K-3, pp. 479–488.

Wright, H.E., Jr, **Kutzbach, J.E.**, **Webb, T.**, **Ruddiman, W.F.**,

Street-Perrott, F.A. and **Bartlein, P.J.** (Eds). (1993) *Global Climates Since the Last Glacial Maximum*. University of Minnesota Press, Minneapolis.

Wright, V.P. (Ed). (1986) *Paleosols*. Princeton University Press, Princeton.

Wright, V.P. and **Marriott, S.B.** (1993) The sequence stratigraphy of fluvial depositional systems: the role of floodplain sediment storage. *Sedim. Geol.*, **86**, 203–210.

Wyroll, K.-H. (1988) Determining the causes of Pleistocene stream aggradation in the central coastal areas of western Australia. *Catena*, **15**, 39–51.

Yang, C.-S. and **Nio, S.-D.** (1993) Application of high resolution sequence stratigraphy to the Upper Rotliegend in the offshore. In: *Siliciclastic Sequence Stratigraphy* (Ed. by P. Weimer and H. W. Posamentier), *Am. Assoc. Petrol. Geol. Mem.*, **58**, 285–316.

Yoshida, S., **Miall, A.D.** and **Willis, A.** (1998) Sequence stratigraphy and marine to nonmarine facies architecture of foreland basin strata, Book Cliffs, Utah, U.S.A. *Am. Assoc. Petrol. Geol. Bull.*, **82**, 1596–1606.

Zaitlin, B.A., **Dalrymple, R.W.** and **Boyd, R.** (1994) The stratigraphic organization of incised-valley systems associated with relative sea-level change. In: *Incised-Valley Systems: Origin and Sedimentary Sequences* (Ed. by R. W. Dalrymple, R. Boyd and B. A. Zaitlin), *Spec. Publ. Soc. Econ. Paleont. Miner.*, **51**, 45–60.

Zaleha, M.J. (1997) Fluvial and lacustrine palaeoenvironments of the Miocene Siwalik Group, Khaur area, northern Pakistan. *Sedimentology*, **44**, 349–368.

Zhang, Z., **Sun, K.** and **Yin, J.** (1997) Sedimentology and sequence stratigraphy of the Shanxi Formation (Lower Permian) in the northwestern Ordos Basin, China: an alternative sequence model for fluvial strata. *Sedim. Geol.*, **112**, 123–136.

Sedimentology (2000) **47** (Suppl. 1), 49–61

Dolomite formation and biogeochemical cycles in the Phanerozoic

STEPHEN J. BURNS*, JUDITH A. MCKENZIE† and CRISOGONO VASCONCELOS†
**Geological Institute, University of Bern, CH-3012 Bern, Switzerland (E-mail: burns@geo.unibe.ch)*
†Geological Institute, ETH-Zentrum, CH-8092 Zürich, Switzerland (E-mail: sediment@erdw.ethz.ch; chris@erdw.ethz.ch)

ABSTRACT

Based on present knowledge of the purely chemical controls on the kinetics of massive dolomite formation, the abundance and distribution of dolomite throughout the Phanerozoic remains an enigma, sometimes referred to as the 'dolomite problem'. Comparing dolomite abundance to secular variation in seawater chemistry indicates that some changes in seawater chemistry are more likely to have resulted from extensive dolomitization rather than to have caused it. The recently formulated microbial dolomite model provides the opportunity to view the geological history of dolomite formation from a new perspective. A biogeochemical approach to the 'dolomite problem' reveals a plausible connection between Phanerozoic geochemical cycles and dolomite formation. In particular, periods of more extensive dolomitization broadly correlate with diverse indicators of decreased oxygen levels in the atmosphere and oceans. Lowered oxygen levels would have fostered a more active community of anaerobic microbes, including sulphate-reducing bacteria, which in turn could have led to more extensive dolomitization of marine carbonates.

Keywords Dolomite, bacteria, sulphate reduction, geochemical cycles, Phanerozoic

INTRODUCTION

In September 1991, carbonate sedimentologists gathered in Ortisei, Italy, to attend the Dolomieu Conference on Carbonate Platforms and Dolomitization, which was organized to celebrate the 200th anniversary of Déodat Dolomieu's classic 1791 paper describing the mineral dolomite in detail for the first time (Zenger *et al.*, 1994). The first century of dolomite research was basically one of discovery and the application of physical and chemical principles to explain observations (Fig. 1). The second century saw new discoveries of modern dolomite environments and development of new ideas which were formulated into models. After two centuries of research, however, long held questions concerning the uneven distribution of dolomite in geological time, and the inability to precipitate dolomite in the laboratory under conditions replicating those found in modern dolomite environments, remained enigmatic (McKenzie, 1991).

The Dolomieu Conference brought together many of the experts in the field, and an IAS special publication, featuring a collection of papers summarizing the highlights of the presentations and discussions, subsequently appeared. In particular, this volume contains an excellent introductory paper by the conference organizers which provides an overview and outlines the current state-of-the-art with respect to dolomite and its origins (Purser *et al.*, 1994). With this background, we pondered what new things could be written about dolomite fewer than 10 years later? As we begin the third century of dolomite research, what new information or thoughts could we add to the reams of words already written about the 'dolomite problem'? See for example the references cited in Purser *et al.* (1994).

Perhaps surprisingly, we feel a new prospective to dolomite research does exist, one that has evolved significantly in the last decades (Fig. 1). This new view adds a microbial factor

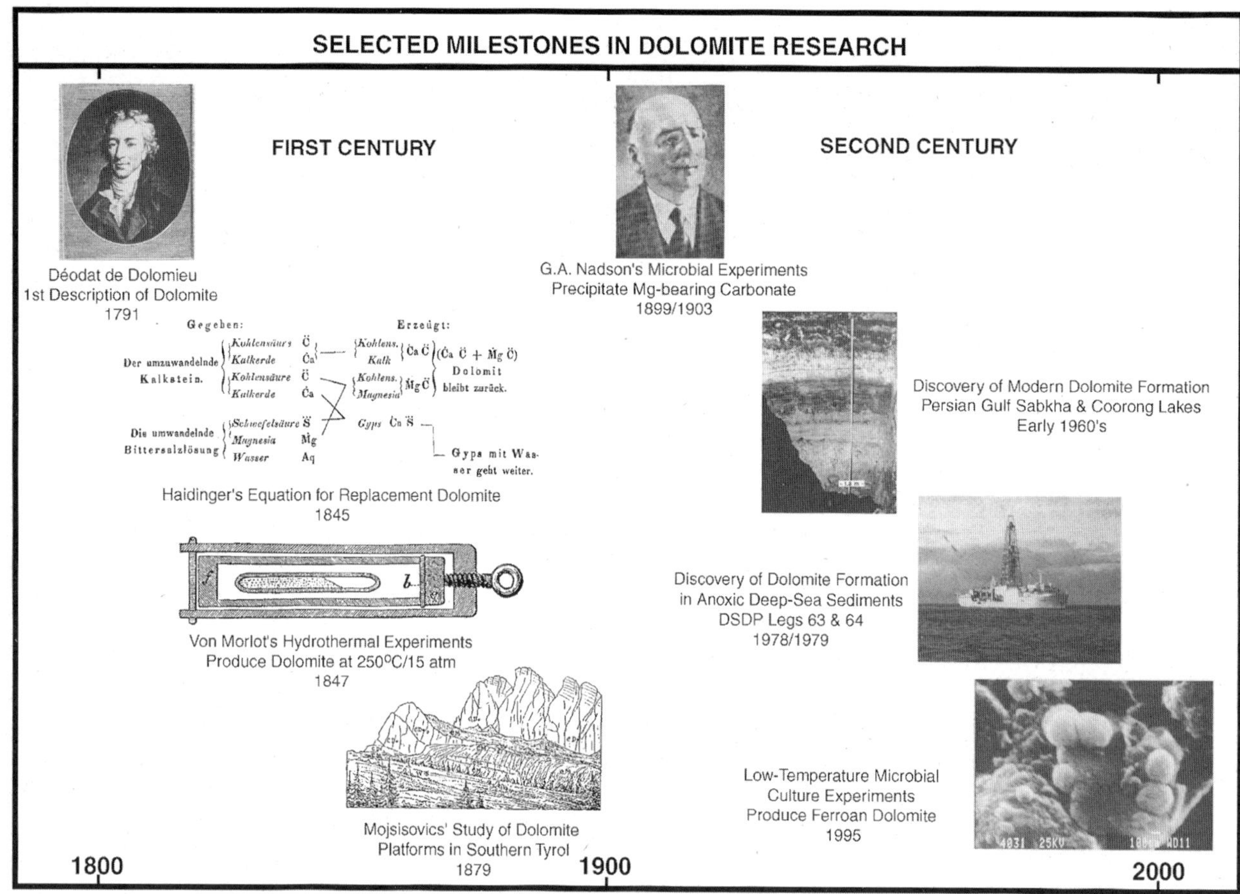

Fig. 1. Illustrated compilation of some selected milestones in the 200-year history of dolomite research. With the first description of dolomite by Dolomieu (1791), the exact chemistry of the mineral was a topic of research beginning with the chemical analysis of Saussure (1792). Based on a common observation that there was a genetic association between beds of dolomite and gypsum, Haidinger (1845) attempted to explain the physical and chemical processes of dolomitization, as shown in his early chemical reaction equations. Subsequently, Von Morlot (1847) reported on experiments based on Haidinger's equations, wherein he produced dolomite under high-temperature and high-pressure conditions. This was probably the first of many attempts in the following century to produce dolomite in the laboratory. The classic area of southern Tyrol, now known as the Dolomite Mountains of northern Italy, was a prime location for geological field work in the nineteenth century (e.g. Mojsisovics, 1879). The second century began with some of the first geomicrobiological experiments by Nadson (1928), which may have produced dolomite. The microbial approach was forgotten with the exciting discoveries of modern dolomite forming under hypersaline conditions beneath the sabkhas of Abu Dhabi (Illing *et al.*, 1965) and in the Coorong lakes of South Australia (Von der Borch, 1965). Beginning in 1968 with the Deep Sea Drilling Project, exploration of deep-sea basins led to the discovery of dolomite forming under anoxic conditions (Pisciotto & Mahoney, 1981; Kelts & McKenzie, 1982). Finally, at the end of the twentieth century, microbial mediation of dolomite precipitation was demonstrated and a new microbial dolomite model evolved (Vasconcelos *et al.*, 1995). Nadson's photograph from Lapo (1987).

to the list of conditions promoting dolomite formation. It first came to the forefront with the discovery of Pleistocene dolomite in organic carbon-rich sediment during deep-sea drilling on the California Margin (DSDP Leg 63, Pisciotto & Mahoney, 1981) and the Gulf of California (DSDP Leg 64, Kelts & McKenzie, 1982) and led to the sulphate reduction model for dolomite precipitation. This model proposes that sulphate ions inhibit dolomite precipitation (Baker & Kastner, 1981). Beginning with the recognition that dolomite does form under anoxic conditions, the potential importance of organic matter and the role of microbial sulphate reduction in the precipitation of dolomite, at least in some environments, have been generally accepted (e.g. Garrison *et al.*, 1984; Burns *et al.*, 1988), although the relationship has been poorly constrained. At the Dolomieu Conference, the participants were in agreement that organic matter was probably an important factor for dolomite precipitation, but the precise function remained elusive (Purser *et al.*, 1994).

Since the appearance of the Dolomieu Conference volume, the organic matter factor has been pursued using a different geomicrobiological approach to understand its importance for dolomite precipitation. Specifically, Vasconcelos *et al.* (1995) reported the results of a bacterial culture experiment in which ferroan dolomite was produced under low-temperature anaerobic conditions, partially replicating those found in a modern dolomite-precipitating environment, Lagoa Verhmelha (Brazil). Because no dolomite was found in sterile control samples, the results of their bacterial culture experiment provide strong evidence for the involvement of bacteria (possibly including sulphate-reducing bacteria) in the precipitation of dolomite. Taking this work a step further, Vasconcelos & McKenzie (1997) proposed a new model for dolomite formation, based on their study of the natural laboratory of Lagoa Verhmelha. Their microbial dolomite model provides a different actualistic system in which to evaluate the geological record of dolomite. For example, Wright (1997) has demonstrated that such an organogenic approach can be successfully applied to evaluate ancient dolomite formations.

The importance of bacteria for the precipitation of dolomite is not, however, a new idea. Beginning in 1899, Nadson (1928) conducted similar bacterial culture experiments in which he used sulphate-reducing bacteria from anoxic surface sediments in a Russian salt lake to produce very small amounts of an authigenic carbonate. A qualitative chemical analysis of the carbonate led him to propose that dolomite might have been precipitated in association with the anaerobes. Half a century later, Neher (1959) reported on an anaerobic bacterial experiment in which dolomite crystals were precipitated. More recently, Folk (1993) has been a strong proponent of the influence of bacteria on mineral formation based on his observations of nanometre-sized spheroids within or upon the faces of crystals, which he proposes are the morphological representations of former bacterial bodies. Although the bacterial link has been previously recognized throughout the second century of dolomite research, albeit rarely reported, the importance of the microbial factor in dolomite formation has been generally neglected and has not been given its rightful consideration within the voluminous literature compiled on the 'dolomite problem'.

With this paper, we re-examine the factors controlling dolomite formation in light of recent developments in dolomite investigations, and apply a microbial approach to the interpretation of the geological record of dolomite formation over the course of the Phanerozoic. We begin by briefly reviewing the microbial model of dolomite formation, as presently understood, based on studies of modern environments and experimental results. Then, we compare secular variation in dolomite abundance through the Phanerozoic to changing environmental conditions over the same time period. This comparison is first made, as has often been done, from a purely physicochemical viewpoint. Finally, we add microbes into the equation to see if they help explain variation in dolomite abundance through time. As we begin a third century of dolomite research, we suggest that major progress will be made in the search for answers to the unresolved questions involving dolomite formation by linking studies of the biosphere with the geosphere. Paraphrasing Nadson (1928), we agree with his statement, 'Understanding the essential role played by this bacterial phenomenon may be the solution to the Dolomite Problem and the problems of the Mg cycle in the ocean.'

THE MICROBIAL DOLOMITE MODEL

Vasconcelos & McKenzie (1997) postulated a new microbial model for dolomite formation based on their study of a coastal lagoon (Lagoa Vermelha), which is located in an unusual hydrological and climatic setting. The region is dominated by a semi-arid microclimate, which leads to extreme hypersalinity in the lagoon during the dry season. Intense evaporation increases the salinity and lowers the water level, permitting the inflow of seawater to supply ions for the microbial processes. During the wet season, precipitation exceeds evaporation, resulting in large variations in salinity, sometimes reaching brackish conditions. Fluctuation between hypersaline and brackish conditions in a short period of time is reflected in the distinctive carbonate mineralogy of the sediment deposited in Lagoa Vermelha. Understanding the annual hydrological pattern in Lagoa Vermelha was important because it provided information about the source and concentration of ions used to form dolomite, as well as the timing of dolomite precipitation. Dolomite apparently precipitates under the most

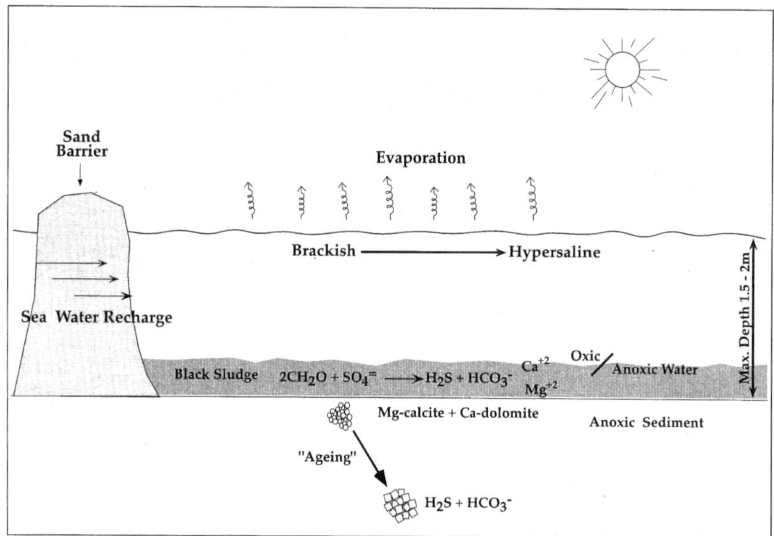

Fig. 2. Schematic diagram illustrating the microbial dolomite model as observed in Lagoa Vermelha (Brazil). See text for discussion. Adapted from Vasconcelos & McKenzie (1997).

hypersaline conditions, whereas high-Mg calcite forms at intermediate salinity and low-Mg calcite during periods with brackish water (Vasconcelos & McKenzie, 1997). The schematic diagram shown in Fig. 2 illustrates the microbial dolomite model as observed in Lagoa Vermelha.

Relatively high productivity in Lagoa Vermelha leads to anoxic conditions at the water–sediment interface and formation of a black, organic carbon-rich sludge. Microbial activity apparently mediates the precipitation of carbonate minerals within the sludge layer (Fig. 2). The occurrence of modern dolomite in this environment implies that microbial activity may influence nucleation of dolomite. Microbial activity in the lagoon is probably directly influenced by the changing water chemistry. The response of the microbial population in Lagoa Vermelha to changes in salinity may be represented by different metabolic processes at the water–sludge interface. Under more extreme hypersaline conditions, sulphate-reducing organisms may become more active.

In the case of dolomite precipitation in Lagoa Vermelha, Mg^{2+} and SO_4^{2-} appear to be essential components. When sulphate-reducing bacteria use SO_4^{2-}, they also take up Mg^{2+} because it forms a strong ionic pair with SO_4^{2-}. In other words, the micro-organisms overcome the kinetic barrier by using the SO_4^{2-} for their metabolism, while at the same time they release Mg^{2+} from the ion pair. On a submicrometre scale, the microbial metabolism saturates the microenvironment around the cell with HCO_3^-, creating conditions favourable for preferential precipitation of dolomite. The microbially nucleated crystals are subsequently buried, where the initially formed Ca-dolomite undergoes an 'ageing' process, whereby inorganic recrystallization occurs to produce a more stoichiometric dolomite (Fig. 2). The microbial dolomite model proposes that a continuous supply of SO_4^{2-} is necessary to maintain the microbial activity required to produce dolomite. In contrast, in a sterile inorganic system, SO_4^{2-} may remain an inhibiting factor to dolomite precipitation, as has been indicated by the experimental work of Baker & Kastner (1981) and Morrow & Ricketts (1988).

The microbial dolomite model offers a new perspective with which to evaluate mechanisms of dolomite formation. The annual cycle in Lagoa Vermelha illustrates the relationship of dolomite precipitation to variation in environmental conditions and implies that the same process might be applied on another time-scale to evaluate global environmental changes.

WHERE DO WE STAND?

Before attempting more broadly to apply a microbial model for dolomite formation to the Phanerozoic history of dolomitization, let us briefly review what is known about dolomite formation. It has long been recognized that almost all large bodies of dolomite '... form by reaction of solutions derived from sea water with previously deposited calcium carbonates' (Lippmann, 1973). All current 'models' of dolomite formation (see

Land, 1985) are primarily physical descriptions of how to move seawater, or modified seawater, through calcium carbonate sediment. Where all such models fail is in describing the chemical controls on why dolomite replaces calcium carbonate sediment in any particular setting, or the kinetic controls on dolomite formation. In 1973, Lippmann recognized that, 'The fact that dolomite is not a normal product of marine carbonate precipitation (and of many experiments) implies the existence of serious kinetic obstacles.' Twenty-five years later, Land (1998) tentatively suggested, '... we all(?) now agree that 'The Dolomite Problem' is one of kinetics.'

Given that dolomite forms primarily by replacement of calcium carbonate, the dolomitization reaction is usually written:

$$2CaCO_3 + Mg^{2+} = CaMg(CO_3)_2 + Ca^{2+}$$

As pointed out by numerous authors, however (e.g. Lippmann, 1973), low-temperature dolomite formation is a dissolution–reprecipitation reaction first requiring the dissolution of calcite and then the reprecipitation of dolomite by the following reaction:

$$Ca^{2+} + Mg^{2+} + 2CO_3^{2-} = CaMg(CO_3)_2$$

This observation is not trivial. Its importance is that observations of what controls dolomite formation in 'marginal' environments, where the overall reaction is not by the replacement of calcium carbonate, will still apply to dolomitization in general. We mention this because, to date, the role of bacteria in dolomite formation has been directly studied only in relatively specialized environments, or in experiments.

When it comes to the question of dolomite kinetics, nearly all of the studies conducted over the past two centuries, and primarily in the past 30 years, have focused on the purely physicochemical controls on rates of dolomite formation. Despite a large number of experimental studies (probably many more than have been reported on in the literature owing to their failure to produce dolomite), there is very little firm evidence for what factors control rates of dolomite precipitation at low temperatures. Leaving bacteria out of the equation for the moment, low-temperature laboratory experiments fail to produce even moderately ordered or stoichiometric dolomite even over quite long time periods. Only at temperatures over about 100 °C, well beyond those expected for synsedimentary dolomite formation, can dolomite be readily precipitated in experiments. At these temperatures, several factors can be shown to influence the rate of dolomitization of precursor carbonate, including solution Mg^{2+}/Ca^{2+} (hereafter Mg/Ca) ratio (Gaines, 1980; Sibley, 1990; Sibley *et al.*, 1994), dissolved sulphate concentration (Baker & Kastner, 1981; Morrow & Ricketts, 1988; Sibley *et al.*, 1994) and degree of supersaturation (Arvidson & Mackenzie, 1997). The importance of these factors to dolomitization at lower temperatures and in natural settings is less certain. All have been cited as promoting dolomite formation in evaporitic settings (Folk & Land, 1975). Sulphate reduction is also thought to lead to dolomite formation in organic carbon-rich, continental margin sediments (Baker & Burns, 1985).

CHANGES IN DOLOMITE ABUNDANCE AND SEAWATER CHEMISTRY THROUGH THE PHANEROZOIC

If we are to attempt to examine the role bacteria may or may not have on dolomite formation in general, that is outside of special, organic carbon-rich environments, we need first to look at how dolomite formation has varied through geological time and how that variation might relate to changes in seawater and atmospheric chemistry. The Phanerozoic is marked by large-scale cyclical changes in climate, sea-level and marine chemistry (Fischer, 1984). These cycles, primarily driven by changes in the rate of seafloor spreading and hydrothermal circulation at ocean ridge systems, affected, among many things, the deposition and cycling of carbonate rocks. For example, the mineralogy of primary calcium carbonate directly precipitating from seawater has alternated between aragonite and calcite (Sandberg, 1983) over the course of the Phanerozoic (Fig. 3). Calcium carbonate mineralogy is related to first-order sea-level variations, with calcite precipitated during periods of high sea-level and rapid seafloor spreading, and aragonite during periods of low sea-level and slow seafloor spreading.

Two explanations have been offered for these relationships, both related to changes in atmospheric or ocean chemistry brought about by changing seafloor spreading rate. First, rapid seafloor spreading could lead to increased

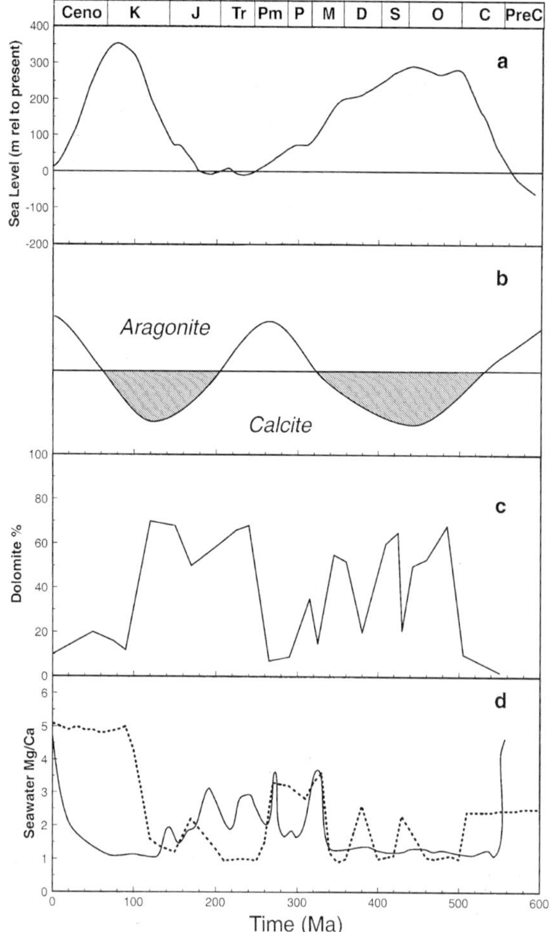

Fig. 3. Phanerozoic cycles of: (a) sea level (after Vail *et al.*, 1977); (b) mineralogy of primary calcium carbonate precipitates from seawater (after Sandberg, 1983); (c) dolomite abundance as per cent total carbonate rock (after Given & Wilkinson, 1987); and (d) seawater Mg/Ca from Hardie, 1996 (solid curve) and Wilkinson & Algeo, 1989 (dashed curve).

atmospheric and surface ocean pCO_2 levels, which would decrease carbonate solubility and favour calcite deposition over aragonite (Sandberg, 1985). Alternatively, rapid spreading should be accompanied by an increased flux of seawater through the global ocean ridge system and thereby decrease the Mg/Ca ratio of seawater (Mackenzie & Pigott, 1981; Hardie, 1996). Experimental work shows that the Mg/Ca ratio controls calcium carbonate mineralogy, and indicates that quite large decreases in seawater Mg/Ca are necessary to yield calcite as the primary carbonate mineral (Morse *et al.*, 1997).

The relative abundance of dolomite as compared with total carbonate rock has also varied cyclically through the Phanerozoic (Fig. 3c). Given & Wilkinson (1987) re-evaluated existing data on the composition of Phanerozoic carbonates. They showed that the relative variation in the distribution of dolomite during the last 570 Myr displays large peaks at the Early Ordovician/Early Silurian and the Early Cretaceous (Fig. 3c). This distribution implies that there is not a systematic increase in dolomite with age, but two discrete periods of significant dolomite formation during the Phanerozoic. Not all agree that this pattern faithfully represents the ratio of dolomite to calcite deposited over time (e.g. see Zenger, 1989). For the purposes of discussion, however, we accept it as representative of the first-order changes in dolomite abundance. The reason for the observed variations in dolomite occurrence remains unclear. Various authors (Mackenzie & Pigott, 1981; Wilkinson *et al.*, 1985; Wilkinson & Given, 1986; Mackenzie & Agegian, 1989; Wilkinson & Algeo, 1989) pointed out that periods of significant dolomite occurrence may be related to variation in atmosphere–hydrosphere environmental conditions. For example, Mackenzie & Morse (1992) postulated that '... conditions for early dolomitization ... are best met during extended times of high sea-level when atmospheric CO_2 levels are high and the saturation state of seawater with respect to carbonate minerals relatively low.'

As mentioned, one proposed cause for changes in the mineralogy of primary calcium carbonates is changes in the Mg/Ca ratio of seawater. Although for many years it was generally accepted that seawater chemistry has remained relatively constant for the past several billion years, this idea has since been challenged (Wilkinson & Algeo, 1989; Spencer & Hardie, 1990). Spencer & Hardie (1990) and Hardie (1996) suggested that secular variations in the mineralogies of potash evaporites require large changes in seawater chemistry, especially the Ca/SO_4 and Mg/Ca ratios. Because potash evaporite mineralogy also appears to be related to the first-order sea-level cycles, Spencer & Hardie (1990) and Hardie (1996) suggested that seawater chemistry changes are driven by changes in ocean ridge hydrothermal fluxes (see Holland *et al.*, 1996, for an opposing view). Quite large changes in seawater Mg/Ca could also be caused by variations in the rate of dolomite formation (Wilkinson & Algeo, 1989). Figure 3d shows two estimations of how seawater Mg/Ca has varied through the Phanerozoic. The curves are fairly similar in overall shape though the proposed mechanisms for the changes are somewhat different. Both curves show Mg/Ca ratios of close to 1 for the period from approximately the late Cambrian to early Mississippian, higher values during the Mississippian/Pennsylvanian and Permian, lower values for much of the Mesozoic, and increasing Mg/Ca either in the Cretaceous (Wilkinson & Algeo, 1989) or Cenozoic (Spencer & Hardie, 1990). In the model of Spencer & Hardie (1990) and Hardie (1996), seawater Mg^{2+} concentration remains relatively constant over time, and changes in Mg/Ca result from large changes in seawater Ca^{2+} concentration caused by changing mid-ocean ridge hydrothermal fluxes. In the model of Wilkinson & Algeo (1989), changes in Mg/Ca result primarily from changing seawater Mg^{2+} concentration caused by changing amounts of dolomite formation.

With an eye on the variations outlined above, we ask the following general question. Can the secular variation in seawater chemistry be used to explain why dolomite has more easily formed during some periods of Earth history than during others? We will first consider this question from a purely physicochemical approach based on our, admittedly limited, knowledge of dolomite reaction kinetics. We will then see if there might be a relationship between changing atmospheric and seawater chemistry and a biological/bacterial influence on dolomite formation.

DOLOMITIZATION AND SEAWATER Mg/Ca

One of the tenets of the dolomitization literature is that dolomite replacement of calcite is promoted by a high seawater Mg/Ca ratio. Estimations of the chemistry of dolomitizing fluids based on studies of naturally occurring dolomites have led many authors to suggest that high Mg/Ca ratios, often together with increased salinity, promote dolomite formation (e.g. Folk & Land, 1975). High-temperature experiments have also shown that the rate of dolomite replacement of calcium carbonate minerals increases with increased solution Mg/Ca (e.g. Gaines, 1980; Lumsden *et al.*, 1995). Comparing the secular variations of seawater chemistry and dolomite abundance, however, suggests that either the seawater Mg/Ca ratio has little effect on dolomite formation, or even that dolomite replacement of calcium carbonate is enhanced by lowered Mg/Ca. Figure 3 shows that the periods during which dolomite makes up the highest percentage of carbonate rock coincide with periods of decreased seawater Mg/Ca. As mentioned, different authors have suggested different mechanisms for decreased Mg/Ca, either changes in the mid-ocean ridge hydrothermal flux (Spencer & Hardie, 1990; Hardie, 1996), or dolomite formation itself (Wilkinson & Algeo, 1989). Regardless of the mechanism, much more dolomite appears to form during periods of decreased seawater Mg/Ca (Given & Wilkinson, 1987).

From a purely thermodynamic viewpoint, the boundary between the dolomite and calcite stability fields is determined solely by the Mg/Ca ratio, as can also be seen by writing the dolomitization reaction:

$$2CaCO_3 + Mg^{2+} = CaMg(CO_3)_2 + Ca^{2+}$$

where, because the activities of the pure solids are equal to 1:

$$k_{sp(cal-dol)} = (Ca^{2+})/(Mg^{2+})$$

The equilibrium solubility product of this reaction is equal to the solubility of calcite squared over the solubility of dolomite:

$$k_{sp(cal-dol)} = (k_{sp-cal})^2/k_{sp-dol}$$

While the equilibrium solubility product for calcite is well established, that for dolomite at low temperatures is not. Typically, equilibrium solubility products for calcite and dolomite of around $10^{-8\cdot35}$ and $10^{-18\cdot22}$ are used to establish the stability boundary between calcite and dolomite (Lippmann, 1973). These values yield a Ca/Mg ratio at equilibrium of $10^{1\cdot52}$ or a Mg/Ca ratio of 0·03. The k_{sp-dol} value used by Lippmann, and most other values found in the literature, are for ordered stoichiometric dolomite. Most dolomite, however, initially forms as Ca-rich, 'protodolomite' with less than perfect ordering. The precise relationship between dolomite solubility and either stoichiometry or ordering is not known, though both disorder and nonstoichiometry will increase the solubility. Helgeson *et al.* (1978), based on thermodynamic criteria, proposed a k_{sp-dol} value of $10^{-16\cdot52}$ for completely disordered dolomite. Long-term dissolution experi-ments by Land (1998) indicated a larger k_{sp-dol}, around $10^{-15\cdot1}$ for dolomite with a composition of $Ca_{0\cdot58}Mg_{0\cdot42}(CO_3)$, although whether equilibrium was fully reached is not certain. A larger dolomite solubility product, for example 10^{-17}, would yield an equilibrium Mg/Ca ratio of 0·5 for the boundary between calcite and dolomite stability. This value is not far from the proposed Mg/Ca ratio of seawater for the periods of the Phanerozoic during which there was extensive dolomite formation, and it is not difficult to imagine that during these periods the sea-water value was fixed by equilibrium with dolomite (Wilkinson & Algeo, 1989). At the very least, decreased seawater Mg/Ca ratios seem to have had no inhibiting effect on dolomite formation.

Is it possible, however, that lowered Mg/Ca ratios might actually enhance dolomite formation? The saturation state of seawater with respect to calcite has also been suggested as a factor in dolomite formation (Mackenzie & Morse, 1992). Although the thermodynamic drive for dolomite to replace calcite in seawater remains whether calcite is supersaturated in seawater or not, mechanistically seawater saturated with calcite presents a problem because the process must proceed via the first step of calcite dissolution, which clearly will not occur under conditions of supersaturation. During periods of lower seawater Mg/Ca, the kinetic inhibition on calcite precipitation might disappear, so that the oceans will remain close to saturation with calcite. The first step toward dolomitization, calcite dissolution, would therefore more readily occur. There are a number of flaws in such an argument. First, there is no reason why dolomite could not form directly by replacing aragonite. Experiments of dolomitization of aragonite often proceed faster than similar experiments with calcite as a precursor carbonate (Katz & Matthews, 1977). Also, if dolomite precipitation can be initiated, dolomite precipitation itself will lower seawater saturation with respect to both calcite and dolomite, eventually to the point at which calcite is undersaturated.

DOLOMITE AND SEAWATER SULPHATE

How might changes in seawater chemistry affect other factors that are suspected to control dolomite reaction kinetics? Dissolved sulphate concentrations have been shown to inhibit dolomite replacement of calcium carbonate (Baker & Kastner, 1981), at least at low sulphate concentrations. It has also been proposed that either very low or very high sulphate concentrations could promote dolomite formation (Brady *et al.*, 1996). Sulphate is very effectively removed from seawater during hydrothermal circulation at mid-ocean ridges, and therefore there could be significant changes in seawater sulphate concentrations associated with the first-order Phanerozoic cycles that might have an effect on dolomitization. Both Spencer & Hardie (1990) and Holland *et al.* (1996) evaluated changes in sulphate concentration associated with changing rates of hydrothermal fluxes. In the Spencer and Hardie model, increases in the hydrothermal flux large enough to drive the Mg/Ca ratio down to 1 or below resulted in about a 25% decrease in the seawater sulphate concentration. In the calculations of Holland *et al.*, a doubling of the hydrothermal flux yielded about a 33% decrease in sulphate concentration. Based on comparison with experimental results of how sulphate concentration affects dolomite precipitation, these changes appear to be too small to have brought about a significant change in the rate of dolomite formation.

DOLOMITE SATURATION STATE IN SEAWATER

Theoretical thermodynamic arguments and high-temperature experiments suggest that the rate of dolomite formation should increase with increasing saturation state (Arvidson & Mackenzie, 1997). The seawater models of Spencer & Hardie (1990) and Holland *et al.* (1996) can be used very roughly to calculate how the ion activity product (IAP) for dolomite during periods of extensive dolomite formation compares with that of the modern ocean. Calculated seawater Ca^{2+}, Mg^{2+} and HCO_3^- (alkalinity) concentrations in both Spencer & Hardie's and Holland *et al.*'s models suggest that the IAP was reduced by at least a factor of 2 compared with that of the modern ocean. Furthermore, the periods of extensive dolomite formation were probably also periods of increased atmospheric pCO_2 (Sandberg, 1985; Mackenzie & Morse, 1992). The effect of increased pCO_2 on the seawater carbonate system would be to further reduce the carbonate ion concentration and therefore the IAP of dolomite and the theoretical rate of dolomite formation.

ATMOSPHERIC pCO_2 AND DOLOMITE FORMATION?

It has also been suggested that the conditions for synsedimentary dolomite formation were best met during times when increased atmospheric pCO_2 levels lowered the saturation state of seawater with respect to carbonate minerals (Mackenzie & Pigott, 1981; Machel & Mountjoy, 1986; Mackenzie & Morse, 1992). The same mechanism has been used to explain 'calcite seas' (Sandberg, 1985). This entire concept, however, does not stand up to close scrutiny. In a system of calcite and pure water, calcite equilibrium is just as easily reached at low or high pCO_2. Low pCO_2 does not lead to supersaturation. Perhaps a better natural test is the observation that at present low pCO_2 (on geological time-scales), calcite has absolutely no trouble precipitating from waters with low Mg/Ca ratios. It does so every summer in hardwater lakes throughout the world.

The modern ocean reaches aragonite supersaturation not because of low pCO_2, but because high Mg/Ca inhibits calcite precipitation (Morse *et al.*, 1997, and references therein). Slow calcite precipitation rates in seawater relative to processes that push the saturation state toward supersaturation, such as photosynthetic drawdown of CO_2 and temperature-related degassing, allow the calcium carbonate ion activity product to reach aragonite saturation. The latter precipitates only because precipitation of the former is slow. Whether or not the saturation state of seawater with respect to calcite has a role in dolomitization is not clear, but, if atmospheric pCO_2 level has no role in determining calcite saturation, neither does it have a role in dolomite formation.

In summary, the proposed secular variations in seawater Mg/Ca, sulphate concentration, saturation state or pCO_2 do little to help explain variations in dolomite abundance. According to our best understanding of the effects of all of these factors on dolomite reaction kinetics, seawater chemistry during 'calcite seas' intervals, those during which dolomite formed more rapidly, should have led to lower rates of dolomite formation, not higher. Yet, despite seawater composition that should have retarded rates of dolomitization, it would appear that some unknown factor(s) allowed much more rapid dolomitization of calcium carbonate sediments. In fact, extensive dolomite formation itself probably changed seawater chemistry to the point at which the Mg/Ca ratio and saturation level of seawater were much closer to dolomite equilibrium than is the case for the modern ocean. The changes in ocean chemistry observed or modelled during Phanerozoic supercycles are more easily explained as a response to more rapid dolomite formation (Wilkinson & Algeo, 1989) rather than the other way around. The question remains why?

A BIOGEOCHEMICAL APPROACH TO SECULAR VARIATIONS IN DOLOMITE

Finally, let us see if the microbial model for dolomite formation, developed from a rather specialized modern environment and test-tube experiments, has any application to dolomitization on geological time-scales. The main component of the model is the presence of sulphate-reducing bacteria, requisite anaerobes. Therefore, in addition to secular variation in the parameters discussed in the preceding section, we need to consider additional parameters that reflect the overall oxidative state of the ocean/atmosphere system. Such a consideration may yield insights into whether particular periods may have favoured environments conducive to sulphate-reducing bacteria and fostered their activity.

As mentioned, the distribution of dolomite during the last 570 Ma displays large peaks at the Early Ordovician/Middle Silurian and the Early Cretaceous (Figs 3 and 4). Variation in the intensity of hydrothermal circulation through mid-ocean ridges that drive changes in sea-level and seawater chemistry outlined above also result in other changes in the chemistry of the ocean/atmosphere system. With regard to sulphate-reducing bacteria, the most important of these will be factors that alter the free oxygen concentration of the atmosphere and dissolved oxygen concentration in the ocean. To a large degree, varying atmospheric oxygen level will be recorded as variation in the carbon and sulphur isotopic composition of chemical sediments in the geological record (Garrels & Perry, 1974; Kump & Garrels, 1986; Berner & Canfield, 1989; Carpenter & Lohmann, 1997; Petsch & Berner, in press). Figure 4 shows variations in: (a) dolomite distribution, (b) $\delta^{34}S$ values of marine evaporites and $\delta^{13}C$ values of marine carbonates, (c) modelled atmospheric O_2 levels and (d) ironstone occurrences during the Phanerozoic. Notably, there is an overall negative correlation between the abundance of dolomite (Fig. 4a) and the modelled atmospheric O_2 levels (Fig. 4c), implying that periods during which seawater oxygen levels are low might be more conducive to massive dolomite formation.

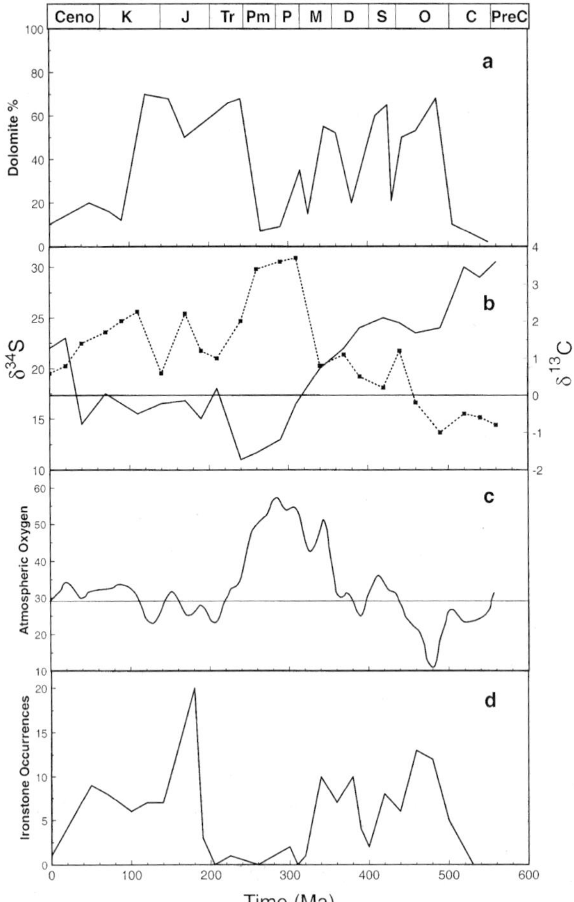

Fig. 4. Phanerozoic cycles of: (a) dolomite abundance; (b) $\delta^{13}C$ of marine carbonates (dashed line) and $\delta^{34}S$ of sulphate (solid line) in marine evaporites (after Lindh, 1983); (c) modelled atmospheric O_2 levels in 10^{18} moles (from Lasaga, 1989; straight line is modern value); and (d) occurrences of oolitic ironstones (after Van Houton & Bhattacharyya, 1982).

Another rough indicator of the overall oxic state of the oceans is the number of ironstone occurrences throughout the Phanerozoic (Fig. 4c). As ironstone occurrences require that iron be transported to the site of deposition in the ferrous state, the process of ironstone formation should also be promoted by low oxygen conditions. Furthermore, the reduction of ferric to ferrous iron in Earth surface systems is also a microbially mediated process carried out by bacteria found in oxygen-depleted environments. Significantly, the Phanerozoic pattern of oolitic ironstone occurrences is strikingly similar to that of percentage dolomite in carbonate rocks.

Based on the secular relationship between indicators of ocean/atmosphere oxygen levels and dolomite abundance, we propose that the primary factor controlling widespread dolomite formation is the free atmospheric oxygen concentration which would in turn control ocean oxygen concentrations. Periods of decreased oxygen levels would allow the environmental conditions necessary for microbial sulphate reduction to be much more easily achieved. The greater global extent of sulphate-reducing conditions in carbonate sediments would then enhance the rates of replacement of calcium carbonate sediments (mainly calcite during these periods) by dolomite. We do not wish to imply that all massive dolomite is necessarily produced by microbes. Perhaps they only initiate the process, or provide nucleation sites for further nonmicrobial dolomite formation. Many elements of the connection between microbes and dolomite remain to be brought to light.

Returning to the connections between secular variation in sea-level and ocean chemistry, how might all of these elements be tied together, and how might they control dolomite formation? Fig. 4b shows that there is, in general, a negative correlation between $\delta^{34}S$ values of marine evaporites and $\delta^{13}C$ values of marine carbonates during the Phanerozoic. Walker (1986) discussed the sulphur fluxes associated with seafloor hydrothermal systems and the impact of these fluxes on this negative correlation between $\delta^{13}C$ and $\delta^{34}S$ values, burial of organic matter, and O_2 content of seawater. His speculative mechanism to explain this association combines a set of processes that might cause a synchronous variation in the fluxes of sedimentary carbon and sulphur with the hydrothermal flux through mid-ocean ridges. He suggested that hydrothermal activity leads to a large flux of sulphide into the deep sea. The excess sulphide decreases the oxygen content of the ocean and atmosphere, promoting the development of anoxic conditions in parts of the deep sea. This anoxia promotes the preservation of organic matter and the deposition of sulphide minerals in seafloor sediments.

Carpenter & Lohmann (1997) reached a similar conclusion that seafloor hydrothermal systems are responsible for the negative correlation between $\delta^{13}C$ and $\delta^{34}S$. They modelled the flux of sulphur, as well as Fe, Mn, CH_4 and H_2, from the seafloor hydrothermal systems and concluded that these fluxes control burial of organic carbon through the consumption of O_2 during inorganic and organically mediated oxidation reactions. Therefore, conditions of elevated seafloor hydrothermal activity alone or together with modified oceanographic circulation, such as sluggish circulation and/or stratification, could produce widespread ocean anoxia and subsequent burial of organic matter. Because the abundance of dolomite is positively correlated with periods of high seafloor hydrothermal activity (i.e. periods of higher sea-level) and low O_2 in seawater (Figs 3 and 4), the nonuniformity of its distribution during the Phanerozoic must be re-examined in terms of alterations in the C–O–S system on the seafloor.

The dolomite peak in the early Phanerozoic is associated with a dramatic increase in the $\delta^{34}S$ value of marine sulphates (Fig. 4d). Such an increase in $\delta^{34}S$ values is interpreted as an indication of times of increased microbial reduction of sulphate to sulphide in carbonaceous sediments, which occurs with a large isotopic fractionation (Claypool *et al.*, 1980). Thus, the period of major dolomite formation in the early Phanerozoic can be correlated with widespread reducing conditions and consequently high sulphate reduction activity. In the Early Cretaceous, however, the dolomite peak is not marked by a similar $\delta^{34}S$ increase in marine sulphates, but the value remains relatively low (Fig. 4d). From the geological record, the Early Cretaceous is recognized as a time of widespread anoxia (Weissert, 1989), which seems to contradict the above interpretation of the variations in $\delta^{34}S$ value of marine sulphates. To explain this discrepancy, Walker (1986) proposes that there was accelerated weathering of sedimentary sulphide in the Early Cretaceous that would lead to an increased influx of isotopically lighter sulphate to the oceans, thus diluting the effect of increased microbial reduction with widespread anoxia. Secular changes in the values of $\delta^{34}S$, as well as $\delta^{13}C$, are therefore not necessarily indicative of periods of major dolomite formation. Weissert (1989) also concluded that there must have been accelerated weathering in the Early Cretaceous based on the increased delivery of siliciclastic material to the oceans.

According to Jannasch (1995), the most significant recent discovery in biological oceanography was the abundant biological activity and specific ecological role of micro-organisms in seafloor hydrothermal systems. The interaction of these micro-organisms with hydrothermal fluids produces the primary food supply for the vent communities. The flux of energy from the geothermal source via emitted reduced chemical species is microbially mediated toward the production of organic carbon as a form of biochemical energy. Periods of higher intensity of hydrothermal circulation during the Phanerozoic could result in cycles of increased microbial activity. Therefore, we propose that the relationship between microbial activity and periods of high seafloor hydrothermal activity must be analysed in order to understand how the biogeochemical cycle might be influential in producing periods of increased massive dolomite formation.

CONCLUSION

Massive dolomite formation appears to be intrinsically related to the redox state of the oceans. If so, the microbial community associated with seafloor hydrothermal systems, together with the rate of geochemical fluxes through mid-ocean ridges, may be the ultimate factors controlling massive dolomite formation on a global scale. The magnitude of the processes occurring in this biosphere/geosphere system and their importance for biogeochemical cycles remains an unknown quantity to be evaluated with future research in the 21st century. In particular, exploration of the microbial biosphere beneath the seafloor will undoubtedly contribute fundamentally new information relevant to our understanding of dolomite formation from which new dolomite models will evolve.

REFERENCES

Arvidson, F.S. and **Mackenzie, F.T.** (1997) Tentative kinetic model for dolomite precipitation rate and its application to dolomite distribution. *Aq. Geochem.*, **2**, 273–298.

Baker, P.A. and **Burns, S.J.** (1985) Occurrence and formation of dolomite in organic-rich continental margin sediments. *Amer. Assoc. Petrol. Geol. Bull.*, **69**, 1917–1930.

Baker, P. and **Kastner, M.** (1981) Constraints on the formation of sedimentary dolomite. *Science*, **213**, 214–216.

Berner, R.A. and **Canfield, D.F.** (1989) A new model for atmospheric oxygen over Phanerozoic time. *Amer. J. Sci.*, **289**, 333–361.

Brady, P.V., **Krumhansl, J.L.** and **Ppenguth, H.W.** (1996) Surface complexation clues to dolomite growth. *Geochim. Cosmochim. Acta*, **60**, 727–731.

Burns, S.J., **Baker, P.A.** and **Showers, W.J.** (1988) The factors controlling the formation and chemistry of dolomite in organic-rich sediments: Miocene Drakes Bay Formation, California. In: *Sedimentology and Geochemistry of Dolostones* (Ed. by V. Shukla and P. A. Baker), *Spec. Publ. Soc. Econ. Paleont. Mineral.*, **43**, 41–52.

Carpenter, S.J. and **Lohmann, K.C.** (1997) Carbon isotope ratios of Phanerozoic marine cements: Re-evaluating the global carbon and sulfur systems. *Geochim. Cosmochim. Acta*, **61**, 4831–4846.

Claypool, G.E., **Holser, W.T.**, **Kaplan, I.R.**, **Sakai, H.** and **Zak, I.** (1980) The age of sulfur and oxygen isotopes in marine sulfate and their mutual interpretation. *Chem. Geol.*, **28**, 199–260.

de Dolomieu, D. (1791) Sur un genre de pierres calcaires trés peu effervescentes avec les acides et phosphorescentes par la collision. *J. Physique*, **39**, 3–10.

Fischer, A.G. (1984) The two Phanerozoic supercycles. In: *Catastrophies in Earth History* (Ed. by W. A. Berggren and J. A. Vancouvering), pp. 129–148. Princeton University Press, Princeton.

Folk, R.L. (1993) SEM imaging of bacteria and nannobacteria in carbonate sediments and rocks. *J. Sedim. Petrol.*, **63**, 990–999.

Folk, R.L. and **Land, L.** (1975) Mg/Ca ratio and salinity: two controls over crystallization of dolomite. *Amer. Assoc. Petrol. Geol. Bull.*, **59**, 60–68.

Gaines, A.M. (1980) Dolomitization kinetics: recent experimental studies. In: *Concepts and Models of Dolomitization* (Ed. by D. H. Zenger, and R. L. Ethington), *Soc. Econ. Paleon. Mineral. Spec. Publ.*, **28**, 603–613.

Garrison, R.E., **Kastner, M.** and **Zenger, D.H.**, eds. (1984) Dolomites of the Monterey Formation and Other Organic-Rich Units. *Spec. Publishers Soc. Econ. Paleont. Mineral, Pacific Section*, **41**.

Garrels, R.M. and **Perry, E.A. Jr.** (1974) Cycling of carbon, sulfur, and oxygen through geologic time. In: *The Sea*, Vol. 5 (Ed. by E. D. Goldberg), pp. 303–336. John Wiley & Sons, New York.

Given, R.K. and **Wilkinson, B.H.** (1987) Dolomite abundance and stratigraphic age: Constraints on rates and mechanisms of Phanerozoic dolostone formation. *J. Sedim. Petrol.*, **57**, 1068–1078.

Haidinger, W. (1845). Ueber die Pseudomorphosen und ihre anogene und katogene Bildung. *Abhandlungen der k. böhm. Gesellschaft der Wissenschaften. Prague, Ser. 5.*, **3**, 231–259.

Hardie, L.A. (1996) Secular variation in seawater chemistry: an explanation for the coupled secular variation in the mineralogies of marine limestones and potash evaporites over the past 600 m.y. *Geology*, **24**, 279–283.

Helgeson, H.C., **Delany, J.M.**, **Nesbitt, H.W.** and **Bird, D.K.** (1978) Summary and critique of the thermodynamic properties of rock forming minerals. *Amer. J. Sci.*, **278-A**, 1–229.

Holland, H.D., **Horita, J.** and **Seyfried, W.E. Jr.** (1996) On the secular variations in the composition of Phanerozoic marine potash evaporites. *Geology*, **24**, 993–996.

Illing, L.V., **Wells, A.J.** and **Taylor, J.C.M.** (1965) Penecontemporary dolomite in the Persian Gulf. In: *Dolomitization and Limestone Diagenesis* (Ed. by L. C. Pray and R. C. Murray), *Soc. Econ. Paleont. Mineral. Spec. Publ.*, **13**, 89–111.

Jannasch, H.W. (1995) Microbial interactions with hydrothermal fluids. In: *Seafloor Hydrothermal Systems: Physical, Chemical, Biological, and Geological Interactions* (Ed. by S. E. Humphris, R. A. Zierenberg, L. S. Mullineaux and R. E. Thomson), *Amer. Geophys. Union Monogr.*, **91**, pp. 273–296.

Katz, A. and **Matthews, A.** (1977) The dolomitization of $CaCO_3$: an experimental study at 252–295 C. *Geochim. Cosmochim. Acta*, **41**, 297–308.

Kelts, K. and **McKenzie, J.A.** (1982) Diagenetic dolomite formation in Quaternary anoxic diatomaceous muds of DSDP Leg 64, Gulf of California. In: *Initial Reports of the Deep Sea Drilling Project* (Ed. by J. R. Curray, D. G. Moore, *et al.*), 64, Part 2, pp. 553–569. US Government Printing Office, Washington, DC.

Kump, L.R. and **Garrels, R.M.** (1986) Modeling atmospheric O_2 in the global sedimentary redox cycle. *Amer. J. Sci.*, **286**, 336–360.

Land, L.S. (1985) The origin of massive dolomite. *J. Geol. Education*, **33**, 112–125.

Land, L.S. (1998) Failure to precipitate dolomite at 25°C from dilute solution despite 1000-fold oversaturation after 32 years. *Aq. Geochem.*, **4**, 361–368.

Lapo, A.V. (1987). *Traces of Bygone Biospheres*, 2nd edn. Mir Publishers, Moscow.

Lasaga, A.C. (1989) A new approach to the isotopic modelling of the variation of atmospheric oxygen through the Phanerozoic. *Amer. J. Sci.*, **298**, 411–435.

Lindh, T.B. (1983) *Temporal variations in the* ^{13}C *and* ^{34}S *and global sedimentation during the Phanerozoic.* M. S. Thesis, University of Miami.

Lippmann, F. (1973). *Sedimentary Carbonates.* Springer-Verlag, Berlin.

Lumsden, D.N., **Morrison, J.W.** and **Lloyd, R.V.** (1995) The role of iron and Mg/Ca ratio in dolomite synthesis at 192 C. *J. Geol.*, **103**, 51–61.

Machel, H.G. and **Mountjoy, E.W.** (1986) Chemistry and environments of dolomitization – a reappraisal. *Earth Sci. Rev.*, **23**, 175–222.

Mackenzie, F.T. and **Agegian, C.** (1989) Biomineralization and tentative links to plate tectonics. In: *Plants and Animals* (Ed. by R. E. Crick), pp. 11–28. Plenum Press, New York.

Mackenzie, F.T. and **Morse, J.W.** (1992) Sedimentary carbonate through Phanerozoic time. *Geochim. Cosmochim. Acta.*, **56**, 3281–3295.

Mackenzie, F.T. and **Pigott, J.D.** (1981) Tectonic controls of Phanerozoic sedimentary rock cycling. *J. Geol. Soc. London*, **138**, 183–196.

McKenzie, J.A. (1991) The dolomite problem: an outstanding controversy. In: *Controversies in Modern Geology: Evolution of Geological Theories in Sedimentology, Earth History and Tectonics* (Ed. by D. W. Müller, J. A. McKenzie and H. Weissert), pp. 37–54. Academic Press, London.

Mojsisovics von Mojsvar, E. (1879) *Die Dolomit-Riffe Von Südtirol. und Venetien.* Alfred Hölder, Vienna.

Morrow, D.W. and **Ricketts, B.D.** (1988) Experimental investigation of sulfate inhibition of dolomite and its mineral analogues. In: *Sedimentology and Geochemistry of Dolostones* (Ed. by V. Shukla and P. A. Baker), *Spec. Publ. Soc. Econ. Paleont. Miner.*, **43**, 27–38.

Morse, J.W., **Wang, Q.** and **Tsio, M.Y.** (1997) Influences of temperature and Mg: Ca ratio on $CaCO_3$ precipitates from seawater. *Geology*, **25**, 85–87.

Nadson, G.A. (1928) Beitrag zur kenntnis der bakteriogenen kalkablagerungen. *Arch. Fuer Hydrobiologie*, **19**, 154–164.

Neher, J. (1959) Bakterien in tieferliegenden gesteinslagen. *Eclog. Geol. Helv.*, **52**, 619–625.

Petsch, S.T. and **Berner, R.A.** (in press) Coupling the long-term geochemical cycles of carbon, phosphorus, sulfur and iron: the effect on atmospheric O_2 and the isotopic records of carbon and sulfur. *Amer. J. Sci.*

Pisciotto, K.A., **Mahoney, J.J.** (1981) Isotopic survey of diagenetic carbonates. In: *Initial Reports of the Deep Sea Drilling Project* (Ed. by R. Yeats, B. Haq, *et al.*), pp. 595–609. US Government Printing Office, Washington, DC.

Purser, B.H., **Tucker, M.E.** and **Zenger, D.H.** (1994) Problems, progress and future research concerning dolomites and dolomitization. In: *Dolomites: a Volume in Honour of Dolomieu* (Ed. by B. Purser, M. Tucker and D. Zenger), *Spec. Publ. Int. Assoc. Sedimentol.*, **21**, 3–20.

Sandberg, P.A. (1983) An oscillating trend in Phanerozoic nonskeletal carbonate mineralogy. *Nature*, **305**, 19–22.

Sandberg, P.A. (1985) Nonskeletal aragonite and pCO_2 in the Phanerozoic and Proterozoic. In: *The Carbon Cycle and Atmospheric CO_2: Natural Variations Archean to Present* (Ed. by E. T. Sundquist and W. Broecker), *Amer. Geophys. Union Monogr.*, **32**, 585–594.

de Saussure, N.-T. (1792) Analyse de la dolomie. *J. Physique*, **40**, 161–173.

Sibley, D.F. (1990) Unstable to stable transformations during dolomitization. *J. Geol.*, **98**, 739–748.

Sibley, D.F., **Nordeng, S.H.** and **Borkowski, M.** (1994) Dolomitization kinetics in hydrothermal bombs and natural settings. *J. Sedim. Res.*, **A64**, 630–637.

Spencer, R.J. and **Hardie, L.A.** (1990) Control of seawater composition by mixing of river waters and mid-ocean ridge hydrothermal brines. In: *Fluid–Mineral Interactions: a Tribute to H. P. Eugster* (Ed. by R. J. Spencer and I.-M. Chou), *Geochem. Soc. Spec. Publ.*, **19**, 409–419.

Vail, P.R., **Mitchum, R.M. Jr**, **Todd, R.G.**, **Vidmer, J.M.**, **Thompson, S.**, **Sangree, J.B.**, **Bubb, J.W.** and **Hatfield, W.G.** (1977) Seismic stratigraphy – application to hydrocarbon exploration. *Amer. Assoc. Petrol. Geol. Mem.*, **26**, 49–212.

Van Houton, F.B. and **Bhattacharyya, D.P.** (1982) Phanerozoic oolitic ironstones-geologic record and facies. *Annu. Rev. Earth Planet. Sci.*, **10**, 441–458.

Vasconcelos, C., **McKenzie, J.A.**, **Bernasconi, S.**, **Grujic, D.** and **Tien, A.J.** (1995) Microbial mediation as a possible mechanism for natural dolomite formation at low temperatures. *Nature*, **377**, 220–222.

Vasconcelos, C. and **McKenzie, J.A.** (1997) Microbial mediation of modern dolomite precipitation and diagenesis under anoxic conditions (Lagoa Vermelha, Rio de Janeiro, Brazil). *J. Sedim. Res.*, **67**, 378–390.

Von der Borch, C.C. (1965) The distribution and preliminary geochemistry of modern carbonate sediments of the Coorong area, South Australia. *Geochim. Cosmochim. Acta*, **29**, 781–799.

Von Morlot, A. (1847) Ueber Dolomit und seine künstliche Darstellung aus Kalkstein. In: *Naturwissenschaftliche Abhandlungen* (Ed. by W. Haidinger), **1**, pp. 305–315.

Walker, J.C.G. (1986) Global geochemical cycles of carbon, sulfur and oxygen. *Mar. Geol.*, **70**, 159–174.

Weissert, H. (1989) C-Isotope stratigraphy, a monitor of paleoenvironmental change: a case study from the early Cretaceous. *Surveys Geophysics*, **10**, 1–69.

Wilkinson, B.H. and **Algeo, T.J.** (1989) Sedimentary carbonate record of calcium magnesium cycling. *Amer. J. Sci.*, **289**, 1158–1194.

Wilkinson, B.H. and **Given, R.K.** (1986) Secular variation in abiotic marine carbonates: Constraints on Phanerozoic atmospheric carbon dioxide contents and oceanic Mg/Ca ratios. *J. Geol.*, **94**, 321–324.

Wilkinson, B.H., **Owen, R.M.** and **Carroll, A.R.** (1985) Submarine hydrothermal weathering, global eustasy and carbonate polymorphism in Phanerozoic marine oolites. *J. Sedim. Petrol.*, **55**, 171–183.

Wright, D.T. (1997) An organogenic origin for widespread dolomite in the Cambrian Eilean Dubh Formation, Northwestern Scotland. *J. Sedim. Res.*, **67**, 54–64.

Zenger, D.H. (1989) Dolomite abundance and stratigraphic age: Constraints on rates and mechanisms of Phanerozoic dolostone formation. *J. Sedim. Petrol.*, **59**, 162–164.

Zenger, D.H., **Bourrouilh-le January, F.G.** and **Carozzi, A.V.** (1994) Dolomieu and the first description of dolomite. In: *Dolomites: a Volume in Honour of Dolomieu* (Ed. by B. Purser, M. Tucker and D. Zenger), *Spec. Publ. Int. Assoc. Sedimentol.*, **21**, 21–28.

Sedimentology (2000), **47** (Suppl. 1), 62–94

The structure and fluid mechanics of turbidity currents: a review of some recent studies and their geological implications

BEN KNELLER and CLARE BUCKEE
School of Earth Sciences, University of Leeds, Leeds LS2 9JT, UK (E-mail: ben@earth.leeds.ac.uk; buckee@earth.leeds.ac.uk)

ABSTRACT

The literature on the structure and behaviour of gravity currents is reviewed, with emphasis on some recent studies, and with particular attention to turbidity currents, though reference is also made to comparable behaviour in pyroclastic flows. Questions of definition are discussed, in particular the distinction between dense currents, which may deposit *en masse*, and more dilute currents. High-density dispersions may exist as a discrete, independently moving layer beneath a more dilute flow, as the basal part of a continuous density distribution or possibly as a transient depositional layer. Existing theory appears inadequate to explain the behaviour of some high-density dispersions. Surge-type currents are contrasted with quasi-steady currents, which may be generated by a variety of mechanisms including direct feed by rivers in flood. Such fluvially generated currents provide one means of generating currents with reversing buoyancy. Geologically significant turbidity currents are impractical for direct study owing to their large scale and (often) destructive nature. Small-scale laboratory currents offer a wealth of insights into turbidity current behaviour. This paper summarizes recent experimental studies that focus on the physical structure of gravity currents, with emphasis on the velocity and turbulence structure, the vertical density distribution and the stability of stratification. Preliminary quantification of the turbulence structure (including controls on turbulent entrainment, turbulent kinetic energy, Reynolds stresses and turbulence production) has been facilitated by recent technological developments that have allowed the measurement of instantaneous fluctuations in both velocity and concentration. Laboratory models, however, generally involve substantial simplification, and require compromises in some parameters to achieve adequate scaling of the parameters of most interest. Mathematical modelling also provides important insights into turbidity current behaviour. We discuss various approaches to modelling, ranging from simple hydraulic equations to systems of partial differential equations that explicitly treat conservation of momentum, fluid and sediment mass, and turbulent kinetic energy. The application for which the model is designed (i.e. to calculate mean head velocity or to create an instantaneous two-dimensional contour plot of downstream velocity in a current) determines the complexity of the mathematical model required. The behaviour of suspension currents around topography is complex and depends upon the relative height of the topography, and upon the density and velocity structure of the current. Many interactions with topography are well described by the internal Froude number, Fr_i. Both reflection and deflection of currents may occur on the upstream side of topography, depending upon Fr_i. On the downstream side of topography, flow separation, lee waves or hydraulic jumps may occur.

Keywords Froude number, gravity current, hyperpycnal, Richardson number, stratification, topography, turbulence.

INTRODUCTION

Density currents are produced where gravity acts upon a density difference between one fluid and another (as such they should probably be called buoyancy currents). In the case of suspension currents, such as turbidity and pyroclastic currents, the density excess is provided by suspended solid particles. Such currents are major agents of sediment transport on land, in lakes, seas and oceans. They are important not only for the potential environmental hazards they pose (reservoir sedimentation, submarine cable breakage, pollutant dispersal and volcanic hazards; see reviews by Middleton, 1993, and Simpson, 1997) but also because turbidite systems are volumetrically the most significant clastic accumulations in the deep sea (Normark *et al.*, 1993), and include the world's largest sedimentary bodies (e.g. the Bengal Fan, Bouma *et al.*, 1985). Turbidite systems also form a significant component of the stratigraphic record, and include many of the world's most important hydrocarbon reservoirs (Weimer & Link, 1991).

This review focuses mainly on selected aspects of turbidity current structure and behaviour, and emphasizes some recent studies. In particular, we discuss the density, velocity and turbulence structure of turbidity currents, attempts to measure and model them, and the behaviour of turbidity currents interacting with topography. We also highlight some of the controversies surrounding turbidity current behaviour. Many of the observations and arguments (and indeed debates) that we report are equally valid for other types of suspension current such as pyroclastic currents. We build upon the review of sediment deposition by turbidity currents by Middleton (1993), which reviews much of the turbidity current literature to that point, and also upon that of Simpson (1997), which deals with gravity currents in general. A recent review by Huppert (1998) on quantitative modelling of suspension currents emphasizes the application of box models (see below) to simple laboratory systems and more complex environmental currents. Pyroclastic flows and surges have recently been reviewed by Freundt & Bursik (1998) and Wohletz (1998), respectively.

Definition of turbidity currents and pyroclastic currents

Turbidity currents are part of a continuum of sediment-gravity flows (Middleton & Hampton, 1976). Turbidity currents are traditionally defined as those sediment-gravity flows in which sediment is suspended by fluid turbulence (Sanders, 1965; Middleton & Hampton, 1973; Lowe, 1982; Middleton, 1993; Simpson, 1997). However, the term 'turbidity current' was adopted to describe a natural phenomenon whose exact nature is often unclear, and in which the sediment support mechanism(s) is commonly impossible to determine in nature. Moreover, a variety of sediment support mechanisms (e.g. turbulence, hindered settling, dispersive pressure owing to grain interactions) may operate simultaneously within a sediment gravity flow. Defining suspension currents by specifying the grain support mechanism is thus problematic (see discussion below and in Mulder *et al.*, 1997a) as it may confuse essentially descriptive terminology (e.g. 'turbidity current') with interpretations or assertions of physical process (e.g. 'turbulent') that cannot yet be supported by observation in all cases. Nonetheless, it is probable that turbulence is the primary or sole grain support mechanism at least in dilute currents, i.e. those with fractional grain concentrations of the order of a few per cent or less (the exact concentration being dependent on the balance of factors that determine the flow Reynolds number).

Definitions are further complicated by a still very incomplete understanding of the turbulence structure of suspension currents, and the considerable confusion in the geological literature between the term *turbulent* (i.e. disturbed by eddies) and *turbid* (i.e. opaque with sediment); McCave & Jones's (1988) description of nonturbulent turbidity currents does not necessarily imply a contradiction. We define a suspension current as '*flow induced by the action of gravity upon a (fluidal) turbid mixture of fluid and (suspended) sediment, by virtue of the density difference between the mixture and the ambient fluid*'. Within this definition, it is accepted that suspension (the 'holding up' of grains above the bed) may involve grain support mechanisms other than fluid turbulence, i.e. it is broader in meaning than 'turbulent suspension'. A turbidity current is a suspension current in which the interstitial fluid is a liquid (generally water); a pyroclastic current is one in which the interstitial fluid is gas. We use the terms 'density current' and 'gravity current' interchangeably to describe any buoyancy-driven current whether the density difference is due to particles or not.

The role of turbulence in suspension currents during both transport and deposition has been the source of considerable debate in both the sedimentological and volcanological communities. The argument has focused on whether particulate material (sediment or pyroclasts) is deposited from high-concentration currents *en masse* or grain-by-grain, and has tended to link the state of the current during deposition (and thus the resulting depositional style) with the longer-term transport regime that preceded it. One view is that suspension currents of relatively high concentration exist as non-Newtonian, essentially plug-like, and perhaps nonturbulent masses, supported largely by matrix strength and/or grain-to-grain interactions (e.g. Shanmugam *et al.*, 1995; Shanmugam, 1996; Miller & Smith, 1977; Wilson, 1985, 1997). These non-Newtonian currents may be overlain by an independently moving dilute cloud, from which they are separated by a more-or-less discrete interface (e.g. Postma *et al.*, 1988; Mulder *et al.*, 1997a; see also Mohrig *et al.*, 1998). According to this view, the deposit of the basal layer may represent a 'frozen' current, formed more or less instantaneously by frictional or cohesive immobilisation.

Another view holds that the transport regime is more expanded, and, even at relatively high particle concentrations, only the basal part of the current has low levels of turbulence; in these cases there is a continuous, rather than stepped, density profile (e.g. Druitt, 1992; Branney & Kokelaar, 1992; Kneller & Branney, 1995; Hiscott *et al.*, 1997). In this view, the transport regime is dominated by turbulent suspension, and deposition occurs by progressive aggradation. Much debate has focused on the use of geological criteria to differentiate between the deposits of different process regimes. The uncertainties involved in this approach are evident from the fact that contrasting models have been applied to the interpretation of the same deposits, including massive deep-water sands (e.g. Jackfork Group; Shanmugam & Moiola, 1995; Slatt *et al.*, 1997) and ignimbrites (e.g. Taupo ignimbrite; Wilson, 1997; Dade & Huppert, 1996, 1997). There may, in nature, be a continuum between the two types of transport regime, though it is harder to envisage a continuum between progressive and *en masse* deposition. At least some high-concentration, debris-flow-like depositional regimes arise by reconcentration during collapse of a more dilute current (e.g. Vrolijk & Southard, 1997), possibly during lofting of buoyant interstitial fluid (see below). Thus a transient, late-stage, debris-flow-like transport regime may arise during collapse of a current in which the long-range transport regime was more dilute, and perhaps dominated by fluid turbulence.

The empirical relation for the effective viscosity of clay-free dispersions of sand was given by Davidson *et al.* (1977) as:

$$\mu_s/\mu = (1 - 1{\cdot}35C)^{-2{\cdot}5} \qquad (1)$$

where μ_s is the apparent viscosity of the clay-free dispersion, μ is the molecular viscosity of water and C is the fractional grain concentration. This relation suggests that even at concentrations of 45% by volume (approaching the maximum concentration for a fluidized dispersion of roughly uniform grain size), the effective viscosity is about an order of magnitude greater than that of water. Utilizing the criterion for turbulence in Newtonian fluids of:

$$\mathrm{Re} = \frac{(\rho_c \overline{U} d)}{\mu_s} = 2000 \qquad (2)$$

(where ρ_c is the mean current density, $\overline{U}$ is the depth-averaged current velocity and d is the current thickness), a high-concentration (45% by volume), noncohesive particulate current ('cohesionless debris flow') should rapidly become fully turbulent when the product of velocity and thickness exceeds about $3 \times 10^{-2}\,\mathrm{m^2\,s^{-1}}$. Moreover, recent experimental work suggests that a clay content of several per cent may be necessary to induce debris-flow-like behaviour in sandy dispersions moving under the influence of gravity (Marr *et al.*, in press). Thus, on the face of it, long-range (many kilometres) nonturbulent transport of cohesionless dispersions seems unlikely on purely mechanical grounds. However, Hallworth & Huppert (1998), reporting experimental results from high-concentration particle-driven gravity currents, describe behaviour in suspensions whose initial volume fraction exceeds 0·3 that is qualitatively different from that in suspensions of lower concentration. These flows formed rapidly moving, densely compacted currents that came to an abrupt halt, forming deposits of fairly constant thickness with an abrupt 'snout'. The dense flow was overlain by an independently moving dilute flow, formed by turbulent mixing at the upper boundary of the dense layer (compare Mohrig *et al.*, 1998), which produced a more extensive, thin deposit. Hallworth & Huppert (1998) were unable to explain these phenomena quantitatively using existing theory. The process observed in these

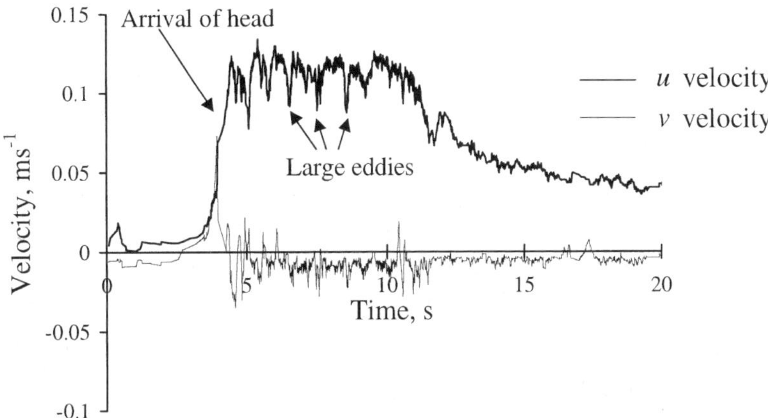

Fig. 1. Time series of downstream (u) and vertical (v) velocity measured in an experimental, lock-exchange brine current, using laser Doppler anemometry (from Kneller *et al.*, 1997).

experiments is analogous to that envisaged by Mulder *et al.* (1997a) for the 1979 event in the Var submarine canyon, off south-east France. The stepped thickness distribution of Hallworth & Huppert's (1998) experimental deposits is strikingly similar to that reported by Tokuhashi (1979) in a study of thickness distributions of Mio-Pliocene turbidite sandstones in Japan.

INITIATION AND EVOLUTION OF TURBIDITY CURRENTS

Normark & Piper (1991) reviewed a variety of initiation mechanisms for turbidity currents, including virtually instantaneous events such as seismogenic slumping (e.g. Heezen & Ewing, 1952; Morgenstern, 1967; Weaver *et al.*, 1992; García & Hull, 1994) that generate surge-type currents. Such currents have their experimental analogue in the finite-volume currents generated by the release of a lock gate, and may consist of a more-or-less clearly differentiated head, body and tail, with a general decrease in both mean velocity and sediment concentration backwards from the head (e.g. Felix, in press). Typically, the velocity of the current passing a fixed point will very rapidly reach its maximum value as the head arrives, after which flow will wane during passage of the body and tail (e.g. Fig. 1). However, even surge-type currents may pass through periods of relatively steady flow. For example, the 1929 Grand Banks turbidity current, which was triggered by slope failure following an earthquake (Cochonat & Piper, 1995), continued for a minimum of 2–3 h (Piper *et al.*, 1988; see also Piper *et al.*, 1999a), and was sufficiently steady to develop large-scale bedforms in the fan valley (Hughes-Clarke *et al.*, 1990). Sustained (steady or quasi-steady) underflows may also be generated during volcanic eruptions and the consequent remobilization of unconsolidated material (e.g. Lipman & Mullineaux, 1981; Kokelaar, 1992), or by seismically triggered subaerial sliding within the drainage basin (e.g. Syvitski & Schafer, 1996). Steady or quasi-steady currents may have an element of inherent instability owing to internal waves, the generation and impact of eddies and longer term 'surging' (e.g. Lambert *et al.*, 1976) that may be expressed in the deposit (e.g. Duringer *et al.*, 1991).

Several analytical and numerical models of turbidity currents have produced solutions for steady flow (e.g. Chu *et al.*, 1979), but the application of steady flow models to the interpretation of deposits has not been generally popular, despite evidence in deposits for steady or quasi-steady flow (Kneller, 1995).

The fluvial connection

Many turbidite systems, especially fan systems, display obvious connections with rivers. The growth of turbidite systems occurs when fluvial systems can discharge directly to the shelf break. This direct link is generally associated with sea-level lowstands (Mutti, 1985; Shanmugam *et al.*, 1985; Mutti & Normark, 1991; Posamentier *et al.*, 1991; Normark *et al.*, 1993) but may also occur during highstands if rates of delta progradation are sufficiently high (Burgess & Hovius, 1998). The initiation of turbidity currents is generally ascribed to remobilization of sediment deposited at the shelf break or in the canyon head (e.g. Kolla & Perlmutter, 1993; Mulder & Syvitski, 1996). However, turbidity currents (that may be sustained for periods of days or weeks) can be generated directly by fluvial discharge into lakes (e.g. Lambert *et al.*, 1976; Sturm & Matter, 1978; Weirich, 1986; Lambert & Giovanoli, 1988), and into the ocean, where they are linked to high sediment load in the source fluvial system. Examples occur on fjord delta-fronts (Prior & Bornhold, 1990; Zeng *et al.*, 1991; Carlson *et al.*, 1992; Phillips & Smith, 1992; see also Nemec, 1990) and in more open marine systems, notably off the Yellow River, the Zaire (Congo) River and the Santa Clara river that supplies the Hueneme Fan on the California Borderland (Heezen *et al.*, 1964; Shepard & Emery, 1973; Reynolds, 1987; Wright *et al.*, 1988, 1990; Piper *et al.*, 1999b). The Yellow River generates seasonal silty hyperpycnal flow (a turbidity current formed directly from dense river effluent; Wright *et al.*, 1990), and the Zaire River supplies a turbid underflow directly to the head of the Zaire Canyon that may flow down onto the fan surface (Eisma & Kalf, 1984; van Weering & van Iperen, 1984) as well as possibly sourcing less frequent sandy turbidity currents.

Rivers that generate hyperpycnal flows at the present day are generally those with high hinterland relief and exceptionally high suspended loads (e.g. the Var, which drains the south-western French Alps; Mulder *et al.*, 1997b), enhanced where the available sediment is fine-grained such as the loess carried by the Yellow River. Mulder & Syvitski (1995) calculated the likely interval between hyperpycnal flow events for a selection of the world's river systems, and conclude that almost half could generate hyperpycnal flows with a return period of 100 years or less. This is probably an underestimate since they implicitly assumed hyperpycnal flow generation only when the depth-averaged suspended sediment concentration led to an effluent density exceeding that of sea water. However, sediment-charged rivers show considerable density stratification, and many flood events might produce both plumes and hyperpycnal flows when the depth-averaged effluent density is less than that of sea water (see, for example, McLeod *et al.*, 1999). Even with this caveat, it is likely that the rivers that supply large fans can never generate hyperpycnal flows under the current (Holocene) hydrological regime. However, most of these fans have been inactive throughout the Holocene. Damuth *et al.* (1988) suggested that sinuous channels on the Amazon Fan, which developed during late Pleistocene lowstands, may have been formed by continuous underflows (see also Hay, 1987).

The association of turbidity currents with flood events does not necessarily imply hyperpycnal river effluent, even when the turbidity currents are associated with bursts of low salinity (Zeng *et al.*, 1991). Rapid deposition at the river mouth during prolonged floods may lead to oversteepening and repeated failure of delta slopes, generating turbidity currents that include a proportion of the brackish pore water of the delta sediments.

Ignitive flows

Both sustained and surge-type fine-grained underflows may also serve as triggering mechanisms for larger sandy flows; where such currents flow into canyons they may become autosuspensive, which is to say that they are self-sustaining (Pantin, 1979; or ignitive, *sensu* Parker, 1982), and may entrain sediment that has previously been introduced into the canyon by littoral drift, storms or smaller turbidity currents. Pantin (in press) has recently provided the first experimental evidence of autosuspension. Canyon-flushing associated with surge-type currents initiated by slope failures may produce currents whose final volume may be several times that of the portion of the slope that has failed (e.g. Grand Banks, Piper & Aksu, 1987; Piper *et al.*, 1988; Cochonat & Piper, 1995; Madeira Abyssal Plain, Masson, 1994). Ignitive flows do not require that the triggering current is sandy, as noted by García & Parker (1993) who demonstrated experimentally the entrainment of sediment by saline currents, and concluded that 'currents of silty mud could also entrain substantial amounts of sand and carry it to deep water'.

Currents with reversing buoyancy

The behaviour of currents with buoyant interstitial fluid (such as currents with warm, fresh or brackish interstitial water entering the sea, pyroclastic currents flowing into water, or lofting pyroclastic currents) has been investigated exerimentally by Sparks *et al.* (1993), Carey *et al.* (1996) and Hürzeler *et al.* (1996), and numerically by Hürzeler *et al.* (1995) using a k–ε turbulence model (see below). McLeod *et al.* (1999) have experimentally investigated the behaviour of such currents on entering the sea, including the effect of flow stratification in splitting the current into a dense underflow and a buoyant plume. The front speed of underflows with buoyant interstitial fluid decreases more rapidly than that of currents in which the interstitial fluid has the same density as the ambient. These underflows ultimately come to a halt as sedimentation results in a reversal of buoyancy (the bulk density drops below that of the ambient fluid), and the current lifts off, the point of lift-off remaining constant for a constant discharge. The lofted fluid carries fine sediment with it, thus potentially decoupling the coarse and fine sediment fractions. The lofted fluid forms a plume that rises to a level of neutral buoyancy (if in a stratified environment) or to the water surface, and spreads as a gravity interflow or overflow. Sediment falling from the plume produces a widespread fall-out deposit, termed hemiturbidite by Stow & Wetzel (1990) describing what they interpreted as the consequences of flow lofting owing to temperature differences between warm interstitial water (introduced from shallow depths as part of the current) and cold ambient water of the deep ocean.

Lofting pyroclastic currents may produce debris-flow-like deposits with abrupt fronts (e.g. Lipman & Mullineaux, 1981), and it is tempting to suggest that the deposits of hyperpycnal flows should have similar properties (i.e. somewhat debrite-like) produced by late-stage reconcentration as the current collapses. Indeed, the deposits of the youngest (latest Pleistocene) lobes of the Mississippi fan (Twichell *et al.*, 1995) have geometries intriguingly similar to those of some lofted pyroclastic currents.

APPROACHES TO UNDERSTANDING SUSPENSION CURRENTS

Prediction of erosion by turbidity currents, and of the distribution of turbidite deposits, such as their extent, thickness and grain size distribution, requires an understanding of the mechanisms of sediment transport and deposition, which are in turn dependent on the fluid dynamics of the currents. Middleton (1993) states that 'a better understanding [of turbidity current dynamics] is required to account for the many diverse phenomena reported from modern environments and ancient turbidite systems'. However, the dynamics of suspension currents are highly complex owing to feedback between turbulence and the suspended sediment. Consequently, a greatly simplified approach is generally taken when modelling suspension currents, commonly by depth averaging.

Furthermore, laboratory experiments investigating the dynamics of gravity currents have established that the distribution of turbulence and shear stresses is significantly different from that in open-channel flows (see section on turbulence structure below and Kneller *et al.*, 1997, 1999; Buckee *et al.*, in press). The turbulence structure affects the way in which sediment is suspended and transported and hence the way in which turbidity current behaviour and sediment deposition is modelled.

Since Kuenen (1937) and Kuenen & Migliorini (1950) first demonstrated experimentally that turbidity currents were the most probable agent of transport of clastic sediment into the deep oceans, subsequent research has followed three main paths: the study of turbidite systems, laboratory experiments and theoretical modelling. Turbidite systems are beyond the scope of this paper and the reader is referred to reviews by Mutti & Normark (1991), Mutti (1992) and Normark *et al.* (1993), and articles in Weimer *et al.* (1994) and Pickering *et al.* (1995). The extreme complexity of most turbidite systems, and indeed individual turbidite beds, has precluded the development of quantitative models of turbidity current behaviour inferred solely from their deposits.

Study of the mechanics of turbidity currents has been limited by the difficulties implicit in studying them in the natural environment. Turbidity currents in the oceans may reach velocities of tens of metres per second (e.g. Mulder *et al.*, 1997a) and heights of hundreds of metres (e.g. Heezen & Ewing, 1952) and even rather modest currents have damaged or destroyed equipment deployed for the purposes of studying them (e.g. Zeng *et al.*, 1991). Consequently, most of what is known about large natural turbidity currents (i.e. those significant in terms of sediment transfer to the deep sea) is inferred from indirect sources, such as submarine cable breaks and heights of deposits above submarine valley floors (e.g. Piper & Savoye, 1993). Direct measurements of pyroclastic currents, with the exception of frontal velocities, are more hazardous still. A few smaller, natural turbidity currents have been visualized or monitored (Hay *et al.*, 1982; Chikita, 1989, 1990; Zeng *et al.*, 1991), but our knowledge of natural suspension currents remains fragmentary. Small-scale laboratory experiments therefore offer one of the best means of studying suspension current dynamics. Advances in experimental technology have increased our understanding from broad descriptions of gravity current morphology (e.g. Middleton, 1966b; Allen, 1985; Simpson, 1997) to the structure of turbulence in these currents (Kneller *et al.*, 1997, 1999; Parsons, 1998; Best *et al.*, in press; Buckee *et al.*, in press).

Mathematical models can also provide significant insights into current dynamics. Analytical solutions have been proposed for some aspects of gravity current behaviour (e.g. Chu *et al.*, 1979), but in the long term numerical techniques present the best hope of understanding and predicting three-dimensional turbidity current processes and deposits. Turbidity currents are, however, highly complex phenomena; they are nonuniform, unsteady, nonlinear, free boundary flows driven by a nonconservative density difference (i.e. the density difference varies as sediment is eroded or deposited) and a combination of body and pressure forces (Allen, 1985). In pyroclastic currents one must also consider the effects of a nonconservative interstitial fluid phase (due to exsolution of magmatic volatiles, substrate boiling and combustion), and volume and pressure changes (due to heating of admixed air, and gas phase compressibility, although this can probably be neglected in currents with low Mach numbers; Furbish, 1997). In most cases there are more variables than governing equations – the so-called closure problem – and the models rely upon simplifying assumptions in order to achieve a closure. The accuracy of the individual models thus depends upon the choice and validity of the assumptions made. Experimental results provide a means of constraining some of these variables as well as providing a test for such models.

EXPERIMENTS ON TURBIDITY CURRENTS

Scaling

Mathematical models provide a valuable method for understanding and predicting the consequences of the complex relationships and feedback mechanisms involved in turbidity currents. However, physical data from field observations, or more practically from experiments, are still required in order to test the simplifying assumptions necessary in such models. The advantages of using experimental models are offset by the difficulties arising from the necessary simplifications involved in scaling. Scaling of natural currents to a laboratory level is a well-documented issue in fluid dynamics (see, e.g. Middleton & Southard, 1984; French, 1986). In a recent review paper, Peakall *et al.* (1996) outline four approaches to the scaling of physical models: 1 : 1 replicas of the field prototype; Froude number similarity; distorted scale modelling; and unscaled experimental analogues. The large scale of turbidity currents in the oceans that produce significant sedimentary deposits (e.g. current thicknesses of 10^1–10^2 m and velocities of 10^0–10^1 m s^{-1}; see, e.g. Heezen & Ewing, 1952) makes it impossible to study these experimentally at a 1 : 1 scale; consequently, scaled laboratory experiments present the best way to study turbidity currents. The results of unscaled experimental analogue models of large turbidity currents (e.g. Alexander & Morris, 1994), although qualitatively informative, are difficult to apply quantitatively. Distorted scale experiments, for example where unrealistically high slopes are used to obtain appropriate bed shear stresses (e.g. Postma *et al.*, 1988) may be the only way of reproducing some aspects of the prototype, but clearly necessitate some circumspection in application of the results.

Froude scale modelling

Froude scale modelling is based on a similarity approach in which the current is fully characterized by a series of dimensionless variables. As long as the values of the dimensionless variables in laboratory currents are known to be comparable with those of the natural current, the experiment is adequately scaled with respect to the parameters included in that variable, allowing modelling of large-scale phenomena in the laboratory (e.g. Middleton, 1966a). Dimensional analysis depends upon identification of the controlling variables, which are then grouped into a smaller number of dimensionless parameters such as the flow Reynolds number (Re), which reflects the ratio of inertial to viscous forces, and the Froude number (Fr), which reflects the ratio of inertial to gravitational forces acting on a fluid flow. Currents that share the same values of Re and Fr are said to be dynamically similar. In the case of Froude scale modelling, Fr takes the same value as the prototype, but Re is relaxed (making the assumption that the effects of viscosity can be neglected if the current is fully turbulent). Froude scale models are only appropriate for turbidity current modelling where the model is fully turbulent ($Re > 2000$) which is sometimes difficult to achieve in the laboratory, and even then assumes that turbulence is self-similar across the range of scales represented by the model and prototype. In gravity currents, the densiometric Froude number

(Fr′) is used, since the action of gravity depends upon the fractional density difference between the current and the ambient fluid:

$$Fr' = \overline{U}/(dg')^{1/2} \quad (3)$$

where $\overline{U}$ is the depth-averaged current velocity, d is some characteristic length scale (normally the current thickness in gravity currents), and the reduced gravity $g' = g(\Delta\rho/\rho_a)$, where g is the acceleration due to gravity, $\Delta\rho$ is the density difference between the current and the ambient fluid and ρ_a is the density of the ambient fluid.

Scaling sediment-laden currents

Experimental studies of sediment-laden turbidity currents must also be scaled to the natural system using a dimensionless settling velocity (making the assumption that settling velocity adequately describes the particle hydrodynamics), that is the ratio of terminal settling velocity of the sediment grains to some velocity scale that is considered characteristic of the current (Middleton, 1966a; Laval *et al.*, 1988). Electrostatic forces may affect very fine sediments (such as chalk, charcoal or clay) which have the correctly scaled settling velocity, but this problem can be solved by using glass beads or silica flour (e.g. Parker *et al.*, 1987; García, 1993). However, all fine sediments experience capillary forces once they have settled, and these can only be reduced by using larger grain sizes. Better sediment analogues can be obtained by using grains of reduced density (e.g. Middleton, 1966b,c), but the consequent reduction in the bulk density of the current necessitates an increase in sediment concentration. Changing sediment concentration or grain size affects particle behaviour, which may no longer scale with the whole current behaviour (Middleton, 1966a; Peakall *et al.*, 1996). Many experimental studies substantially simplify the system by neglecting sediment deposition and erosion, and use brine (or other dense solutions) to create the density difference (Laval *et al.*, 1988). However, while saline currents are believed to be dynamically similar to fine-grained turbidity currents (Stacey & Bowen, 1988a), the vertical density structure (and hence current dynamics) may be considerably different in coarse-grained turbidity currents.

Review of experimental work

Table 1 shows the majority of experimental work carried out on the dynamics of gravity currents (thus excluding a substantial set of experiments primarily concerned with sediment deposition from lock-exchange currents; see, e.g. Alexander & Morris, 1994; Kneller & McCaffrey, 1995; Gladstone *et al.*, 1998). Table 1 highlights the way in which technological advances have led to increasingly detailed studies of gravity current dynamics. Initial experimentation focused on the anatomy of gravity currents, especially the head (e.g. Keulegan, 1957; Middleton, 1966b; Britter & Simpson, 1978). More recently, the development of facilities in which steady currents can be generated has allowed the body of experimental currents to be examined (e.g. Parker *et al.*, 1987; García, 1993, 1994). However, flow measurements were made primarily using intrusive equipment (e.g. micropropellors), which disrupt the flow, and so only mean flow properties rather than the turbulence structure were determined. In the last three years (see Table 1), the application of several new experimental techniques, e.g. laser Doppler anemometry (LDA) in combination with refractive index (RI) matching, acoustic Doppler velocimetry (ADV) and ultrasonic Doppler velocity profiling (UDVP), has allowed the preliminary quantification of the turbulence structure both in the head and in the body of laboratory gravity currents (e.g. Kneller *et al.*, 1997, 1999; Parsons, 1998; Best *et al.*, in press; Buckee *et al.*, in press). There are still technical limitations to this work, specifically the difficulty in measuring instantaneous density fluctuations, and the lack of high-resolution velocity data in particulate currents (LDA has not successfully been used in sediment-laden density currents, and UDVP measures at frequencies < 30 Hz, too low for a full determination of the turbulence spectrum; Parsons, 1998).

THE ANATOMY OF GRAVITY CURRENTS

The head

Gravity currents are described as having a well-defined head, body and in some cases a tail. The dynamics of the head are important because they may set a boundary condition for the current as a whole (Simpson & Britter, 1979). Both Allen (1971) and Middleton (1993) suggest that the head is a 'locus of erosion', and therefore important sedimentologically. The head has an overhanging 'nose' (Fig. 2), a result of the no-slip condition at the lower boundary (Britter & Simpson, 1978) and frictional resistance at the

Table 1. Summary of experimental studies on gravity currents.

Author(s)	Type of gravity current	Focus	Experimental facility	Velocity measurement	Density measurement	Key words
Kuenen (1937)	Sediment-laden	Deposit	Modified lock-exchange	n/a	n/a	Deposits
Keulegan (1957)	Solute (Saline)	Head	Lock-exchange	n/a	n/a	Anatomy of the head, simple laws
Ellison & Turner (1959)	Solute (Saline)	Body	Continuous	Thermistors (hot-wire)	Sampling	Mean flow properties, stratification, entrainment
Middleton (1966a)	Solute & sediment-laden	Head	Continuous & lock-exchange	n/a	n/a	Anatomy and motion of the head
Middleton (1966b)	Solute & sediment-laden	Head, body	Continuous	Slow-motion photography	n/a	Simple laws
Britter & Simpson (1978)	Solute (Saline)	Head	Arrested head	Dye streak measurements & hot-film probe	Conductivity meter (no results shown)	Mean flow properties, mixing rate, Kelvin–Helmholtz billows
Simpson & Britter (1979)	Solute (Saline)	Head	Arrested head	Dye streak measurements & hot-film probe	Conductivity meter (no results shown)	Mean flow properties, mixing rate, Kelvin–Helmholtz billows, lobes & clefts
Parker *et al.* (1987)	Sediment-laden	Body	Continuous	Micropropellor	Siphon sampling	Mean flow properties, sediment entrainment & deposition, entrainment of ambient
García (1993)	Solute & sediment-laden	Body	Continuous	Micropropellor	Optical probe	Mean flow properties, hydraulic jump, deposits
García & Parker (1993)	Solute & sediment-laden	Body	Continuous	Micropropellor	Optical probe	Mean flow properties, sediment entrainment
García (1994)	Sediment-laden	Body	Continuous	Micropropellor	Siphon sampling	Mean flow properties, sediment deposition.
Altinakar *et al.* (1996)	Sediment-laden	Body	Modified lock-exchange	Micropropellor	Siphon sampling	Mean flow properties
García & Parsons (1996)	Solute (Saline)	Head	Arrested head	Indirectly from arrested head facility	n/a	Mixing rate
Hallworth *et al.* (1996)	Solute	Head	Lock-exchange	n/a	n/a	Entrainment into the head
Kneller *et al.* (1997)	Solute (RI matched)	Head, body	Lock-exchange	LDA, video	n/a	Mean flow properties, turbulence, reflections
Parsons (1998)	Solute (Saline)	Head	Arrested head	ADV, laser-induced fluorescence (LIF)	Conductivity probe	Turbulence, entrainment of ambient, mixing rate
Kneller *et al.* (1999)	Solute (RI matched)	Head, body	Lock-exchange	LDA, video	n/a	Mean flow properties, turbulence
Best *et al.* (in press)	Sediment-laden	Head, body	Modified lock-exchange	UDVP, video	n/a	Mean flow properties, turbulence, reflections
Buckee *et al.* (in press)	Solute (RI matched)	Body	Continuous	LDA	Conductivity probe	Mean flow properties, turbulence, stratification

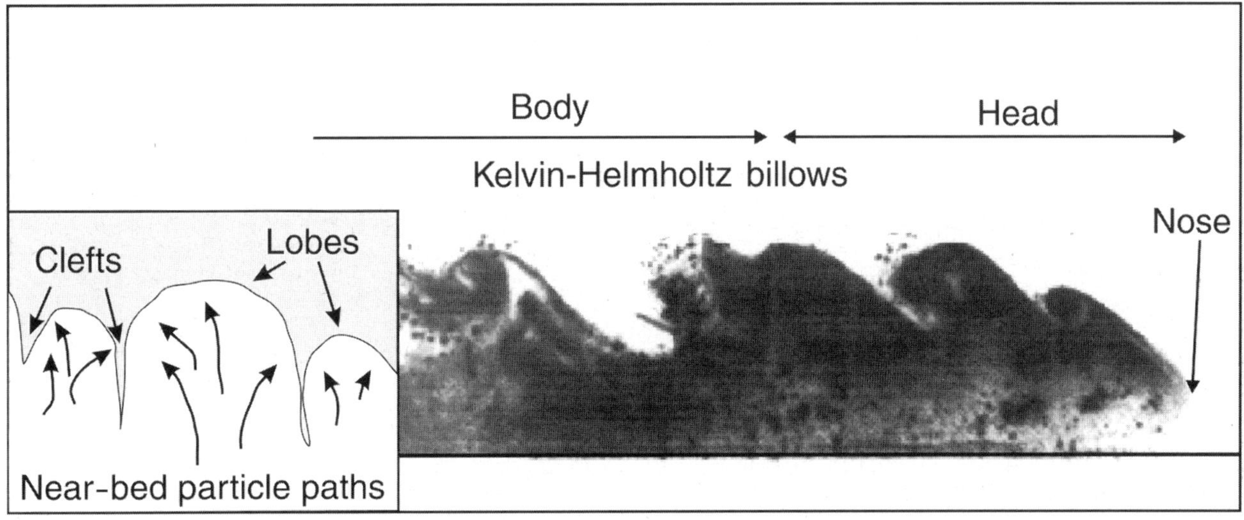

Fig. 2. Quasi two-dimensional image of the head and body of a saline gravity current illustrating well-developed Kelvin–Helmholtz billows (side view); modified from Simpson (1969). Inset shows schematic view of lobes and clefts seen from below.

upper boundary. At the rear of the head a series of transverse vortices (Fig. 2) are present, identified as Kelvin–Helmholtz instabilities (Britter & Simpson, 1978; see below).

Middleton (1966b) performed the first comprehensive set of experiments on the heads of brine currents and turbidity currents in the laboratory. He investigated the effects of slope on the velocity and shape of the head, finding that for low slopes (angles < 2·3°) the head velocity is adequately described by Keulegan's (1957) formula in which the head velocity is independent of the slope. Further experiments suggest that the dimensionless head velocity is only weakly dependent on slope (for slopes of 5–90°) because, with increasing slope, higher gravitational forces are counteracted by increased frictional resistance at the upper boundary owing to increased rates of ambient entrainment (Britter & Linden, 1980). The downstream velocity in the body of the current, which does depend on the slope, has been shown to be up to 30–40% faster than the head velocity (Middleton, 1966b,c; Kneller *et al.*, 1997, 1999; Best *et al.*, in press). Consequently, the head height increases with slope, as the body velocity increases and material moves more rapidly into the head (Hopfinger & Tochon-Danguy, 1977; Britter & Linden, 1980; Simpson, 1997).

Mixing of the current with the ambient fluid is an important process that occurs at the head, primarily by the detraining of dense fluid out of the back of the head in a series of transverse vortices (e.g. Allen, 1971; Britter & Simpson, 1978; Simpson & Britter, 1979). Britter & Simpson (1978) used slow-motion pictures to trace the shapes of the billows at the upper boundary, showing that the ratio of billow amplitude to wavelength was the same as for Kelvin–Helmholtz billows. These billows can grow in wavelength and amplitude to about the size of the head, and collapse as dense fluid is detrained into a 'mixing region' of approximately the same thickness as the billow amplitude (Britter & Simpson, 1978; Simpson & Britter, 1979). Middleton (1966b) investigated the motion of fluid and particles within, and in front of, the head, showing that flow lines diverge within the head and that the head displaces the ambient fluid. Further experiments on the displacement of ambient fluid by the head of the current led to the description of the 'lobe and cleft' structure at the base of the head, attributed to a gravitational instability formed when less dense fluid is over-run by, and incorporated into, the head (e.g. Simpson, 1969, 1972; Allen, 1971). Simpson (1969) visualized the lobe and cleft structure (Fig. 2), demonstrating that the clefts extend a substantial distance back from the head. Recent work, however, shows that lobe spacing scales with the flow thickness and that the development of the lobe and cleft structures is probably attributable to a secondary instability associated with Kelvin–Helmholtz vortex breakdown (Schowalter *et al.*, 1994; Parsons, 1998). Simpson & Britter (1979) used dye tracers to visualize the ambient fluid being

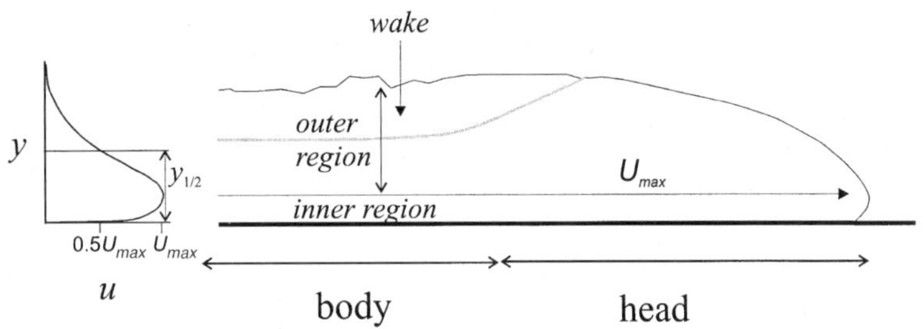

Fig. 3. Schematic diagram of the head and body of a gravity current, showing a typical downstream velocity profile.

entrained into the head at its base. They calculated a flux of light ambient fluid moving under the head and found it accounted for only 1% of the total mixing, the rest occurring at the upper boundary.

The body

Few physical descriptions of the body of gravity currents exist in the literature. Ellison & Turner (1959) describe the body as a region of steady downstream velocity which has a thin, dense layer of fluid near the base of the current which, with increasing downstream velocity, mixes with the ambient fluid at the upper boundary as an 'irregular succession of large eddies'. Similarly, Britter & Simpson (1978) and Simpson & Britter (1979) used visual observations and measurements to divide gravity currents, behind the head, into two distinct regions; the body, which is the lower dense layer, and a region of less dense, mixed fluid that has been mixed out of the head of the current. A similar structure was inferred by Kneller *et al.* (1999) based on the effect of mixing on the refractive index of brine currents. Some controversy surrounds the nature of the mixed region above the dense layer; it has been suggested that this is not strictly part of the gravity current but should be described as a zone of 'clouded water' entrained by the underflow (Middleton, 1966b; based in part on Kuenen, 1951; Blanchet & Villatte, 1954).

MEAN FLOW PROPERTIES OF GRAVITY CURRENTS

Downstream velocity profile

Figure 3 shows a typical vertical profile of the downstream velocity and serves to define some of the terms used. The shape of the downstream velocity profile in experimental and natural gravity currents (Chikita, 1990; Altinakar *et al.*, 1996; Simpson, 1997) is similar to the velocity profile in turbulent plane wall jets (Launder & Rodi, 1983). Gravity currents (and wall jets) are described as having an inner and outer region divided by the velocity maximum. The inner, wall-bounded region has a positive velocity gradient and is generally less than half the thickness of the outer region, which has a negative velocity gradient. The height of the velocity maximum is controlled by the ratio of the drag forces at the upper and lower boundaries (Middleton, 1966c; Kneller *et al.*, 1997), and in many experimental currents is found to occur at about 0·2–0·3 of the height of the current (Altinakar *et al.*, 1996; Kneller *et al.*, 1997, 1999; Best *et al.*, in press). For gravity currents travelling over a rough bed (e.g. turbidity currents over bedforms or gravel beds) the height of the velocity maximum will be raised compared with those that are dominated by drag at the upper boundary (as in supercritical currents, e.g. García, 1993; Buckee *et al.*, in press) which tends to lower the level of the velocity maximum. Despite this variability, it is possible to collapse velocity profiles from many different experimental currents (Fig. 4), by using a characteristic length scale, $y_{1/2}$ (defined as the distance between the bed and the height in the outer region at which the downstream velocity is half the maximum downstream velocity, see Fig. 3; Buckee *et al.*, in press).

Concentration profiles

Gravity currents are density stratified (i.e. there is a vertical gradient in concentration), having a dense basal layer of fluid and sediment (or dense fluid), with a less dense, more homogeneous mixed region above. A number of models for the distribution of density in gravity currents have been proposed (see discussion in Peakall *et al.*, in press; Fig. 5). Initial two-layer models (e.g. Middleton, 1966b, 1993; Simpson

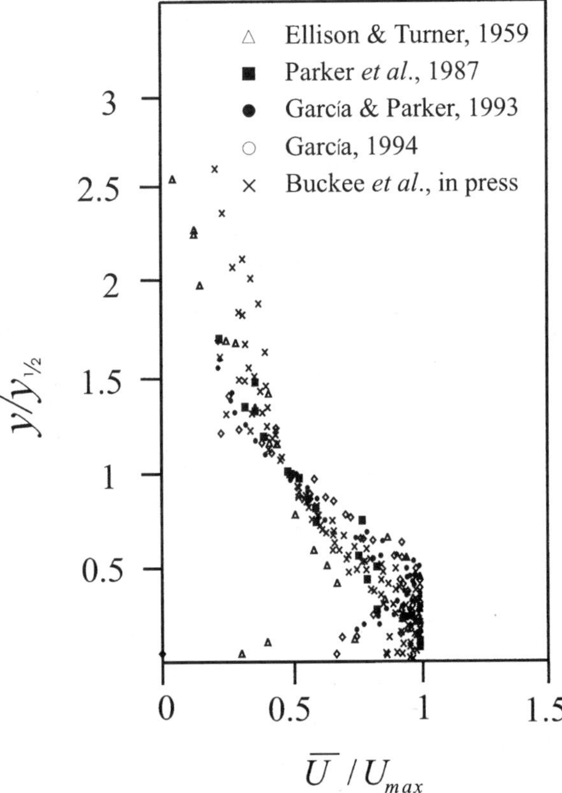

Fig. 4. Collapse of downstream velocity data using characteristic height $y_{1/2}$ (defined as the height in the outer region at which the downstream velocity is half the maximum velocity, see Fig. 3). The data represent both supercritical and subcritical currents.

& Britter, 1979) were based on visual observations of saline currents (e.g. Middleton, 1966b; Britter & Simpson, 1978; Simpson & Britter, 1979) and measurements in clay suspension currents (e.g. Blanchet & Villatte, 1954) and saline currents (e.g. Britter & Simpson, 1978; Simpson & Britter, 1979). The inflection point in the density distribution was observed to occur well above the level of the velocity maximum, and divided the 'body' of the current from the 'mixed' fluid detrained from the head (Fig. 5a). However, concentration measurements from both experimental sediment-laden currents (e.g. García, 1993, 1994; Altinakar *et al.*, 1996) and saline currents (e.g. Parsons, 1998; Buckee *et al.*, in press) demonstrate that the density structure is somewhat more complex. Two main types of sediment concentration profile have been observed (Peakall *et al.*, in press). A smooth profile (Fig. 5b) is commonly seen in low-concentration, weakly depositional currents (Altinakar *et al.*, 1996; García, 1990, 1994) and in saline gravity currents (Ellison & Turner, 1959; Buckee *et al.*, in press). These currents are nonetheless highly stratified, with a density gradient that is greatest near the base of the current and decreases rapidly around the level of the velocity maximum. The second class of density distribution has a 'stepped-concentration' profile (Fig. 5c) and is commonly observed

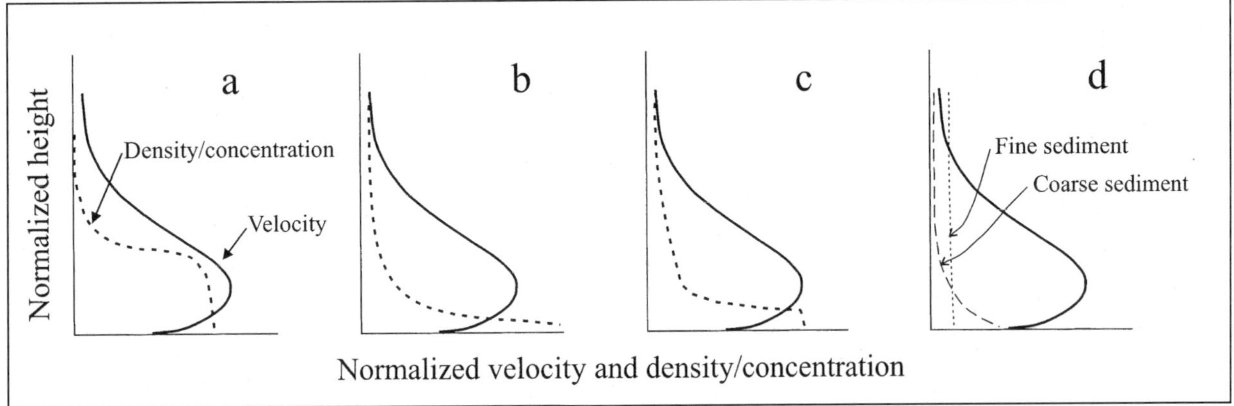

Fig. 5. Schematic diagram showing various characteristic density/concentration profiles (dashed lines) in density currents; the same downstream velocity profile (solid line) has been used in each case for reference. In each case the y-axis represents a normalized height and the x-axis normalized concentration and velocity. (a) A two-layer model type concentration profile, dividing the flow into a constant density lower region (the current) and an upper region of fluid detrained from the head (e.g. Britter & Simpson, 1978; Simpson & Britter, 1979; Middleton, 1993). (b) A smooth profile, characteristic of low-concentration, weakly depositional flows (e.g. Altinakar *et al.*, 1996; García, 1990, 1994). (c) A stepped concentration profile observed in erosional flows (e.g. García, 1993). (d) A Rouse-type distribution of sediment grain-sizes observed in turbidity currents (e.g. García, 1994), in which coarse material is concentrated towards the lower part of the flow whereas fine-grained material is more evenly distributed throughout the depth of the flow (modified from Peakall *et al.*, in press).

in erosional currents (García, 1993) or currents interpreted to have a high entrainment rate at the upper boundary (Peakall *et al.*, in press). Experiments in which vertical grain size distributions have been measured (García, 1994) show that fine-grained material is more uniformly distributed in the vertical than the coarse material, which tends to become concentrated in the lower part of the current. Theoretical and experimental studies of turbidity currents (Stacey & Bowen, 1988a,b; García & Parker, 1993; García, 1994; Altinakar *et al.*, 1996), and limited field data (e.g. Chikita, 1989; Normark, 1989; Zeng & Lowe, 1997), suggest that they have patterns of vertical sediment distribution rather similar to those of shear flows, in which the vertical sediment concentration profile obeys a power law distribution (given by Rouse, 1937; see Middleton & Southard, 1984) with an exponent that is directly proportional to the ratio of the current shear velocity, U_*, to the grain settling velocity, U_s (Middleton & Southard, 1984). Relatively low values of U_*/U_s predict high near-bed concentrations of suspended sediment, decaying rapidly upwards (Fig. 5d). This has two important consequences; firstly, any polydisperse (i.e. multiple grain-size) current must be stratified in terms of both density and grain-size, particularly if it is depositional or close to being so; secondly, the grain-size range near the base of the current is broader than that higher in the current.

TURBULENCE STRUCTURE OF GRAVITY CURRENTS

The head: temporal variations, vertical structure and entrainment

Data on the turbulence structure in the head of a gravity current have been presented by Kneller *et al.* (1997, 1999) and Parsons (1998). Time series of instantaneous downstream velocity clearly record the arrival of the head and the passage of large, low-frequency Kelvin–Helmholtz billows superimposed on a period of quasi-steady motion (Kneller *et al.*, 1997, 1999; Fig. 1). Instantaneous velocities, associated with large eddies, were found to exceed the maximum mean downstream velocity by up to 50%. Similarly, large instantaneous peaks in Reynolds stress were recorded. Two areas of high negative Reynolds stress were identified – at the top of the head, related to the Kelvin–Helmholtz billows, and beneath the nose of the current, attributed to the entrainment of over-ridden ambient fluid which is buoyant within the current (Kneller *et al.*, 1999). The turbulence structure in gravity current heads, in which Fr is always less than one, is dominated by shearing at the upper interface. This is reflected in the distributions of turbulent kinetic energy and density fluctuations, both of which have a maximum at the upper boundary of the head (Kneller *et al.*, 1997, 1999; Parsons, 1998; Best *et al.*, in press).

García & Parsons (1996) and Parsons (1998), demonstrated experimentally that mixing in gravity current heads is Reynolds number dependent; dimensionless mixing rates decrease with decreasing Reynolds number. In large (high-Re) currents, which are fully turbulent, entrainment is primarily due to secondary instabilities (i.e. Kelvin–Helmholtz vortex breakdown) in the upper part of the current. In small currents (i.e. low-Re currents in which all the turbulent subranges are not present, for instance the laboratory experiments of Britter & Simpson, 1978), entrainment due to secondary instabilities (i.e. Kelvin–Helmholtz vortex breakdown) is less significant as viscous effects become more important. Care must be taken in laboratory currents therefore to ensure that the Reynolds number is sufficiently high (> 2000) to minimize such viscous effects as they may compromise the scaling approach taken. Hallworth *et al.* (1996) examined entrainment into the head of lock-exchange currents, using a pH neutralization technique whereby the alkali current entrained acidic ambient fluid. The amount of dilution in the head could be quantified using pH indicator solution. Hallworth *et al.* (1996) suggest that in lock-exchange currents, total entrainment at a given distance from the source is independent of the reduced gravity and related primarily to the initial volume. Experiments on gravity currents travelling over a free surface (i.e. the water–air interface) led Hallworth *et al.* (1996) to the conclusion that approximately two-thirds of the total entrainment into the head, for currents over a rigid surface, occurs by overriding of ambient fluid. These conclusions are in contrast to most other results which show that entrainment is a function of the densiometric Froude number (e.g. Ellison & Turner, 1959) and hence the initial reduced gravity. Furthermore, many experimental studies have shown that entrainment at the lower boundary

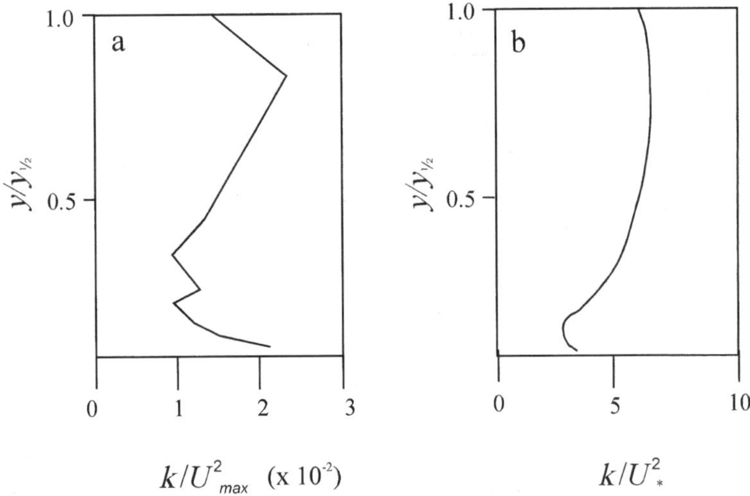

Fig. 6. A comparison of an experimental profile of turbulent kinetic energy (k) normalized by the maximum downstream velocity (U_{max}) squared (Buckee *et al.*, in press) and a theoretical profile of k normalized by the square of the shear velocity (U_*) adapted from Brørs & Eidsvik (1993). The experimental profile (a) is less smooth than the theoretical profile (b) owing to poorer vertical resolution. However, the results do compare well; maximum normalized turbulent kinetic energy occurs at $y/y_{1/2} = 0{\cdot}8$ in both cases. Brørs & Eidsvik (1993) quote maximum k/U^2_{max} of the order of 10^{-2} for flows which are comparable to the experimental gravity currents in Buckee *et al.* (in press).

by overriding of ambient fluid is relatively insignificant (e.g. Simpson & Britter, 1979; Parsons & García, 1995; Parsons, 1998; see above). The contradictory results presented in Hallworth *et al.* (1996) probably reflect the significant differences in the turbulence structure of gravity currents flowing over fixed (i.e. the bed) and free surfaces (i.e. the water–air interface), and the increased influence of viscosity in low-*Re* (<2000) currents.

The body: temporal variations

Time series of downstream velocity in the body of quasi-steady gravity currents reveal the presence of large coherent structures that advect with the current (Kneller *et al.*, 1997; Best *et al.*, in press). Instantaneous downstream velocities may be up to 40% higher than the maximum mean downstream velocity in the body (Buckee *et al.*, in press), and therefore equivalent to, or higher than, instantaneous velocities in the head. This result suggests that the body of the current may play a significant role in sediment entrainment, as such high turbulent velocities imply high Reynolds stresses (Kneller *et al.*, 1997) and hence high potential erosion rates. Spectral analysis of velocity time series (Kneller *et al.*, 1999) shows that low-frequency (<10 Hz) motions dominate the energy budget at all depths in the current.

The body: vertical structure

Laboratory experiments show that turbulence intensities are highest at the top of the gravity current (e.g. Kneller *et al.*, 1997, 1999; Best *et al.*, in press), and related to large-scale shearing and mixing at the upper boundary. Low turbulence intensities are observed around the level of the velocity maximum, or the high density gradient near the base of the current (Best *et al.*, in press; Buckee *et al.*, in press). Similarly, turbulent kinetic energy is low at around the level of the velocity maximum (Kneller *et al.*, 1999; Buckee *et al.*, in press). In currents with high levels of shear at the upper boundary (e.g. lock-exchange currents, supercritical currents) maximum turbulent kinetic energy occurs towards the top of the current (Kneller *et al.*, 1999; Buckee *et al.*, in press; Fig. 6). However, in subcritical currents without a counterflow of ambient fluid (owing to replacement of fluid in the lock), maximum turbulent kinetic energy occurs towards the base of the current and Reynolds stresses at that point are high and positive, reflecting an upward transfer of momentum (Buckee *et al.*, in press), with the level of zero Reynolds stress being displaced above the velocity maximum. Negative Reynolds stresses in the upper part of the current are associated with mixing of the ambient fluid downward into the current. Significant instantaneous variations may occur in the vertical Reynolds stress profile however, reflecting flow unsteadiness and the passage of large eddies (Kneller *et al.*, 1997).

Energy budget in gravity currents

Buckee *et al.* (in press) calculated rates of turbulent kinetic energy production in steady experimental gravity currents from shearing of the mean flow and from buoyancy flux. The results showed that the highest rates of turbulence production occurred due to shearing near the lower boundary, where the velocity gradient is highest. The results also suggested that turbulence production by shear is more significant in currents that are more highly density stratified (e.g. currents in which vertical mixing is inhibited; subcritical currents) while turbulence production by buoyancy flux is more important in currents that are less stratified (e.g. currents in which there is more significant vertical mixing; supercritical currents). The study by Buckee *et al.* (in press) was limited, however, by poor vertical resolution for the calculation of velocity gradients and the fact that the entire energy budget could not be resolved owing to lack of spatial resolution in the downstream direction.

DENSITY STRATIFICATION IN GRAVITY CURRENTS

Definitions

A stable density stratification is one in which the overall density decreases upwards (Turner, 1973). However, a current that is stratified on a large scale may, on closer examination, reveal poorly stratified regions as well as well-stratified regions. The gradient Richardson number (Ri_g) is calculated with respect to local density gradients and highlights regions that are less stably stratified than the whole:

$$Ri_g = \frac{-g d\rho_y / d_y}{\rho_0 (dU_y/dy)^2} \tag{4}$$

where ρ_0 is some reference density (normally taken as that of the ambient fluid) and U_y is the mean downstream velocity at height y in the flow. A widely accepted criterion for the stability of stratification is that if $Ri_g > 0{\cdot}25$ locally, then the stratification is sufficiently strong to significantly inhibit vertical mixing at that level (Turner, 1973). However, the presence of a stable stratification does not necessarily imply the absence of turbulence. A stable stratification tends to inhibit vertical mixing, and an external energy supply is therefore required to generate turbulence and prevent its rapid dissipation (e.g. Tritton, 1988).

Stability of stratification in gravity currents

García (1993) and Buckee *et al.* (in press) calculated Ri_g above the velocity maximum in both subcritical and supercritical experimental gravity currents. Typically, supercritical currents were found to have Ri_g in the range 0·13–0·23, below the critical value of 0·25, whereas for subcritical currents Ri_g had a range between 0·54 and 0·84. The difference in stability between the two flow regimes reflects the greater degree of mixing that occurs at the upper boundary of supercritical currents (Ellison & Turner, 1959). Furthermore, Buckee *et al.* (in press) calculated vertical profiles of Ri_g for two experimental currents (Fig. 7), showing that the strength of the stratification varies with height in gravity currents. Stacey & Bowen (1988a,b) calculate theoretical Ri_g profiles for supercritical currents which are in reasonable agreement with the data of Buckee *et al.* (in press) for the region above the velocity maximum. At the level of the velocity maximum, where the velocity gradient is zero, calculated Ri_g goes to infinity. In currents with a smooth sediment concentration profile, the stratification below the velocity maximum is generally stable. However, when the concentration profile is stepped (e.g. Fig. 5c) there may be a region of less stable stratification close to the bed.

THEORETICAL MODELS OF GRAVITY CURRENT BEHAVIOUR

Mathematical modelling of gravity currents takes a range of forms from simple hydraulic equations and box models to highly complex turbulence models (see Table 2 and below). Similarly, mathematical models may be used for a wide range of purposes, from providing estimates of current velocity and thickness to predicting turbidite geometries and grain size distributions,

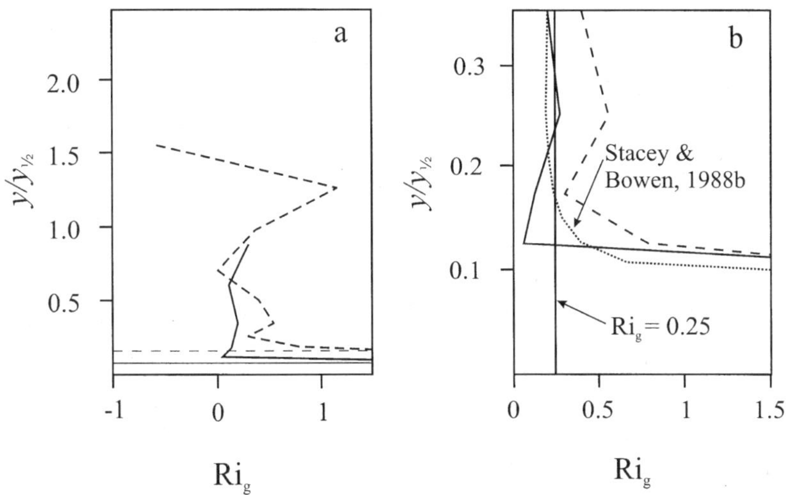

Fig. 7. Calculated gradient Richardson numbers for experimental gravity currents ($Fr > 1$, solid line and $Fr < 1$, dashed line) taken from Buckee *et al.* (in press). (a) The horizontal lines represent the level of the velocity maximum for each current. In (b) a detail of (a) is shown co-plotted with a theoretical curve for Ri_g (dotted line) for a current in which $Fr > 1$ (adapted from Stacey & Bowen, 1988b). It should be noted that the curves derived from experimental data are less smooth than the theoretical curve owing to their poorer vertical resolution. The line $Ri_g = 0{\cdot}25$ highlights the value above which the flow is locally stably stratified.

to modelling the vertical structure of turbulence in a gravity current. In order to model gravity current dynamics, a series of equations must be solved: the conservation of fluid mass, the conservation of sediment mass and the conservation of momentum (known as the Navier–Stokes equation). In addition, an equation for conservation of turbulent energy may be employed. The Navier–Stokes equation is a nonlinear partial differential equation which can be solved for turbulent currents, such as gravity currents, by the application of simplifying assumptions (e.g. Tennekes & Lumley, 1972; Tritton, 1988).

Simple models

The first models of the hydraulic behaviour of turbidity currents were one-equation models based on a modified form of the Chézy equation for steady, uniform flow, which can be derived from the Navier–Stokes equations (Komar, 1977; Stacey & Bowen, 1988a):

$$\overline{U} = \sqrt{\frac{8g'Sd}{f_b + f_i}} \tag{5}$$

where $\overline{U}$ is the depth-averaged mean current velocity, S is the slope ($\sin \alpha$), d is the thickness of the current, g' is the reduced gravity and f_b and f_i are the Darcy–Weisbach friction coefficients at the lower and upper boundaries of the current, respectively (Daly, 1936; Kuenen, 1952; Hinze, 1960; Middleton, 1966b,c, 1993; Middleton & Southard, 1984). It has been suggested that for large turbidity currents $f_b + f_i$ is approximately 0·01 (Middleton & Southard, 1984), where $f_i > f_b$, and that both terms are likely to be functions of the densiometric Froude number (Middleton, 1966a; Chikita, 1989). Modified Chézy equations have frequently been used to model the bulk flow properties of natural turbidity currents (e.g. Kuenen, 1952; Bowen *et al.*, 1984; Kirwan *et al.*, 1986; Mulder *et al.*, 1998). Some of these models have been extended in order to study turbidity current evolution (e.g. on the Navy submarine fan, Bowen *et al.*, 1984) and erosion and deposition (e.g. in the Saguenay fjord, Mulder *et al.*, 1998).

Box models

The primary utility of box models is to provide a simple and rapid way to model the rate of propagation of finite-volume, low-concentration, surge-type currents and their broad deposit characteristics. Box models are most effective for modelling currents with relatively low concentrations of small heavy particles which are transported by, and sedimented from, turbulent currents (turbidity currents) (for a review see

Table 2. Summary of mathematical models of gravity currents.

Author(s)	Model	Comments
Kuenen (1952)	Chézy equation	Calculated properties of the 1929 Grand Banks current
Ellison & Turner (1959)	Depth-averaged	Used in conjunction with laboratory experiments to examine entrainment of ambient fluid at the upper boundary of saline currents
Kirwan *et al.* (1960)	General analytical solution	Calculated evolution of the 1929 Grand Banks current and the Orleansville current in order to model the origin of the currents
Hinze (1960)	Analytical solution in which eddy viscosity is assumed to be constant in the vertical direction	Models density and velocity profiles above the velocity maximum. Predicts velocity and flow thickness for the Grand Banks current in reasonable agreement with other researchers
Chu *et al.* (1979)	Depth-averaged	Examines the effect of changing certain variables (sediment exchange with the bed, slope, ambient entrainment and lower boundary drag) on flow evolution
Bowen *et al.* (1984)	Modified Chézy equation, including terms for ambient entrainment and Coriolis force	Modelled flow evolution of turbidity currents on the Navy submarine fan
Parker *et al.* (1986)	Depth-averaged (3- and 4-equation models)	Examines the theory of self-acceleration (or autosuspension). The 4-equation model appears to predict self-acceleration realistically
Stacey & Bowen (1988a,b)	Algebraic turbulence closure, eddy viscosity and eddy diffusivity vary vertically with the gradient Richardson number (with a linear bridge across the velocity maximum)	Models vertical profiles of velocity, density and gradient Richardson number Results compare well with laboratory experiments The second paper uses the model to examine self-acceleration and flows bearing multiple grain sizes
Eidsvik & Brørs (1989)	k–ε turbulence closure model	Predicts vertical distribution of velocity, density, turbulent kinetic energy and eddy viscosity Results compare well with experimental data. Predicts self-acceleration
Brørs & Eidsvik (1992)	Reynolds stress turbulence closure model	Predicts vertical distribution of velocity, density, turbulent kinetic energy and eddy viscosity Results compare well with experimental data
Dade & Huppert (1994)	Box model	Scaling arguments used to apply the model to the deposits of 2D (channelized) turbidites Predicts run-out length and deposit thickness
Zeng & Lowe (1997)	Depth-averaged, and accounts for vertical distribution of sediment	Models flow evolution and deposit characteristics
Kubo *et al.* (1998)	Box model	Compares model results with palaeocurrent velocities calculated from grain size data from an extensive turbidite in the Miocene–Pliocene Kiyusomi Formation, Japan
Mulder *et al.* (1998)	Modified Chézy equation, including terms for ambient entrainment and sediment exchange with the bed	Modelled flow evolution of a 28-day flood in the Saguenay fjord in 1663 Compared model deposit characteristics (run-out distance, thickness, grain size) with real data
Felix (in press)	Mellor-Yamada 2½ turbulence closure model, includes influence of particles on turbulence	Models vertical and downstream distribution of velocity, density, turbulent kinetic energy and eddy viscosity. Results compare well with experimental data. Models turbidite deposits (thickness, grain size distributions)

Huppert, 1998). Box models are not based directly on the Navier–Stokes equations; instead the current is modelled as an evolving series of rectangles with no horizontal or vertical variation in current properties (Huppert, 1998). The rate of propagation is controlled by the Froude number of the head, which is assumed to be constant in the range 0·7–0·8 (e.g. Middleton, 1993; Huppert, 1998). In this way a dimensional form of the solution can be found which can be solved analytically. Box models simulate currents which are very similar to lock-exchange experiments and achieve good results (except near the point of initiation) when the model results are compared with the deposits of strongly depositional lock-exchange currents (Bonnecaze *et al.*, 1993; Huppert, 1998). These models are adequate for modelling such simple cases, but have rarely been tested against natural deposits owing to the difficulty in constraining initial conditions. Dade & Huppert (1994, 1995) have made use of box models to simulate a known deposit (Black shell turbidite) and calculate initial current volumes and sediment concentrations. Kubo *et al.* (1998) tested Dade & Huppert's (1995) box models and Bonnecaze *et al.*'s (1995) axisymmetric model against well-constrained data from an extensive turbidite bed of the Miocene–Pliocene Kiyosumi Formation, Japan. Using grain size data to calculate palaeocurrent velocities, Kubo *et al.* (1998) showed that the box models significantly underestimated current velocity. This may be partly because box models fail to account for the difference between the velocity of the head and that of fluid within the body.

An inherent problem with box models that also applies to other depth-averaged models (see below) is the assumption that the vertical sediment concentration profile is uniform (as a result of vertical mixing by turbulence); in fact any polydisperse current must be density stratified (Fig. 5d) and this assumption fails. Similarly, box models do not explicitly account for turbulence, or for entrainment of ambient fluid at the upper boundary, both of which are fundamental. Furthermore, Piper *et al.* (1999a) show that even a current generated by slope failure (e.g. the 1929 Grand Banks current), which might be regarded as a surge-type current, may produce 'sustained flow over many hours' which cannot be simulated by a box model. Perhaps neither the deposits produced in lock-exchange experiments nor the results of box models realistically reflect the patterns of deposition beneath more complex natural turbidity currents.

More complex depth-averaged models

In turbidity current modelling, the Navier–Stokes equations are commonly solved in a depth-averaged form. The assumption is made that there are no significant vertical variations in the properties of the current and therefore the governing equations are integrated over the height of the current. This assumption results in one mean value for each variable (such as velocity and sediment concentration) at each point downstream. As it is clear that the flow properties of turbidity currents vary dramatically with height, some researchers impose vertical profiles (or shape-factors) for velocity and sediment concentration (Turner, 1973; Pantin, 1979; Parker *et al.*, 1987; García, 1994). While this approach does take into account some significant properties of the current, the shape-factor is generally taken as constant in these numerical experiments while in natural turbidity currents both velocity and sediment concentration profiles may evolve with time. Such models are therefore quasi-two-dimensional, not truly two-dimensional as the vertical variation in properties is not actually modelled. Depth-averaged models account for turbulence by employing an empirically derived function for the entrainment of ambient fluid at the upper boundary (e.g. see Ellison & Turner, 1959; Fukushima *et al.*, 1985). Depth-averaged models are generally solved numerically (i.e. by discretizing the governing equations, e.g. Parker *et al.*, 1986; Zeng & Lowe, 1997). However, by making many simplifying assumptions (such as steady flow), the depth-averaged Navier–Stokes equations may be reduced to terms that may be solved analytically (see Stacey & Bowen, 1988a). While complex numerical solutions may require considerable computational expense, they may be more appropriate than analytical solutions for gravity currents which are typically highly unsteady.

In the turbidity current literature, three-equation depth-averaged models are most common, in which the equations for fluid mass, sediment mass and fluid momentum are solved simultaneously (Ellison & Turner, 1959; Fietz & Wood, 1967; Chu *et al.*, 1979; Parker *et al.*, 1986; García, 1994; Zeng & Lowe, 1997). In four-equation models, conservation of turbulent energy is also taken into account (Johnson, 1962; Parker *et al.*, 1986). Most depth-averaged models have been used to study aspects of the evolution of turbidity currents. Ellison & Turner (1959) developed an early model for studying entrainment of ambient fluid at the upper boundary of saline currents. Their theoretical results were combined with experimental data to produce a semi-empirical entrainment function and a series of equations that could be solved to predict current velocity and thickness. Chu *et al.* (1979) modelled flow evolution, examining the transition from supercritical flow on a slope to subcritical flow on the basin floor via a hydraulic jump, and the effect of changing variables such as slope, sediment exchange with the bed and entrainment at the upper boundary on this development. Self-acceleration (or auto-suspension after Pantin, 1979) has been studied theoretically in a similar way (Parker *et al.*, 1986); a four-equation model linking sediment entrainment to mean turbulent energy appeared to give realistic results. A recent model (Zeng & Lowe, 1997) predicts current development and sediment deposition, producing detailed models of deposit thickness and grain-size distribution in both the downstream and the vertical directions.

Depth-averaged models, while not actually modelling the fluid dynamic processes within turbidity currents, can produce relatively accurate predictions of current evolution and deposit characteristics which provide valuable insights for geologists concerned with ancient turbidity current deposits.

Models incorporating turbulence

A second approach to solving the Navier–Stokes equation for turbulent currents is to use a model that accounts for turbulence and hence can resolve vertical current properties. This can be achieved analytically; for instance, Hinze (1960) reduced the Navier–Stokes equations to model a steady current with constant eddy viscosity in the vertical profile. Similarly, Stacey & Bowen (1988a) assumed eddy viscosity and eddy diffusivity were constant in the vertical direction and produced an analytical solution for unsteady currents. However, they demonstrated that this approach is of limited value since the velocity and sediment concentration profiles are decoupled in the model, which is unrealistic. Stacey & Bowen (1988a) also developed a numerical model in which eddy diffusivity and eddy viscosity vary vertically as a function of the gradient Richardson number. This model produced vertical velocity and sediment concentration profiles that compared well with Ellison & Turner's (1959) experimental data, and gradient Richardson number profiles that compare well with Buckee *et al.* (in press; Fig. 7). Several more complicated turbulence closure models have been applied to turbidity currents (for a review of turbulence closure models see Rodi, 1980). Eidsvik & Brørs (1989) used a k–ε model (Rodi, 1980), and Brørs & Eidsvik (1992) used a Reynolds stress model; both models produce realistic results when compared with experimental data sets (e.g. Fig. 6). Interestingly, some features predicted by these models, such as low turbulent kinetic energy at the level of the velocity maximum, were initially dismissed by the authors as unrealistic but have since been observed in laboratory experiments (Parsons, 1998; Kneller *et al.*, 1999; Best *et al.*, in press; Buckee *et al.*, in press; Fig. 6). More recently, Felix (in press) has adapted the Mellor–Yamada level 2½ closure model (Mellor & Yamada, 1982) to include the influence of suspended sediment on turbulence and to model current development in both the vertical and the downstream directions. This approach allows a unique two-dimensional representation of the properties of the whole current (Fig. 8), including the head, and consequently can predict spatial variability (in terms of thickness and grain size) of turbidite deposits.

A turbulence-based model is appropriate for a situation in which there is significant vertical variation in the current properties, and hence is ideal for use with turbidity currents, and may lead to better understanding of turbidity current dynamics (Stacey & Bowen, 1988a). A limitation of the use of turbulence models is that the methods are currently very costly in terms of computer power and time. In many cases the required model output (such as grain size distributions or deposit thickness) does not warrant the computational power required, given the uncertainties over initial conditions of natural currents in the environment, and a more simple depth-averaged model would serve the purpose in modelling.

EFFECTS OF TOPOGRAPHY ON TURBIDITY CURRENTS

In any system in which the topography encountered by sediment gravity flows is not completely flat, the topography exerts a potentially major influence on deposition (e.g. Miller & Smith, 1977; Pickering *et al.*, 1989; Fisher, 1990; Thornburg *et al.*, 1990; Apps *et al.*, 1994), either

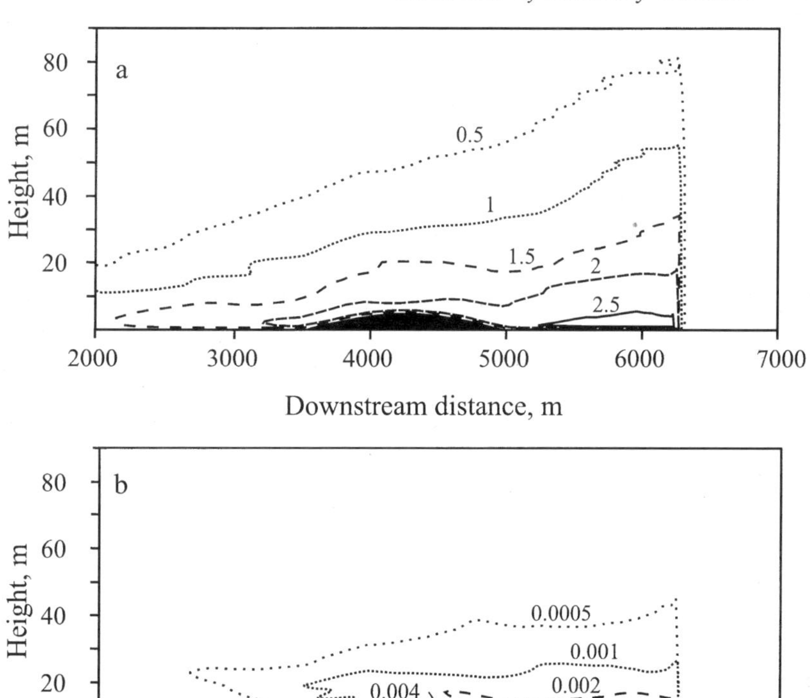

Fig. 8. A 2D time-slice of a turbidity current travelling from left to right, (a) downstream velocity (in m s^{-1}) and (b) total volume concentration fraction (adapted from Felix, in press). The model current is initiated as a static suspension (1000 m × 50 m) at 3% volume concentration, composed of three grain sizes in equal proportions. This image is taken 2000 s after initiation, after the current has passed over an obstacle of 100 m × 5 m. Notice the high velocity core near the front of the current in (a) and the general decrease in mean velocity and sediment concentration backwards from the head.

by controlling the nonuniformity of the currents or by confining and 'ponding' them, either partially, as in a valley, or completely (e.g. Wilson & Walker, 1985; Branney & Kokelaar, 1992; Rothwell *et al.*, 1992; Hodgson *et al.*, 1992; Prather *et al.*, 1999). The behaviour of turbidity currents and pyroclastic currents around obstructing topography varies with the forward velocity of the current, the obstacle height, the current density and, most significantly, the density stratification within the current (Long, 1953, 1955, 1970; Valentine, 1987; Grue, 1992; Lawrence, 1993; Castro & Snyder, 1993; Lane-Serff *et al.*, 1995; see also Muck & Underwood, 1990; Alexander & Morris, 1994). This topographic interaction has major implications for the spatial distribution of sediment in the deep sea, for the interactions of unconfined currents with intrabasinal highs and basin margins, and for channellized currents with channel margins and levées. It is also relevant to hazard mitigation in volcanic areas.

Run-up heights

For obstacles that are much larger than the current, oceanographic data suggest that run-up distances in nature may be many hundreds of metres (Dolan *et al.*, 1989; Muck & Underwood, 1990; Lucchi & Camerlenghi, 1993). Similarly, pyroclastic currents may surmount topographic ridges several hundreds of metres high (e.g. Miller & Smith, 1977; Wilson & Walker, 1985). According to Lane-Serff *et al.* (1995), a finite volume of fluid related to the head may surmount the obstacle if H_* (the ratio of the obstacle height to the current body thickness) is less than 4 or 5, or if the obstacle height is less than 1·5 times the height of the head of the current, according to Muck & Underwood (1990). Both these studies reported the results of experiments with saline currents. However, it seems likely that in relatively poorly stratified currents, the maximum run-up height is probably dependent on the bulk Froude number of the current:

$$h_{\max} = \frac{1}{2}\left(\frac{\overline{U}^2}{g'}\right) \tag{6}$$

(Rottman *et al.*, 1985). For strongly stratified currents, Kneller & McCaffrey (1999) developed Allen's (1985) energy balance arguments to suggest an expression for maximum run-up

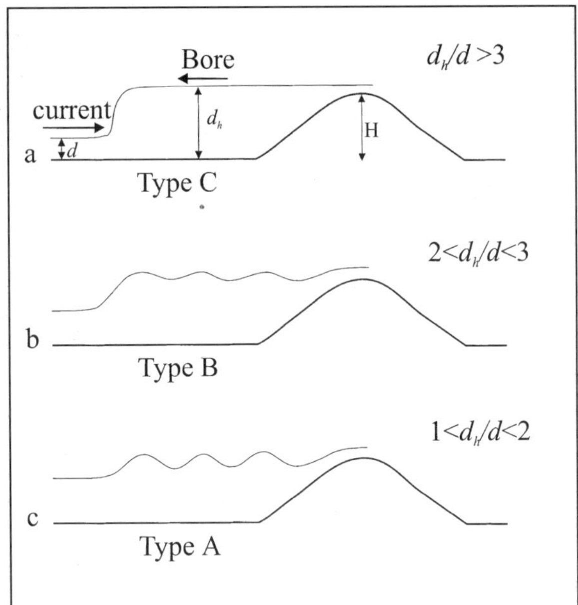

Fig. 9. Nomenclature of three types of bores defined by Rottman & Simpson (1989) (modified from Edwards, 1993).

height that takes account of both the velocity and the density profiles of the current:

$$h_{max} = y + \frac{\rho_y u_y^2 (1 - E)}{2g\Delta\rho_y} \tag{7}$$

where u_y is the downstream component of velocity at initial height y, g is the acceleration due to gravity, $\Delta\rho_y$ is the density difference between the fluid at initial height y and the ambient fluid, and E is the fractional energy loss due to friction. The value of y that yields the highest run-up of any parcel of fluid in the current is given by:

$$\frac{d}{dy}\left[y + \frac{\rho_y u_y^2 (1 - E)}{2g\Delta\rho_y}\right] = 0 \tag{8}$$

h_{max} is thus a function of the velocity and density profiles, and cannot be generalized in systems in which the density stratification is significant – as in most sediment-laden systems.

The blocking of turbidity currents by topography

Complete blocking is when none of the current gets over the top of the topography. For two-dimensional systems, i.e. where the obstacle axis is perpendicular to flow and effectively infinite in extent, Rottman *et al.* (1985) give the criterion for complete blocking as:

$$\frac{\overline{U}}{g'd} = 0.5(\mathrm{H}_* - 1)^2\left(\frac{\mathrm{H}_* + 1}{\mathrm{H}_*}\right) \tag{9}$$

Partial blocking occurs where the relative obstacle height is less and part of the forward current continues over the obstacle. Woods *et al.* (1998) provide an analysis of the criteria for full and partial blocking of unstratified currents by two-dimensional topography, based on the concept of critical flow velocity (u_c, Eq. 11) and critical flow thickness (h_c, Eq. 13) that are used to nondimensionalize the mean flow velocity ($\overline{U}$) and obstacle height (H), thus:

$$U_o = \frac{\overline{U}}{u_c} \tag{10}$$

where

$$u_c = (g'Q)^{1/3} \tag{11}$$

(where U_0 is the nondimensional downstream velocity and Q is the volume flux), and,

$$D_0 = \mathrm{H}/h_c \tag{12}$$

where

$$h_c = Q/u_c \tag{13}$$

where D_0 is the nondimensional height. The minimum height for complete blocking is dependent on U_0 and D_0, and is equal to the thickness of the bore generated upstream (Woods *et al.*, 1998, and see below), which is $\sim 2{\cdot}2h_c$ for critical currents. The situation is more complex for three-dimensional topography; flow probably goes around the obstacle when the densiometric Froude number is much less than 1, but stratification plays an important role (see below).

The effect of blocking is to generate a disturbance upstream of the obstacle, which consists of an internal bore (a more or less abrupt downstream increase in current thickness with an accompanying drop in velocity) that migrates upstream. Where the current is completely blocked, the fluid velocities within the bore may be negative (i.e. upstream) and it therefore constitutes a current 'reflection' (Edwards, 1993; Best *et al.*, in press).

Kneller *et al.* (1991), Edwards (1993) and Kneller (1995) have shown that in three-dimensional situations, these 'reflections' propagate perpendicular to the reflecting surface. The bore is often referred to as a moving internal hydraulic jump, though it does not necessarily separate regions of supercritical and subcritical flow (defined as regions where the densiometric Froude number is, respectively, greater than or less than unity).

The bore may take different forms (Fig. 9) determined by the ratio of the height of the bore to the thickness of the forward current through which it moves (Fig. 9; Wood & Simpson, 1984; Edwards, 1993; Edwards *et al.*, 1994; Best *et al.*, in press; see also Simpson, 1997; Rottman & Simpson, 1989). This ratio in turn depends on the speed of the forward current, the current density and the upstream slope of the obstacle (Edwards *et al.*, 1994). The strongest, Type C in the nomenclature of Edwards *et al.* (1994), consists of an undercutting turbulent gravity current head that generates a bore within the overlying dense fluid (Fig. 9a). Considerable mixing with the ambient fluid may occur at the upper boundary of a Type C bore (see, e.g. Best *et al.*, in press). Type B, *sensu* Edwards *et al.* (1994) is an undular bore, and has a rather smoother front with distinct undulations behind it (Fig. 9b). Lastly, Type A bores consist of a series of internal solitary waves (Fig. 9c). These types form a continuous spectrum. To date, no investigation has examined the effect of varying density stratification on the type of bore produced, but it seems likely that strong bores are associated with stronger stratification.

Multiple palaeocurrent directions within single beds have been reported from many turbidite systems, both ancient (Marjanac, 1990; Kneller *et al.*, 1991; Edwards, 1993) and modern (Pickering *et al.*, 1992), and are probably present in many more. These multiple-palaeocurrent directions have been interpreted as 'reflection' of the turbidity current from topography at the basin margin; comparable interactions with topography have also been described for pyroclastic currents (Valentine *et al.*, 1992; Woods *et al.*, 1998). In some cases, the 'reflections' are more or less diametrically opposed to the forward current (e.g. Ellis, 1982; Pickering & Hiscott, 1985), while in many cases they are roughly perpendicular. Often the changes in current direction occur in association with abrupt reversals in grading (Pickering & Hiscott, 1985), and these reversals or repetitions have been used to imply current reflection.

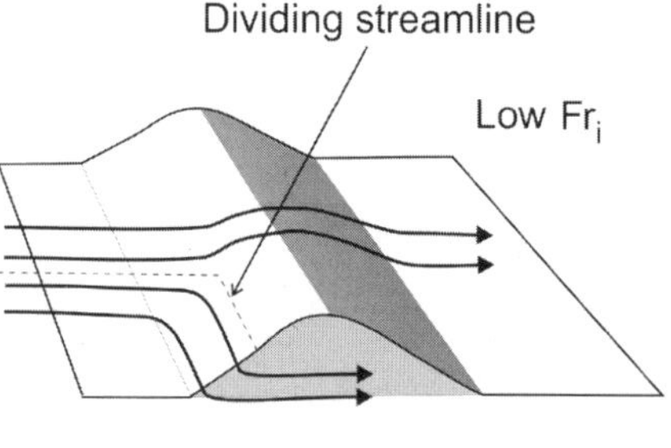

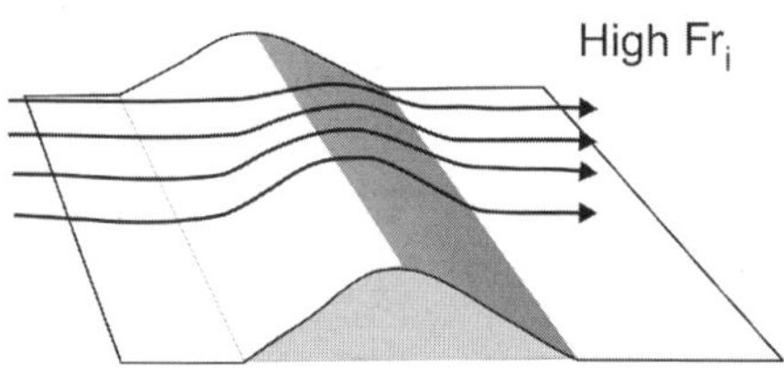

Fig. 10. Effect of internal Froude number (Fr_i) on behaviour of flows upstream of 3D topography; dividing streamline forms at low Fr_i.

Kneller *et al.* (1997) discussed the possibility that internal solitary waves of Type A may be responsible for the creation of wave ripples in deep water – as undoubtedly occurs (Ricci Lucchi, 1981; Fujioka *et al.*, 1989). In lock-exchange experiments, surge-type currents generate only a short train of 3–4 waves; it is not clear whether a sustained current would create a more continuous wave train. It is unlikely that such a short wave train could generate ripples from a flat bed, but they would readily modify the form of pre-existing current ripples. Particle velocities within a solitary wave are given by:

$$U = \frac{Hg'(h+H)^{0.5}}{3h} \tag{14}$$

(Allen, 1982, p. 35) where h is the thickness of the lower fluid layer (in this case the residual current), H is the wave height and g' is reduced gravity $g[\Delta\rho/\rho_a]$. Edwards *et al.* (1994) demonstrated that H shows some dependence upon the slope of the obstruction at which the wave was generated. Conservative values of g' (density differences of a few per cent) and h and H (of the order of a few metres) yield estimates of velocity and bed shear stress capable of moving sand in traction.

A number of examples in the rock record consist of symmetrical ripples associated with what in all other respects look like deep-water sequences (Ricci Lucchi, 1981; Kneller *et al.*,

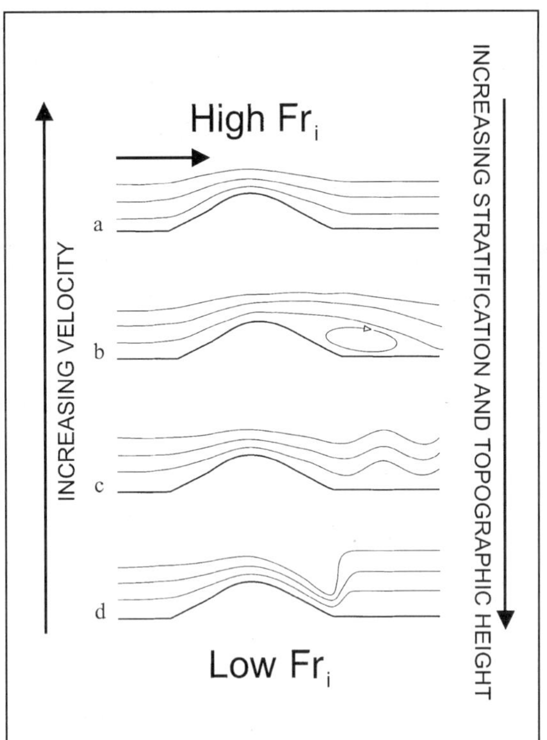

Fig. 11. The effect of internal Froude number (Fr_i) on the behaviour of flows downstream of topography.

1991); at least some of these may have been deposited well below storm wave base. Deep-sea photographs of symmetrical ripples in water depths of thousands of metres (e.g. Fujioka *et al.*, 1989), possibly formed by this mechanism, cast doubt on the reliability of wave ripples as water-depth indicators.

Stratified flow around 3D topography

Where the relative height of the obstacle is small (of the same order as the inner region of the current), the behaviour of the current around three-dimensional topography can probably be treated in the same way as stratified boundary layer flows. In these circumstances, behaviour both upstream and downstream of topography can be related to an internal Froude number (see, e.g. Baines, 1995), generally defined as:

$$\mathrm{Fr_i} = \frac{\overline{U}}{\mathrm{NH}} \qquad (15)$$

where $\overline{U}$ is the current velocity, H is the obstacle height, or some more general length scale (y_s) such as the dimensionless depth (η) over which a known density difference ($\Delta\rho_k$) occurs (Valentine, 1987):

$$y_s = \frac{\Delta\rho_s d}{(\mathrm{d}\rho/\mathrm{d}\eta)} \qquad (16)$$

where $\eta = (y/d)$. N is a parameter known as the Brunt–Väisälä frequency, or simply the buoyancy frequency, which is the frequency of unforced gravity waves on the density stratification, given by:

$$\mathrm{N}^2 = \left(g\frac{\partial\rho/\partial y}{\rho_a}\right) \qquad (17)$$

where g is the acceleration due to gravity, $\partial\rho/\partial y$ is the vertical component of the density gradient, and ρ_a is a reference density (normally taken as the density of the ambient fluid).

Hunt & Snyder (1980) used an internal Froude number criterion to describe strongly stratified atmospheric flow over hills (see also Baines, 1979; Snyder *et al.*, 1985). For small values of Fr_i, a critical plane ('dividing streamline' of Hunt & Snyder, 1980) exists within the current, above which fluid particles have sufficient kinetic energy to move up and over obstructing topography, whereas in the denser, lower regions of the current, fluid particles have insufficient kinetic energy to rise up the topography, and are deflected around it. The height of this dividing plane, H_s, is approximated by:

$$\mathrm{H_s} = \mathrm{H}\,(1 - \theta \mathrm{Fr_i}) \qquad (18)$$

where θ is constant with a value of about 1 (Hunt & Snyder, 1980), although it has been assigned values from 0·7 to 2·0, and may depend on the shape of the obstacle (Baines, 1979; Snyder *et al.*, 1985). For small values of Fr_i, all the fluid below the dividing plane moves around the barrier, while that above this level moves over it in an essentially two-dimensional pattern (Fig. 10). For values of Fr_i greater than $1/\theta$ there exists no dividing plane, and fluid at all levels moves over the obstacle; where there are gaps between obstacles, only the fluid opposite the gap moves through it, i.e. there is effectively no deflection (Baines, 1979).

Downstream effects of topography on gravity current behaviour

The downstream behaviour of a current is also dependent upon the internal Froude number

(Fig. 11), although the threshold values of Fr_i separating these different behaviours varies with the form of the topography in three dimensions. In the simplest case, where the topography is small enough that Fr_i is very large (>>1; H_* likely to be less than ~0·1), simple divergence of streamlines occurs on the downstream side of the topography (Long, 1955). This is likely to be the case over small downstream steps in topography.

At slightly lower Fr_i (Fig. 11b), flow separation occurs downstream of the topography (Lawrence, 1993), in exactly the same fashion as occurs downstream of ripples and dunes in shear flows. The probable velocity and density scales in turbidity currents mean that flow separation is likely in similar situations, i.e. downstream of scours and large bedforms such as dunes or gravel waves, assuming currents are of sufficiently low density that turbulence is not suppressed near the bed. The higher velocities of pyroclastic currents lead to flow separation on a much larger scale beneath blasts and surges (e.g. Lipman & Mullineaux, 1981).

At still lower Fr_i (larger obstacle sizes and density gradients, lower velocities; Fig. 11c), flow over the lee side becomes supercritical (Castro & Snyder, 1993), and lee-waves occur downstream of the obstacle, whose wavelength is a function of the internal Froude number (see also Huppert & Miles, 1969). Flood (1988) suggested a lee-wave origin for fine-grained sediment waves, generally having wavelengths of 500–3000 m and heights of 10–100 m (McCave & Tucholke, 1986; see also Blumsack & Weatherly, 1989), that occur in many slope, base of slope and levee settings. A lee-wave model satisfactorily explains the commonly observed up-current migration of mud waves. Additionally, in some cases these waves can plausibly be related to turbidity current activity in the lee of topography, e.g. mud-waves on the Var sedimentary ridge formed by turbidity current overspill from the adjacent fan valley (cf. Savoye *et al.*, 1993; who suggest an antidune origin), and sediment waves on the right-hand levée of the Hueneme fan channel (Piper *et al.*, 1999b), in both cases apparently fed by hyperpycnal flow. Wherever such waves appear to migrate upslope, it is likely they have formed beneath gravity-driven currents.

Hydraulic jumps

At the lowest values of Fr_i a hydraulic jump occurs downstream of the obstacle (Fig. 11d), where fluid goes through an abrupt downstream deepening; this is effectively a breaking lee wave. For three-dimensional cases, downstream wave breaking is more likely for objects with a larger width-to-length ratio (Castro & Snyder, 1993).

Several authors have described the existence and mechanics of internal hydraulic jumps in stratified or multilayer experimental systems (Long, 1953; Yih & Guha, 1955; Wood & Simpson, 1984), and the effects of such jumps on sediment deposition from both steady (García & Parker, 1989; García, 1993) and highly unsteady experimental turbidity currents (Alexander & Morris, 1994; see below). The possibility of hydraulic jumps in natural systems has been postulated by Middleton (1970), amongst others (see also Menard, 1964; Komar, 1971, 1975; Hand, 1974, 1975), and the analytical model of steady turbidity current flow of Chu *et al.* (1979) suggested the existence of a hydraulic jump between the uniform flow of the slope or channel ($Fr' > 1$) and the nonuniform flow of the basin plain ($Fr' < 1$). Experimental data from García & Parker (1989) and García (1993) suggest that while increased bedload deposition coincides with the jump (i.e. where shear velocity falls), the zone of enhanced deposition of suspended load is displaced slightly downstream of the jump. This may be a function of the lag-time for material to settle through the current (i.e. a function of U_*/U_s), or the decay time for the additional turbulence (and thus high Reynolds stress) associated with the jump. A number of authors invoke hydraulic jumps at slope breaks or channel mouths to explain sediment bypassing (e.g. Mutti & Normark, 1991; Mutti, 1992) possibly associated with scour (e.g. Kenyon *et al.*, 1995; Vicente-Bravo & Robles, 1995; see also Shor *et al.*, 1990). If the buoyancy frequency, N, varies vertically, several different effects may occur in different parts of the current, for example flow separation may occur on the downstream-facing slope of the topography, with a hydraulic jump forming further downstream, immediately beyond the break in slope.

Implications for sedimentation

Where turbidity currents interact with large obstacles and are partially blocked, they will inevitably experience a massive and rapid decrease in both competence and capacity associated with the upstream jump, and sedimentation is likely to occur here. This is supported by the observations of Alexander &

Morris (1994), Kneller (1995) and Kneller & McCaffrey (1995) that show marked localization of deposition associated with the jump, despite the sudden increase in turbulence that would be expected there. In rapidly waning currents, the jump may remain almost stationary where the current strikes the obstacle at a high angle; however, in steadier currents the jump – and associated deposition – might migrate upstream, producing a less localized deposit. Rapidly waning currents may also produce higher energy jumps (Type C or B) since they propagate through the thinning tail of the forward current and so the ratio of jump height to forward current thickness progressively increases.

Deposition may similarly be concentrated in the lee of the obstacle in association with the downstream hydraulic jump, possibly producing an abrupt downstream thickening in the deposit. However, sedimentation from currents that remain subcritical even over the lee side of the obstacle (slower, thicker or denser currents) might occur over the entire lee slope of the obstacle where flow separation occurs. Deposition beneath supercritical flow on the lee side of obstacles may produce antidune stratification, whose preservation in the sediment thus need not imply that the current was supercritical except on the downstream facing slope of the obstacle.

DISCUSSION

The debate over the nature and importance of high-density dispersions dates back at least to the introduction of the term 'fluxoturbidite' (e.g. Carter, 1975), and seems set to continue until experimental studies and theory together can elucidate their mobility and manner of deposition. Distinction between dense and dilute flows on the basis of their deposits alone is a hazardous business; the deposits themselves are likely to reflect only transient conditions in the depositional regime (only the lowermost part of the current above progressively aggrading deposits), and may say nothing directly about the transport regime. At any rate, existing theory seems inadequate to explain the behaviour of some highly mobile dense dispersions, and arguments based solely on the geological interpretation of deposits may be inadequate to resolve issues of process. The link between textural features of the deposit on the one hand and process on the other is somewhat indirect and in many cases poorly understood. Some geological features that may allegedly be used to differentiate between different processes are susceptible to misinterpretation. For example, normal grain-size grading in a turbidite does not necessarily arise in the same way as that produced by static settling of a multisize dispersion in the absence of any effective bed shear stress (thus depending on the vertical structure of the dispersion), but by a temporal decline in bed shear stress at a point beneath a waning (but still moving) current – hence the presence of current-induced structures in many turbidites. Such grading thus depends on the longitudinal structure of the current, which is likely to be dependent on the initiation mechanism and downstream evolution of the flow.

The relative importance of hyperpycnal flow compared with other initiation mechanisms is also unclear, and is likely to vary with the configuration of the shelf and drainage basin. The narrowing of shelves during sea-level lowstands may militate in favour of hyperpycnal flow in general, though Mulder & Syvitski (1996) considered that the changing balance of discharge and suspended sediment load reduced the likelihood of hyperpycnal flow in any given river system. Recognition of ancient deposits produced by hyperpyncal flow is still conjectural. Modelling may help to reduce the uncertainty in the interpretation of deposits, but the compromises involved in both experimental and mathematical approaches mean that the results may describe only limited aspects of the prototype. Experimental studies are also limited by the physical difficulties involved in making the required observations. Many model studies necessarily involve a certain lack of geological realism that must be taken into account when applying the results to natural systems. However, experimental observations on currents can help alert us to factors that cannot generally be neglected, such as the existence of density stratification and variations in its stability. In this context, it is again worthy of note that stable stratification does not imply the absence of turbulence.

The use of bulk (i.e. depth-averaged) parameters, such as overall flow Richardson number or even flow Reynolds number, may be inappropriate under many circumstances. Many currents are better characterized by parameters that capture the vertical density and velocity structure, such as gradient Richardson number and internal Froude number. The vertical velocity and turbulence structure of turbidity currents is fundamental to the manner in which sediment is distributed within the current, and thus affects many aspects of deposition, especially where the presence of topography means that different levels in the current are in contact with the bed. This includes depositional topography such as the levées of submarine channels. The stratification of currents flowing down levéed channels is instrumental in partitioning different grain-sizes of sediment between the channel axis and the levée (e.g. Peakall *et al.*, in press, and references therein). However, Fr_i also determines the degree of overspill on channel bends from the height of the dividing streamline with respect to the levée crest (Fig. 10) and the behaviour of the portion of the current that overtops the levée (Fig. 11).

Despite differences in structure between suspension currents and open channel flows, expressions such as the law-of-the-wall (when suitably modified for stratified flow; Turner, 1973), and the quadratic stress law are probably reasonable approximations when treating processes such as near-bed sediment distribution, bedload transport and erosion, which are dominated by the near-bed region of the current, since the velocity structure of the inner region of a suspension current is similar to that of a shear flow (Kneller *et al.*, 1999).

Technological advances have facilitated increasingly useful models of density current behaviour, both experimentally and numerically. Experiments show that the turbulence structure in density currents is highly complex. Further work is required, particularly with respect to sediment-laden currents and the interaction between the turbulence structure and suspended particles. Turbulence-based numerical models appear to predict realistically the dynamic structure of gravity currents (e.g. Stacey & Bowen, 1988a,b; Eidsvik & Brørs, 1989; Brørs & Eidsvik, 1992; Felix, in press) and hence are more likely to be useful in modelling turbidite deposition. Experimental and numerical models are, however, limited by the lack of testing against environmental data, owing to the large scale of natural turbidity currents and the inherent difficulties in comparing model output with turbidite deposits for which the initial conditions are largely or wholly unknown. The future surely lies in a mix of experimental and theoretical modelling, along with increasingly quantitative field studies. Ultimately, however, there is no substitute for direct observation; full instrumentation of natural currents in the deep sea remains one of the most exciting prospects for the future.

ACKNOWLEDGEMENTS

An article of this nature requires invidious choices as to what to include and what to omit. We apologize to those readers who may feel that we have erred in our judgement, and to those who feel that their work has been overlooked or misrepresented. Many people have shared their ideas with us, but we thank especially Jeff Peakall and Henry Pantin, and also Bill McCaffrey and Maarten Felix, all of whom helped improve the manuscript substantially. We are grateful to Gerry Middleton and an anonymous reviewer for sympathetic, constructive and prompt reviews of an imperfect initial manuscript, and to Jim Best for his patient and thorough editing.

NOMENCLATURE

- C Fractional sediment concentration
- d Flow thickness
- d_h Mean height of the bore waveform
- D_o Nondimensionalized height
- E Fractional energy loss due to friction
- f_b Darcy–Weisbach friction coefficient at the lower boundary
- f_i Darcy–Weisbach friction coefficient at the upper boundary
- Fr Froude number
- Fr′ Densiometric Froude number
- $\mathrm{Fr_i}$ Internal Froude number
- g Acceleration due to gravity
- g' Reduced gravity
- h Thickness of the lower fluid layer
- h_c Critical flow thickness
- h_{max} Maximum run-up height
- H Obstacle height
- $\mathrm{H_*}$ Ratio of obstacle height to current body thickness
- $\mathrm{H_s}$ Height of critical dividing plane
- H Wave height
- k Turbulent kinetic energy
- N Brunt–Väisälä frequency
- Q Volume flux
- Re Flow Reynolds number
- $\mathrm{Ri_g}$ Gradient Richardson number
- S Slope
- u Instantaneous downstream velocity
- u_c Critical flow velocity
- U_y Component of downstream velocity at initial height y
- U_* Shear velocity
- U_s Grain settling velocity

U Particle velocity within a wave
$\overline{U}$ Depth-averaged downstream velocity
U_{max} Maximum mean downstream velocity
U_o Nondimensionalized downstream velocity
U_y Mean downstream velocity at height y
v Instantaneous vertical velocity
$y_{1/2}$ Distance between the bed and the level in the outer region at which the downstream velocity is half the maximum downstream velocity
y_s Length scale in internal Froude number
y Height
θ Constant
$\Delta\rho$ Density difference between the current and the ambient fluid ($\rho_c - \rho_a$)
$\Delta\rho_s$ A known density difference
$\Delta\rho_y$ Density difference between the fluid at height y and the ambient fluid
η Dimensionless depth, y/d
ρ_a Density of the ambient fluid
ρ_c Mean current density
ρ_o Reference density
ρ_y Current density at initial height y
μ Molecular viscosity of water
μ_s Apparent viscosity of a clay-free dispersion

REFERENCES

Alexander, J. and **Morris, S.** (1994) Observations on experimental, non-channelized, high-concentration turbidity currents and variations in deposits around obstacles. *J. Sedim. Res.*, **64**, 899–909.

Allen, J.R.L. (1971) Mixing at turbidity current heads, and its geological implications. *J. Sedim. Petrol.*, **41**, 97–113.

Allen, J.R.L. (1982) *Sedimentary Structures: Their Character and Physical Basis*, Vols I & II. Elsevier. Amsterdam.

Allen, J.R.L. (1985) *Principles of Physical Sedimentology.* George Allen and Unwin, London.

Altinakar, M.S., Graf, W.H. and **Hopfinger, E.J.** (1996) Flow structure in turbidity currents. *J. Hydraul. Res.*, **34**, 713–718.

Apps, G.M., Peel, F.J., Travis, C.J. and **Yielding, C.A.** (1994) Structural controls on Tertiary deep water deposition in the northern Gulf of Mexico. In: *Submarine Fans and Turbidite Systems* (Ed. by P. Weimer, A. H. Bouma and B. F. Perkins), pp. 1–7. SEPM, Gulf Coast Section.

Baines, P.G. (1979) Observations of stratified flow past three–dimensional barriers. *J. Geophys. Res.*, **84**, 7834–7838.

Baines, P.G. (1995). *Topographic Effects in Stratified Flows.* Cambridge University Press Monographs on Mechanics, Cambridge, UK.

Best, J., Peakall, J. and **Kirkbride, A.** (in press) Mean flow and turbulence structure of sediment-laden gravity currents: new insights using ultrasonic Doppler velocity profiling. In: *Sediment Transport and Deposition by Particulate Gravity Currents* (Ed. by B. C. Kneller, W. D. McCaffrey, J. Peakall and T. Druitt), *Spec. Publ. Int. Assoc. Sediment.*

Blanchet, C. and **Villatte, H.** (1954) Experimental studies of density currents in a glass sided flume. *U.N. Econ. Comm. for Asian and Far East, Water Res. Ser., Flood Control Ser.*, U.N. Ser, no. **9**, ST/ECAFE/Ser. F/9.

Blumsack, S.L. and **Weatherly, G.L.** (1989) Observations of the nearby flow and a model for the growth of mudwaves. *Deep Sea Res.*, **36**, 1327–1339.

Bonnecaze, R.T., Hallworth, M.A., Huppert, H.E. and **Lister, J.R.** (1995) Axisymmetric particle-driven gravity currents. *J. Fluid Mech.*, **294**, 93–121.

Bonnecaze, R.T., Huppert, H.E. and **Lister, J.R.** (1993) Particle-driven gravity currents. *J. Fluid Mech.*, **250**, 339–369.

Bouma, A.H., Normark, W.R. and **Barnes, N.E.** (1985) *Submarine Fans and Related Turbidite Systems.* Springer-Verlag. New York.

Bowen, A.J., Normark, W.R. and **Piper, D.J.W.** (1984) Modelling of turbidity currents on Navy submarine fan, California continental borderland. *Sedimentology*, **31**, 169–185.

Branney, M.J. and **Kokelaar, B.P.** (1992) A reappraisal of ignimbrite emplacement: progressive aggradation and changes from particulate to non-particulate flow during emplacement of high-grade ignimbrite. *Bull. Volcanol.*, **54**, 504–520.

Britter, R.E. and **Linden, P.F.** (1980) The motion of the front of a gravity current travelling down an incline. *J. Fluid Mech.*, **91**, 531–543.

Britter, R.E. and **Simpson, J.E.** (1978) Experiments on the dynamics of a gravity current head. *J. Fluid Mech.*, **88**, 223–240.

Brørs, B. and **Eidsvik, K.J.** (1992) Dynamic Reynolds stress modelling of turbidity currents. *J. Geophys. Res.*, **97**, 9645–9652.

Buckee, C.M., Kneller, B.C. and **Peakall, J.** (in press) Turbulence structure in steady, solute-driven gravity currents. In: *Sediment Transport and Deposition by Particulate Gravity Currents* (Ed. by B. C. Kneller, W. D. McCaffrey, J. Peakall and T. Druitt), *Spec. Publ. Int. Assoc. Sediment.*

Burgess, P.M. and **Hovius, N.** (1998) Rates of delta progradation during highstands: consequences for timing of deposition in deep-marine systems. *J. Geol. Soc. Lond.*, **155**, 217–222.

Carey, S., Sigurdsson, H., Mandeville, C. and **Bronto, S.** (1996) Pyroclastic flows and surges over water: an example from the 1883 Krakatau eruption. *Bull. Volcanol.*, **57**, 493–511.

Carlson, P.R., Powell, R.D. and **Phillips, A.C.** (1992) Submarine sedimentary features on a fjord delta front, Queen inlet, Glacier Bay, Alaska. *Can. J. Earth Sci.*, **29**, 565–573.

Carter, R.M. (1975) A discussion and classification of subaqueous mass transport with particular reference to grain flow, slurry-flow and fluxoturbidites. *Earth Sci. Rev.*, **11**, 145–177.

Castro, I.P. and **Snyder, W.H.** (1993) Experiments on wave breaking in stratified flow over obstacles. *J. Fluid Mech.*, **255**, 195–211.

Chikita, K. (1989) A field study on turbidity currents initiated from spring run-offs. *Water Resources Res.*, **25**, 257–271.

Chikita, K. (1990) Sedimentation by river-induced turbidity currents; field measurements and interpretation. *Sedimentology*, **37**, 891–905.

Chu, F.H., Pilkey, W.D. and **Pilkey, O.H.** (1979) An analytical study of turbidity current steady flow. *Mar. Geol.*, **33**, 205–220.

Cochonat, P. and **Piper, D.J.W.** (1995) Source area of sediments contributing to the 'Grand Banks' 1929 turbidity current. In: *Atlas of Deep Water Environments; Architectural Style in Turbidite Systems* (Ed. by K. T. Pickering, R. N. Hiscott, N. H. Kenyon, F. Ricci-Lucchi and R. D. A. Smith), 12–13. Chapman & Hall, London.

Dade, W.B. and **Huppert, H.E.** (1994) Predicting the geometry of channelized deep-sea turbidites. *Geology*, **22**, 645–648.

Dade, W.B. and **Huppert, H.E.** (1995) Modelling suspension-driven gravity surges. *Sedimentology*, **42**, 453–471.

Dade, W.B. and **Huppert, H.E.** (1996) Emplacement of the Taupo ignimbrite by a dilute turbulent flow. *Nature*, **381**, 509–512.

Dade, W.B. and **Huppert, H.E.** (1997) Emplacement of the Taupo ignimbrite – Reply. *Nature*, **385**, 307–308.

Daly, R.A. (1936) Origin of submarine canyons. *Am. J. Sci., 5th Ser.*, **31**, 401–420.

Damuth, J.E., Flood, R.D., Kowsmann, R.O., Belderson, R.H. and **Gorini, M.A.** (1988) Anatomy and growth pattern of Amazon deep-sea fan as revealed by long-range side-scan sonar (GLORIA) and high resolution seismic studies. *AAPG Bull.*, **72**, 885–911.

Davidson, J.F., Harrison, D. and **Carvalho, G.D.** (1977) On the liquid like behaviour of fluidized beds. *Ann. Rev. Fluid Mech.*, **9**, 55–86.

Dolan, J., Beck, C. and **Ogawa, Y.** (1989) Upslope deposition of extremely distal turbidites: an example from the Tiburon Rise, west-central Atlantic. *Geology*, **17**, 990–994.

Druitt, T. (1992) Emplacement of the 18 May, 1980 lateral blast deposit east-northeast of Mount St. Helens, Washington. *Bull. Volcanol.*, **54**, 554–572.

Duringer, P., Paicheler, J.C. and **Schneider, J.L.** (1991) Un courant d'eau continu – peut-il generer des turbidites? Resultats d'experimentations analogiques (Can a continuous water flow generate turbidites? – results from analogue experiments). *Mar. Geol.*, **99**, 231–246.

Edwards, D.A. (1993). *Turbidity Currents: Dynamics, Deposits and Reversals*, Lecture Notes in Earth Sciences, no. 44. Springer-Verlag, Berlin.

Edwards, D.A., Leeder, M.R., Best, J.L. and **Pantin, H.M.** (1994) On experimental reflected density currents and the interpretation of certain turbidites. *Sedimentology*, **41**, 437–461.

Eidsvik, K.J. and **Brørs, B.** (1989) Self-accelerated turbidity current prediction based upon (k–ε) turbulence. *Cont. Shelf Res.*, **9**, 617–627.

Eisma, D. and **Kalf. J.** (1984) Dispersal of Zaire River suspended material in the estuary and the Angola Basin. *Neth J. Sea Res.*, **17**, 385–411.

Ellis, D. (1982) *Palaeohydrodynamics and computer simulation of turbidites in the Marnoso-Arenacea, northern Apennines, Italy*. PhD thesis, University of St Andrews.

Ellison, T.H. and **Turner, J.S.** (1959) Turbulent entrainment in stratified flows. *J. Fluid Mech.*, **6**, 423–448.

Felix, M. (in press) A two-dimensional numerical model for a turbidity current. In: *Sediment Transport and Deposition by Particulate Gravity Currents* (Ed. by B. C. Kneller, W. D. McCaffrey, J. Peakall and T. Druitt), *Spec. Publ. Int. Assoc. Sediment.*

Fietz, T.R. and **Wood, I.R.** (1967) Three dimensional density current. *J. Hydraul. Div., Proc. ASCE*, **93**, 1–23.

Fisher, R.V. (1990) Transport and deposition of a pyroclastic surge across an area of high relief: the 18 May 1980 eruption of Mount St Helens, Washington. *Geol. Soc. Am. Bull.*, **102**, 1038–1054.

Flood, R.D. (1988) A lee wave model for deep-sea mudwave activity. *Deep Sea Res.*, **35**, 973–983.

French, R.H. (1986) *Open-channel Hydraulics*. McGraw-Hill.

Freundt, A. and **Bursik, M.** (1998) Pyroclastic flow transport mechanisms. In: *From Magma to Tephra* (Ed. by A. Freundt and M. Rosi), 173–245. Elsevier, Amsterdam.

Fujioka, K., Watanabe, M. and **Kobayashi, K.** (1989) Deep-sea photographs of the northwestern Pacific ocean – an invitation to deep-sea environment. *Bull. Ocean Res. Inst.*, **27**.

Fukushima, Y., Parker, G. and **Pantin, H.M.** (1985) Prediction of ignitive turbidity currents in Scripps submarine canyon. *Mar. Geol.*, **67**, 55–81.

Furbish, D.J. (1997) *Fluid Physics in Geology*. Oxford University Press. New York.

García, M.H. (1990) *Depositing and eroding sediment-driven flows: turbidity currents*. PhD thesis, University of Minnesota, St. Anthony Falls Hydraulic Laboratory.

García, M.H. (1993) Hydraulic jumps in sediment-driven bottom currents. *J. Hydraul. Eng.*, **119**, 1094–1117.

García, M.H. (1994) Depositional turbidity currents laden with poorly sorted sediment. *J. Hydraul. Eng.*, **120**, 1240–1263.

García, M.H. and **Parker, G.** (1989) Experiments on hydraulic jumps in turbidity currents near a canyon-fan transition. *Science*, **245**, 393–396.

García, M.H. and **Parker, G.** (1993) Experiments on the entrainment of sediment into suspension by a dense bottom current. *J. Geophys. Res.*, **98**, 4793–4807.

García, M.H. and **Parsons, J.D.** (1996) Mixing at the front of gravity currents. *Dyn. Atmos. Oceans*, **24**, 197–205.

García, M.O. and **Hull, D.M.** (1994) Turbidites from giant Hawaiian landslides: results from Ocean Drilling Program site 842. *Geology*, **22**, 159–162.

Gladstone, C., Phillips, J.C. and **Sparks, R.S.J.** (1998) Experiments on bidisperse, constant-volume gravity currents; propagation and sediment deposition. *Sedimentology*, **45**, 833–843.

Grue, J. (1992) Non-linear water waves at a submerged obstacle or bottom current. *J. Fluid Mech.*, **244**, 455–476.

Hallworth, M.A. and **Huppert, H.E.** (1998) Abrupt transitions in high-concentration, particle-driven gravity currents. *Phys. Fluids*, **10**, 1083–1087.

Hallworth, M.A., Huppert, H.E., Phillips, J.C. and **Sparks, R.S.J.** (1996) Entrainment into two-dimensional and axisymmetric turbulent gravity currents. *J. Fluid Mech.*, **308**, 289–311.

Hand, B.M. (1974) Supercritical flow in density currents. *J. Sedim. Petrol.*, **44**, 637–648.

Hand, B.M. (1975) Supercritical flow in density currents: reply. *J. Sedim. Petrol.*, **45**, 750–753.

Hay, A.E. (1987) Turbidity currents and submarine channel formation in Rupert Inlet, British Columbia. 2. The roles of continuous and surge-type flow. *J. Geophys. Res.*, **92**, 2883–2900.

Hay, A.E., Burling, R.W. and **Murray, J.W.** (1982) Remote acoustic detection of a turbidity current surge. *Science*, **217**, 833–835.

Heezen, B.C. and **Ewing, M.** (1952) Turbidity currents and submarine slumps, and the 1929 Grand Banks earthquake. *Am. J. Sci.*, **250**, 849–873.

Heezen, B.C., Menzies, R.J., Schneider, E.D., Ewing, W.M. and **Granelli, N.C.L.** (1964) The Congo Submarine Canyon. *Am. Ass. Petrol. Geol. Bull.*, **48**, 1126–1149.

Hinze, J.O. (1960) On the hydrodynamics of turbidity currents. *Geol. Mijnb.*, **39**, 18–25.

Hiscott, R.N., Pickering, K.T., Bouma, A.H., Hand, B.M.,

Kneller, B.C., Postma, G. and **Soh, W.** (1997) Basin-floor fans in the North Sea: sequence stratigraphic models vs sedimentary facies: discussion. *Am. Ass. Petrol. Geol. Bull.*, **81**, 662–665.

Hodgson, N.A., Farnsworth, J. and **Fraser, A.J.** (1992) Salt-related tectonics, sedimentation and hydrocarbon plays in the Central Graben, North Sea, UKCS. In: *Exploration Britain; Geological Insights for the Next Decade* (Ed. by R. F. P. Hardman), *Geol. Soc. Lond. Spec. Publ.*, **67**, 31–63.

Hopfinger, E.J. and **Tochon-Danguy, J.C.** (1977) A model study of powder snow avalanches. *J. Glaciol.*, **19**, 343–356.

Hughes-Clarke, J.E.H., Shor, A.N., Piper, D.J.W. and **Mayer, L.A.** (1990) Large-scale current-induced erosion and deposition in the path of the 1929 Grand Banks turbidity current. *Sedimentology*, **37**, 613–629.

Hunt, J.C.R. and **Snyder, W.H.** (1980) Experiments on stably and neutrally stratified flow over a model three-dimensional hill. *J. Fluid Mech.*, **96**, 671–704.

Huppert, H.E. (1998) Quantitative modelling of granular suspension flows. *Phil. Trans. R. Soc. Lond. A*, **356**, 2471–2496.

Huppert, H.E. and **Miles, J.W.** (1969) Lee waves in a stratified flow. Part 3. Semi-elliptical obstacle. *J. Fluid Mech.*, **55**, 481–496.

Hürzeler, B.E., Ivey, G.N. and **Imberger, J.** (1995) Spreading model for a turbidity-current with reversing buoyancy from a constant-volume release. *Mar. Freshwater Res.*, **46**, 393–408.

Hürzeler, B.E., Imberger, J. and **Ivey, G.N.** (1996) Dynamics of turbidity current with reversing buoyancy. *J. Hydraul. Eng.*, **122**, 230–236.

Johnson, M.A. (1962) Turbidity currents. *Sci. Prog.*, **50**, 257–273.

Kenyon, N.H., Millington, J., Droz, L. and **Ivanov, M.K.** (1995) Scour holes in a channel-lobe transition zone, Rhone Cone. In: *Atlas of Deep Water Environments; Architectural Style in Turbidite Systems* (Ed. by K. T. Pickering, R. N. Hiscott, N. H. Kenyon, F. Ricci-Lucchi and R. D. A. Smith), 212–215. Chapman & Hall, London.

Keulegan, G.H. (1957) An experimental study of the motion of saline water from locks into fresh water channels. *US. Natl. Bur. Stand., Report* **5168**.

Kirwan, A.D. Jr, Doyle, L.J., Bowles, W.D. and **Brooks, G.R.** (1986) Time-dependent hydrodynamic models of turbidity currents analyzed with data from the Grand Banks and Orleansville events. *J. Sedim. Petrol.*, **56**, 379–386.

Kneller, B.C. (1995) Beyond the turbidite paradigm: physical models for deposition of turbidites and their implications for reservoir prediction. In: *Characterisation of Deep Marine Clastic Systems* (Ed. by A. Hartley and D. J. Prosser), *Geol. Soc. London Spec. Publ.*, **94**, 29–46.

Kneller, B.C. and **Branney, M.J.** (1995) Sustained high density turbidity currents and the deposition of thick massive sands. *Sedimentology*, **42**, 607–616.

Kneller, B.C., Bennett, S.J. and **McCaffrey, W.D.** (1997) Velocity and turbulence structure of gravity currents and internal solitary waves: potential sediment transport and the formation of wave ripples in deep water. *Sediment. Geol.*, **112**, 235–250.

Kneller, B.C., Bennett, S.J. and **McCaffrey, W.D.** (1999) Velocity structure, turbulence and fluid stresses in experimental gravity currents. *J. Geophys. Res., Oceans.*, **104**, 5281–5291.

Kneller, B.C., Edwards, D., McCaffrey, W. and **Moore, R.** (1991) Oblique reflection of turbidity currents. *Geology*, **19**, 250–252.

Kneller, B.C. and **McCaffrey, W.D.** (1995) Modelling the effects of salt-induced topography on deposition from turbidity currents. In: *Salt, Sediment and Hydrocarbons* (Ed. by C. J. Travis, H. Harrison, M. R. Hudeac, B. C. Vendeville, F. J. Peel and R. F. Perkins), 137–145. SEPM Gulf Coast Section, Houston.

Kneller, B.C. and **McCaffrey, W.D.** (1999) Depositional effects of flow non-uniformity and stratification within turbidity currents approaching a bounding slope: deflection, reflection and facies variation. *J. Sedim. Res.*, **69**, 980–991.

Kokelaar, B.P. (1992) Ordovician marine volcanic and sedimentary record of rifting and volcanotectonism: Snowdon, Wales, United Kingdom. *Geol. Soc. Am. Bull.*, **104**, 1433–1455.

Kolla, V. and **Perlmutter, M.A.** (1993) Timing of turbidite sedimentation on the Mississippi fan. *Am. Ass. Petrol. Geol. Bull.*, **77**, 1129–1141.

Komar, P.D. (1971) Hydraulic jumps in turbidity currents. *Geol. Soc. Am. Bull.*, **82**, 1477–1488.

Komar, P.D. (1975) Supercritical flow in density currents: a discussion. *J. Sedim. Petrol.*, **45**, 747–749.

Komar, P.D. (1977) Computer simulation of turbidity current flow and the study of deep-sea channels and fan sedimentation. In: *The Sea; Ideas and Observations on Progress in the Study of the Seas*, Vol. 6, *Marine Modeling* (Ed. by E. D. Goldberg, I. N. McCave, J. J. O'Brien and J. H. Steele), 603–621. John Wiley and Sons, New York.

Kubo, Y., Masuda, F., Tokuhashi, S. and **Sakai, T.** (1998) Spatial variation in paleocurrent velocities estimated from a turbidite bed of the Mio-Pliocene Kiyosumi Formation in Boso Peninsula, Japan. *J. Geol. Soc. Jpn.*, **104**, 359–364.

Kuenen, P.H. (1937) Experiments in connection with Daly's hypothesis on the formation of submarine canyons. *Leidse Geo. Meded.*, **8**, 327–355.

Kuenen, P.H. (1951) Properties of turbidity currents of high density. In: *Turbidity Currents and the Transportation of Coarse Sediments to Deep Water*, (Ed. by J. L. Hough), *SEPM Spec. Publ.*, **2**, 14–33.

Kuenen, P.H. (1952) Estimated size of the Grand Banks turbidity current. *Am. J. Sci.*, **250**, 874–884.

Kuenen, P.H. and **Migliorini, C.I.** (1950) Turbidity currents as a cause of graded bedding. *J. Geol.*, **58**, 91–127.

Lambert, A. and **Giovanoli, F.** (1988) Records of riverborne turbidity currents and indications of slope failures in the Rhone deltas of Lake Geneva. *Limnol. Oceanogr.*, **33**, 458–468.

Lambert, A.M., Kelts, K.R. and **Marshall, N.F.** (1976) Measurements of density underflows from Walensee, Switzerland. *Sedimentology*, **23**, 87–105.

Lane-Serff, G.F., Beal, L.M. and **Hadfield, T.D.** (1995) Gravity current flow over obstacles. *J. Fluid Mech.*, **292**, 9–53.

Launder, B.E. and **Rodi, W.** (1983) The turbulent wall jet – measurements and modelling. *Ann. Rev. Fluid Mech.*, **15**, 429–459.

Laval, A., Cremer, M., Beghin, P. and **Ravenne, C.** (1988) Density surges: two-dimensional experiments. *Sedimentology*, **35**, 73–84.

Lawrence, G.A. (1993) The hydraulics of steady two-layer flow over a fixed obstacle. *J. Fluid Mech.*, **254**, 605–633.

Lipman, P.W. and **Mullineaux, D.R.** (1981) The 1980 eruptions of Mount St. Helens, Washington. *US Geological Survey Professional Paper*, **1250**.

Long, R.R. (1953) Some aspects of the flow of stratified fluids (I). A theoretical investigation. *Tellus*, **5**, 42–57.

Long, R.R. (1955) Some aspects of the flow of stratified fluids (III). Continuous density gradients. *Tellus*, **7**, 341–357.

Long, R.R. (1970) Blocking effects in flow over obstacles. *Tellus*, **22**, 471–480.

Lowe, D.R. (1982) Sediment gravity flows: II. Depositional models with special reference to the deposits of high-density turbidity currents. *J. Sedim. Petrol.*, **52**, 279–297.

Lucchi, R. and **Camerlenghi, A.** (1993) Upslope turbiditic sedimentation on the southeastern flank of the Mediterranean ridge. *Boll. Oceanol. Teorica Ed Applicata*, **11**, 3–25.

Marjanac, T. (1990) Reflected sediment gravity flows and their deposits in the flysch of Middle Dalmatia, Yugoslavia. *Sedimentology*, **37**, 921–930.

Marr, J.D.G., Harff, P.A., Shanmugam, G. and **Parker, G.** (in press) Experiments on subaqueous sandy debris flows: the role of clay and water content in flow dynamics and depositional structures. *Geol. Soc. Am. Bull.*

Masson, D.G. (1994) Late Quaternary turbidity current pathways to the Madeira Abyssal Plain and some constraints on turbidity current mechanisms. *Basin Res.*, **6**, 1–17.

McCave, I.N. and **Jones, K.P.N.** (1988) Deposition of ungraded muds from high-density non-turbulent turbidity currents. *Nature*, **333**, 250–252.

McCave, I.N. and **Tucholke, B.E.** (1986) Deep-current controlled sedimentation in the western North Atlantic. In: *The Geology of North America*, Vol. M, *the Western North Atlantic Region* (Ed. by P. R. Vogt and B. E. Tucholke), 1117–1126. Geological Society of America, Boulder.

McLeod, P., Carey, S. and **Sparks, R.S.J.** (1999) Behaviour of particle-laden flows into the ocean: experimental simulation and geological implications. *Sedimentology*, **46**, 523–537.

Mellor, G.L. and **Yamada, T.** (1982) The development of a turbulence closure model for geophysical fluid problems. *Rev. Geophys. Space Phys.*, **20**, 851–875.

Menard, H.W. (1964) *Marine Geology of the Pacific*. McGraw-Hill, New York.

Middleton, G.V. (1966a) Small-scale models of turbidity currents and the criterion for auto-suspension. *J. Sedim. Petrol.*, **36**, 202–208.

Middleton, G.V. (1966b) Experiments on density and turbidity currents; [Part 1], Motion of the head. *Can. J. Earth Sci.*, **3**, 523–546.

Middleton, G.V. (1966c) Experiments on density and turbidity currents; [Part 2], Uniform flow of density currents. *Can. J. Earth Sci.*, **3**, 627–637.

Middleton, G.V. (1970) Experimental studies related to problems of flysch sedimentation. In: *Flysch Sedimentology in North America* (Ed. by J. Lajoie), *Spec. Pap. Geol. Assoc. Can.*, **7**, 253–272.

Middleton, G.V. (1993) Sediment deposition from turbidity currents. *Ann. Rev. Earth Planet. Sci.*, **21**, 89–114.

Middleton, G.V. and **Hampton, M.A.** (1973) Sediment gravity flows: mechanisms of flow and deposition. In: *Turbidites and Deep Water Sedimentation. Short Course Notes*, 1–38. SEPM (Pacific Section), Los Angeles.

Middleton, G.V. and **Hampton, M.A.** (1976) Subaqueous sediment transport and deposition by sediment gravity flows. In: *Marine Sediment Transport and Environmental Management* (Ed. by D. J. Stanley and D. J. P. Swift), 197–218. Wiley, New York.

Middleton, G.V. and **Southard, J.B.** (1984) *Mechanics of Sediment Movement. SEPM. Short Course*, **3**, 401pp.

Miller, T.P. and **Smith, R.L.** (1977) Spectacular mobility of ash flows around Aniakchak and Fisher calderas, Alaska. *Geology*, **5**, 173–176.

Mohrig, D., Whipple, K.X., Hondzo, M., Ellis, C. and **Parker, G.** (1998) Hydroplaning of subaqueous debris flows. *Geol. Soc. Am. Bull.*, **110**, 387–394.

Morgenstern, N.R. (1967) Submarine slumping and the initiation of turbidity currents. In: *Marine Geotechnique* (Ed. by A. F. Richards), 189–220. University of Ilinois Press, Urbana, Illinois.

Muck, M.T. and **Underwood, M.B.** (1990) Upslope flow of turbidity currents: a comparison among field observations, theory, and laboratory methods. *Geology*, **18**, 54–57.

Mulder, T. and **Syvitski, J.P.M.** (1995) Turbidity currents generated at river mouths during exceptional discharges to the world oceans. *J. Geol.*, **103**, 285–299.

Mulder, T. and **Syvitski, J.P.M.** (1996) Climatic and morphologic relationships of rivers: implications of sea-level fluctuations on river loads. *J. Geol.*, **104**, 509–523.

Mulder, T., Savoye, B. and **Syvitski, J.P.M.** (1997a) Numerical modelling of a mid-sized gravity flow: the 1979 Nice turbidity current (dynamics, processes, sediment budget and seafloor impact). *Sedimentology*, **44**, 305–326.

Mulder, T., Savoye, B. and **Syvitski, J.P.M.** (1997b) Des courants de turbidité hyperpycnaux dans la tête du canyon du Var ? Données hydrologiques et observations de terrain. *Oceanol. Acta*, **20**, 607–626.

Mulder, T., Syvitski, J.P.M. and **Skene, K.I.** (1998) Modeling of erosion and deposition by turbidity currents generated at river mouths. *J. Sedim. Res.*, **A68**, 124–137.

Mutti, E. (1985) Turbidite systems and their relations to depositional sequences. In: *Provenance of Arenites* (Ed. by G. G. Zuffa), 65–93. Reidel Publishing Company, Dortrecht.

Mutti, E. (1992). *Turbidite Sandstones*. Agip, San Donato Milanese, Parma.

Mutti, E. and **Normark, W.R.** (1991) An integrated approach to the study of turbidite systems. In: *Seismic Facies and Sedimentary Processes of Submarine Fans and Turbidite Systems* (Ed. by P. Weimer and M. H. Link), pp. 75–106. Springer-Verlag, New York.

Nemec, W. (1990) Aspects of sediment movement on steep delta slopes. In: *Coarse-Grained Deltas* (Ed. by A. Colella and D. B. Prior), *Spec. Publ. Int. Assoc. Sedimentol.*, **10**, 29–73.

Normark, W.R. (1989) Observed parameters for turbidity-current flow in channels, Reserve Fan, Lake Superior. *J. Sedim. Petrol.*, **59**, 423–431.

Normark, W.R. and **Piper, D.J.W.** (1991) Initiation processes and flow evolution of turbidity currents: implications for the depositional record. In: *From Shoreline to Abyss* (Ed. by R. H. Osborne), *SEPM Spec. Publ.*, **46**, 207–230.

Normark, W.R., Posamentier, H. and **Mutti, E.** (1993) Turbidite systems: state of the art and future directions. *Rev. Geophys.*, **B31**, 91–116.

Pantin, H.M. (1979) Interaction between velocity and effective density in turbidity flow: phase-plane analysis, with criteria for autosuspension. *Mar. Geol.*, **31**, 59–99.

Pantin, H.M. (in press) Experimental evidence for autosuspension. In: *Sediment Transport and Deposition by Particulate Gravity Currents* (Ed. by B. C. Kneller, W. D. McCaffrey, J. Peakall and T. Druitt), *Spec. Publ. Int. Assoc. Sediment.*

Parker, G. (1982) Conditions for the ignition of catastrophically erosive turbidity currents. *Mar. Geol.*, **46**, 307–327.

Parker, G., Fukushima, Y. and **Pantin, H.M.** (1986) Self-accelerating turbidity currents. *J. Fluid Mech.*, **171**, 145–181.

Parker, G., García, M.H., Fukushima, Y. and **Yu, W.** (1987) Experiments on turbidity currents over an erodible bed. *J. Hydraul. Res.*, **25**, 123–147.

Parsons, J.D. (1998) *Mixing mechanisms in density intrusions.* PhD Thesis, University of Illinois at Urbana-Champaign.

Parsons, J.D. and **García, M.H.** (1995) Flow structure and mixing characteristics in saline gravity current fronts. *Hydrosystems Report* **45**. University of Illinois, Urbana-Champaign.

Peakall, J., Ashworth, P. and **Best, J.** (1996) Physical modelling in fluvial geomorphology: principles, applications and unresolved issues. In: *The Scientific Nature of Geomorphology* (Ed. by B. L. Rhoads and C. E. Thorn), 221–253. John Wiley and Sons, Chichester.

Peakall, J., McCaffrey, W.D. and **Kneller, B.C.** (in press) A process model for the evolution, morphology and architecture of sinuous submarine channels. *J. Sed. Res.*

Phillips, A.C. and **Smith, N.D.** (1992) Delta slope processes and turbidity currents in prodeltaic submarine channels, Queen Inlet, Glacier Bay, Alaska. *Can. J. Earth Sci.*, **29**, 93–101.

Pickering, K.T. and **Hiscott, R.N.** (1985) Contained (reflected) turbidity currents from the Middle Ordovician Cloridorme Formation, Quebec, Canada: an alternative to the antidune hypothesis. *Sedimentology*, **32**, 373–394.

Pickering, K.T., Hiscott, R.N. and **Hein, F.J.** (1989) *Deep Marine Environments: Clastic Sedimentation and Tectonics.* London. Unwin Hyman.

Pickering, K.T., Hiscott, R.N., Kenyon, N.H., Ricci-Lucchi, F. and **Smith, R.D.A.** (Eds) (1995) *Atlas of Deep Water Environments. Architectural Style in Turbidite Systems.* Chapman & Hall. London.

Pickering, K.T., Underwood, M.B. and **Taira, A.** (1992) Open-ocean to trench turbidity-current flow in the Nankai Trough: flow collapse and reflection. *Geology*, **20**, 1099–1102.

Piper, D.J.W. and **Aksu, A.E.** (1987) The source and origin of the 1929 Grand Banks turbidity current inferred from sediment budgets. *Geo-Mar. Lett.*, **7**, 177–182.

Piper, D.J.W., Cochonat, P. and **Morrison, M.L.** (1999a) The sequence of events around the epicentre of the 1929 Grand Banks earthquake: initiation of debris flow and turbidity current inferred from sidescan sonar. *Sedimentology*, **46**, 79–97.

Piper, D.J.W., Hiscott, R.N. and **Normark, W.R.** (1999b) Outcrop scale acoustic facies analysis and latest Quaternary development of Hueneme and Dume submarine fans, offshore California. *Sedimentology*, **46**, 47–78.

Piper, D.J.W. and **Savoye, B.** (1993) Processes of late Quaternary turbidity current flow and deposition on the Var fan, north-west Mediterranean Sea. *Sedimentology*, **40**, 557–582.

Piper, J.W., Shor, A.N. and **Clarke, J.E.H.** (1988) The 1929 'Grand Banks' earthquake, slump, and turbidity current. In: *Sedimentologic Consequences of Convulsive Events* (Ed. by H. E. Clifton), *Geol. Soc. Am. Special Pap.*, **229**, 77–92.

Posamentier, H.W., Erskine, R.D. and **Mitchum, R.M. Jr** (1991) Models for submarine-fan deposition within a sequence-stratigraphic framework. In: *Seismic Facies and Sedimentary Processes of Submarine Fans and Turbidite Systems* (Ed. by P. Weimer and M. H. Link), 127–136. Springer-Verlag, New York.

Postma, G., Nemec, W. and **Kleinspehn, K.** (1988) Large floating clasts in turbidites: a mechanism for their emplacement. *Sediment. Geol.*, **58**, 47–61.

Prather, B.E., Booth, J.E., Steffens, G.S. and **Craig, P.A.** (1999) Classification, lithologic calibration and stratigraphic succession of seismic facies from intraslope basins, deep water Gulf of Mexico. *AAPG Bull.*, **83**, 701–727.

Prior, D.B. and **Bornhold, B.D.** (1990) The underwater development of Holocene fan deltas. In: *Coarse-Grained Deltas* (Ed. by A. Colella and D. B. Prior), *Spec. Publ. Int. Assoc. Sedimentol.*, **10**, 75–90.

Reynolds, S. (1987) A recent turbidity current event, Hueneme Fan, California: reconstruction of flow properties. *Sedimentology*, **34**, 129–137.

Ricci Lucchi, F. (1981) The Marmoso-Arenacea: a migrating turbidite basin 'oversupplied' by a highly efficient dispersal system. In: *Excursion Guidebook with Contributions on Sedimentology of Some Italian Basins.* (Ed. by F. Ricci-Lucchi), 231–303. International Association of Sedimentologists 2nd Regional Meeting, Bologna.

Rodi, W. (1980) *Turbulence Models and Their Application in Hydraulics – a State of the Art Review.* A.A. Balkema, Delft, The Netherlands.

Rothwell, R.G., Pearce, T.J. and **Weaver, P.P.E.** (1992) Late Quaternary evolution of the Madeira Abyssal Plain, Canary Basin, NE Atlantic. *Basin Res.*, **4**, 103–132.

Rottman, J.W. and **Simpson, J.E.** (1989) The formation of internal bores in the atmosphere: a laboratory model. *Quart. J. Royal Met. Soc.*, **115**, 941–963.

Rottman, J.W., Simpson, J.E. and **Hunt, J.C.R.** (1985) Unsteady gravity current flows over obstacles: some observations and analysis related to the Phase II trials. *J. Hazardous Materials*, **11**, 325–340.

Rouse, H. (1937) Modern conceptions of the mechanics of turbulence. *Trans. Am. Soc. Civ. Eng.*, **102**, 436–505.

Sanders, J.E. (1965) Primary sedimentary structures formed by turbidity currents and related resedimentation mechanisms. In: *Primary Sedimentary Structures and Their Hydrodynamic Interpretation – a Symposium* (Ed. by G. V. Middleton), *SEPM Spec. Publ.*, **12**, 192–219.

Savoye, B., Piper, D.J.W. and **Droz, L.** (1993) PlioPleistocene evolution of the Var deep-sea Fan off the French Riviera. *Mar. Petrol. Geol.*, **10**, 550–571.

Schowalter, D.G., Van Atta, C.W. and **Lasheras, J.C.** (1994) A study of streamwise vortex structure in a stratified shear layer. *J. Fluid Mech.*, **281**, 247–281.

Shanmugam, G. (1996) High-density turbidity currents: are they sandy debris flows? *J. Sedim. Res.*, **66**, 2–10.

Shanmugam, G., Bloch, R.B., Mitchell, S.M., Beamish, G.W.J., Hodgkinson, R.J., Damuth, J.E., Straume, T., Syvertsen, S.E. and **Shields, K.E.** (1995) Basin-floor fans in the North Sea: sequence stratigraphic models vs. sedimentary facies. *AAPG Bull.*, **79**, 477–512.

Shanmugam, G. and **Moiola, R.J.** (1995) Reinterpretation of depositional processes in a classic flysch sequence (Pennsylvanian Jackfork Group), Ouachita Mountains, Arkansas and Oklahoma. *AAPG Bull.*, **79**, 672–695.

Shanmugam, G., Moiola, R.J. and **Damuth, J.E.** (1985) Eustatic control of submarine fan development. In: *Submarine Fans and Related Turbidite Systems.* (Ed. by A. H. Bouma, W. R. Normark and N. E. Barnes), 23–28. Springer-Verlag, New York.

Shepard, F.P. and **Emery, K.O.** (1973) Congo Submarine Canyon fan valley. *AAPG Bull.*, **57**, 1679–1691.

Shor, A.N., Piper, D.J.W., Clarke, J.E. and **Mayer, L.A.** (1990) Giant flute-like scours and other erosional features formed by the 1929 Grand Banks turbidity current. *Sedimentology*, **37**, 631–645.

Simpson, J.E. (1969) A comparison between laboratory and atmospheric density currents. *Quart. J. Royal Met. Soc.*, **95**, 758–765.

Simpson, J.E. (1972) Effects of the lower boundary on the head of a gravity current. *J. Fluid Mech.*, **53**, 759–768.

Simpson, J.E. (1997) *Gravity Currents in the Environment and the Laboratory*, 2nd edn. Cambridge University Press. New York.

Simpson, J.E. and **Britter, R.E.** (1979) The dynamics of the head of a gravity current advancing over a horizontal surface. *J. Fluid Mech.*, **94**, 477–495.

Slatt, R.M., Weimer, P., Stone, C.G., Lowe, D.R., Coleman, J.L. Jr, Bouma, A.H., DeVries, M.B., Stone, C.G.D.A.A.E., Jordan, D.W., Shanmugam, G. and **Moiola, R.J.** (1997) Reinterpretation of depositional processes in a classic flysch sequence (Pennsylvanian Jackfork Group), Ouachita Mountains, Arkansas and Oklahoma; discussions and reply. *AAPG Bull.*, **81**, 449–491.

Snyder, W.H., Thompson, R.S., Eskridge, R.E., Lawson, R.E., Castro, I.P., Lee, J.T., Hunt, J.C.R. and **Ogawa, Y.** (1985) The structure of strongly stratified flow over hills: dividing-streamline concept. *J. Fluid Mech.*, **152**, 249–288.

Sparks, R.S.J., Bonnecaze, R.T., Huppert, H.E., Lister, J.R., Hallworth, M.A., Mader, H. and **Phillips, J.** (1993) Sediment-laden gravity currents with reversing buoyancy. *Earth Planet. Sci. Lett.*, **114**, 243–257.

Stacey, M.W. and **Bowen, A.J.** (1988a) The vertical structure of density and turbidity currents: theory and observations. *J. Geophys. Res.*, **93**, 3528–3542.

Stacey, M.W. and **Bowen, A.J.** (1988b) The vertical structure of turbidity currents and a necessary condition for self-maintenance. *J. Geophys. Res.*, **93**, 3543–3553.

Stow, D.A.V. and **Wetzel, A.** (1990) Hemiturbidite: a new type of deep-water sediment. *Proc. Ocean Drilling Program, Scientific Results*, **116**, 25–34.

Sturm, M. and **Matter, A.** (1978) Turbidites and varves in Lake Brienz (Switzerland): deposition of clastic detritus by density current. In: *Modern and Ancient Lake Sediments* (Ed. by A. Matter and M. E. Tucker), *Spec. Publ. Int. Assoc. Sediment.*, **2**, 147–168.

Syvitski, J.P.M. and **Schafer, C.T.** (1996) Evidence for an earthquake-triggered basin collapse in Saguenay fjord, Canada. *Sediment. Geol.*, **104**, 127–153.

Tennekes, H. and **Lumley, J.L.** (1972) *A First Course in Fluid Turbulence*. MIT Press, Cambridge, Mass.

Thornburg, T.M., Kulm, L.D. and **Hussong, D.M.** (1990) Submarine-fan development in the southern Chile Trench: a dynamic interplay of tectonics and sedimentation. *Geol. Soc. Am. Bull.*, **102**, 1658–1680.

Tokuhashi, S. (1979) Three-dimensional analysis of a large sandy-flysch body, Mio-Pliocene Kiyosumi Formation, Boso Peninsula, Japan. *Memoirs Fac Science, Kyoto University, Series Geol Mineralogy*, **46**, 1–60.

Tritton, D.J. (1988) *Physical Fluid Dynamics*. Oxford University Press, Oxford.

Turner, J.S. (1973) *Buoyancy Effects in Fluids. Cambridge Monographs on Mechanics and Applied Mathematics.* Cambridge University Press, Cambridge.

Twichell, D.C., Schwab, W.C. and **Kenyon, N.H.** (1995) Geometry of sandy deposits at the distal edge of the Mississippi Fan, Gulf of Mexico. In: *Atlas of Deep Water Environments; Architectural Style in Turbidite Systems* (Ed. by K. T. Pickering, R. N. Hiscott, N. H. Kenyon, F. Ricci-Lucchi and R. D. A. Smith), 282–286. Chapman & Hall, London.

Valentine, G.A. (1987) Stratified flow in pyroclastic surges. *Bull. Volcanol.*, **49**, 616–630.

Valentine, G.A., Wohletz, K.H. and **Kieffer, S.W.** (1992) Effects of topography on facies and compositional zonation in caldera-related ignimbrites. *Geol. Soc. Am. Bull.*, **104**, 154–165.

Vicente-Bravo, J.C. and **Robles, S.** (1995) Large-scale mesotopographic bedforms from the Albian Black Flyshe, northern Spain: characterisation, setting and comparisons with recent analogues. In: *Atlas of Deep Water Environments; Architectural Style in Turbidite Systems* (Ed. by K. T. Pickering, R. N. Hiscott, N. H. Kenyon, F. Ricci-Lucchi and R. D. A. Smith), 282–286. Chapman & Hall, London.

Vrolijk, P.J. and **Southard, J.B.** (1997) Experiments on rapid deposition of sand from high-velocity flows. *Geosci. Canada*, **24**, 45–54.

Weaver, P.P.E., Rothwell, R.G., Ebbing, J., Gunn, D. and **Hunter, P.M.** (1992) Correlation, frequency of emplacement and source directions of megaturbidites on the Madeira abyssal plain. *Mar. Geol.*, **109**, 1–20.

van Weering, T.C.E. and **van Iperen, J.** (1984) Fine-grained sediments of the Zaire deep-sea fan, southern Atlantic Ocean. In: *Fine-Grained Sediments: Deep-Water Processes and Facies* (Ed. by D. A. V. Stow and D. J. W. Piper), *Geol. Soc. London Spec. Publ.*, **15**, 95–113.

Weimer, P. and **Link, M.H.** (1991) Global petroleum occurrences in submarine fans and turbidite systems. In: *Frontiers in Sedim. Geol., Seismic Facies and Sedimentary Processes of Submarine Fans and Turbidite Systems* (Ed. by P. Weimer and M. H. Link), 9–67. Springer Verlag, New York.

Weimer, P., Bouma, A.H. and **Perkins, R.F.** (Eds) (1994). *Submarine Fans and Turbidite Systems.* SEPM, Gulf Coast Section, Houston.

Weirich, F. (1986) The record of density-induced under-flows in a glacial lake. *Sedimentology*, **33**, 261–277.

Wilson, C.J.N. (1985) The Taupo eruption, New Zealand II. The Taupo ignimbrite. *Phil. Trans. Royal Soc. London, Series, A*, **314**, 229–310.

Wilson, C.J.N. (1997) Emplacement of Taupo ignimbrite. *Nature*, **385**, 306–307.

Wilson, C.J.N. and **Walker, G.P.L.** (1985) The Taupo eruption, New Zealand I. General aspects. *Phil. Trans. Royal Soc. London A*, **314**, 199–228.

Wohletz, K.H. (1998) Pyroclastic surges and compressible two-phase flow. In: *From Magma to Tephra* (Ed. by A. Freundt and M. Rosi), 247–312. Elsevier, Amsterdam.

Wood, I.R. and **Simpson, J.E.** (1984) Jumps in layered miscible fluids. *J. Fluid Mech.*, **140**, 329–342.

Woods, A.W., Bursik, M.I. and **Bukatov, A.V.** (1998) The interaction of ash flows with ridges. *Bull. Volcanol.*, **60**, 38–51.

Wright, L.D., Wiseman, W.J., Bornhold, B.D., Prior, D.B., Suhayda, J.N., Keller, G.H., Yang, L.S. and **Fan, Y.B.** (1988) Marine dispersal and deposition of Yellow River silts by gravity-driven underflows. *Nature*, **332**, 629–632.

Wright, L.D., Wiseman, W.J. Jr, Yang, Z.S., Bornhold, B.D., Keller, G.H., Prior, D.B. and **Suhayda, J.N.** (1990) Processes of marine dispersal and deposition of suspended silts off the modern mouth of the Juanghe (Yellow) River. *Cont. Shelf Res.*, **10**, 1–40.

Yih, C.S. and **Guha, C.R.** (1955) Hydraulic jump in a fluid system of two layers. *Tellus*, **7**, 358–366.

Zeng, J. and **Lowe, D.R.** (1997) Numerical simulations of turbidity current flow and sedimentation: I. Theory. *Sedimentology*, **44**, 67–84.

Zeng, J., **Lowe, D.R.**, **Prior, D.B.**, **Wiseman, W.J. Jr** and **Bornhold, B.D.** (1991) Flow properties of turbidity currents in Bute Inlet, British Columbia. *Sedimentology*, **38**, 975–996.

Sedimentology (2000) **47** (Suppl. 1), 95–120

Spatial and temporal distribution of diagenetic alterations in siliciclastic rocks: implications for mass transfer in sedimentary basins

S. MORAD*, J. M. KETZER* and L. F. DE ROS†
**Department of Earth Sciences, Uppsala University, 752 36 Uppsala, Sweden (E-mail: sadoon.morad@geo.uu.se; marcelo.ketzer@geo.uu.se)*
†Universidade Federal do Rio Grande do Sul, Instituto de Geociências, Av. Bento Gonçalves, 9500, CEP 91501–970, Porto Alegre, Brazil (E-mail: lfderos@if.ufrgs.br)

ABSTRACT

The spatial and temporal distribution of diagenetic alterations in siliciclastic sequences is controlled by a complex array of interrelated parameters that prevail during eodiagenesis, mesodiagenesis and telodiagenesis. The spatial distribution of near-surface eogenetic alteration is controlled by depositional facies, climate, detrital composition and relative changes in sea-level. The most important eogenetic alterations in continental sediments include silicate dissolution and the formation of kaolinite, smectite, calcrete and dolocrete. In marine and transitional sediments, eogenetic alterations include the precipitation of carbonate, opal, microquartz, Fe-silicates (glaucony, berthierine and nontronite), sulphides and zeolite. The eogenetic evolution of marine and transitional sediments can probably be developed within a predictable sequence stratigraphic context. Mesodiagenesis is strongly influenced by the induced eogenetic alterations, as well as by temperature, pressure and the composition of basinal brines. The residence time of sedimentary sequences under certain burial conditions is of key importance in determining the timing, extent and patterns of diagenetic modifications induced. The most important mesogenetic alterations include feldspar albitization, illitization and chloritization of smectite and kaolinite, dickitization of kaolinite, chemical compaction as well as quartz and carbonate cementation. Various aspects of deep-burial mesodiagenesis are still poorly understood, such as: (i) whether reactions are accomplished by active fluid flow or by diffusion; (ii) the pattern and extent of mass transfer between mudrocks and sandstones; (iii) the role of hydrocarbon emplacement on sandstone diagenesis; and (iv) the importance and origin of fluids involved in the formation of secondary inter- and intragranular porosity during mesodiagenesis. Uplift and incursion of meteoric waters induce telogenetic alterations that include kaolinitization and carbonate-cement dissolution down to depths of tens to a few hundred metres below the surface.

Keywords Siliciclastic rocks, distribution pattern, sequence stratigraphy

INTRODUCTION

Siliciclastic diagenesis occurs in multicomponent geochemical systems and is driven by temperature-related thermodynamic and kinetic disequilibrium between the minerals and pore waters. Additionally, changes in lithostatic and hydrostatic pressures put important constraints on the degree of mechanical and chemical compaction and, consequently, on the porosity and permeability of the sediments and the extent and pattern of fluid flow. Considerable efforts have been made to understand the spatial and temporal distribution of diagenetic alterations in siliciclastic rocks, mainly due to the realization by oil companies that these alterations exert profound controls on reservoir quality. It is now evident that diagenesis causes a complex deviation from the simple progressive trend of declining porosity and permeability vs. depth. This paper aims to review the state-of-the-art of knowledge on the spatial and temporal distribution of diagenetic alterations and to address key controversies, such as: (i) the extent and patterns of mass transfer between mudrocks and sandstones; (ii) the role of organic acids in diagenetic reactions; and (iii) whether mesogenetic reactions are mediated by advective or diffusive mass transfer.

The boundary between eo- and mesodiagenesis (*sensu* Choquette & Pray, 1970; Schmidt & McDonald, 1979) is not precise in terms of burial depths and temperatures. However, in this review, mesogenetic modifications are considered to occur at depths $\geqslant 2$ km and temperatures $\geqslant 70\,°C$, which coincide with the initiation of chemical compaction, clay-mineral transformation reactions, burial quartz cementation and substantial thermal alteration of organic matter. Mesodiagenesis encompasses modifications that occur over a wide range of burial depths and temperatures and is therefore in this paper tentatively subdivided into shallow (2–3 km; T= 70–100 °C) and deep (> 3 km; $T > 100\,°C$). The burial depths at which mesogenetic reactions occur may vary considerably depending on the burial-thermal history of the sequence. In many basins, burial and mesodiagenesis are succeeded by uplift and telodiagenesis mediated by near-surface waters. It is thus obvious that determining the precise timing and burial depths of diagenetic alterations is often difficult to achieve.

SPATIAL DISTRIBUTION OF EOGENETIC ALTERATIONS IN CONTINENTAL SEDIMENTS

The spatial distribution of diagenetic alterations in continental sediments is strongly influenced by climatic conditions, patterns of regional groundwater flow and depositional facies (Fig. 1). Under warm, subhumid to humid conditions (precipitation > 1000 mm yr^{-1}) and in proximal fluvial sediments, active percolation of undersaturated meteoric waters causes the dissolution of detrital silicates (primarily lithic grains, feldspar and mica) and precipitation of kaolinite. The spatial distribution of kaolinite is influenced by the amounts and distribution of unstable detrital silicates and the hydraulic conductivity of the sand body. Grain dissolution and formation of kaolinite are thus most prevalent in laterally persistent, permeable sediments, such as proximal, fluvial channel sands, but less important in fine-grained sandstones and siltstones of crevasse splay and distal fluviolacustrine facies (Figs 1 and 2).

Conversely, under warm, arid to semiarid conditions (precipitation < 200 mm yr^{-1}) and in distal, fluviolacustrine and interdune facies, meteoric-water flux is strongly limited, leading to the formation of smectitic clays, Fe-oxides/oxyhydroxides, sulphates, K-feldspar, zeolites, calcrete, dolocrete and silcrete. The spatial distribution of these cements is related to mineral–water interaction during regional groundwater flow from proximal to distal settings and to evaporative ionic concentration of pore waters (Fig. 1). Groundwater in permeable, proximal fluvial sandstones adjacent to uplifted blocks has elevated $\alpha_{Ca2+}/\alpha_{Mg2+}$ ratios, and hence commonly forms phreatic calcrete (Garcia *et al.*, 1998). Phreatic, concretionary calcrete is frequently elongated in the direction of regional, groundwater flow (Johnson, 1989; Mozley & Davis, 1996). In some cases, elongated concretions precipitate from groundwater flowing along faults (Mozley & Goodwin, 1995). Groundwaters in the distal deposits have high $\alpha_{Mg2+}/\alpha_{Ca2+}$ ratios due to calcrete precipitation in the proximal facies and to evaporative concentration of dissolved salts favouring the precipitation of dolocrete (Fig. 1; Arakel *et al.*, 1990; Madé *et al.*, 1994). Precipitation of dolomite or calcite close to the water table may occur due to subtle variations in pore water composition due to various degrees of mixing between phreatic and vadose waters (Watts, 1980). Sandstones rich in magnesian grains are also likely to favour the precipitation

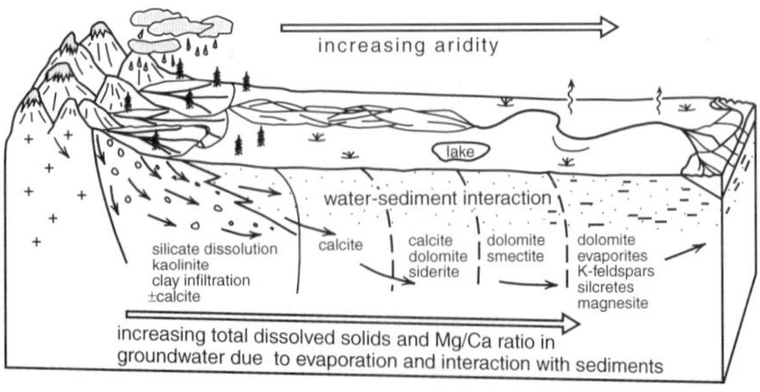

Fig. 1. Block diagram showing the geochemical modifications and flow pattern of regional groundwater as well as the related spatial distribution of eogenetic alterations in proximal-to-distal, continental facies.

of dolocrete (cf. Watts, 1980). In lake-margin environments, phreatic dolocrete forms by the mixing of groundwater with lake brines during periods of strong evaporation and lake-level lowering (Colson & Cojan, 1996). Phreatic calcrete and dolocrete layers may extend over areas of up to 10 × 100 km, such as in Quaternary sediments in Australia (Arakel, 1986), exerting important controls on reservoir compartmentalization and fluid flow. Siderite is a common, yet often minor, diagenetic cement in siliciclastic lacustrine and fluvial overbank sediments (Figs 1 and 2). Eogenetic meteoric siderite has a chemical composition that ranges from near pure, $FeCO_3$ end-member composition (Mozley, 1989) to various extents of enrichment in Mg, Ca and Mn (Morad *et al.*, 1998).

In distal sediments and arid climate, vadose silcrete (composed mainly of opal-A and opal-CT) occurs as concretions and as laterally continuous layers that extend over hundreds of metres, such as in Quaternary and Recent sediments in Australia (van de Graaf, 1983). Groundwater (phreatic) silcrete is composed mainly of quartz overgrowths and micro- to megaquartz and can extend laterally up to several kilometres particularly in distal sediments (Thiry & Milnes, 1991). K-feldspar occurs as overgrowths around detrital feldspars, as pseudomorphic replacement of detrital feldspars and as intergranular crystals in distal fluvial sediments (Fig. 1; Morad *et al.*, 1989). The alteration of distal sediments enriched in felsic volcaniclastic fragments causes a pervasive cementation by microcrystalline K-feldspar, zeolites (clinoptilolite, phillipsite and analcime) and smectite (Sibley, 1978; Surdam & Boles, 1979; De Ros *et al.*, 1994).

SPATIAL DISTRIBUTION OF EOGENETIC ALTERATIONS IN MARINE AND TRANSITIONAL SEDIMENTS

The spatial distribution of eogenetic alterations in marine and transitional sediments is strongly influenced by sea-level changes, depositional facies and the extent of mixing between marine and meteoric waters (Figs 3–5). The climatic conditions, sediment permeability and availability of proper hydraulic head control the magnitude of alterations induced by meteoric waters into transitional and marine sediments below the seafloor. In coastal settings (peritidal and intertidal zones), pore waters are of mixed marine and meteoric origin. Conversely, the diagenesis of sediments submerged on the shelf, slope and basinal plain is mediated overall by diffusive mass transfer in largely stationary marine pore waters. The most common eogenetic alterations in marine and transitional sediments include:

1 Fe-silicates, including berthierine, glaucony and nontronite. Berthierine and glaucony are closely associated with diagenetic phosphate that

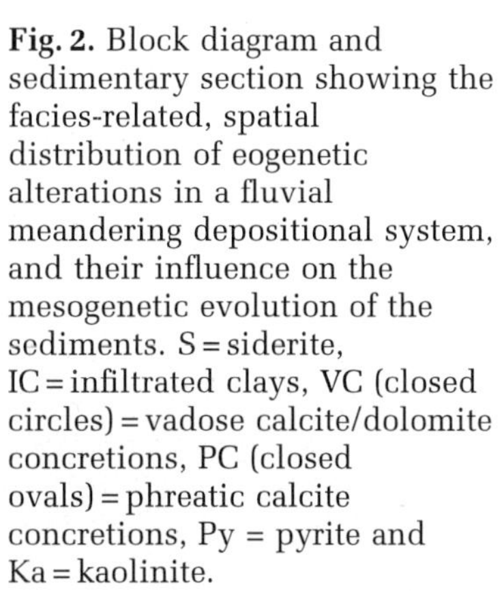

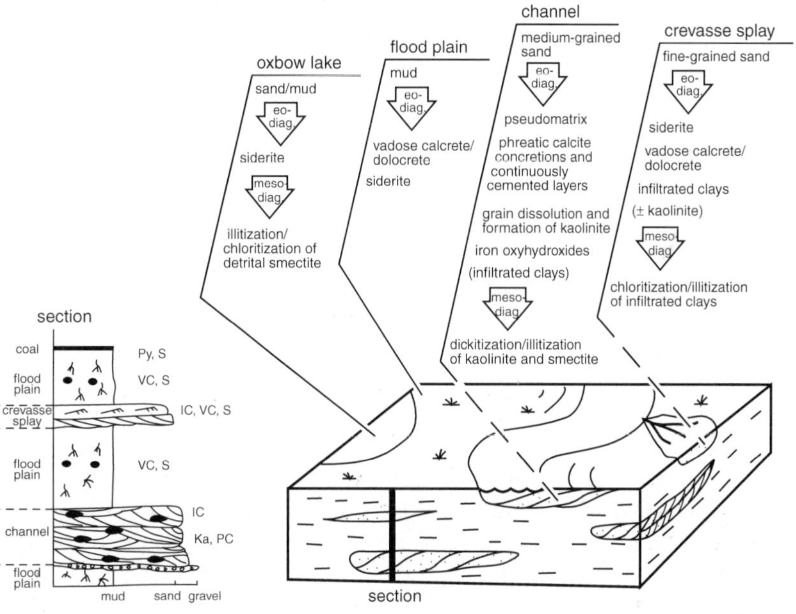

Fig. 2. Block diagram and sedimentary section showing the facies-related, spatial distribution of eogenetic alterations in a fluvial meandering depositional system, and their influence on the mesogenetic evolution of the sediments. S = siderite, IC = infiltrated clays, VC (closed circles) = vadose calcite/dolomite concretions, PC (closed ovals) = phreatic calcite concretions, Py = pyrite and Ka = kaolinite.

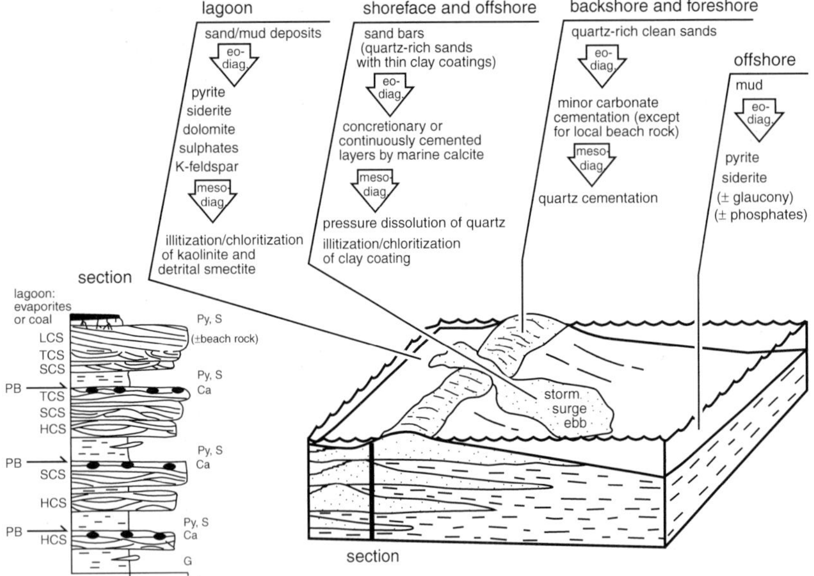

Fig. 3. Block diagram and sedimentary section showing the facies-related, spatial distribution of eogenetic alterations in a wave- and storm-dominated shelf, and their influence on the mesogenetic evolution of the sediments. LCS = low-angle cross-stratification, TCS = trough cross-stratification, SCS = swaley cross-stratification, HCS = hummocky cross-stratification, PB = parasequence boundary, Py = pyrite, S = siderite, Ca (closed ovals) = calcite concretions, and G = glaucony. Calcite cement may also occur in the interbedded, organic-rich mudstones.

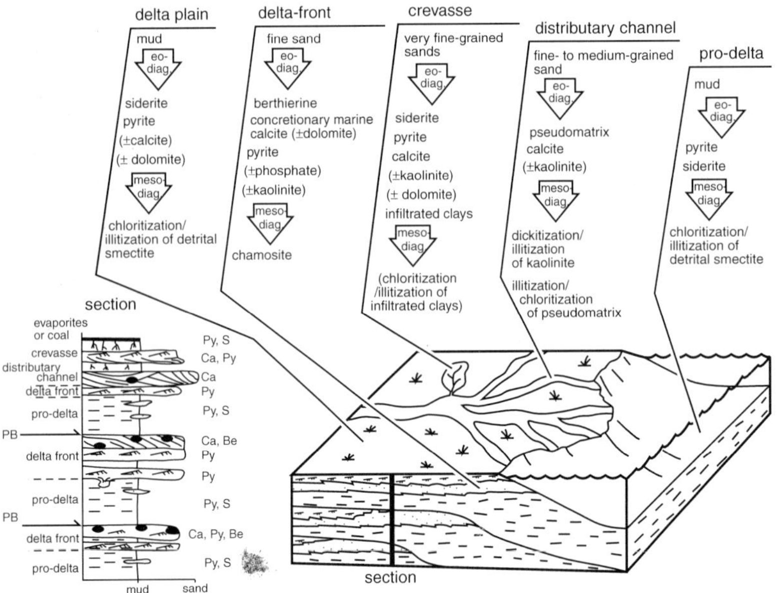

Fig. 4. Block diagram and sedimentary section showing the facies-related distribution of eogenetic alterations in a deltaic depositional system, and their influence on the mesogenetic evolution of the sediments. PB = parasequence boundary, S = siderite, Be = berthierine, Ca (closed ovals) = calcite concretions. Pyrite (Py) occurs throughout most of the sequence.

occurs as concretions in mudrocks and, less commonly, as cement in sandstones. Berthierine (1 : 1 clay mineral; $Fe^{2+}:Fe^{3+} \geqslant 4:1$; e.g. Hornibrook & Longstaffe, 1996) forms in shallow-water, marine and deltaic sediments (Figs 3 and 4), such as in the Niger and Mississippi deltas. Berthierine forms from suboxic, Fe-reducing, marine and brackish pore waters in sediments that are enriched in Fe-oxides/oxyhydroxides and organic matter. Glaucony forms in the outer shelf from suboxic marine pore waters that undergo nitrate reduction immediately below the seafloor (Fig. 3). Low sedimentation rates and slight enrichment in organic matter in this setting prevent a rapid consumption of dissolved oxygen and build up of Fe^{2+} in pore water. In some cases, the formation of glaucony and berthierine is closely linked to the alteration of volcanic-ash layers (Sturesson, 1992), as these layers release considerable amounts of Fe, Mg, K, Si and Al ions upon interaction with seawater. Upwelling of hot

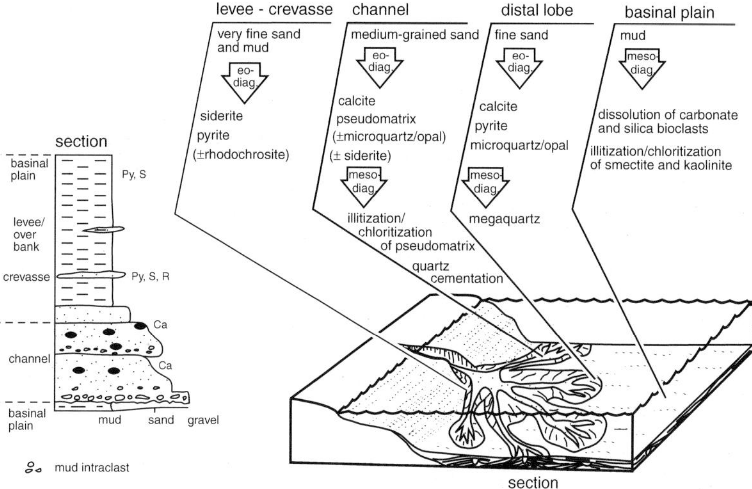

Fig. 5. Block diagram and sedimentary section showing the facies-related, spatial distribution of eogenetic alterations in a marine turbidite depositional system, and their influence on the mesogenetic evolution of the sediments. Py = pyrite, S = siderite, Ca (closed ovals) = calcite/dolomite, R = rhodochrosite.

hydrothermal fluids to the shallow seafloor can lead to the formation of abundant Fe-oxide ooids, such as those found today in the vicinity of volcanic islands in Indonesia (Heikoop *et al.*, 1996). In these areas, a 1-m-thick layer composed of ooidal Fe-oxide formed during the Recent. Upon shallow burial of these ooids in the suboxic zone, the dissolution of Fe-oxyhydroxides may lead to the formation of berthierine (Sturesson *et al.*, 1999). Nontronite [$Fe^{3+}{}_2Si_4O_{10}(OH)_2$], which is closely associated with Mn-oxyhydroxides, forms in oxic pore-waters (dissolved oxygen $\geqslant 2$ mg kg^{-1}) of pelagic, deep-sea sediments that are poor in organic matter. In these deep-sea settings, abundant Fe and Mn are derived from hydrothermal vents near mid-ocean ridges.

2 Sulphides are conspicuous cements in marine and transitional sediments with anoxic, sulphidic pore waters. The amounts of sulphides are greatest in organic matter-rich near-shore sediments and in shelfal settings influenced by upwelling of nutrient-rich marine waters. The sulphides occur as microcrystalline and framboidal pyrite that is disseminated within eogenetic cements and clay matrix and replace micas, rock fragments, bioclasts and intraclasts. The Fe-sulphides in Recent sediments occur as framboidal mackinawite ($FeS_{0\cdot9}$) and greigite (Fe_3S_4). These minerals are metastable and transform rapidly into pyrite during shallow eodiagenesis. The amounts of Fe-sulphides formed are related to diffusibility of sulphate ions from seawater and to the amounts and reactivity of detrital Fe minerals and organic matter (Raiswell, 1982). Slow supply rates of dissolved iron and/or sulphide ions, such as in abyssal-plain sediments, lead to the formation of discrete euhedral pyrite. In some basins, such as the Baltic Sea, microcrystalline alabandite (MnS) and gamma-MnS precipitate in Recent sediments in areas characterized by a smaller degree of input of reactive, detrital Fe- than Mn-oxyhydroxides (Boettcher & Huckriede, 1997).

3 Carbonate cements in marine and transitional sediments are dominated by calcite. Intertidal and lower peritidal sands in tropical and subtropical regions (latitude between ≈30°S and 30°N) are cemented extremely rapidly and extensively by microcrystalline calcite and aragonite leading to the formation of tight beach rocks (Fig. 3). Carbonate precipitation in these coastal settings occurs from seawater due to evaporation and CO_2 degassing (Hanor, 1978). Seawater is moved into beach sands by tidal waves and storm-generated currents.

Carbonate cement in marine and deltaic sandstones occurs as scattered and stratabound concretions at the bottom or top of the bed (Figs 3–5). The coalescence of concretions develops continuously cemented siliciclastic layers that persist over considerable lateral distances ($\geqslant 1\cdot5$ km; Prosser *et al.*, 1993; Klein *et al.*, 1999) which may induce reservoir compartmentalization. According to Bjørkum & Walderhaug (1990), the original distribution of the carbonate bioclasts, which they consider to be the main cement source, and the number of nucleation sites are the primary controls on the spatial distribution and geometry of calcite cement. However, this has not always been found to be the case (e.g. McBride *et al.*, 1995). Concretions are spheroidal, ovoidal or flattened parallel to the bedding planes. Compared with spheroidal concretions, flattened concretions may form due to: (i) permeability anisotropy, with maximum growth occurring parallel to the bedding plane; (ii) growth in the direction of fluid flow (McBride *et al.*, 1994); (iii) greatest flux of ions, due to greater concentration of bioclasts along the bedding plane (Walderhaug & Bjørkum, 1998); or (iv) concretion growth during compaction of the layer (Raiswell, 1971).

Calcite cementation in shallow-marine and deltaic sediments (Figs 3 and 4) occurs by marine and mixed marine–meteoric waters (Simpson & Hutcheon, 1995). Cementation in subtidal and deeper-water environments occurs largely by marine pore water that is slightly modified in composition by increased carbonate alkalinity due to suboxic to anoxic degradation of organic matter. Carbon and calcium ions can also be transferred directly from the seawater by diffusion subsequent to the establishment of chemical potential gradients with low ionic concentrations in pore waters due to calcite precipitation. In addition to seawater, other sources of carbonate cement include the dissolution of biogenic carbonates, micritic carbonate intraclasts and volcanic material (Morad & De Ros, 1994). Bioclastic carbonates, and hence calcite cement, are common in wave- and storm-dominated sands (Fig. 3), turbiditic sediments (Fig. 5) and in tidal channels and point bars. The time required to develop a continuously cemented sandstone layer decreases with an increase in: (i) the number of nucleation sites available for concretion growth and (ii) the ratio of percentage of cement in concretion to percentage of cement source (Wilkinson, 1989). The number of nucleation sites depends, in turn, on the number of sites where selective supersaturation (e.g. bioturbation) occurs and/or the distribution pattern of substrates (i.e. carbonate grains) for calcite precipitation and concretion growth.

Other eogenetic carbonate cements include dolomite, siderite, rhodochrosite and magnesite, which are most common in shelfal and deep-sea muds (Figs 3–5). The precipitation of eogenetic dolomite in marine sediments occurs by mixing between seawater and freshwater, and in anoxic pore waters in which the dissolved sulphate ions were removed by bacterial reduction to sulphides (Baker & Kastner, 1981); this is often referred to as organogenic dolomite, and forms in the bacterial sulphate-reduction and methanogenesis zones. Microcrystalline siderite cement precipitates as concretions in mudrocks and sandstones. Siderite in shallow-water marine sediments is enriched in Ca and Mg (Mozley, 1989) whereas in deep-sea sediments it is enriched in Mn (Chow *et al.*, 1996), due to the derivation of Mn-oxyhydroxide particles from the discharge sites of hydrothermal waters in the vicinity of mid-ocean ridges. Enrichment of deep-sea sediments with Mn-oxyhydroxide also accounts for the formation of authigenic rhodochrosite (Chow *et al.*, 1996). Magnesite that occurs in deep-sea sediments is enriched in Fe and Mn, and forms either by direct precipitation or through the transformation of landfordite (hydrous Mg-carbonate) during shallow-burial and slight increase in temperature ($T \approx 40\,°C$; Matsumoto, 1992).

In some cases, the localized precipitation of marine carbonate cements with extremely low $\delta^{13}C$ values at or immediately below the seafloor is related to an increase in carbonate alkalinity due to the oxidation of methane at sites of gas seepage (Hovland *et al.*, 1987; von Rad *et al.*, 1996). In deep-sea sediments, the dissociation of methane hydrates (clathrates) causes a strong enrichment of authigenic carbonates (e.g. siderite) in ^{12}C and ^{18}O isotopes (Matsumoto, 1990).

4 Precipitation of zeolite, K-feldspar, albite, smectite (commonly Mg- and Fe-rich), opal-A, opal-CT, chalcedony and microquartz as rims in marine sediments that are strongly enriched in volcanic fragments and biogenic silica, such as in deep-water, marine turbidites of arc basins. The dissolution of opaline skeletons of radiolaria, diatoms and sponge spicules results in the formation of grain-coating microquartz (Carvalho *et al.*, 1995; Hendry & Trewin, 1995). Microquartz coatings inhibit the nucleation and growth of syntaxial quartz overgrowths, and hence influence the spatial distribution of porosity and permeability in deep reservoirs (Aase *et al.*, 1996). Eogenetic alteration of volcanic fragments to zeolite induces a rapid loss of permeability. The zeolites are typically phillipsite and clinoptilolite, being formed as a result of a high α_{H4SiO4} in the pore waters at temperatures $\lesssim 35\,°C$ (Boles, 1977).

5 In coastal sabkha settings, siliciclastic sediments are cemented by anhydrite, gypsum, barite, calcite, dolomite and, in a few cases, halite. However, the amount of halite cement in the sedimentary record is probably underestimated due to its dissolution during thin-section preparation. Other cements in coastal sabkha settings include magnesite and quartzine (length-slow fibrous quartz); the latter forms by the replacement of sulphate cements.

RELATIONSHIPS BETWEEN THE SPATIAL DISTRIBUTION OF EOGENETIC ALTERATIONS AND SEQUENCE STRATIGRAPHY

Relative changes in sea-level and rates of sediment supply are the main factors controlling transgression and regression events, which can be predicted within the context of sequence stratigraphy. These events control the position and dynamics of the meteoric, mixed marine/meteoric and marine pore-water zones. Changes in pore-water composition driven by transgression and regression promote various diagenetic reactions. The intensity of these reactions in a specific pore-water zone depends on the duration of cycles of relative sea-level changes. Typically, diagenetic reactions are related to third- and fourth-order sea-level changes (McKay *et al.*, 1995; Taylor *et al.*, 1995; Loomis & Crossey, 1996; Rossi & Cañaveras, 1999), i.e. between 10 Myr and 200 kyr (Vail *et al.*, 1977). Nevertheless, fifth-order cycles are long enough to create recognizable diagenetic alterations, such as cementation of carbonate platforms (Read & Horbury, 1993) and of beach and probably offshore sediments.

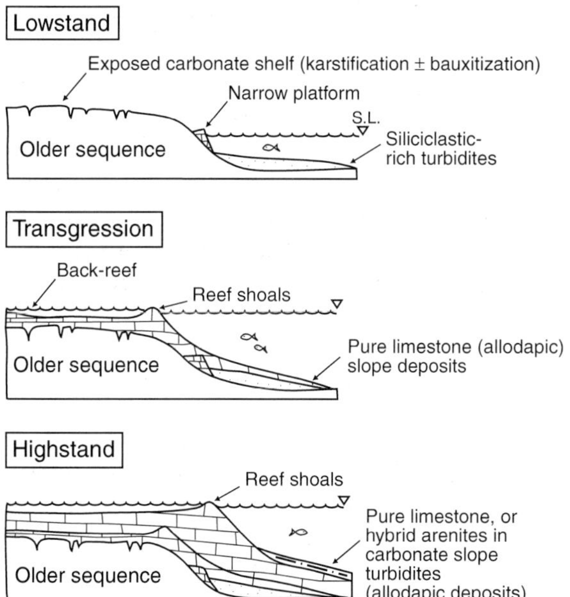

Fig. 6. Schematic relationships between relative sea-level changes and the composition of sediments in deep-water turbidites. Note that transgressive and highstand, deep-water deposits are enriched in carbonate intraclasts (i.e. hybrid arenites), and thus have a greater potential to be tightly cemented by calcite than lowstand, siliciclastic turbidites.

Transgression and regression also influence the detrital composition of sandstones in terms of the proportion and spatial distribution of intra- and extrabasinal grains (Zuffa *et al.*, 1995). Noncarbonate intrabasinal grains are volumetrically important only during times of reduced supply of extrabasinal sediments, such as during transgression. Otherwise, these intrabasinal grains would be diluted by extrabasinal sediments (Garzanti, 1991; MacQuaker *et al.*, 1996). During transgression and highstand times, sea water covers the shelf and carbonate production is at a maximum. Re-sedimentation of shallow-water carbonate sediments into deep-water sites immediately after its deposition on the shelf (allodapic, *sensu* Dolan, 1989) produces pure limestones or hybrid arenites rich in carbonate fragments in transgressive and highstand system tracts (TST and HST; Fig. 6; Dolan, 1989; Fontana *et al.*, 1989). Therefore, such TST and HST deposits have a great potential to be cemented by carbonates. Intraclasts are usually ductile, and can thus cause a rapid destruction of porosity and permeability of hybrid sandstones upon mechanical compaction. Conversely, during exposure of the shelf due to relative sea-level fall or progradation during late HST, carbonate production is at a minimum, chemical weathering dominates over erosion and thus sediments reaching basinal settings are dominantly siliciclastic. These conditions result in cementation and karstification of limestones and possible bauxite formation in the shelf (Bardossy & Combes, 1999). Consequently, there will be little or no intrabasinal carbonate sediment in lowstand systems tract (LST) deposits (Fig. 6), and hence lower potential for carbonate cementation and/or for rapid porosity–permeability loss as a result of mechanical compaction.

Regression and sequence boundaries

Regression is produced by relative sea-level fall (forced regressions; Posamentier *et al.*, 1992) and/or by excessive sediment supply (normal regression). In both cases, the meteoric and mixed marine–meteoric zones migrate basinward. The eogenetic alterations due to forced and normal regressions are similar, but those induced due to forced regressions tend to be localized below the

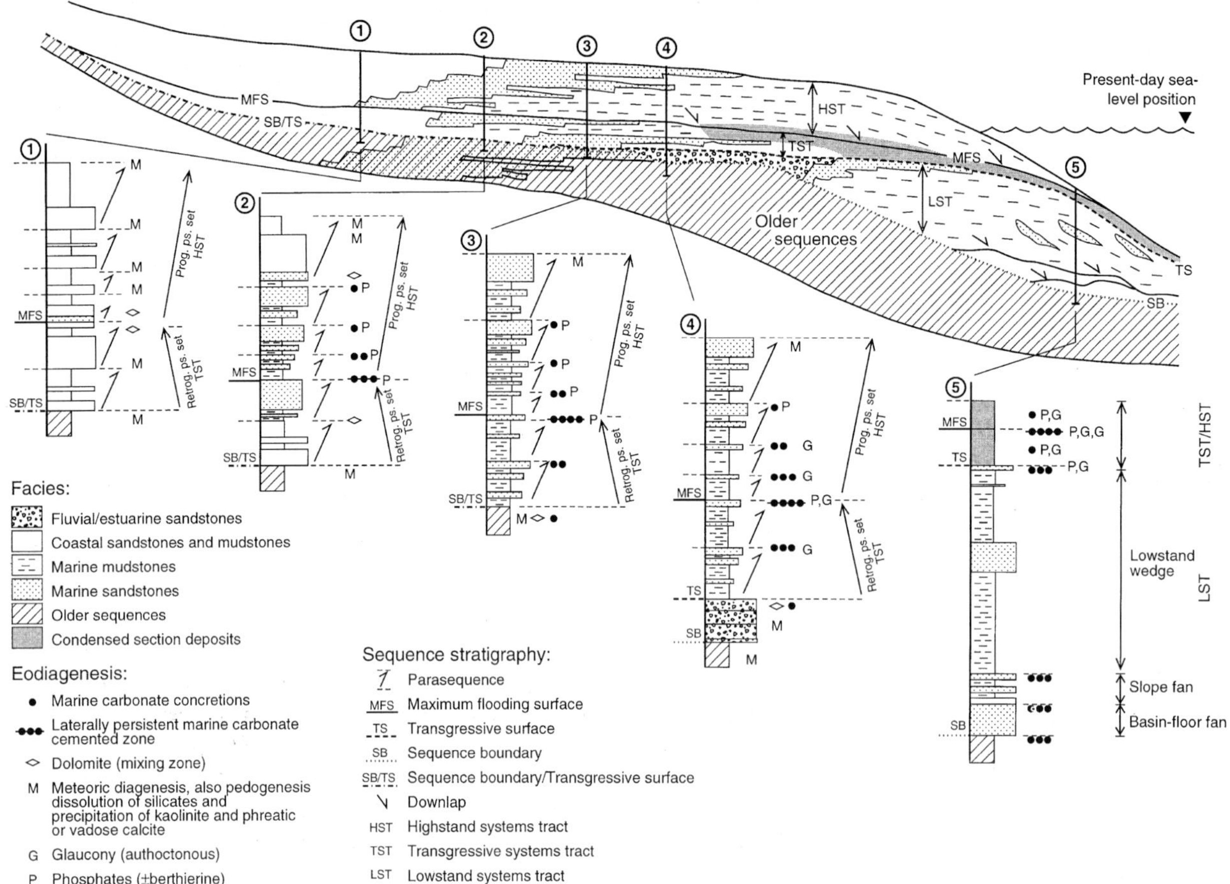

Fig. 7. Examples of the spatial distribution of eogenetic alterations that occur along key sequence stratigraphic surfaces and system tracts. Sequence stratigraphy framework modified from van Wagoner *et al.* (1990).

subaerial exposure surfaces, i.e. unconformities (Fig. 7; van Wagoner *et al.*, 1990). Conversely, eogenetic alterations during normal regression are localized in zones of high sediment supply or relatively low sediment accommodation space, such as during late stages of the HST.

During forced regression and lowstand times, large areas are subaerially exposed on the shelf, leading to an enlargement of the areas of meteoric recharge. The basinward migration of the meteoric zone promotes the flushing of shallow-marine sediments and, in some cases, even deep-water turbidites by meteoric waters (Hayes & Boles, 1992; Carvalho *et al.*, 1995). Meteoric-water flushing during regression results in the dissolution of framework silicates and early marine and mixing-zone cements and in the precipitation of kaolinite and meteoric carbonate cements (Fig. 8; Rossi & Cañaveras, 1999). Carbonate cements that precipitated during regression display a decrease in the Sr, Na and Mg content, as well as lower $\delta^{18}O$ and higher $^{87}Sr/^{86}Sr$ values towards the centre of the pores, indicating progressive geochemical modification of pore-waters towards a meteoric composition (Kaldi & Gidman, 1982; Glasmann *et al.*, 1989; Hart *et al.*, 1992; Morad *et al.*, 1992). Alterations resulting from the incursion of meteoric water are expected to be progressively more intensive towards HST due to the occurrence of a distinct progradation. In the interfluve areas along sequence boundaries or subaerially exposed areas during normal regression, pedogenesis may result in the formation of clay cutans and precipitation of calcretes, dolocretes and silcretes.

Transgression, parasequence boundaries and condensed sections

A relative sea-level rise and increase in accommodation space promotes transgression and, consequently, marine flooding of subaerially exposed sediments with landward shifting of the mixed and marine pore-water zones, and restriction of meteoric-water recharge areas.

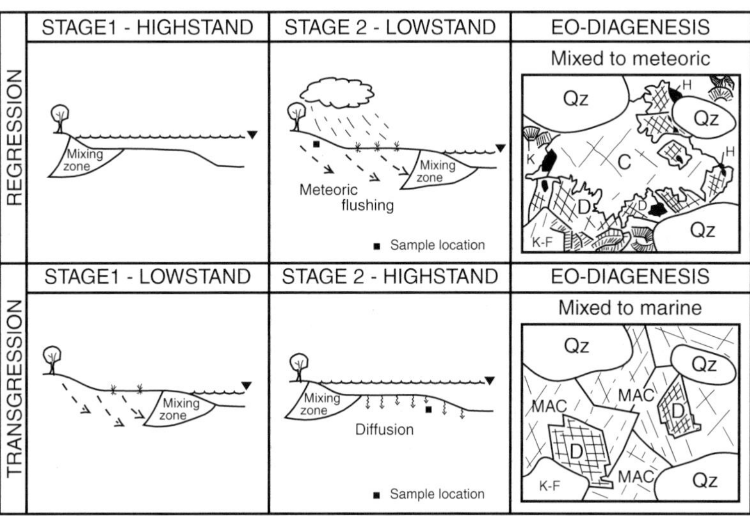

Fig. 8. Diagenetic alterations in siliciclastic sediments during regression and transgression. During regression, earlier formed mixing-zone dolomite cement (D) is dissolved or calcitized through interaction with meteoric water; this process is associated with the formation of meteoric calcite cement (C), iron oxides/oxyhydroxides (H) and kaolinite (K). During transgression, earlier formed mixing-zone dolomite is post-dated by marine calcite cement (MAC) formed by ionic diffusion from the overlying seawater.

When the transgressive surface coincides with the sequence boundary, marine water will infiltrate below flooding surfaces, mixing with meteoric waters and promoting the precipitation of dolomite (Yin & Surdam, 1985; Taylor *et al.*, 1995; Dixon & Davis, 1997). Dolomitization or calcite cementation is also likely to occur close to the MFS in continental sediments when the fluvial bars on top of the TST are reworked by coastal processes (e.g. tides) and in the TS at top of the incised valley deposits (Fig. 7; Taylor *et al.*, 1995).

In the marine pore-water zone, typical diagenetic processes include the precipitation of carbonate concretions at shallow depths below the seafloor (Fig. 8). Ions needed for this cementation are derived by diffusion from the overlying sea water and from the dissolution of aragonitic and high-Mg calcite grains. As ionic diffusion is most efficient over distances $\leqslant 10$ m, marine carbonate cementation is expected to be most common during a long 'residence time' at shallow depths below the seafloor (Kantorowicz *et al.*, 1987; Wilkinson, 1989). Hence, cementation would occur along surfaces of low sedimentation rates, such as parasequence boundaries or MFS and within condensed sections (Fig. 7; Taylor *et al.*, 1995; Loomis & Crossey, 1996). As the 'residence time' is longest along the MFS, it has the highest potential to develop laterally persistent cemented horizons by coalescence of concretions. Considering deep-water turbidites, a rise in relative sea-level induces modifications in the architecture of turbidites, whereby basin-floor fan deposits evolve into slope-fan deposits. This facies transition, which is possibly marked by a down-lap surface, represents a major break in sedimentation, and is thus a potential site for marine carbonate cementation, particularly in sediments enriched in carbonate grains.

Stratigraphic levels corresponding to low sedimentation rates in marine environments also favour the formation of glaucony and phosphates (Fig. 7), which thus occur along parasequence boundaries, TST and lower HST deposits and within condensed section deposits (Loutit *et al.*, 1988; Amorosi, 1995). Glaucony displays an upward increase (from TST) and then decrease (towards HST) in both abundance and maturity (Fig. 7; Amorosi, 1995). The maturation of glaucony, i.e. an increase in K_2O (derived from seawater) and development of a mica-like structure, is thus enhanced by long residence time at shallow depths below the seafloor. Glaucony and phosphate-rich sediments may also occur without a clear association with transgressions. In such sediments, glaucony is typically allochthonous, derived by erosion and reworking of authochthonous glaucony (Baum & Vail, 1988; Amorosi, 1995). Additionally, the formation of glaucony may also be closely linked to the alteration of volcanic-ash layers (Sturesson, 1992). Compared with glaucony, berthierine forms during somewhat higher sedimentation rates in sediments with greater organic-matter content, and hence closer to the strandline. Berthierine and siderite commonly form ooids in ironstones, which mark intervals of low sedimentation rates (e.g. parasequence boundaries and MFS) and condensed sections (Burkhalter, 1995; MacQuaker *et al.*, 1996). The frequency and abundance of glaucony and berthierine are higher in ancient than in Recent sediments, probably due to the relatively low present-day sea-level. A transgression and rise in sea-level induce conditions of upwelling of nutrient-rich seawater into the shelf and formation of organic-rich sediments.

SPATIAL AND TEMPORAL DISTRIBUTION OF MESOGENETIC ALTERATIONS

Subsequent to the main phase of mechanical compaction and the initiation of chemical compaction (burial depths $\gtrsim 2$ km; $T \gtrsim 70$ °C), temperature and basinal brines increasingly control not only the spatial but also the temporal distribution of diagenetic alterations. Basinal brines evolve chemically and isotopically during progressive burial. In addition to the continued, but less important, brittle grain fracturing and plastic deformation of ductile grains, a wide variety of chemical, physical and biochemical mesogenetic modifications are induced in the sediments, including cementation, mineral transformation, dissolution of grains and cements, chemical compaction and overpressuring (Fig. 9). Mineral transformation affects mainly clay minerals and detrital feldspars due to their sensitivity to changes in temperature and pore-water composition. Additionally, the spatial and temporal distribution of mesogenetic alterations is strongly influenced by: (i) the extent and distribution of eogenetic alterations (Figs 1–5); (ii) the residence time at certain *P–T* conditions, indicating that mesogenetic reactions are kinetically controlled; and (iii) alterations in mudrocks, such as thermal maturation of organic matter and formation of organic acids, the dissolution of K-feldspar, biogenic silica and carbonates, as well as the illitization and chloritization of smectite and kaolinite.

Flow patterns and geochemical evolution of formation waters

A subject of long-standing debate and controversy is whether fluids are stagnant, moving sluggishly or actively (by advection and convection) in the deep subsurface. Thus, whether the spatial and temporal distribution of mesogenetic reactions are mediated by advective fluid flow or by

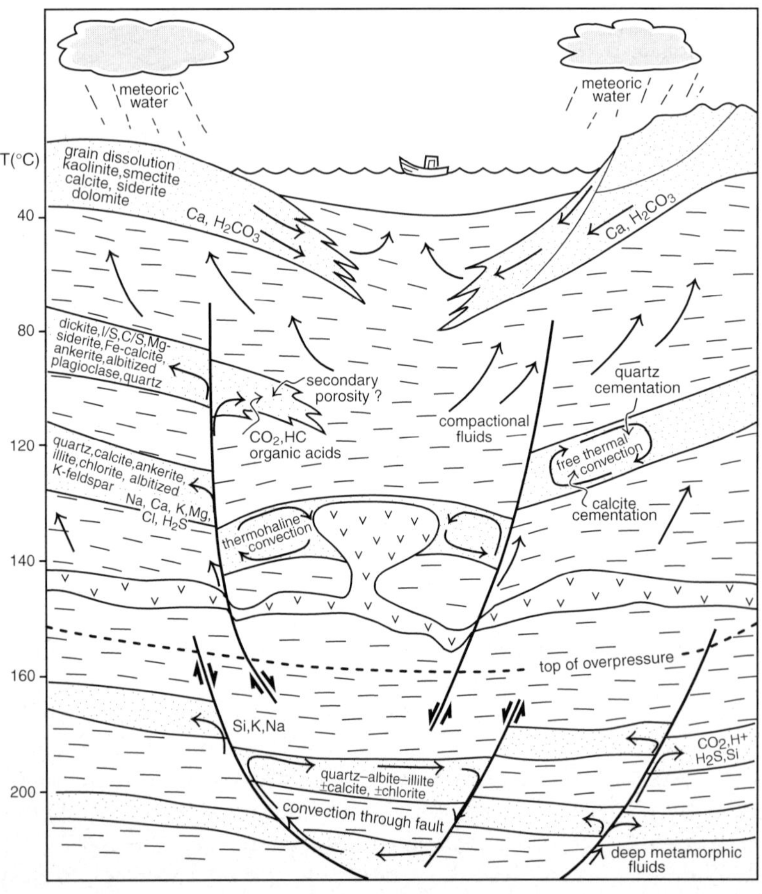

Fig. 9. Spatial and temporal distribution of diagenetic alterations and patterns of fluid flow and mass transfer in a hypothetical basin.

diffusion is uncertain, and depends, among other things, on the burial diagenetic evolution and tectonic setting of the basin. In basins that are tectonically inactive, overpressured or compartmentalized by low-permeability rocks or structural barriers, fluids become stagnant for prolonged times. However, the episodic release of overpressure from mudrocks causes pulses of focused fluid flow along major faults (Fig. 9; Cartwright, 1994). Owing to the low compressibility of water, flow is probably enhanced by simultaneous vertical and lateral stresses. Concrete geochemical and palaeogeothermal evidence for the occurrence of vertical and lateral mass and heat flux along a large, fairly shallow (≈ 2 km) growth fault has been provided by Losh *et al.* (1999). Slow (less than a few mm yr^{-1}) fluid flow occurs by compactional expulsion of waters from mudrocks and by convection. Convection is driven by buoyancy forces associated with temperature- and salinity gradients (Hanor, 1987; Evans & Nunn, 1989). However, so far, there are no unequivocal data supporting the role of fluid convection within a sandstone unit on the spatial and temporal distribution of diagenetic minerals in sedimentary basins. Nevertheless, fluid convection may occur along fault planes, particularly when the basin experiences high heat flux, such as resulting from extensional tectonics. In the absence of fluid convection, mass transfer from mudrocks into sandstones occurs along faults and by cross-formational ionic diffusion. The latter process suggests that the spatial distribution of mesogenetic alterations in sandstones occurs over distances of metres or tens of metres from the mudrocks. At burial depths >3 km, an important factor that favours diffusive mass transfer is the limited amount of pore water available compared with the volume needed to explain common mesogenetic alterations. The spatial distribution of diagenetic reactions is also strongly controlled by the degree of hydrocarbon saturation in the reservoir and by the concentration of dissolved and exsolved gases, such as CO_2. A high degree of hydrocarbon supersaturation impedes fluid flow in reservoirs. In such reservoirs, diagenetic reactions may slow and proceed mainly by diffusion (Worden *et al.*, 1998).

Shallow mesogenetic alterations

The most typical mesogenetic alterations that occur at burial depths of 2–3 km and temperatures of 70–100 °C include (Fig. 9):

1 Precipitation of carbonate cements, including calcite, ankerite, magnesian siderite and ferroan magnesite. In sandstones in which pore-water pH is buffered by aluminosilicate equilibrium or by organic acids, excess CO_2 derived from thermal oxidation of organic matter is eliminated by carbonate precipitation (Smith & Ehrenberg, 1989) in the vicinity of associated organic-rich mudrocks. Increased carbonate alkalinity and consequent carbonate cementation may occur due to the anaerobic or, less commonly, aerobic degradation of oil by deeply infiltrating meteoric waters (Smith, 1996). Hence, the spatial distribution of carbonate cements is closely related to the position of the oil–water interface. Important sources of carbonate ions include associated mudrocks and marls (Souza & Silva, 1998), mainly through the dissolution of biogenic and micritic calcite. Calcium supply from, and hence the intimate spatial distribution of carbonate cement with, these lithologies occurs simultaneously with smectite illitization (Wintsch & Kvale, 1994) and the generation of organic acids. The latter process lowers the pH values of pore waters, and thus enhances biogenic-carbonate dissolution in marls and mudrocks.

2 The precipitation of small amounts of dickite as well as increasing crystal-structural ordering and incipient to pervasive dickitization of kaolinite. Dickitization is a dissolution–precipitation process that occurs in pore waters with low α_{K+}/α_{H+} ratio; otherwise illite forms (Ehrenberg *et al.*, 1993; Morad *et al.*, 1994). Low α_{K+}/α_{H+} ratio in pore fluids can be achieved through input of acids (Morad *et al.*, 1994) and through absence or retardation of K^+ release from minerals (Ehrenberg, 1991). Acidic fluids in the temperature ranges considered can be derived from the thermal maturation of kerogen. Hence, dickitization is expected to be most pervasive in sandstones that are poor in detrital K-feldspars (e.g. in aeolian facies or due to eogenetic dissolution) and/or in the vicinity of sites of influx of acidic waters such as along faults or adjacent to organic-rich mudrocks.

3 Progressive, burial transformation of detrital and eogenetic smectite, first into poorly, and then at greater depths into well-ordered, mixed-layer illite/smectite (I/S) or chlorite/smectite (C/S), respectively, and of berthierine into chamosite. These alterations, which proceed via dissolution/precipitation, rather than by solid-state, layer-conserving reactions (Rask *et al.*, 1997), are volumetrically more important in mudrocks than in sandstones. In sandstones, I/S and C/S form by replacement of mechanically infiltrated, grain-coating smectite, mud intraclasts and volcanic fragments. Infiltrated clays, and hence grain coating chloritic and illitic clay minerals, are common in point-bar and delta-plain sandstone facies.

Thus, illitization of smectite in mudstones consumes K^+ but releases considerable amounts of Si, Mg, Fe, Na and Ca ions that can be used in diagenetic reactions in adjacent sandstones (Boles & Franks, 1979). The release of interlayer water influences the spatial distribution of diagenetic reactions by shifting mineral-water equilibrium by diluting the total solid concentrations in formation waters. Although K^+ can be supplied locally from the concomitant dissolution of K-feldspar (i.e. closed-system diagenesis; Hower *et al.*, 1976), an increasing number of studies suggest that the illitization of smectite in mudrocks may occur in an open system in which an external source of K^+, such as interbedded sandstones, is required (e.g. Furlan *et al.*, 1996; Land *et al.*, 1997). In contrast to illitization, the spatial distribution of mesogenetic alterations in siliciclastic sequences, as a consequence of smectite chloritization in mudrocks, is poorly explored in the literature. Chloritization occurs in sediments that are rich in biotite and intermediate to mafic volcanic rock fragments, as these materials act as sources of Fe and Mg required for the reaction. Hence, diagenetic chlorite in sandstones should be spatially associated with basaltic dykes and sills. Iron is also commonly derived from the reduction of goethite and hematite. Therefore, chlorite formation is often related spatially and temporally to the formation and migration of hydrocarbons, such as in bleached zones close to faults along which such reducing fluids have migrated (Surdam *et al.*, 1993). As illitization and chloritization are processes that involve ionic flux, they are expected to occur more readily as well as at shallower depths in permeable sandstones than in mudrocks. Chang *et al.* (1986) found that corrensite formation begins at a burial depth of 1500 m (60 °C) in sandstones and 2000 m (70 °C) in mudrocks. K–Na-rich smectite is potentially prone to illitization whereas Mg-rich smectite, which forms in sandstones rich in intermediate to mafic volcaniclastic sediments, is prone to chloritization (Chang *et al.*, 1986).

The transformation of berthierine into chamosite, which occurs in shallow-water marine siliciclastic facies, begins at temperatures of ≈ 70 °C, but becomes extensive at temperatures exceeding ≈ 100 °C (Odin, 1985; Jahren & Aagaard, 1989; Hillier & Velde, 1992). The spatial and temporal distribution of mesogenetic alteration of smectitic glaucony is poorly understood. However, smectitic glaucony should be potentially subjected to illitization and chloritization (Wilson & Stanton, 1994).

4 Incipient to extensive replacement of detrital plagioclase by nearly pure, albite end-member, i.e. albitization (Boles, 1982; Morad *et al.*, 1990) releases excess Al^{3+} and Ca^{2+} as by-products. The immobile Al^{3+} ions precipitate as minor dickite, laumontite or mixed-layer clays in the vicinity of albitized grains, whereas Ca^{2+} ions either dissolve in pore waters or precipitate as calcite or laumontite (Boles, 1982; Helmold & Kamp, 1984; Morad *et al.*, 1990). The overall albitization reaction can thus be tentatively written as follows:

$$\underset{\text{anorthite}}{CaAl_2Si_2O_8} + 2SiO_2 + 0{\cdot}5H_2O + Na^+ + H^+ = \underset{\text{albite}}{NaAlSi_3O_8} + 0{\cdot}5\underset{\text{dickite}}{Al_2Si_2O_5(OH)_4} + Ca^{2+} \quad (1)$$

This reaction denotes that albitization influences the anorthite solid solution in plagioclase and is controlled by pH (i.e. $-\log \alpha_{H^+}$) and $\alpha_{Na^+}/\alpha_{Ca^{2+}}$ of the formation waters rather than by the absolute

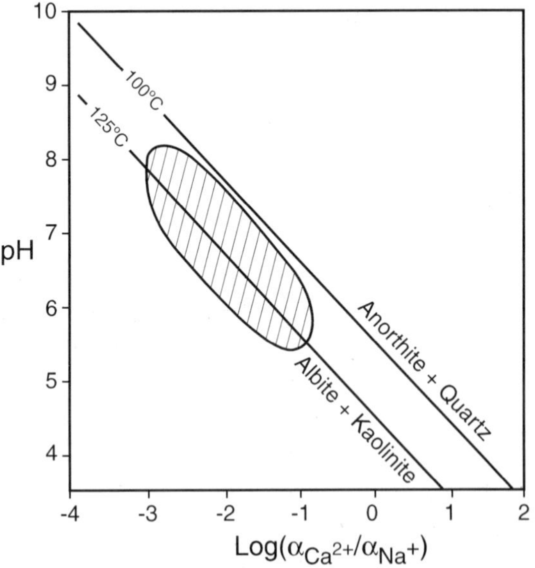

Fig. 10. Equilibrium diagram showing the temperature-related albitization of detrital calcic plagioclase (anorthite) and formation of kaolinite as a by-product. Hatched area represents the composition of most formation waters from North Sea reservoirs (data from Egeberg & Aagaard, 1989; Warren & Smalley, 1994).

Na content (Fig. 10). In North Sea reservoirs, the formation waters are in equilibrium with albite + kaolin rather than with anorthite + quartz at a temperature of 100 °C (Morad *et al.*, 1990; Fig. 10). However, albitization may commence at temperatures as low as ≈ 65 °C (Saigal *et al.*, 1988). A high $\alpha_{Na^+}/\alpha_{Ca^{2+}}$ in the water would be enhanced by the input of Na^+ and/or the precipitation of Ca^{2+} as calcite, which occurs in the presence of very low CO_2 fugacities (Wood, 1994). Illitization of smectite in mudrocks was suggested to be a viable source of Na^+ (Boles, 1982; Aagaard *et al.*, 1990). However, an external source of Na^+ is probably not required as the detrital plagioclase contains significant amounts of sodium (Morad *et al.*, 1990). Plagioclase albitization may thus occur in a closed diagenetic system. Nevertheless, a number of studies have shown that albitization is most pervasive in permeable sandstones (Boles & Ramseyer, 1988; Saigal *et al.*, 1988; Milliken, 1989), indicating that it occurs in open diagenetic systems as well (Aagaard *et al.*, 1990).

Deep mesodiagenesis

As sandstones enter the window of deep mesodiagenesis (≥ 3 km; T > 100 °C), they may dramatically lose their porosity and permeability, unless the sediments are, for example, subjected to overpressuring or to shallow-burial emplacement of hydrocarbon. In some sequences, considerable secondary porosity develops due to the dissolution of framework grains and intergranular cement. The most important mesogenetic alterations include (Fig. 9):

1 Chemical compaction and quartz cementation are the most important porosity-reducing alterations in deeply buried sandstones. Quartz cementation commonly occurs near stylolites and sites of intergranular dissolution. Dissolved silica is transferred by diffusion due to the establishment of a dissolved-silica concentration gradient between the sites of quartz dissolution and precipitation. These sandstones are referred to as silica exporters and importers, respectively (Houseknecht, 1988). Clean, coarse-grained quartz-cemented sandstones (e.g. foreshore sands) are thus expected to be spatially associated with silica-exporting, micaceous, fine-grained sands and sands rich in clay-coated quartz grains (e.g. shoreface sand bar; Wescott, 1983). Oelkers *et al.* (1996) found that the spatial distribution of quartz cement about stylolitic surfaces depends on temperature and on the interstylolite distance. In cores, taken from a depth of 3·6 km, the distribution of quartz cement is independent of distance from the stylolite. Conversely, sandstones at depths of 5 km show a greater concentration of quartz cement near stylolitic surfaces only if the surfaces are closely spaced (about 50 cm).

Additional sources of silica include illitization of kaolinite and smectite reactions and the dissolution of detrital feldspars. Quartz cementation of sandstones may occur due to the transfer of silica released from adjacent mudrocks by the illitization of smectite and kaolinite and by the dissolution of quartz, siliceous bioclasts and volcanic ash (Moncure *et al.*, 1984; Hendry & Trewin, 1995; Gluyas *et al.*, 2000). Quartz cementation may also occur as a result of a decrease in quartz solubility during cooling and mixing of brines with low-salinity pore waters in sandstones near major faults (cf. Hanor, 1996). The frequent zonation of quartz overgrowths observed under cathodoluminescence is, in some cases, due to variations in Al and transitional metal contents (Hogg *et al.*, 1992; Kraishan *et al.*, 2000). This zonation may result from multiple cementation episodes, interrupted by episodes of non-cementation (Hogg *et al.*, 1992). It is, however, uncertain whether or not the cementation episodes are related to pulses of brine upwelling and cooling or reflect geochemical evolution of formation waters due to water–sediment interaction during progressive burial of the sandstones. An

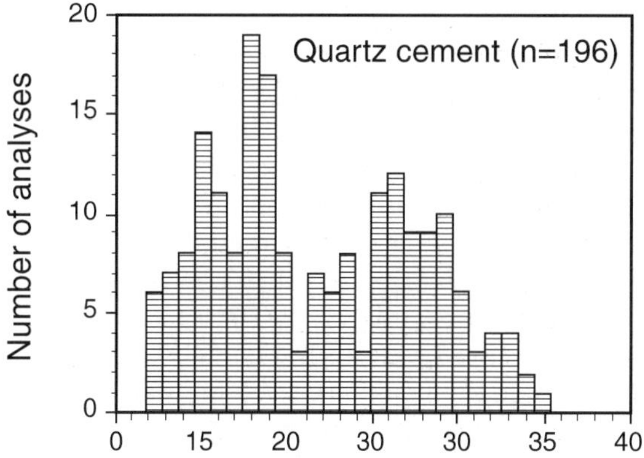

Fig. 11. Histogram showing the ranges and frequency of $\delta^{18}O_{SMOW}$ values of quartz cement (based on data from Lee *et al.*, 1985; McBride *et al.* 1987; Tilley & Longstaffe, 1989; Brint *et al.*, 1991; Girard & Deynoux, 1991; Aplin *et al.*, 1993; Girard & Barnes, 1995; Hervig *et al.*, 1995; Hogg *et al.*, 1995; Summer *et al.*, 1995; Graham *et al.*, 1996; Macaulay *et al.*, 1997; Sullivan *et al.*, 1997; William *et al.*, 1997) in sandstones.

example of the latter is the dissolution of aluminosilicates and increase in Al^{3+} concentrations of formation waters due to complexation by organic acids.

The $\delta^{18}O_{SMOW}$ values of quartz cement reported in the literature vary widely (+12 to +35‰; Fig. 11), and apparently reflect the wide range of precipitation temperatures and/or of oxygen isotopic compositions of the fluids involved. Indeed, fluid-inclusion analyses in quartz overgrowths reveal a wide range of salinities and precipitation temperatures. Quartz with high $\delta^{18}O$ values may have precipitated at low temperatures as groundwater silcretes.

2 Precipitation of finely crystalline, discrete quartz or blocky carbonate cements (e.g. ferroan dolomite, ankerite and calcite in pores that remain after quartz cementation (Boles, 1978; Land & Fisher, 1987). Indeed, carbonate cements that post-date quartz overgrowths are a global phenomena, which may be related to the different temperature-related solubility of the two minerals. With an increase in temperature, carbonate minerals have retrogressive solubility whereas quartz has a progressive solubility. Therefore, during progressive burial, quartz becomes more soluble, whereas carbonates tend to precipitate. Petrographic evidence substantiating this includes the partial replacement of quartz overgrowths by abutting carbonate cement. The precipitation of mesogenetic carbonate cements may also occur due to a decrease of p_{CO2} during release of overpressure, such as during faulting and hydrofracturing in mudrocks, or due to cross-formational fluid flow (Caritat & Baker, 1992; Schulz-Rojahn, 1993). Healing of hydrofractures is followed by another period of pressure build up due to clay dehydration and thermal alteration of organic matter.

3 Illitization of I/S and kaolin. In sandstones, illite occurs mainly as mat-like and fibrous or lath-like rims around framework grains, and as a pore-filling cement. Illite rims form by precipitation from pore waters and by transformation of I/S. The mat-like illite may inherit the habit of precursor clay, such as palygorskite. When illite has originated by transformation of eogenetic infiltrated smectitic coatings, it implies that: (i) its distribution is facies related, and (ii) coatings occur along grain contacts. Conversely, late authigenic illite is absent from such contacts. Thus, intergranular pressure dissolution is enhanced by illitized, originally eogenetic, smectitic coatings rather than by authigenic (neoformed) illite. Oversized illite patches form by mesogenetic illitization of kaolin and argillaceous fragments. The typical book-like stacking crystal habit of kaolin may be preserved during illitization (De Ros, 1998). The illitization of kaolinite also induces a considerable damage to reservoir permeability due to its position in pore throats and to the formation of abundant, discrete quartz cement as a by-product:

$$\underset{\text{kaolin}}{Al_2Si_2O_5(OH)_4} + \underset{\text{K-feldspar}}{KAlSi_3O_8} = \underset{\text{illite}}{KAl_3Si_3O_{10}(OH)_2} + \underset{\text{quartz}}{2SiO_2} + H_2O \quad (2)$$

Illite becomes the dominant clay mineral at temperatures ⩾130 °C (Ehrenberg & Nadeau, 1989; Giles *et al.*, 1992). However, Darby *et al.* (1997) concluded, based on illite dating in reservoir sandstones from the North Sea, that temperature is not the most important factor controlling illite precipitation. Theoretically, illite may form at temperatures as low as ≈50 °C (Bjørkum & Gjelsvik, 1988) under prolonged conditions of low water-to-rock ratio (Huang *et al.*, 1986), high α_{K+}/α_{H+} ratio and relatively low degree of quartz supersaturation in formation waters. Hence, illite formation is kinetically controlled.

Dickite is less susceptible to illitization than kaolinite due to its better-ordered crystal lattice

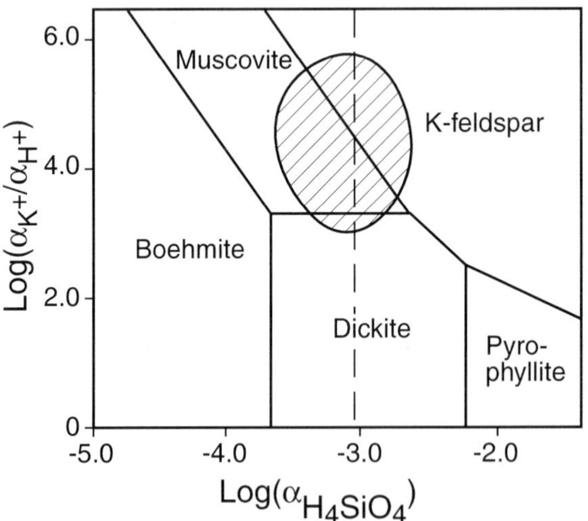

Fig. 12. Stability diagram of the system $K_2O–Al_2O_3–SiO_2$ at 100 °C and 170 bar based on thermodynamic data of SUPCRT; the stability field of dickite is calculated based on data from Zotov *et al.* (1998). Quartz saturation is represented by the dashed line. Hatched area represents the composition of most formation waters from the North Sea reservoirs (data from Egeberg & Aagaard, 1989; Warren & Smalley, 1994).

(Morad *et al.*, 1994). Accordingly, dickite has a wider stability field than kaolinite as revealed by thermodynamic data presented by Zotov *et al.* (1998; Fig. 12). Therefore, extensive dickitization during shallow mesodiagenesis enhances the preservation of reservoir quality as it retards illite formation. Pervasive or complete dickitization of kaolinite occurs at ≈110 °C (Ehrenberg *et al.*, 1993). Hence, during mesodiagenesis, illite and dickite compete in the replacement of kaolinite. Illitization is favoured over dickitization under conditions of low water-to-rock ratio and high $\alpha_{K^+}/\alpha_{H^+}$ ratio. The $\alpha_{K^+}/\alpha_{H^+}$ ratio required to stabilize illite at quartz saturation decreases with increasing temperature (Fig. 13). K^+ is most often believed to be derived internally on the scale of sandstone bed (Ehrenberg, 1991), or from juxtaposed evaporites (Gaupp *et al.*, 1993). As evaporites contain very little or no intercrystalline porosity (i.e. little connate waters), they may supply K^+ due to their dissolution by formation waters (Fig. 9). During a closed-system diagenesis of K-feldspar-poor sandstones, dickite will survive, to be later transformed into nacrite or pyrophyllite when pore waters are characterized by high activity of dissolved silica (Figs 12 and 14) at the boundary between diagenesis and low-grade metamorphism (≈200 °C; Schulz-Rojahn & Phillips, 1989), as follows:

$$\underset{\text{dickite/nacrite}}{Al_2Si_2O_5(OH)_4} + 2H_4SiO_4 = \underset{\text{pyrophyllite}}{2AlSi_2O_5(OH)} + 5H_2O \quad (3)$$

This reaction is fairly uncommon, as most kaolinite and dickite are apparently pervasively transformed into chlorite and illite during mesodiagenesis.

4 Fe-chlorite, which occurs as rims of pseudohexagonal crystals that are arranged perpendicular to the surfaces of framework grains. Such chlorite has a spatial distribution that is opposite to that of quartz cement, as it inhibits the precipitation of quartz overgrowths. Hence, chlorite coatings contribute to porosity and permeability preservation in deep sandstone reservoirs (Pittman *et al.*, 1992; Ehrenberg, 1993). Such reservoirs are, however, characterized by high amounts of irreducible water due to the presence of abundant microporosity in the chlorite, which poses a serious reservoir evaluation problem. It is, so far, poorly understood why chlorite is more efficient in inhibiting the precipitation of quartz overgrowths than illitic clays; the latter clays are commonly partially to completely engulfed by quartz overgrowths (Bonnell *et al.*, 1999). It is possible that the chemistry of fluids that are in equilibrium with chlorite, and not the chlorite itself, somehow retards the nucleation and growth of quartz.

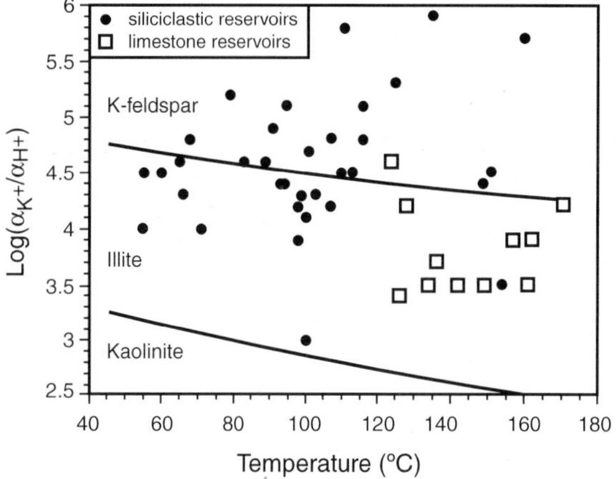

Fig. 13. Diagram showing the equilibrium relationship between North Sea formation waters (data from Egeberg & Aagaard, 1989; Warren & Smalley, 1994) and kaolinite, illite and K-feldspar related to $\log(\alpha_{K^+}/\alpha_{H^+})$ ratio and temperature. Note that the $\log(\alpha_{K^+}/\alpha_{H^+})$ ratios for the siliciclastic reservoirs increase with increasing temperature due to dissolution and albitization of K-feldspars. The $\log(\alpha_{K^+}/\alpha_{H^+})$ ratio needed to illitize kaolinite decreases with increasing temperature.

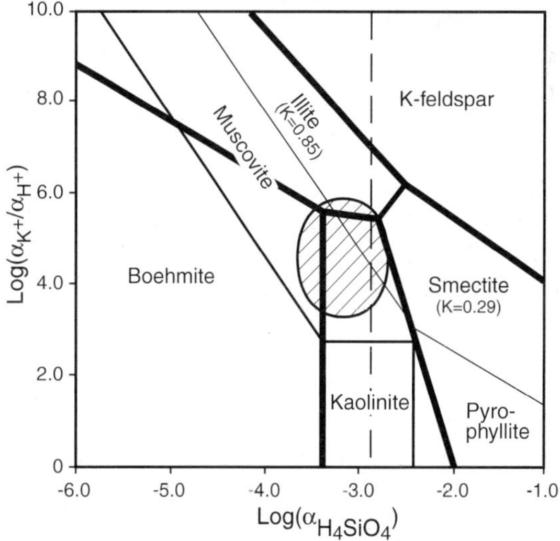

Fig. 14. Stability diagram in the system K_2O–Al_2O_3–SiO_2 at 110 °C based on thermodynamic data of SUPCRT (thin lines; Helgeson *et al.*, 1979) and on data from Aja *et al.* (1991; bold lines). The latter data suggest a dramatic increase in the stability field of kaolinite. Quartz saturation is represented by the dashed line. Hatched area represents the composition of most formation waters from the North Sea reservoirs (data from Egeberg & Aagaard, 1989; Warren & Smalley, 1994).

Chlorite fringes form by precipitation from pore waters, by grain replacement and by the progressive transformation of eogenetic berthierine or S/C coatings (Moraes & De Ros, 1990; Longstaffe *et al.*, 1992; Ehrenberg, 1993; Humphreys *et al.*, 1994). Thus, although chlorite rims reveal an early diagenetic, precompactional habit, it is a deep-burial, high-temperature mineral (Jahren & Aagaard, 1989). In contrast to mesogenetic neoformed chlorite, chloritized smectite or berthierine coatings are continuous along the intergranular grain contact. Nevertheless, the fringe-like habit of chlorite develops only on grain surfaces that face an open pore space, and not at grain contacts. At such contacts, the precursor smectite preserves its tangential alignment around the grains.

In sandstones containing pore-lining chlorite, the biotite, mud intraclast, volcanic rock fragments and, less commonly, feldspars are replaced by chlorite too (Morad, 1990; Pittman *et al.*, 1992; Hillier, 1994; Humphreys *et al.*, 1994). Biotite chloritization is accompanied by the consumption of H^+ from pore waters and release of K^+ and Si^{4+}, leading to the creation of a microenvironment characterized by elevated pH, $\alpha_{K^+}/\alpha_{H^+}$ and $\alpha_{H_4SiO_4}$ in water films within the biotite. This microenvironment enhances the precipitation of numerous authigenic minerals, such as K-feldspar and carbonates (Boles & Johnson, 1984). Micro-crystalline, mesogenetic K-feldspar is also closely associated with illite and chlorite that replace mud intraclasts (Morad *et al.*, 1994). Grain-rimming and replacive chlorites thus form within a similar depth–temperature interval to illite, and are most abundant at temperatures $\gtrsim 100\,°C$ (Jahren & Aagaard, 1989). However, illite usually covers, and hence post-dates, chlorite, probably reflecting an increase in $\alpha_{K^+}/\alpha_{H^+}$ with increasing temperatures of formation waters (Fig. 13).

5 Albitization of detrital K-feldspars is a dissolution–precipitation process rather than a simple, solid-state diffusional exchange of K by Na. At this burial stage, the remaining less-calcic plagioclase grains will be albitized too. The albitization of K-feldspar is spatially and temporally linked with and provides K^+ needed for illitization of kaolinite in sandstones (Morad *et al.*, 1990). Indeed, the dissolution and albitization of K-feldspar are governed kinetically by the consumption of K^+ in the illitization of kaolinite (Aagaard *et al.*, 1990). Upon fixation of K^+ in illite, albitization is thus controlled by pH, activity of Na^+ in formation waters and temperature (Fig. 15). The α_{Na^+} and $\alpha_{Na^+}/\alpha_{H^+}$ required to stabilize albite–illite–quartz at the expense of K-feldspar and kaolinite decreases considerably with increase in burial depth and temperature (Fig. 15). Pervasive to complete illitization of kaolinite and albitization of K-feldspars in North Sea sandstones occurs at temperatures $\gtrsim 130\,°C$ (Bjørlykke *et al.*, 1986).

In some deeply buried sandstones ($T \gtrsim 110\,°C$), albitization is accompanied by cementation and replacement of the feldspar by coarse laumontite (Helmold & Kamp, 1984; Merino *et al.*, 1997). In order to stabilize laumontite, the formation waters in these sandstones should be characterized by extremely low p_{CO_2} (Crossey *et al.*, 1984). In volcaniclastic sandstones, both albitization and laumontite formation occur at relatively shallow depths and low temperatures ($\approx 50\,°C$; Surdam & Boles, 1979; Wopfner *et al.*, 1991; Remy, 1994). This implies that laumontite forms in a wide temperature range (≈ 50–$200\,°C$; Boles & Coombs, 1977).

6 Precipitation of pyrite and, less commonly, Cu–Pb–Zn sulphides occurs during thermal sulphate reduction and invasion of the reservoirs by metal-rich brines (Siebert, 1985; Metcalfe *et al.*, 1994). Variable amounts of barite and ankerite are associated with these sulphides. Diagenetic ores composed of this mineral assemblage are referred to as Mississippi Valley-type ores which form by large-scale, tectonically driven fluid flow, such as in foreland basins. As fluid flow is mostly focused, these minerals are distributed mainly along fractures and faults.

Spatial and temporal distribution of mesogenetic secondary porosity

Various patterns of spatial and temporal distribution of secondary porosity that result from the dissolution of framework grains (mainly feldspars

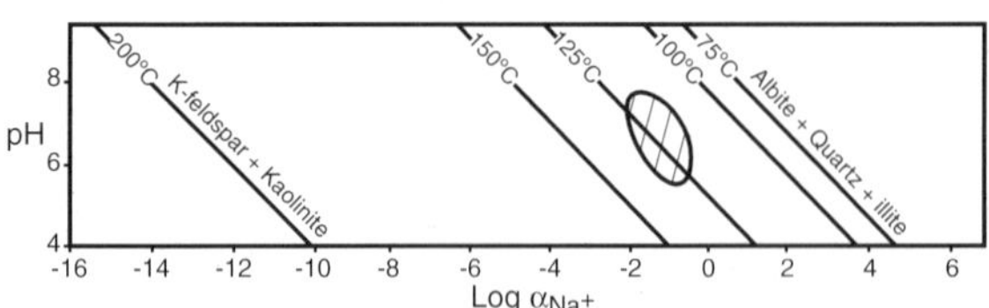

Fig. 15. Stability diagram for the reaction 2K-feldspar + 2·5kaolinite + Na^+ = albite + 2illite + 2quartz + 2·5H_2O + H^+, showing that the stability field of K-feldspar + kaolinite decreases successively with increasing temperature. Hatched area represents the composition of most formation waters from the North Sea reservoirs (data from Egeberg & Aagaard, 1989; Warren & Smalley, 1994).

and rock fragments) and cements (mainly calcite) are reported in deeply buried sandstones. However, determination of the timing and burial depths at which mineral dissolution occurs is difficult. This is particularly true for siliciclastic sequences that have undergone complex burial–uplift histories. Additionally, the degree of preservation vs. destruction of eogenetic secondary porosity during mesodiagenesis is poorly understood, and the distinction between primary and secondary intergranular porosity is fraught with uncertainty. It appears that the presence of relatively small amounts of rigid, evenly distributed, intergranular eogenetic cement, such as dolomite rhombs and quartz overgrowths, can limit or prevent mechanical compaction during deep burial, even in sandstones that are rich in ductile lithic grains (Souza *et al.*, 1995). Indeed, mouldic pores outlined by, for example, clay minerals and feldspar overgrowths are common over a wide range of burial depths. However, primary and secondary porosities decline fairly sharply with depth, except for hydrocarbon-saturated (Swarbrick, 1994) and overpressurred sandstones (Hall, 1994). Other conditions that favour porosity preservation include a short period of deep burial (Sombra & Chang, 1997) and mechanically stable framework-grains (e.g. quartz, chert and feldspars).

Despite the uncertainties, it is argued in the literature that secondary porosity can account for many high-quality, deep reservoirs such as those in the US Gulf Coast (e.g. Taylor & Land, 1996). As undersaturated meteoric waters are unlikely to penetrate siliciclastic sequences to depths greater than 3 km, particularly in overpressured and actively subsiding basins, mineral dissolution has been suggested to be accomplished mainly by organic acids and CO_2 released during thermal maturation of organic matter and oil generation in source mudrock (Fig. 9). Organic acids (mainly carboxylic) form also by the interaction of water with hydrocarbons, mineral oxidation of organic matter and by bacterial degradation of organic matter and hydrocarbons (MacGowan & Surdam, 1988). Generation of CO_2 also occurs by thermal reduction of sulphate ions which are derived from the dissolution of evaporite deposits.

Organic acid anions, which occur in variable concentration in formation waters depending on temperature, type of organic matter in the source rocks, lithology and redox reactions (Carothers & Kharaka, 1978; MacGowan & Surdam, 1988; Shock, 1988), may enhance the solubility of carbonates and aluminosilicates by complexing Si, Al, Ca and Mg (Surdam *et al.*, 1984, 1989). Maximum concentrations of organic acid anion occur at ≈ 80–120 °C. Over this temperature range, the pH of formation waters is buffered by carboxylic acid anions. At temperatures ⩾ 100 °C, the carboxylic acids begin to progressively decarboxylate, resulting in an increase in p_{CO2}. If pH is buffered by silicates, the increase in p_{CO2} at temperatures of 100–120 °C will cause dissolution of carbonate cements.

Carboxylic and carbonic acids may account for secondary porosity, particularly in sandstones adjacent to hydrocarbon source rocks due to the dissolution of calcite cement and, less commonly, of aluminosilicates (Moncure *et al.*, 1984; Surdam *et al.*, 1984; Surdam & Yin, 1995). It was shown experimentally that ionic solubility is enhanced substantially through complexation by carboxylic compounds. For instance, acetate complexes with Ca, Mg and Al and oxalate complexes with Si (Bennett & Siegel, 1987). However, mass-balance calculations and the usually low measured concentrations of organic acids in formation waters, as well as the absence of a marked increase in Al and Si concentrations, raise some doubts about the efficiency of these acids in creating secondary porosity in sandstone reservoirs (Giles *et al.*, 1995). Nevertheless, relatively high concentrations of Al (up to 6 mg L^{-1}) have been reported in formation waters from Tertiary sediments in the San Joaquin basin, USA (Fisher & Boles, 1990) and Palaeozoic–Cenozoic sediments in the Tarim Basin, China (Chufuang *et al.*, 1997). Acetate, which is a dominant species, may reach concentrations of up to 1000–3000 mg L^{-1} in some oil-field brines (Carothers & Kharaka, 1978; Fisher, 1987), whereas oxalate concentrations only reach 100 mg L^{-1}, probably due to the precipitation of Ca-oxalate (Manning *et al.*, 1992). Additionally, organic acids may react with clay minerals and carbonates, and thus be neutralized while still in the source mudrock. Obviously, despite the importance of organic acids deduced based on theory and experimental work, their influence on the spatial and temporal distribution of diagenetic alterations is still a controversial topic. However, the formation-water samples so far analysed reflect compositions that post-date the peak of thermal maturation of organic matter and hydrocarbon formation. Consequently, the peak of maximum organic-acid concentrations in formation waters and direct evidence for their active role in influencing the spatial distribution of mesogenetic alterations have not been observed.

Other sources of acidic fluid during mesodiagenesis include low-grade metamorphic reactions (so-called reverse weathering) in deep regimes of sedimentary basins (Taylor & Land, 1996). As these fluids are believed to move by focused flow, the spatial distribution of resulting secondary porosity is strongly associated with major, deep fault zones. In some rare cases, large amounts of CO_2, which can be derived from the contact metamorphism of carbonate rocks, can be emplaced in sandstone reservoirs, such as in the Norphlet Formation, Pisgah Anticline, Mississippi, USA (Studlick *et al.*, 1990).

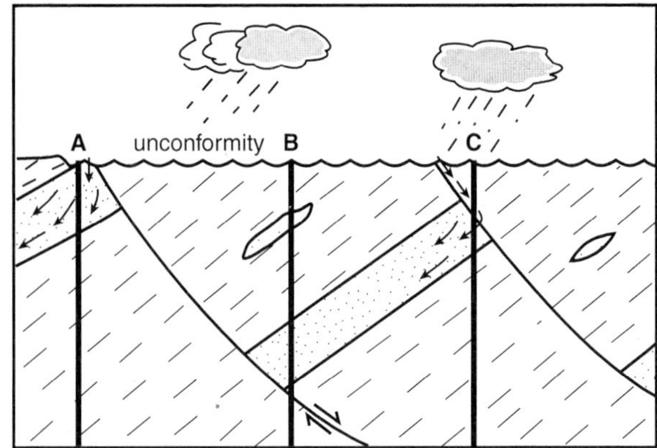

Fig. 16. Hypothetical diagram showing the possible ways by which meteoric telodiagenesis occurs below an unconformity surface. Different distribution patterns of telogenetic alterations (e.g. kaolinite content) occur below the surface in each of the wells A, B and C. Telogenetic alterations in well A are systematically related to the unconformity surface. Alterations are halted in well B due to the occurrence of a thick mudstone layer, but meteoric water flow may occur along the tilted sandstone layer. In well C, alterations occur near a fault.

SPATIAL AND TEMPORAL DISTRIBUTION OF TELOGENETIC ALTERATIONS

Telodiagenesis occurs when sedimentary sequences that have undergone eo- and mesogenetic modifications are uplifted and subjected to pressure rebound and the influence of meteoric waters. Factors controlling the spatial and temporal distribution and the extent of meteoric telodiagenesis below unconformity surfaces include: (i) hydraulic head, which depends on the degree of uplift above sea-level; (ii) climate; (iii) degree of porosity and permeability loss during eo- and mesodiagenesis; (iv) presence vs. absence of thick, impermeable mudrock layers which would halt the infiltration of meteoric water; (v) presence of faults, which extend to the surface and act as conduits for meteoric water; (vi) duration of the exposure; and (vii) the mineral composition (Fig. 16). Although meteoric waters were suggested to penetrate to depths of 2 km or more in sedimentary basins, the thickness of pervasive telogenetic alterations varies, often from tens to hundreds of metres (Emery *et al.*, 1990; Shanmugam, 1990).

Therefore, spatially and temporally extensive telogenetic meteoric alterations are expected to occur around uplifted blocks in humid climates and in mineralogically immature sandstones with good hydraulic conductivity. Telogenetic alterations are better preserved if the rate of uplift/erosion is lower than the rate of meteoric alteration (Bjørkum *et al.*, 1990, 1993), and are expected to increase systematically upwards when the sandstones are exposed at the unconformity surfaces (Fig. 17). As the incursion of meteoric water along faults bounding uplifted blocks occur to depths of a few hundreds of metres, the spatial and temporal distribution of telogenetic modifications are not necessarily systematically related to unconformity surfaces (Fig. 16).

The most important telogenetic modifications promoted by meteoric flushing in siliciclastic sequences include:

1 Dissolution of unstable framework grains, such as feldspars, mica, rock fragments, mud intraclasts and heavy minerals (Smale, 1989; Emery *et al.*, 1990). Under humid conditions, the dissolution of detrital aluminosilicates is typically accompanied by the precipitation of kaolinite. Unlike eogenetic kaolinite, telogenetic kaolinite post-dates quartz overgrowths and is closely associated with dissolved eo- and mesogenetic carbonate cement. The dissolution of K-feldspars and removal of K below unconformities are evidenced by the progressive increase in the Th/K ratio in natural gamma-ray logs (Emery *et al.*, 1990).

It is generally agreed that telodiagenesis enhances the porosity and permeability of sandstones at shallow depths below the unconformity surface due to the dissolution of framework grains and cements (Fig. 17; Giles & De Boer, 1990; Ghosh, 1997). This enhancement is most evident under humid climatic conditions even when

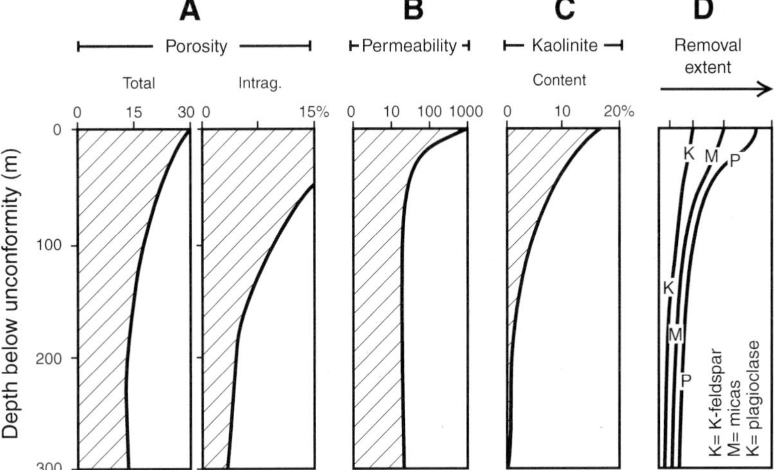

Fig. 17. Distribution patterns of porosity–permeability (e.g. Shanmugan, 1990) and kaolinite contents due to telodiagenesis below an unconformity surface under a humid climate. An increasing order of dissolution of plagioclase, micas and K-feldspar occurs towards the unconformity surface.

silicate dissolution is accompanied by precipitation of kaolinite (Garcia *et al.*, 1998). Sandstones with normal porosity of < 10% and permeability of < 10 mD can reach porosity and permeability values up to 25% and 1000 mD, respectively, close to unconformities (Fig. 17; Shanmugam, 1990).

2 Alteration of eo- and mesogenetic cements, such as the dissolution of calcite, dolomite, sulphate and feldspar overgrowths and the oxidation of ferroan carbonates, chlorite and pyrite (Stanton, 1986; Morad *et al.*, 1995; Garcia *et al.*, 1998). Sequential mineral dissolution by meteoric waters occurs in the upper 200 m below the unconformity surface (Fig. 17; Thrailkill, 1968; Shanmugam, 1988; Shanmugan, 1990; Hayes & Boles, 1992). Dissolution of calcite cement is most important in humid climates. Conversely, calcite dissolution under semi-arid conditions is followed by supersaturation and precipitation. Tightly cemented sandstones, particularly with ferroan carbonates, are not normally subjected to significant dissolution.

3 Pedogenesis and clay infiltration along the unconformity surface occur under semi-arid climatic conditions, in some cases accompanied by the formation of duricrusts like calcrete, silcrete and ferricrete (Shanmugam, 1988). Telogenetically infiltrated clays cover mesogenetic cements, such as quartz overgrowths.

CONCLUDING REMARKS

The complex array of parameters that control the spatial and temporal distribution of diagenetic alterations in siliciclastic sequences include: (a) tectonic setting and thermal–burial history of the basin; (b) origin and geochemical evolution of formation waters; (c) depositional environment and facies; (d) climatic conditions; (e) the physical and chemical properties of framework grains; (f) relative changes in sea-level; (g) amounts and reactivity of organic matter; and (h) timing of hydrocarbon emplacement and the degree of hydrocarbon saturation. Despite the complexity and wide variations of the diagenetic systems, attempts have been made to develop global models for the diagenetic evolution of siliciclastic sequences. This paper has demonstrated that the spatial distribution of eogenetic alterations can be constrained based on variations in depositional facies and climatic conditions, and within a predictive, sequence stratigraphic context. Many of the conceptual and numerical diagenetic models so far developed are biased, as they rely heavily on the huge amounts of data collected from two extensively studied hydrocarbon provinces, namely the US Gulf Coast and the North Sea. The challenge ahead is to develop reliable, predictive, three-dimensional conceptual and numerical models that consider, on a world-wide basis, as much surface and subsurface data about the spatial and temporal distribution of diagenetic alterations as possible. Furthermore, successful models should be developed based on: (1) improved thermodynamic and kinetic databases on organic and inorganic phases and reactions; (2) a combination of 'microscopic' observations (paragenetic relationships and fluid-inclusion data) and larger, basin-wide observations of the distribution of diagenetic alterations; and (3) if diagenetic models are to be useful in hydrocarbon exploration and exploitation, multidisciplinary research involving seismic interpreters, petrophysicists and reservoir engineers should be carried out.

Among the key questions that have to be answered before successful models can be developed are: (1) How can we distinguish between mesogenetic alterations that are mediated by diffusion from those mediated by active fluid flow? In others words, do mesogenetic reactions occur in open or closed systems in terms of mass and heat transfer? What are the scales of mass transfer? And what is the precise role of faults in mass and heat transfer? (2) What is the relationship between the thermal maturation of organic matter and the timing of formation and emplacement of hydrocarbons in reservoirs and the spatial and temporal distribution of mesogenetic alterations? (3) Can the formation-water samples so far collected from sedimentary basins be used as the basis for a global understanding of the spatial and temporal distribution of mesogenetic alterations? In other words, is the present a key to the past?

ACKNOWLEDGEMENTS

We thank Dr Peter Mozley for inviting us to contribute to this millennium issue. S.M. thanks the Swedish Natural Sciences Research Council (NFR) and J.M.K. and L.F.DR. thank the Brazilian Research Council (CNPq) for financial support. Comments and suggestions made by Drs I. S. Al-Aasm, É. Brosse, P. Mozley, J. Schulz-Rojahn, K. Ramseyer, K. G. Taylor and R. Worden helped us improve the manuscript.

REFERENCES

Aagaard, P., Egeberg, P.K., Saigal, G.C., Morad, S. and **Bjørlykke, K.** (1990) Diagenetic albitization of detrital K-feldspar in Jurassic, Lower Cretaceous and Tertiary clastic reservoir rocks from offshore Norway. II. Formation-water chemistry and kinetic considerations. *J. Sedim. Petrol.*, **60**, 575–581.

Aase, N.E., Bjørkum, P.A. and **Nadeau, P.H.** (1996) The effect of grain-coating microquartz on preservation of reservoir porosity. *Bull. Am. Assoc. Petrol. Geol.*, **80**, 1654–1673.

Aja, S.U., Rosenberg, P.E. and **Kittrick, J.A.** (1991) Illite equilibria in solutions: I Phase relationships in the system K_2O–Al_2O_3–SiO_2–H_2O between 25 and 250°C. *Am. Miner.*, **55**, 1353–1364.

Amorosi, A. (1995) Glaucony and sequence stratigraphy: a conceptual framework of distribution in siliciclastic sequences. *J. Sedim. Res.*, **B65**, 419–425.

Aplin, A.C., Warren, E.A., Grant, S.M. and **Robinson, A.G.** (1993) Mechanisms of quartz cementation in North Sea reservoir sandstone: constraints from fluid composition. In: *Diagenesis and Basin Development* (Ed. by A. Horbury and A. G. Robinson), *AAPG Studies in Geology*, 36, pp. 7–23. AAPG, Tulsa.

Arakel, A.V. (1986) Evolution of calcrete in paleodrainages of Lake Napperby area, central Australia. *Paleogeog. Paleoclim. Paleoecol.*, **54**, 283–303.

Arakel, A.V., Jacobson, G. and **Lyons, W.** (1990) Sediment–water interaction as a control on geochemical evolution of playa lake systems in the Australian arid interior. *Hydrobiologia*, **197**, 1–12.

Baker, P.A. and **Kastner, M.** (1981) Constraints on the formation of sedimentary dolomite. *Science*, **213**, 214–216.

Bardossy, G. and **Combes, P.J.** (1999) Karst bauxites: interfingering of deposition and palaeoweathering. In: *Palaeoweathering, Palaeosurfaces and Related Continental Deposits* (Ed. by M. Thiry and R. Simon-Coincon), *Spec. Publ. Int. Assoc. Sediment.*, **27**, 189–206.

Baum, G.R. and **Vail, P.R.** (1988) Sequence stratigraphic concepts applied to Paleogene outcrops, Gulf and Atlantic basins. In: *Sea-Level Changes: an Integrated Approach* (Ed. by C. K. Wilgus, B. S. Hastings, C. G. S. C. Kendall, H. W.Posamentier, C. A. Ross and J. C. van Wagoner), *Spec. Publ. Soc. Econ. Paleont. Miner.*, **42**, 309–327.

Bennett, P. and **Siegel, D.I.** (1987) Increased solubility of quartz in water due to complexing by organic compounds. *Nature*, **326**, 684–686.

Bjørkum, P.A. and **Gjelsvik, N.** (1988) An isochemical model for formation of authigenic kaolinite, K-feldspar and illite in sediments. *J. Sedim. Petrol.*, **58**, 506–511.

Bjørkum, P.A., Knarud, R. and **Bergan, M.** (1993) How important is the late Cimmerian unconformity in controlling formation of kaolinite in sandstones of the North Sea?. In: *Diagenesis and Basin Development* (Ed. by A. Horbury and A. G. Robinson), *AAPG Studies in Geology*, 36, pp. 261–269. AAPG, Tulsa.

Bjørkum, P.A., Mjøs, R., Walderhaug, O. and **Hurst, A.** (1990) The role of the Late Cimmerian unconformity for the distribution of kaolinite in the Gullfaks Field, northern North Sea. *Sedimentology*, **37**, 396–406.

Bjørkum, P.A. and **Walderhaug, O.** (1990) Lateral extent of calcite-cemented zones in shallow marine sandstones. In: *North Sea Oil and Gas Reservoirs II* (Ed. by A. T. Buller, E. Bjerg, O. Hjelmeland, J. Kleppe, O. Torsæter and J. O. Aasen), pp. 331–336. Graham and Trotman, London.

Bjørlykke, K., Aagaard, P., Dypvik, H., Hastings, D.S. and **Harper, A.S.** (1986) Diagenesis and reservoir properties of Jurassic sandstones from the Haltenbanken area, offshore mid Norway. In: *Habitat of Hydrocarbons on the Norwegian Continental Shelf* (Ed. by Spencer, *et al.*), pp. 275–286. Graham and Trotman, London,.

Boettcher, M.E. and **Huckriede, H.** (1997) First occurrence and stable isotope composition of authigenic gamma -MnS in the central Gotland Deep (Baltic Sea). *Mar. Geol.*, **137**, 201–205.

Boles, J. (1977) Zeolites in deep-sea sediments. In: *Mineralogy and Geology of Natural Zeolites* (Ed. by F. A. Mumpton), *Miner. Soc. Am. Short Course Notes*, 4, pp. 137–163, Southern Priority Company, Blacksburg, VA.

Boles, J.R. (1978) Active ankerite cementation in the subsurface Eocene of southwest Texas. *Contrib. Miner. Petrol.*, **68,** 13–22.

Boles, J.R. (1982) Active albitization of plagioclase, Gulf Coast Tertiary. *Am. J. Sci.*, **282,** 165–180.

Boles, J.R. and **Coombs, D.S.** (1977) Zeolite facies alteration of sandstones, Southland Syncline, New Zealand. *Am. J. Sci.*, **77,** 982–1012.

Boles, J.R. and **Franks, S.G.** (1979) Clay diagenesis in Wilcox sandstones of southwest Texas: implications of smectite diagenesis on sandstone cementation. *J. Sedim. Petrol.*, **49,** 55–70.

Boles, J.R. and **Johnson, S.K.** (1984) Influence of mica surfaces on pore water pH. *Chem. Geol.*, **43,** 303–317.

Boles, J.R. and **Ramseyer, K.** (1988) Albitization of plagioclase and vitrinite reflectance as paleothermal indicators, San Joaquin Basin, California. In: *Studies of the Geology of the San Joaquin Basin* (Ed. by S. A. Graham), pp. 129–139. SEPM.

Bonnell, L.M., Warren, E.A. and **Lander, R.H.** (1999) The timing of illitization, Haltenbanken, Mid-Norway. *AAPG Annual Convention, San Antonio, Texas, Abstract with Programs*, p. A14.

Brint, J.F., Hamilton, P.J., Haszeldine, R.S., Fallick, A.E. and **Brown, S.** (1991) Oxygen isotopic analysis of diagenetic quartz overgrowths from the Brent sands: a comparison of two preparation methods. *J. Sedim. Petrol.*, **61,** 527–533.

Burkhalter, R.M. (1995) Ooidal ironstones and ferruginous microbialites: origin and relation to sequence stratigraphy (Aalenian and Bajocian, Swiss Jura mountains). *Sedimentology*, **42,** 57–74.

Carothers, W.W. and **Kharaka, Y.K.** (1978) Aliphatic acid anions in oil-field waters – implications for origin of natural gas. *Bull. Am. Assoc. Petrol. Geol.*, **62,** 2441–2453.

Caritat, P.D. and **Baker, J.C.** (1992) Overpressure release, cross-formational porewater flow and diagenesis. In: *Water–Rock Interaction* (Ed. by Y. K. Kharaka and A. S. Maest), pp. 1161–1164. Balkema, Rotterdam.

Cartwright, J.A. (1994) Episodic basin-wide fluid expulsion from geopressured shale sequences in the North Sea basin. *Geology*, **22,** 447–450.

Carvalho, M.V.F., De Ros, L.F. and **Gomes, N.S.** (1995) Carbonate cementation patterns and diagenetic reservoir facies in the Campos Basin Cretaceous turbidites, offshore eastern Brazil. *Mar. Petrol. Geol.*, **12,** 741–758.

Chang, H.K., Mackenzie, F.T. and **Schoonmaker, J.** (1986) Comparisons between the diagenesis of dioctahedral and trioctahedral smectite, Brazilian offshore basins. *Clays Clay Min.*, **34,** 407–423.

Choquette, P.W. and **Pray, L.C.** (1970) Geologic nomenclature and classification of porosity in sedimentary carbonates. *Bull. Am. Ass. Petrol. Geol.*, **54,** 207–250.

Chow, N., Morad, S. and **Al-Aasm, I.S.** (1996) Origin of authigenic carbonates in Eocene to Quaternary sediments from the Arctic and Norwegian-Greenland Sea. In: *Proceedings of the Ocean Drilling Program, Scientific Results* (Ed. by A. Myhre, J. Thiede, J. Firth, W. F. Ruddiman and L. Johnson), 151, pp. 415–434. US Government Printing Office, Washington.

Chufuang, C., Bowen, M., Wei, L. and **Fangang, Z.** (1997) Water–rock interaction in Tarim Basin: constraints from oilfield water geochemistry. *Chinese J. Geochem.*, **16,** 289–303.

Colson, I. and **Cojan, I.** (1996) Groundwater dolocretes in a lake-marginal environment: an alternative model for dolocrete formation in continental settings (Danian of the Provence Basin, France). *Sedimentology*, **43,** 175–188.

Crossey, L.J., Frost, B.R. and **Surdam, R.C.** (1984) Secondary porosity in laumontite-bearing sandstones. In: *Clastic Diagenesis* (Ed. by R. C. Surdam and D. A. McDonald), *Mem. Am. Assoc. Petrol. Geol.*, **37,** 225–237.

Darby, D., Wilkinson, M., Fallick, A.E. and **Haszeldine, R.S.** (1997) Illite dates record deep fluid movements in petroleum basins. *Petrol. Geosci.*, **3,** 133–140.

De Ros, L.F. (1998) Heterogeneous generation and evolution of diagenetic quartzarenites in the Silurian-Devonian Furnas Formation of the Parana Basin, southern Brazil. *Sediment. Geol.*, **116,** 99–128.

De Ros, L.F., Sgarbi, G.N.C. and **Morad, S.** (1994) Multiple authigenesis of K-feldspar in sandstones: evidence from the Cretaceous Areado Formation, São Francisco Basin, central Brazil. *J. Sedim. Res.*, **A64,** 778–787.

Dixon, R.J. and **Davis, J.M.** (1997) A hydrogeologic analysis of the mixing-zone model of dolomitization. *Abstracts with Programs. Geol. Soc. Am.*, **29,** 475.

Dolan, J.F. (1989) Eustatic and tectonic controls on deposition of hybrid and siliciclastic/carbonate basinal cycles: discussion with examples. *Bull. Am. Assoc. Petrol. Geol.*, **73,** 1233–1246.

Egeberg, P.K. and **Aagaard, P.** (1989) Origin and evolution of formation waters from oil fields on the Norwegian Shelf. *Appl. Geochem.*, **4,** 131–142.

Ehrenberg, S.N. (1991) Kaolinized, potassium-leached zones in the contacts of the Garn Formation, Haltenbanken, mid-Norwegian continental shelf. *Mar. Petrol. Geol.*, **8,** 250–269.

Ehrenberg, S.N. (1993) Preservation of anomalously high porosity in deeply buried sandstones by grain-coating chlorite: examples from the Norwegian continental shelf. *Bull. Am. Assoc. Petrol. Geol.*, **77,** 1260–1286.

Ehrenberg, S.N., Aagaard, P., Wilson, M.J., Fraser, A.R. and **Duthie, D.M.L.** (1993) Depth-dependent transformation of kaolinite to dickite in sandstones of the Norwegian Continental shelf. *Clay Miner.*, **28,** 325–352.

Ehrenberg, S.N. and **Nadeau, P.H.** (1989) Formation of diagenetic illite in sandstones of the Garn Formation, Haltenbanken area, mid-Norwegian continental shelf. *Clay Miner.*, **24,** 233–253.

Emery, D., Myers, K.J. and **Young, R.** (1990) Ancient subaerial exposure and freshwater leaching in sandstones. *Geology*, **18,** 1178–1181.

Evans, D.G. and **Nunn, J.A.** (1989) Free thermohaline convection in sediments surrounding a salt column. *J. Geophys. Res.*, **94,** 12413–12422.

Fisher, J.B. (1987) Distribution and occurrence of aliphatic acid anions in deep subsurface waters. *Geochim. Cosmochim. Acta*, **51,** 2459–2468.

Fisher, J.B. and **Boles, J.R.** (1990) Water–rock interaction in Tertiary sandstones, San Joaquin, California, USA. *Chem. Geol.*, **82,** 83–101.

Fontana, D., Zuffa, G.G. and **Garzanti, E.** (1989) The interaction of eustacy and tectonism from provenance studies of the Eocene Hecho Group Turbidite Complex (Eocene-Central Pyrenees, Spain). *Basin Res.*, **2,** 223–237.

Furlan, S., Clauer, N., Chaudhuri, S. and **Sommer, F.** (1996) K transfer during burial diagenesis in the Mahakam Delta Basin (Kalimantan, Indonesia). *Clays Clay Min.*, **44,** 157–169.

Garcia, A.J.V., Morad, S., De Ros, L.F. and **Al-Aasm, I.S.** (1998) Paleogeographical, paleoclimatic and burial history controls on the diagenetic evolution of Lower Cretaceous

Serraria sandstones in Sergipe-Alagoas Basin, NE Brazil. In: *Carbonate Cementation in Sandstones* (Ed. by S. Morad), *Spec. Publ. Int. Assoc. Sediment.*, **26**, 107–140.

Garzanti, E. (1991) Non-carbonate intrabasinal grains in arenites: their recognition, significance, and relationship to eustatic cycles and tectonic setting. *J. Sedim. Petrol.*, **61**, 959–975.

Gaupp, R., **Matter, A.**, **Platt, J.**, **Ramseyer, K.** and **Walzebuck, J.** (1993) Diagenesis and fluid evolution of deeply buried Permian (Rotliegende) gas reservoirs, northwest Germany. *Bull. Am. Assoc. Petrol. Geol.*, **77**, 1111–1128.

Ghosh, S.K. (1997) Porosity distribution in relation to unconformities and its implications. *Annual Meeting Abstracts, American Association of Petroleum Geologists and Society of Economic Palaeontologists and Mineralogists*, **6**, 40.

Giles, M.R. and **De Boer, R.B.** (1990) Origin and significance of redistributional secondary porosity. *Mar. Petrol. Geol.*, **6**, 378–397.

Giles, M.R., **De Boer, R.B.** and **Marshall, J.D.** (1995) How important are organic acids in generating secondary porosity in the subsurface?. In: *Organic Acids in Geological Processes* (Ed. by E. D. Pittman and M. D. Lewan), pp. 447–470. Springer-Verlag, Berlin.

Giles, M.R., **Stevenson, S.**, **Martin, S.V.**, **Cannon, S.J.C.**, **Hamilton, P.J.**, **Marshall, J.D.** and **Samways, G.M.** (1992) The reservoir properties and diagenesis of the Brent Group: a regional perspective. In: *Geology of the Brent Group* (Ed. by A. C. Morton, R. S. Haszeldine, M. R. Giles and S. Brown), *Spec. Publ. Geol. Soc.*, **61**, 289–327.

Girard, J.P. and **Barnes, D.A.** (1995) Illitization and paleothermal regimes in the Middle Ordovician St. Peter Sandstone, Central Michigan Basin: K–Ar, oxygen isotope, and fluid inclusion data. *Bull. Am. Assoc. Petrol. Geol.*, **79**, 49–69.

Girard, J.P. and **Deynoux, M.** (1991) Oxygen isotope study of diagenetic quartz overgrowths from the Upper Proterozoic quartzites of western Mali, Taoudeni Basin: implications for conditions of quartz cementation. *J. Sedim. Petrol.*, **61**, 406–418.

Glasmann, J.R., **Lundegard, P.D.**, **Clark, R.A.**, **Penny, B.K.** and **Collins, I.D.** (1989) Geochemical evidence for the history of diagenesis and fluid migration: Brent Sandstones, Heather Field, North Sea. *Clay Miner.*, **24**, 255–264.

Gluyas, J., **Garland, C.**, **Oxtoby, N.** and **Hogg, A.** (2000) Quartz cement: the Miller's Tale. In: *Quartz Cementation of Oilfield Sandstones* (Ed. by R. H. Worden and S. Morad), *Spec. Publ. Int. Assoc. Sediment.*, in press.

Graham, C.M., **Valley, J.W.** and **Winter, B.L.** (1996) Ion microprobe analysis of $^{18}O/^{16}O$ in authigenic and detrital quartz in the St. Peter Sandstone, Michigan Basin and Wisconsin Arch., USA: contrasting diagenetic histories. *Geochim. Cosmochim. Acta*, **60**, 5101–5116.

Hall, P.L. (1994) Physical and chemical aspects of the development of overpressuring in sedimentary environments. *Clay Miner.*, **29**, 425–437.

Hanor, J.S. (1978) Precipitation of beachrock cements: mixing of marine and meteoric waters vs. CO2-degassing. *J. Sedim. Petrol.*, **48**, 489–501.

Hanor, J.S. (1987) Kilometre-scale thermohaline overturn of pore waters in the Lousiana Gulf Coast. *Nature*, **327**, 501.

Hanor, J.S. (1996) Variations in chloride as a driving force in siliciclastic diagenesis. In: *Siliciclastic Diagenesis and Fluid Flow: Concepts and Applications* (Ed. by L. J. Crossey, R. Loucks and M. W. Totten), *Spec. Publ. Soc. Econ. Paleont. Miner.*, **55**, 3–12.

Hart, B.S., **Longstaffe, F.J.** and **Plint, A.G.** (1992). Evidence for relative sea-level changes from isotopic and elemental composition of siderite in the Cardium Formation, Rocky Mountain Foothills. *Bull. Can Petrol. Geol.*, **40**, 52–59.

Hayes, M.J. and **Boles, J.R.** (1992) Volume tric relations between dissolved plagioclase and kaolinite in sandstones: implications for aluminum mass transfer in the San Joaquin Basin, California. In: *Origin, Diagenesis, and Petrophysics of Clay Minerals in Sandstones* (Ed. by D. W. Houseknecht and E. D. Pittman), *Spec. Pub. Soc. Econ. Paleont. Miner.*, **47**, 111–123.

Heikoop, J.M., **Tsujita, C.J.**, **Risk, M.J.**, **Tomascik, T.** and **Mah, A.J.** (1996) Modern iron ooids from a shallow-marine volcanic setting: Mahengetang, Indonesia. *Geology*, **24**, 759–762.

Helgeson, H.C., **Delany, J.M.**, **Nesbitt, H.W.** and **Flowers, G.C.** (1979) Summary and critique of the thermodynamic properties of rock-forming minerals. *Am. J. Sci.*, **278A**, 299pp.

Helmold, K.P. and **Kamp, P.C.V.** (1984) Diagenetic mineralogy and controls on albitization and laumontite formation in Paleogene arkoses, Santa Ynez mountains, California. In: *Clastic Diagenesis* (Ed. by D. A. McDonald and R. C. Surdam), *Mem. Am. Assoc. Petrol. Geol.*, **37**, 239–276.

Hendry, J.P. and **Trewin, N.H.** (1995) Authigenic quartz microfabrics in Cretaceous turbidites: evidence for silica transformation processes in sandstones. *J. Sedim. Res.*, **A65**, 380–392.

Hervig, R.L., **Williams, L.B.**, **Kirkland, I.K.** and **Longstaffe, F.J.** (1995) Oxygen isotope microanalyses of diagenetic quartz: possible low temperature occlusion of pores. *Geochim. Cosmochim. Acta*, **59**, 2537–2543.

Hillier, S. (1994) Pore-lining chlorites in siliciclastic reservoir sandstones: electron microprobe SEM and XRD data, and implications for their origin. *Clay Miner.*, **29**, 665–679.

Hillier, S. and **Velde, B.** (1992) Chlorite interstratified with a 7 Angstrom mineral: an example from offshore Norway and possible implications for the interpretation of the composition of diagenetic chlorites. *Clay Miner.*, **27**, 475–486.

Hogg, A.J.C., **Sellier, E.** and **Jourdan, A.J.** (1992) Cathodoluminescence of quartz cements in Brent Group sandstones, Alwyn South, UK North Sea. In: *Geology of the Brent Group* (Ed. by A. C. Morton, R. S. Haszeldine, M. R. Giles and S. Brown), *Spec. Publ. Geol. Soc.*, **61**, 421–440.

Hogg, A.J.C., **Pearson, M.J.** and **Fallick, A.E.** (1995) An integrated thermal and isotopic study of the diagenesis of the Brent Group, Alwyn South, U.K. North Sea. *Appl. Geochem.*, **10**, 531–546.

Hornibrook, E.R.C. and **Longstaffe, F.J.** (1996) Berthierine from the Lower Cretaceous Clearwater Formation, Alberta, Canada. *Clays Clay Min.*, **44**,1–21.

Houseknecht, D.W. (1988) Intergranular pressure solution in four quartzose sandstones. *J. Sedim. Petrol.*, **58**, 228–246.

Hovland, M., **Talbot, M.R.**, **Qvale, H.**, **Olaussen, S.** and **Aasberg, L.** (1987) Methane-related carbonate cements in pockmarks of the North Sea. *J. Sedim. Petrol.*, **57**, 881–892.

Hower, J., **Eslinger, E.V.**, **Hower, M.** and **Perry, E.A.** (1976) Mechanism of burial metamorphism of argillaceous sediments. I. Mineralogical and chemical evidence. *Geol. Soc. Am. Bull.*, **87**, 725–737.

Huang, W.L., **Bishop, A.M.** and **Brown, R.W.** (1986) The effect of fluid/rock ratio on feldspar dissolution and illite formation under reservoir conditions. *Clay Miner.*, **21**, 585–601.

Humphreys, B., **Kemp, S.J.**, **Lott, G.K.**, **Bermanto,**

Dharmayanti, D.A. and **Samsori, I.** (1994) Origin of grain-coating chlorite by smectite transformation: an example from Miocene sandstones, North Sumatra back-arc basin, Indonesia. *Clay Miner.*, **29**, 681–692.

Jahren, J.S. and **Aagaard, P.** (1989) Compositional variations in diagenetic chlorites and illites, and relationships with formation-water chemistry. *Clay Miner.*, **24**, 157–170.

Johnson, M.R. (1989) Palaeogeographic significance of oriented calcareous concretions in the Triassic Katberg Formation, South Africa. *J. Sedim. Petrol.*, **59**, 1008–1010.

Kaldi, J. and **Gidman, J.** (1982) Early diagenetic dolomite cements: examples from the Permian Lower Magnesian Limestone of England and the Pleistocene carbonates of the Bahamas. *J. Sedim. Petrol.*, **52**, 1073–1085.

Kantorowicz, J.D., Bryant, I.D. and **Dawans, J.M.** (1987) Controls on the geometry and distribution of carbonate cements in Jurassic sandstones: Bridport sands, southern England and Viking Group, Troll Field, Norway. In: *Diagenesis of Sedimentary Sequences* (Ed. by J. D. Marshall), *Spec. Publ. Geol. Soc.*, **36**, 103–118.

Klein, J.S., Mozley, P., Campbell, A. and **Cole, R.** (1999) Spatial distribution of carbon and oxygen isotopes in laterally extensive carbonate-cemented layers: implications for mode of growth and subsurface identification. *J. Sedim. Res.*, **69**, 184–191.

Kraishan, G.M., Rezaee, M.R. and **Worden, R.H.** (2000) Trace element composition of quartz cement: a key to reveal origin of silica. In: *Quartz Cementation of Oilfield Sandstones* (Ed. by R. H. Worden and S. Morad), *Spec. Publ. Int. Assoc. Sediment.*, in press.

Land, L.S. and **Fisher, R.S.** (1987) Wilcox sandstone diagenesis, Texas Gulf Coast: a regional isotopic comparison with the Frio Formation. In: *Diagenesis of Sedimentary Sequences* (Ed. by J. D. Marshall), *Spec. Publ. Geol. Soc.*, **36**, 219–235.

Land, L.S., Mack, L.E., Milliken, K.L. and **Lynch, F.L.** (1997) Burial diagenesis of argillaceous sediments, south Texas Gulf of Mexico sedimentary basin: a reexamination. *Geol. Soc. Am. Bull.*, **109**, 2–15.

Lee, M., Aronson, J.L. and **Savin, S.M.** (1985) K/Ar dating of the time of gas emplacement in Rotliegendes sandstone, Netherlands. *Bull. A,. Assoc. Petrol. Geol.*, **69**, 1381–1385.

Longstaffe, F.J., Tilley, B.J., Ayalon, A. and **Connolly, C.A.** (1992) Controls on porewater evolution during sandstone diagenesis, Western Canada sedimentary basin: an oxygen isotope perspective. In: *Origin, Diagenesis, and Petrophysics of Clay Minerals in Sandstones* (Ed. by D. W. Houseknecht and E. D. Pittman), *Spec. Publ. Soc. Econ. Paleont. Miner.*, **47**, 13–34.

Loomis, J.L. and **Crossey, L.J.** (1996) Diagenesis in a cyclic, regressive siliciclastic sequence: the point Lookout Sandstone, San Juan Basin, Colorado. In: *Siliciclastic Diagenesis and Fluid Flow: Concepts and Applications,* (Ed. by L. J. Crossey, R. Loucks and M. W. Totten), *Spec. Publ. Soc. Econ. Paleont. Miner.*, **55**, 23–36.

Losh, S., Eglinton, L., Schoell, M. and **Wood, J.** (1999) Vertical and lateral fluid flow related to a large growth fault, south Eugene Island block 330 field, offshore Louisiana. *Bull. Am. Ass. Petrol. Geol.*, **83**, 244–276.

Loutit, T.S., Hardenboll, J., Vail, P.R. and **Baum, G.R.** (1988). Condensed sections: the key to age dating and correlation of continental margin sequences. In: *Sea-Level Changes: an Integrated Approach* (Ed. by C. K. Wilgus, B. S. Hastings, C. G. S. C. Kendall, H. W. Posamentier, C. A. Ross and J. C. van Wagoner), *Spec. Publ. Soc. Econ. Paleont. Miner.*, **42**, 183–216.

Macaulay, C.I., Boyce, A.J., Fallick, A.E. and **Haszeldine, R.S.** (1997) Quartz veins record vertical flow at a graben edge: Fulmar Oil Field, Central North Sea. *Bull. Am. Assoc. Petrol. Geol.*, **81**, 2024–2035.

MacGowan, D.B. and **Surdam, R.C.** (1988) Difunctional carboxilic acids in oilfield waters. *Org. Geochem.*, **12**, 245–259.

MacQuaker, J.H.S., Taylor, K.G., Young, T.P. and **Curtis, C.D.** (1996) Sedimentological and geochemical controls on ooidal ironstones and 'bone-bed' formation and some comments on their sequence-stratigraphical significance. In: *Sequence Stratigraphy in British Geology* (Ed. by S. P. Hesselbo and D. N. Parkinson), *Spec. Publ. Geol. Soc.*, **103**, 97–107.

Madé, B., Ben Baccar, M. and **Fritz, B.** (1994) Geochemical modelling of diagenetic reactions: a thermodynamic and kinetic approach. *Miner. Magazine*, **58A**, 549–550.

Manning. A.A.C., Gestsdóttir, K. and **Rae, E.I.** (1992) Feldspar dissolution in the presence of organic acid anions under diagenetic conditions: an experimental study. In: *Advances in Organic Geochemistry* (Ed. by J. R. Maxwell, C. B. Eckardt, S. R. Larter and D. A. C. Manning), *Org. Geochem.*, **19**, 483–492.

Matsumoto, R. (1990) Vuggy carbonate crust formed by hydrocarbon seepage on the continental shelf of Baffin Island, northeast Canada. *Geochem. J.*, **24**, 143–158.

Matsumoto, R. (1992) Diagenetic dolomite, calcite, rhodochrosite, magnesite, and landsfordite from Site 799, Japan Sea – Implications for depositional environments and the diagenesis of organic-rich sediments. In: *Proceedings of the Ocean Drilling Program, Scientific Results* (Ed. by K. A. Pisciotto, J. C. Ingle, M. T. von Breymann and J. Barron), 127/128, pp. 75–98. US Government Publishing Office, Washington.

McBride, E.F., Land, L.S. and **Mack, L.E.** (1987) Diagenesis of eolian and fluvial feldspathic sandstones, Norphlet Formation (Upper Jurassic), Rankin County, Mississippi, and Mobile County, Alabama. *Bull. Am. Ass. Petrol. Geol.*, **71**, 1019–1034.

McBride, E.F., Milliken, K.L., Cavazza, W., Cibin, U., Fontana, D., Picard, M.D. and **Zuffa, G.G.** (1995) Heterogeneous distribution of calcite cement at the outcrop scale in Tertiary sandstones, northern Apennines, Italy. *Bull. Am. Assoc. Petrol. Geol.*, **79**, 1044–1063.

McBride, E.F., Picard, M.D. and **Folk, R.L.** (1994) Oriented concretions, Ionian coast, Italy: evidence of groundwater flow direction. *J. Sedim. Res.*, **A64**, 535–540.

McKay, J.L., Longstaffe, F.J. and **Plint, A.G.** (1995) Early diagenesis and its relationship to depositional environment and relative sea-level fluctuations (Upper Cretaceous Marshybank Formation, Alberta and British Columbia). *Sedimentology*, **42**, 161–190.

Merino, E., Girard, J.-P., May, M.T. and **Ranganathan, V.** (1997) Diagenetic mineralogy, geochemistry, and dynamics of Mesozoic arkoses, Hartford Rift Basin, Connecticut, U.S.A. *J. Sedim. Res.*, **67**, 212–224.

Metcalfe, R., Rochelle, C.A., Savage, D. and **Higgo, J.W.** (1994) Fluid–rock interactions during continental red bed diagenesis: implications for theoretical models of mineralization in sedimentary basins. In: *Geofluids: Origin, Migration and Evolution of Fluids in Sedimentary Basins* (Ed. by J. Parnell), *Spec. Publ. Geol. Soc.*, **78**, 301–324.

Milliken, K.L. (1989) Petrography and composition of authigenic feldspars, Oligocene Frio Formation, South Texas. *J. Sedim. Petrol.*, **59**, 361–374.

Moncure, G.K., **Lahann, R.W.** and **Siebert, R.M.** (1984) Origin of secondary porosity and cement distribution in a sandstone/shale sequence from the Frio Formation (Oligocene). In: *Clastic Diagenesis* (Ed. by D.A.McDonald and R.C.Surdam), *Mem. Am. Assoc. Petrol. Geol.*, **37**, 151–161.

Morad, S. (1990) Mica alteration reactions in Jurassic reservoir sandstones from the Haltenbanken area, offshore Norway. *Clays Clay Miner.*, **38**, 584–590.

Morad, S., **Al-Aasm, I.S.**, **Longstaffe, F.J.**, **Marfil, R.**, **De Ros, L.F.**, **Johansen, H.** and **Marzo, M.** (1995) Diagenesis of a mixed siliciclastic/evaporitic sequence of the Middle Muschelkalk (Middle Triassic), the Catalan Coastal Range, NE Spain. *Sedimentology*, **42**, 749–768.

Morad, S., **Ben Ismail, H.**, **De Ros, L.F.**, **Al-Aasm, I.S.** and **Serrhini, N.-E.** (1994) Diagenesis and formation water chemistry of Triassic reservoir sandstones from southern Tunisia. *Sedimentology*, **41**, 1253–1272.

Morad, S., **Bergan, M.**, **Knarud, R.** and **Nystuen, J.P.** (1990) Albitization of detrital plagioclase in Triassic reservoir sandstones from the Snorre Field, Norwegian North Sea. *J. Sedim. Petrol.*, **60**, 411–425.

Morad, S. and **De Ros, L.F.** (1994) Geochemistry and diagenesis of stratabound calcite cement layers within the Rannoch Formation of the Brent Group, Murchison Field, North Viking Graben (northern North Sea) – comment. *Sedim. Geol.*, **93**, 135–141.

Morad, S., **De Ros, L.F.**, **Nystuen, J.P.** and **Bergan, M.** (1998) Carbonate diagenesis and porosity evolution in sheet-flood sandstones: evidence from the Middle and Lower Lunde Members (Triassic) in the Snorre Field, Norwegian North Sea. In: *Carbonate Cementation in Sandstones* (Ed. by S. Morad), *Spec. Publ. Int. Assoc. Sediment.*, **26**, 53–85.

Morad, S., **Marfil, R.** and **De La Peña, J.A.** (1989) Diagenetic K-feldspar pseudomorphs in the Triassic Buntsandstein sandstones of the Iberian Range, Spain. *Sedimentology*, **36**, 635–650.

Morad, S., **Marfil, R.**, **Al-Aasm, I.S.** and **Gomez-Gras, D.** (1992) The role of mixing-zone dolomitization in sandstone cementation: evidence from the Triassic Buntsandstein, the Iberian Range, Spain. *Sedim. Geol.*, **80**, 53–65.

Moraes, M.A.S. and **De Ros, L.F.** (1990) Infiltrated clays in fluvial Jurassic sandstones of Recôncavo Basin, northeastern Brazil. *J. Sedim. Petrol.*, **60**, 809–819.

Mozley, P.S. (1989) Relation between depositional environment and the elemental composition of early diagenetic siderite. *Geology*, **17**, 704–706.

Mozley, P.S. and **Davis, M.** (1996) Relationship between oriented calcite concretions and permeability correlation structure in an alluvial aquifer, Sierra Ladrones Formation, New Mexico. *J. Sedim. Res.*, **66**, 11–16.

Mozley, P.S. and **Goodwin, L.B.** (1995) Patterns of cementation along a Cenozoic normal fault: a record of paleoflow orientations. *Geology*, **23**, 539–542.

Odin, G.S. (1985) Significance of green particles (glaucony, berthierine, chlorite) in arenites. In: *Provenance of Arenites* (Ed. by G. G. Zuffa), NATO-ASI Series C: Mathematical and Physical Sciences, 148, pp. 279–307. Reidel, Dordrecht.

Oelkers, E.H., **Bjørkum, P.A.** and **Murphy, W.M.** (1996) A petrographic and computational investigation of quartz cementation and porosity reduction in North Sea sandstones. *Am. J. Sci.*, **296**, 420–452.

Pittman, E.D., **Larese, R.E.** and **Heald, M.T.** (1992) Clay coats: occurrence and relevance to preservation of porosity in sandstones. In: *Origin, Diagenesis, and Petrophysics of Clay Minerals in Sandstones* (Ed. by D. W. Houseknecht and E. D. Pittman), *Spec. Publ. Soc. Econ. Paleont. Miner.*, **47**, 241–264.

Posamentier, H.W., **Allen, G.P.**, **James, D.P.** and **Tesson, M.** (1992) Forced regressions in a sequence stratigraphic framework: concepts, examples, and exploration significance. *Bull. Am. Assoc. Petrol. Geol.*, **76**, 1687–1709.

Prosser, D.J., **Daws, J.A.**, **Fallick, A.E.** and **Williams, B.P.J.** (1993) Geochemistry and diagenesis of stratabound calcite cement layers within the Rannoch Formation of the Brent Group, Murchison Field, North Viking Graben (northern North Sea). *Sedim. Geol.*, **87**, 139–164.

Raiswell, R. (1971) The growth of Cambrian and Liassic concretions. *Sedimentology*, **17**, 147–171.

Raiswell, R. (1982) Pyrite texture, isotopic composition and the availability of iron. *Am. J. Sci.*, **282**, 1244–1263.

Rask, J.H., **Bryndzia, L.T.**, **Braunsdorf, N.R.** and **Murray, T.E.** (1997) Smectite illitization in Pliocene-age Gulf of Mexico mudrocks. *Clays Clay Miner.*, **45**, 99–109.

Read, J.F. and **Horbury, A.D.** (1993) Eustatic and tectonic controls on porosity evolution beneath sequence-bounding unconformities and parasequence disconformities on carbonate platforms. In: *Diagenesis and Basin Development* (Ed. by A. Horbury and A. G. Robinson), *AAPG Studies in Geology*, 36, pp. 155–197. AAPG, Tulsa.

Remy, R.R. (1994) Porosity reduction and major controls on diagenesis of Cretaceous-Paleocene volcaniclastic and arkosic sandstone, Middle Park Basin, Colorado. *J. Sedim. Res.*, **A64**, 797–806.

Rossi, C. and **Cañaveras, J.C.** (1999) Pseudospherulitic fibrous calcite in paleo-groundwater, unconformity-related diagenetic carbonates (Paleocene of the Áger Basin and Miocene of the Madrid Basin, Spain). *J. Sedim. Res.*, **69**, 224–238.

Saigal, G.C., **Morad, S.**, **Bjørlykke, K.**, **Egeberg, P.K.** and **Aagaard, P.** (1988) Diagenetic albitization of detrital K-feldspars in Jurassic, Lower Cretaceous, and Tertiary clastic reservoirs from offshore Norway, I. Textures and origin. *J. Sedim. Petrol.*, **58**, 1003–1013.

Schmidt, V. and **McDonald, D.A.** (1979) The role of secondary porosity in the course of sandstone diagenesis. In: *Aspects of Diagenesis* (Ed. by P. A. Scholle and P. R .Schluger), *Spec. Publ. Soc. Econ. Paleont. Miner.*, **29**, 175–207.

Schulz-Rojahn, J.P. (1993) Calcite-cemented zones in the Eromanga Basin: clues to petroleum migration and entrapment? *Aust. Petrol. Explor. Explor. Ass. J.*, **33**, 63–76.

Schulz-Rojahn, J.P. and **Phillips, S.E.** (1989) Diagenetic alteration of Permian reservoir sandstones in the Nappamerri Trough and adjacent areas, southern Cooper Basin. In: *The Cooper and Euromanga Basins, Australia* (Ed. by B. J. O'Neil), Proceedings, Adelaide, Australia, pp. 629–645. Society of Petroleum Engineers and Australian Society of Exploration Geophysicists.

Shanmugam, G. (1988) Origin, recognition, and importance of erosional unconformities in sedimentary basins. In: *New Perspectives in Basin Analysis* (Ed. by K. L. Kleinspehn and C. Paola), Chapter 5, pp. 83–108. Springer-Verlag, New York.

Shanmugam, G. (1990) Porosity prediction in sandstones using erosional unconformities. In: *Prediction of Reservoir Quality Through Chemical Modelling* (Ed. by I. D. Meshri and P. J. Ortoleva), *Mem. Am. Assoc. Petrol. Geol.*, **49**, 1–23.

Shock, E.L. (1988) Organic acid metastability in sedimentary basins. *Geology*, **16**, 886–890.

Sibley, D.F. (1978) K-feldspar cement in the Jacobsville sandstone. *J. Sedim. Petrol.*, **48**, 983–986.

Siebert, R.M. (1985) The origin of hydrogen sulfide, elemental sulfur, carbon dioxide, and nitrogen in reservoirs. *Timing of Siliciclastic Diagenesis: Relationship to Hydrocarbon Migration*, pp. 30–31. Sixth Annual Research Conference, Gulf Coast Section, SEPM, Society of Economic Palaeontologists and Mineralogists.

Simpson, G. and **Hutcheon, I.** (1995) Pore-water chemistry and diagenesis of the modern Fraser River Delta. *J. Sedim. Res.*, **A65**, 648–655.

Smale, D. (1989). Leaching of heavy minerals above and below the Mid-Cretaceous Unconformity in the Ohuriawa Gorge area of the Waipara River, North Canterbury, New Zealand. *J. Sedim. Petrol.*, **59**, 1011–1021.

Smith, J.S. (1996) Origins of calcite and pyrite cements in the Carrizo Sandstone on the flank of Butler salt dome, East Texas Basin; petrographic and isotopic evidence for fluid circulation near salt domes. *MSc Thesis*, University of Texas, Austin.

Smith, J.T. and **Ehrenberg, S.N.** (1989) Correlation of carbon dioxide abundances with temperature in clastic hydrocarbon reservoirs: relationships to inorganic chemical equilibrium. *Mar. Petrol. Geol.*, **6**, 129–135.

Sombra, C.L. and **Chang, H.K.** (1997) Burial history and porosity evolution of Brazilian Upper Jurassic to Tertiary sandstone reservoirs. In: *Reservoir Quality Prediction in Sandstones and Carbonates* (Ed. by J. A. Kupecz, J. G. Gluyas and S. Bloch), *Mem. Am. Assoc. Petrol. Geol.*, **69**, 79–90.

Souza, R.S. and **Silva, C.M.** (1998) Origin and timing of carbonate cementation of the Namorado Sandstone (Cretaceous), Albacora Field, Brazil: implications for oil recovery. In: *Carbonate Cementation in Sandstones* (Ed. by S. Morad), *Spec. Publ. Int. Assoc. Sediment.*, **26**, 309–325.

Souza, R.S., De Ros, L.F. and **Morad, S.** (1995) Dolomite diagenesis and porosity preservation in lithic reservoirs: Carmópolis Member, Sergipe-Alagoas Basin, northeastern Brazil. *Bull. Am. Assoc. Petrol. Geol.*, **79**, 725–748.

Stanton, P.T. (1986) Sedimentology, diagenesis, and reservoir potential of the Pennsylvanian Tyler Formation, central Montana. *MSc Thesis*, University of Colorado, Boulder.

Studlick, J.R.J., Shew, R.D., Basye, G.L. and **Ray, J.R.** (1990) A giant carbon dioxide accumulation in the Norphlet Formation, Pisgah Anticline, Mississippi. In: *Sandstone Petroleum Reservoirs* (Ed. by J. H. Barwis, J. G. McPherson and J. R. J. Studlick), *Casebooks in Earth Science*, pp. 181–203. Springer-Verlag, New York.

Sturesson, U. (1992) Volcanic ash: the source material for Ordovician chamosite ooids in Sweden. *J. Sedim. Petrol.*, **62**, 1084–1094.

Sturesson, U., Heikoop, J.M. and **Risk, M.J.** (1999) Modern and Palaeozoic iron ooids – a similar volcanic origin. *Sedim. Geol.*, in press.

Sullivan, M.D., Macaulay, C.I., Fallick, A.E. and **Haszeldine, R.S.** (1997) Imported quartz cement in aeolian sandstone grew from water of uniform composition but has complex zonation. *Terra Nova*, **9**, 237–241.

Summer, N.S., Ayalon, A. and **Kolodny, Y.** (1995) Oxygen isotope systematics of a sandstone-quartzite transition in contact zones. *Israel J. Earth Sci.*, **44**, 169–181.

Surdam, R.C., Boese, S.W. and **Crossey, L.J.** (1984) The chemistry of secondary porosity. In: *Clastic Diagenesis* (Ed. by R. C. Surdam and D. A. McDonald), *Mem. Am. Assoc. Petrol. Geol.*, **37**, 127–149.

Surdam, R.C. and **Boles, J.R.** (1979) Diagenesis of volcanic sandstones. In: *Aspects of Diagenesis* (Ed. by P. A. Scholle and P. R. Schluger), *Spec. Publ. Soc. Econ. Paleont. Miner.*, **26**, 227–242.

Surdam, R.C., Dunn, T.L., Heasler, H.P. and **Macgowan, D.B.** (1989) Porosity evolution in sandstone/shale systems. In: *Short Course on Burial Diagenesis* (Ed. by I. E. Hutcheon), pp. 61–133. Miner. Assoc. Canada.

Surdam, R.C., Jiao, Z.S. and **Macgowan, D.B.** (1993) Redox reactions involving hydrocarbons and mineral oxidants: a mechanism for significant porosity enhancement in sandstones. *Bull. Am. Ass. Petrol. Geol.*, **77**, 1509–1518.

Surdam, R.C. and **Yin, P.** (1995) Organic acids and carbonate stability, the key to predicting positive porosity anomalies. In: *Organic Acids in Geological Processes* (Ed. by E. D. Pittman and M. D. Lewan), pp. 398–448. Springer-Verlag, Berlin.

Swarbrick, R.E. (1994) Reservoir diagenesis and hydrocarbon migration under hydrostatic palaeopressure conditions. *Clay Miner.*, **29**, 463–473.

Taylor, K.G., Gawthorpe, R.L. and **van Wagoner, J.C.** (1995) Stratigraphic control on laterally persistent cementation, Book Cliff, Utah. *J. Sedim. Res.*, **69**, 225–228.

Taylor, T.R. and **Land, L.S.** (1996) Association of a allochthonous waters and reservoir enhancement in deeply buried Miocene sandstones: Picaroon Field, Corsair Trend, Offshore Texas. In: *Siliciclastic Diagenesis and Fluid Flow: Concepts and Applications* (Ed. by L. J. Crossey, R. Loucks and M. W. Totten), *Spec. Publ. Soc. Econ. Paleont. Miner.*, **55**, 37–48.

Thiry, M. and **Milnes, A.R.** (1991) Pedogenic and groundwater silcretes at Stuart Creek opal field, South Australia. *J. Sedim. Petrol.*, **61**, 111–127.

Thrailkill, J. (1968) Chemical and hydrologic factors in the excavation of limestones caves. *Geol. Soc. Am. Bull.*, **79**, 19–46.

Tilley, B.J. and Longstaffe, F.J. (1989) Diagenesis and isotopic evolution of porewaters in the Alberta Deep Basin: the Falher Member and Cadomin Formation. *Geochim. Cosmochim. Acta*, **53**, 2529–2546.

Vail, P.R., Mitchum, R.M. and **Thompson, S.** (1977) Seismic stratigraphy and global changes of sea-level, part 3: relative changes of sea-level from coastal onlap. In: *Seismic Stratigraphy – Applications to Hydrocarbon Exploration* (Ed. by C. E. Payton), *Mem. Am. Assoc. Petrol. Geol.*, **26**, 63–81.

van de Graaff, W.J.E. (1983) Silcrete in Western Australia: geomorphological settings, textures, structures, and their genetic implications. In: *Residual Deposits: Surface Related Weathering Processes and Materials* (Ed. by R. C. L. Wilson), 11, pp. 159–166. Blackwell, Oxford.

van Wagoner, J.C., Mitchum, R.M., Campion, K.M. and **Rahmanian, V.D.** (1990) Siliciclastic sequence stratigraphy in well logs, cores, and outcrops: concepts for high-resolution correlation of time and facies. *AAPG Methods in Exploration Series, 7*. AAPG, Tulsa.

von Rad, U., Roesch, H., Berner, B., Geyh, M., Marchig, V. and **Schulz, H.** (1996) Authigenic carbonates derived from oxidized methane vented from the Makran accretionary prism off Pakistan. *Mar. Geol.*, **136**, 55–77.

Walderhaug, O. and **Bjørkum, P.A.** (1998) Calcite cement in shallow marine sandstones: growth mechanisms and geometry. In: *Carbonate Cementation in Sandstones* (Ed. by S. Morad), *Spec. Publ. Int. Assoc. Sediment.*, **26**, 179–192.

Warren, E.A. and **Smalley, P.C.** (1994) North Sea Formation Water Atlas. *Mem. Geol. Soc.*, **15**.

Watts, N.L. (1980) Quaternary pedogenic calcretes from the Kalahari (south Africa): mineralogy, genesis and diagenesis. *Sedimentology*, **27**, 661–686.

Wescott, W.A. (1983) Diagenesis of Cotton Valley sandstone (Upper Jurassic), east Texas: implications for tight gas formation pay recognition. *Bull. Am. Assoc. Petrol. Geol.*, **67**, 1002–1013.

Wilkinson, M. (1989). Discussion: evidence for surface reaction-controlled growth of carbonate concretions in shales. *Sedimentology*, **36**, 951–953.

William, L.B., **Hervig, L.** and **Dutton, S.P.** (1997) Constraints on paleofluid compositions in the Travis Peak Formation, East Texas: Evidence from microanalyses of oxygen isotopes in diagenetic quartz. In: *Basin-Wide Diagenetic Patterns* (Ed. by I. P. Montañez, J. M. Gregg and K. L. Shelton), *Spec. Publ. Soc. Econ. Paleont. Miner.*, **57**, 269–280.

Wilson, M.D. and **Stanton, P.T.** (1994). Diagenetic mechanism of porosity and permeability reduction and enhancement. In: *Reservoir Quality Assessment and Prediction in Clastic Rocks* (Ed. by M. D. Wilson), *SEPM Short Course*, **30**, 59–118.

Wintsch, R.P. and **Kvale, C.M.** (1994) Differential mobility of elements in burial diagenesis of siliciclastic rocks. *J. Sedim. Res.*, **A64**, 349–361.

Wood, J.R. (1994) Geochemical Models. In: *Reservoir Quality Assessement and Prediction in Clastic Rocks* (Ed. by M. D. Wilson), *SEPM Short Course*, **30**, 23–40.

Wopfner, H., **Markwort, S.** and **Semkiwa, P.M.** (1991) Early diagenetic laumontite in the Lower Triassic Manda Beds of the Ruhuhu Basin, southern Tanzania. *J. Sedim. Petrol.*, **61**, 65–72.

Worden, R.H., **Oxtoby, N.H.** and **Smalley, P.C.** (1998) Can oil emplacement stop quartz cementation in sandstones? *Petrol. Geosci.*, **4**, 129–137.

Yin, P. and **Surdam, R.C.** (1985) Naturally-enhanced porosity and permeability in the hydrocarbon reservoirs of the Gippsland Basin, Australia. *First Wyoming Enhanced Oil Recovery Symposium, Laramie, WY*, pp. 79–109. EOR Institute, University of Wyoming.

Zotov, A., **Mukhamet-Galeev, A.** and **Schott, J.** (1998) An experimental study of kaolinite and dickite relative stability at 150–300 °C and the thermodynamic properties of dickite. *Am. Miner.*, **83**, 516–524.

Zuffa, G.G., **Cibin, U.** and **Di Giulio, A.** (1995) Arenite petrography in sequence stratigraphy. *J. Geol.*, **103**, 451–459.

Sedimentology (2000) **47** (Suppl. 1), 121–178

Quantitative models of sedimentary basin filling

CHRIS PAOLA
Department of Geology & Geophysics and St. Anthony Falls Laboratory, University of Minnesota, Minneapolis, MN 55455, USA (E-mail: cpaola@tc.umn.edu)

ABSTRACT

Quantitative modelling of the filling of sedimentary basins was begun in earnest in the 1960s. Dozens of themes and variations have been proposed since then, and have yielded an abundance of idealized stratigraphic patterns as functions of both imposed changes and basin properties. Post-plate-tectonic modelling began with 'rigid-lid' models, which show the stratigraphic signature of subsidence variation. This work introduced the connection between stratigraphy and the rheology of the lithosphere. Rigid-lid models are the simplest type of geometric model, in which the sediment surface is assigned prescribed geometries, usually corresponding to different depositional environments. These can reproduce many aspects of overall stratal geometry but are formally restricted to relatively long timescales, for which quasi-steady surface topography can be assumed. So-called dynamic models attempt to represent the morphodynamics of the sediment surface by abstracting and averaging short-term transport processes. Most of the dynamic models proposed to date can be seen as special cases of a single general morphodynamic equation.

The most important result of the first wave of quantitative basin-filling models is that even relatively simple models can produce reasonable stratal patterns. We now have a wide array of tools for exploring scenarios, searching for general behaviours and effects, and making initial quantitative predictions. We have also learned that basin response to external forcing as recorded in stratigraphy can be as sensitive to the characteristics of the basin as to the forcing.

The main brake on the development of basin modelling is not computing power but lack of methods and data for testing the models we have already developed. Physical experiments, which are only just beginning, are one means of doing this. Experimental stratigraphy is a bridge to quantitative field tests, which will require collaboration among academic researchers from a wide range of areas, and between academia and industry, on projects of greater scale and degree of integration than we have seen so far. The advancement of quantitative sedimentary geology will also require significant changes in the way the subject is taught, at all levels.

Keywords Basin, sedimentation, stratigraphy, quantitative model.

'The sediments are a sort of epic poem of the Earth.' – Rachel Carson

INTRODUCTION

Goals and scope of the paper

The 'epic poem of the Earth' is written in a language that we still do not entirely understand. So far, interpretation of the language of sediments has been done with scant recourse to mathematics, the common language of science. However, this is changing. As our science matures, we find that application of quantitative techniques lets us ask sharper questions, makes our hypotheses less ambiguous and hence easier to test, rules out some apparently plausible explanations and suggests new ones that would never have occurred to our unaided intuition. In the last 20 years or so this analytical approach has led to a bloom of quantitative models of the filling of sedimentary basins.

Much has been written about the supposed goals and motivations for modelling in the Earth sciences, but in the end the goal of modelling is the same as that of all other aspects of science: to generate insight. Models are simply expressions of our ideas about how things work. All of us use models, no matter how inductive or empirical we think we are. Models of one form or another have been used since geology began, to guide our observations and to help us make sense of what we see and measure. What has changed is that to our collection of descriptive models we have added more analytical, quantitative models based on physical, chemical and biological principles. This paper is a survey of what has been done in theoretical and physical modelling of sedimentary systems at large scales, what we have learned from it and what some of the main challenges are now.

It is worth clarifying some terms before going any further. I will follow common usage if not etymology and use 'stratigraphy' for the study of sedimentary sequences at large scales (roughly formations and greater), and 'sedimentology' for the study of sediments and sedimentary rocks at smaller scales. 'Sedimentary geology' includes both. 'Physical modelling' means experimental studies of sedimentary processes, at either full or reduced scale. 'Theoretical modelling' means any form of modelling based on quantitative analysis. Theoretical models become 'numerical models' when they involve equations that must be solved numerically. In some cases numerical models may use equations that are supposed to embody real processes but are not directly derived from differential equations; this subcategory of numerical models is often called 'rule-based'. In the extreme case, rule-based models may be semiquantitative or even qualitative even though they are implemented numerically.

This paper is a review of quantitative stratigraphic models. For the most part, this means that its scope is restricted to models of sedimentation on scales for which subsidence is important. However, the focus is on basin filling, not subsidence itself. And, although deep-marine sedimentation is of great importance in modelling sedimentary systems, aspects of this are covered in the accompanying review by Kneller & Buckee (2000) and so will not be discussed here.

The development of quantitative basin analysis

Space does not permit a lengthy historical review of the evolution of the quantitative approach to understanding sedimentary systems. However, as this paper was being written the sedimentary–geology community learned with sorrow of the death of Francis Pettijohn. Pettijohn was one of the founders of modern basin analysis, and his work has greatly affected the research reviewed here. His approach was most comprehensively expressed in his well-known books (Krumbein & Pettijohn, 1938; Pettijohn, 1957; Pettijohn & Potter, 1964; Pettijohn *et al.*, 1973; Potter & Pettijohn, 1977), of which the latter two are the most relevant to basin analysis.

Francis Pettijohn was an outspoken advocate of a pragmatic, field-orientated approach to basin study. He doubtless would not have approved of many of the more esoteric aspects of the work described in this paper. Nonetheless, his approach was also consistently analytical. The Pettijohn school placed great importance on reconstructing spatial transport patterns via palaeocurrents, a topic that is currently not in vogue but that will return to play a major role in modelling as we move from two to three dimensions. An early example of quantitative analysis developed by Pettijohn and some of his students involved the exponential model proposed by Sternberg (1875) for downstream fining in fluvial systems. The idea was that by fitting this model to grain-size data from ancient systems, one could extrapolate upstream to estimate how far away the source area was (Schlee, 1957; Pelletier, 1958; Yeakel, 1962; Meckel, 1967). The estimate has considerable uncertainty but is nonetheless better than having no estimate at all, both still typical features of applications of quantitative models to ancient systems. The Sternberg fining model was one of the first quantitative models developed in fluvial geomorphology. Although essentially empirical, it was supported by a simple physical theory that added to its appeal. Its application by Pettijohn's group nicely illustrates the close connection between geomorphic and stratigraphic modelling. The basic logical structure – combining a quantitative model with careful measurements from rocks and general geological constraints to derive new information about the depositional system – is also still recognizable in the application of models to the stratigraphic record today.

The first modern stratigraphic model was proposed by Sloss (1962). It was conceptual, rather than quantitative, but it was formulated in a careful and disciplined way that made it as well suited as a basis for quantitative modelling as for conceptual refinement. Sloss argued that the shape of a body of sedimentary rock is controlled by four variables:

Q: the quantity of sediment supplied per unit time, equivalent in current models to the supply sediment flux $Q_s(\mathbf{x}_0, \mathbf{x}_1, \ldots)$ where $\mathbf{x}_0, \mathbf{x}_1, \ldots$, are the locations of sediment sources at the edges of the basin.

R: the 'receptor value', the rate of subsidence relative to base level. This is now called 'accommodation' and is equivalent in current models to $\sigma + \partial H/\partial t$, where σ is subsidence rate, H base level and t time.

D: the dispersal, a measure of the total effect of all the sediment transport processes, expressed as a rate. This is represented in various forms in the work discussed below but is closest to the sediment flux distribution within the depositional system, $q_s(x, y)$.

M: a measure of the composition and texture of the sediment supply. This could be represented in various ways depending on the detail required, but fundamentally it is the probability density $f_{D0}(D)$ of grain size D in the sediment supply.

On the basis of this model, Sloss worked out a series of scenarios for transgression and regression (Fig. 1). His generalized formulation implied a broad view of the causes of sedimentary signals. This view was to some extent lost in the early stages of sequence stratigraphy, with its exclusive emphasis on sea-level (a restriction to which Sloss himself later objected; Sloss, 1991). Modern basin-filling models, however, very much reflect this spirit of searching for a complete, fundamental set of independent controlling variables for stratal geometry.

The conceptual branch that grew from Sloss's ideas blossomed in the 1970s in the form of sequence stratigraphy. The quantitative branch took longer to bear fruit; the results are the subject of this paper. An early attempt to formalize and quantify Sloss's conceptual framework was presented by Schwarzacher (1966), who proposed a one-dimensional sedimentation model (a first-order, linear, ordinary differential equation) based on the hypothesis that the sedimentation rate under water increases with depth. This is how some aspects of carbonate sedimentation are modelled today. The most important and prescient result of Schwarzacher's analysis was that his model evolved toward a steady-state solution in which sedimentation rate balances accommodation (subsidence plus eustatic sea-level rise).

The first modern-style dynamic model of basin filling is, interestingly enough, a model of evaporite sedimentation proposed by Briggs & Pollack (1967). All the elements are there: analysis from first principles to derive a model; a target field area (the Michigan Basin) to provide appropriate palaeogeographical constraints; numerical solution of the equations; and (naturally) state-of-the-art computer graphics (Fig. 2a). It is interesting that after this early work, evaporite sedimentation has all but disappeared from quantitative basin-filling models (though for an exception, see Waltham (1992)).

Both Sloss's paper and the Briggs–Pollack evaporite model figure as inspirations for the now classic book by Harbaugh & Bonham-Carter (1970). In a single stroke, this book showed the way to build computational stratigraphic models. In addition to general solution methods and algorithms, it presents a remarkable range of models (Fig. 2b): diffusion in one and two dimensions, potential flow, delta evolution, evaporite precipitation and carbonate deposition (not to mention palaeoecology and hydrothermal ore formation). Stochastic effects are included, with an emphasis on Markov chains. A number of the sedimentation algorithms had been developed earlier by Harbaugh and were published beginning in 1966 (Harbaugh, 1966) by the Kansas Geological Survey in its Computer Contribution series, one of the most farsighted publication series ever produced by a regional geological survey. It would be difficult to exaggerate the contribution that Harbaugh and Bonham-Carter's work made to the development of quantitative stratigraphy. If it can be criticized, it would only be for being slightly ahead of its time: the missing ingredient was plate tectonics.

The development of plate tectonics obviously had an enormous influence on sedimentary geology as on most aspects of Earth science – it is no accident that one of the crucial papers that began the modern tectonic era (Dietz, 1963) was a reinterpretation of the classical geosyncline. Thus the connection between plate tectonics and stratigraphy was established immediately. The advent of plate-tectonic theory also had a more subtle, long-term effect on thinking in the Earth sciences: it led to a great flowering of geomechanics. The realization that

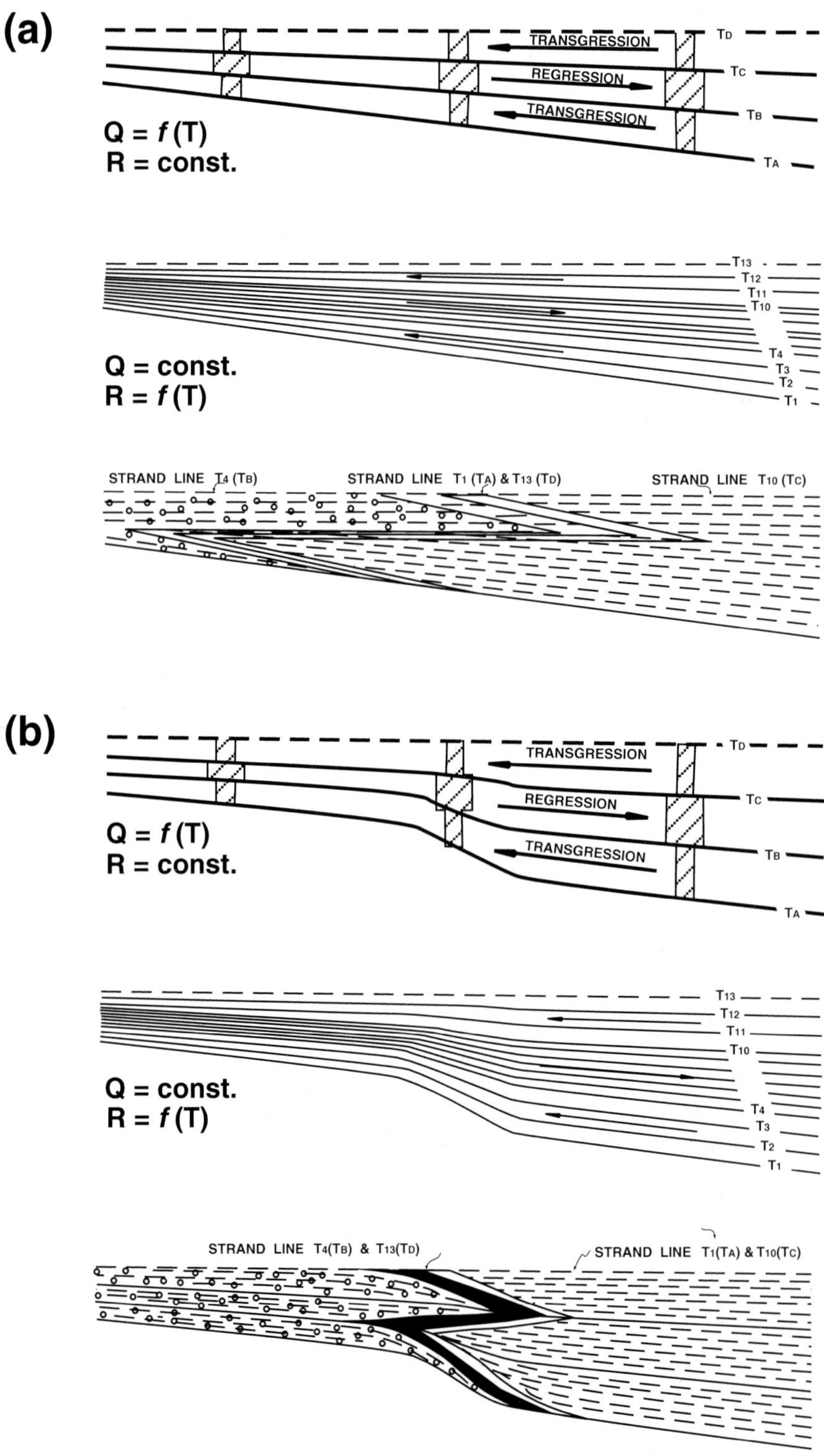

Fig. 1. Stratigraphic cycles as modelled by Sloss (1962). In each scenario, the transgression–regression pattern (bottom) can be produced either by changing sediment supply (top) or relative sea-level (middle). The cross-hatched areas in the top panels represent sediment supply, and the lines in the middle panels represent equal time intervals. (a) Linear increase in subsidence; (b) nonlinear increase in subsidence.

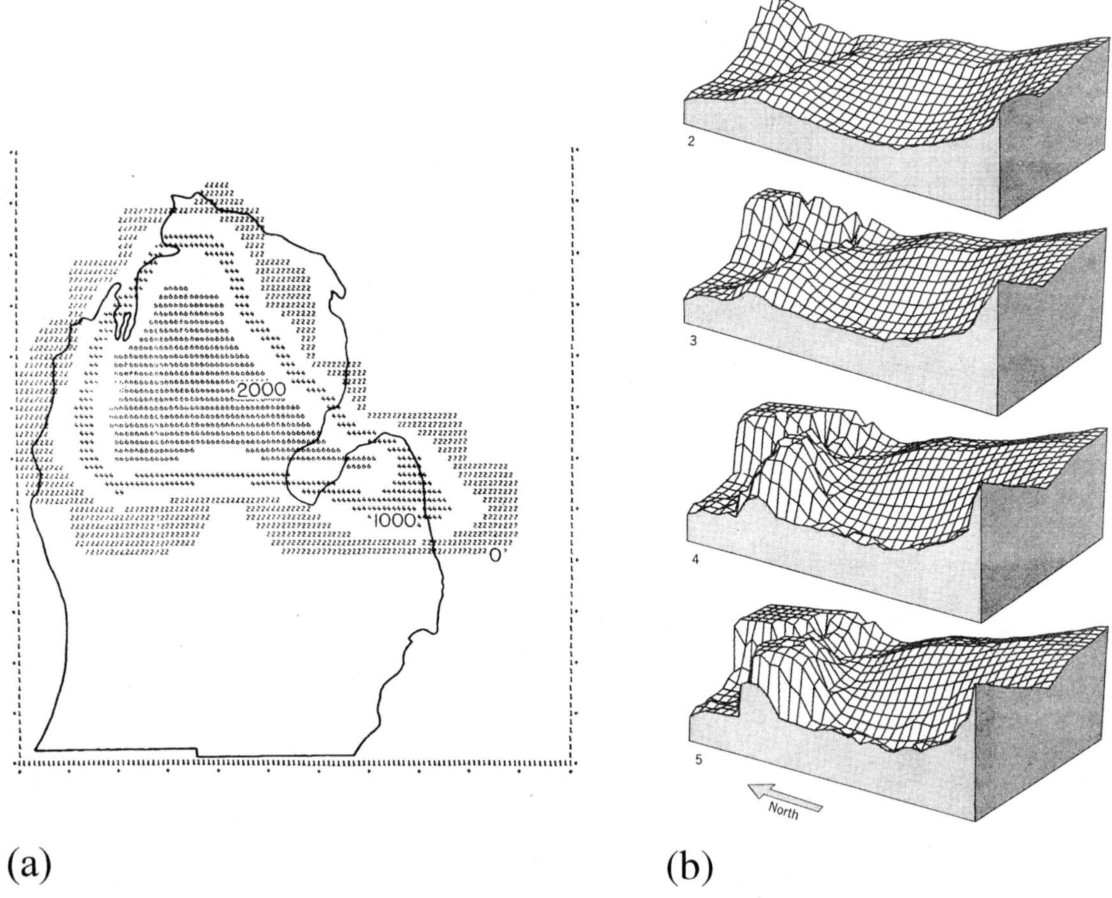

Fig. 2. The state of modelling of basin sedimentation in the late 1960s. (a) Evaporite thickness in the Michigan basin, reprinted with permission from Briggs & Pollack (1967; © 1967, American Association for the Advancement of Science, contour interval is in feet). (b) Three-dimensional image of carbonate bank migration from Harbaugh & Bonham-Carter (1970; © 1970, reprinted by permission of John Wiley & Sons, Inc.). The block is 64 km long.

so many apparently unrelated geological phenomena could be unified in the light of a bold, simple mechanical model inspired new efforts to understand other aspects of geodynamics in terms of simple mechanical analogues. In sedimentary geology, the most dramatic effect of this growth of 'mechanism' was the advent of quantitative models for basin subsidence. These, in turn, allowed development of quantitative models for the geometry of the strata filling the basin.

The crucial paper that launched the revolution in stratigraphic modelling was published in 1978. As it happens 1978 saw the publication of several papers that together make it a turning point in the analysis of sedimentary systems. Here we depart briefly from following scientific themes to focus on that year.

1978 – the true beginning of the third millennium

It is a commonplace that decades as defined by a particular zeitgeist often do not begin and end on exact calendar boundaries. Compared with a few years' leeway for decades, it does not seem unreasonable to define the start of a new millennium for sedimentary geology 22 years early, in 1978. Consider the following developments.

Physically based subsidence models

These were developed in the geophysics community. In a very brief paper, McKenzie (1978) provided the first quantitative model for thermal subsidence following extension. This model followed from Sleep's (1971) observation that the time history of basement elevation in a passive margin is equivalent to the spatial distribution of crustal elevation about an ocean spreading centre. McKenzie then proposed that passive-margin subsidence can be modelled in terms of an instantaneous extension followed by conductive cooling and shrinking of the stretched lithospheric column. A series of papers elaborating on this model soon followed, but McKenzie's basic idea remains at the centre of analysis of extensional subsidence. Flexural subsidence was first modelled quantitatively by Haxby *et al.* (1976). Beaumont (1978) built on this by proposing a physical model for basin subsidence on viscoelastic lithosphere, i.e. lithosphere whose rheology is elastic-dominated on short timescales and viscous-dominated on long ones. This work introduced the idea that stratigraphic patterns may bear the signature of the rheology of the underlying lithosphere. A few years later, Jordan (1981) and Beaumont (1981) proposed elastic and viscoelastic (respectively) first-order models for foreland basins. Later we will look at how all of these subsidence models were coupled to surface-evolution models to build stratigraphy.

Shoreline migration

It had been assumed implicitly since the problem was first considered that palaeoshoreline as observed in the rock record is a direct recorder of relative sea-level variation. Pitman (1978) provided the first formal analysis of the relation of shoreline to sea-level in the light of tectonic subsidence, and came up with the surprising result that shoreline need not be directly coupled to relative sea-level. Pitman's result is shown in Fig. 3. The key assumptions are: steady rigid-beam subsidence; constant coastal-plain slope (in space and time); sea-level variation around a steady long-term fall (the target interval was mid-Cretaceous through Miocene time); and, most important, zero sedimentation rate at the shoreline. With these assumptions, Pitman was able to derive a first-order linear differential equation for the lateral motion of the shoreline. In the long-term limit, the model approaches a steady state in which the shoreline is at the point on the coastal plain where the subsidence rate is equal to the rate of eustatic fall. The timescale over which this happens is a 'tectonic time constant' that depends on basin length, subsidence rate and the slope of the coastal plain. For eustatic variations with periods that are long compared with this time constant, the shoreline responds in a quasi-equilibrium manner. Shoreline and sea-level are out of phase by 90°, with maximum transgression corresponding to maximum rate of sea-level rise, and vice versa. For typical basin geometries, the time constant is of the order of 10^4–10^6 years, well within the range of known eustatic forcing mechanisms.

Angevine (1989) extended Pitman's analysis to both short-term and long-term cycles (relative to the same tectonic time constant) and showed that the model implies that shoreline response to long-term sea-level change is not only phase shifted but also strongly attenuated (Fig. 3). Angevine also showed that in the high-frequency limit, the shoreline behaves in the expected manner, varying in-phase with sea-level with an amplitude set by the amplitude of sea-level variation and the slope of the coastal plain. Thus sedimentary basins could be thought of as high-pass filters for sea-level changes.

Pitman's counter-intuitive result sparked considerable research, including both further theoretical analyses and attempts to verify the phase-shift behaviour from field data. Beyond its obvious importance to interpreting sea-level

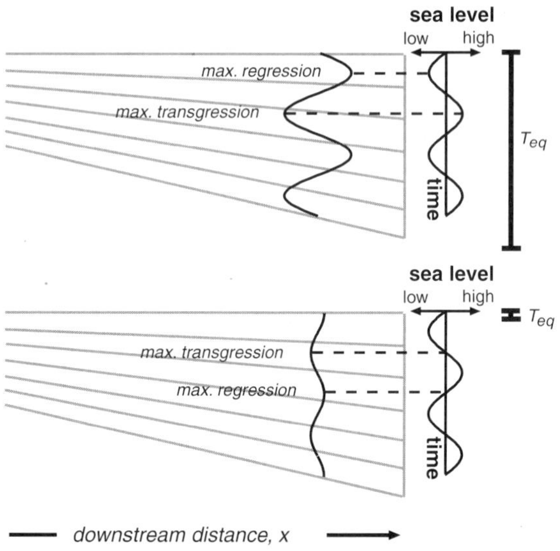

Fig. 3. Sketch of high- and low-frequency shoreline behaviour as predicted by Pitman (1978) and Angevine (1989). T_{eq} is the basin time constant. Transport is from left to right, and the position of shoreline is superimposed on equally spaced time lines for the high-frequency (top panel) and low-frequency (bottom panel) cases. Note the change in both phase and amplitude of the shoreline response for the two cases.

history from the stratigraphic record, Pitman's model was a breakthrough in bringing a measure of mathematical sophistication to modelling a stratigraphic signal (the marine–nonmarine facies boundary). It involved an oversimplification of the mechanics of coastal-plain sedimentation, but its real importance lies in having opened up a new set of questions that previously had not even been considered.

Alluvial architecture

Building on earlier qualitative work (Allen, 1965; Allen, 1974), Leeder (1978) proposed the first physically based quantitative model for alluvial architecture. Leeder envisaged a fluvial channel belt of fixed cross-sectional area avulsing at a constant mean frequency to random locations within a fluvial basin. Each avulsion removes any previously deposited floodplain fines, which fill all space not occupied by channel-belt deposits. The result most frequently cited from Leeder's initial work is that, all else being held constant, periods of rapid subsidence produce cross-sectional panels with lower density and connectivity of channel-belt deposits than periods of slow subsidence.

Leeder's elegant and intuitively appealing approach provoked numerous refinements and variations, as well as field tests, which are discussed below. One of the key assumptions, that the avulsion rate is independent of sedimentation rate, was later shown experimentally to be incorrect (Bryant *et al.*, 1995), and current architecture models allow avulsions to be determined by sedimentation or imposed externally (Mackey & Bridge, 1995). None of this diminishes the fundamental contribution of Leeder's model in establishing the quantitative connection between avulsion dynamics and the mesoscale architecture of fluvial deposits.

River-channel width

The problem of what sets the width of rivers has long been one of the most fundamental questions in fluvial geomorphology. In engineering studies of modern river systems, the width is usually specified, but in modelling ancient systems over long timescales, it must be determined from basic principles. Parker (1978a,b) proposed the first width models for channels with active sediment transport. For sand-bed rivers (Parker, 1978a), the width is set by a condition in which bank erosion is balanced by a lateral turbulent flux of suspended sediment away from the channel centre. For bedload-dominated (gravel) channels (Parker, 1978b), there is no bank erosion. Lateral drag slows the near-bank flow, allowing for weak bedload transport in the channel centre. These papers provided an independent, physically based means of coupling channel width to other variables (e.g. sediment flux, slope), thus providing a closed system of equations for dynamic river evolution.

It is worth noting that this group of landmark papers includes contributions from geophysics, civil engineering and stratigraphy. The growth of analytical sedimentary geology has, above all, been sparked by a confluence of ideas from different fields.

The growth of quantitative, physically based stratigraphic models

The publication of the initial subsidence models provoked a flurry of work examining the implications of various subsidence scenarios for stratal geometry. Again, much of the initial work was done in the geophysics community. The novelty and enormous potential of this approach meant that work on large-scale stratigraphic modelling soon overshadowed analytical study of smaller-scale aspects of sedimentary systems. The initial emphasis was on rheology and lithosphere dynamics, and on predicting large-scale stratal relationships that could be identified in seismic sections. In this context, the sediment fill was generally represented by a horizontal surface; variable stratal geometry was produced by the transport and rotation of palaeosurfaces by the deforming basement. Following fluid-mechanics usage, we will term models with a fixed horizontal sediment surface 'rigid-lid' approximations.

An improvement on the stratigraphic side was made when these geodynamic models were coupled to more realistic representations of the sediment surface. The next step beyond a horizontal surface is to assign prespecified geometries (usually planes with fixed slopes) to surfaces associated with various environments (e.g. fluvial, or submarine). Models that do this are called 'geometric models'. The most recent step in the evolution of basin models has been to represent the dynamics of the sediment surface explicitly. Various approaches have been taken to this, which we will review in the next section.

This line of work has led us to the present state of large-scale modelling of sedimentary systems. Major collections that give a sense of the state of basin modelling are Cross (1990), Franseen *et al.* (1991), Slingerland *et al.* (1994) and Harbaugh *et al.* (1999). The second and fourth of these were produced under the auspices of the Kansas Geological Survey, continuing the role it played in launching quantitative stratigraphic modelling via its Computer Contribution series. An excellent synthesis of modern basin analysis, including extensive discussion of quantitative models, is presented in Allen & Allen (1990). The end of the twentieth century must be one of the most exciting times in history to be in sedimentary geology: joining the ever-expanding armoury of field techniques and conceptual models we now have a wide array of quantitative models. In the next sections, a review is presented of some of the main classes of these, and some of their principal results.

A FIELD GUIDE TO THEORETICAL MODELS

Because basin modelling is in a period of vigorous growth, people are still experimenting with a wide variety of approaches. Models in current use can be classified along the following nonexclusive axes.

Geometric vs. dynamic models

A *geometric model* is one in which the sediment surface is represented by one or more surfaces with predetermined geometry. The simplest kind of geometric model is a rigid-lid model, in which the surface is a fixed horizontal plane. In this case all of the variation in stratal geometry is associated with subsidence. Additional realism is gained by assigning slopes to the sediment surface, which may vary to reflect, for instance, fluvial vs. marine-shelf environments. A geometric model can be adjusted to conserve sediment mass and to allow for variation in sediment supply, a major improvement over rigid-lid models. Geometric models can produce many major features of stratal geometry, including unconformities and plausible time–space sections, and they are computationally simple. They offer a high degree of flexibility but incorporate few process-based constraints on what real transport systems can and cannot do. They also miss stratal features in which the dynamics of the sediment surface plays a role, for example unconformities that develop when fluvial slopes change in response to a change in water supply (Fig. 4). This shortcoming is remedied in *dynamic models*, in which the surface evolves according to some representa-

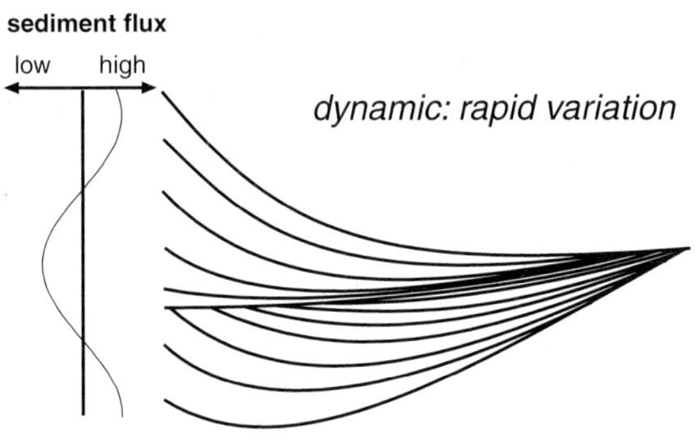

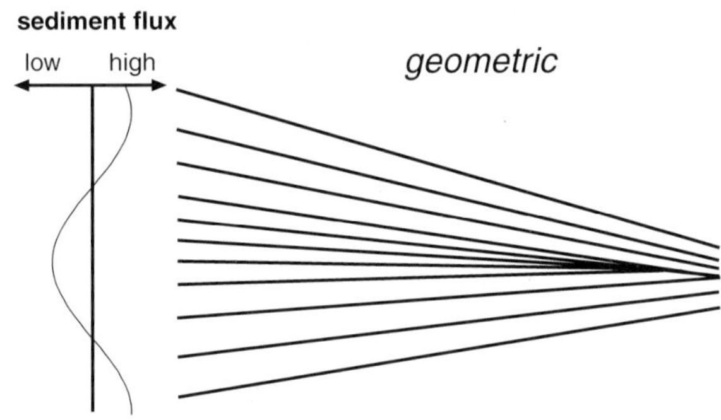

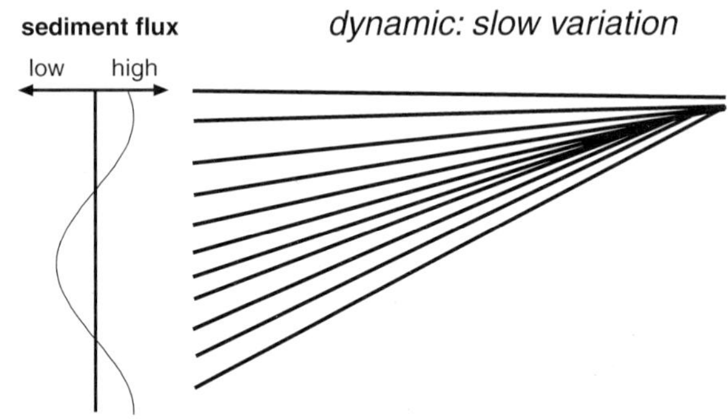

Fig. 4. Sketch showing how geometric and dynamic fluvial models respond differently to a change in sediment supply. In all cases, subsidence is a rigid beam with hinge to the right. A single cycle of variation in sediment supply is imposed. The only role that timescale plays in the geometric case is to determine the total amount of subsidence (rotation) during the cycle. In the dynamic case, rapid changes strongly affect the sediment surface. A proximal unconformity is produced by changes in the initial surface slope. However, slow changes produce a quasi-equilibrium response in which the sediment surface plays a minor role. Hence the stratigraphy produced by the geometric model more closely resembles that for the slow-variation dynamic case.

tion of how sediment and transport and deposition interact with subsidence to produce surface topography. Most 'dynamic' models of sediment surfaces are technically kinematic (i.e. involve only length and time) rather than dynamic (involving length, time and mass). However, if they have been derived carefully, force balances play a role in their origin if not their final form. Dynamic basin models are also sometimes called 'process-based' models.

One-dimensional vs. two-dimensional vs. three-dimensional models

The first dimension in stratigraphy is the vertical, and one-dimensional models are a natural starting point for thinking about vertical sequences. Two-dimensional models add the horizontal dimension parallel to the mean transport direction. The development of well-founded one- and two-dimensional models is an essential first step in basin modelling, even though many natural systems are three-dimensional. Two-dimensional models that attempt to account partially for three-dimensional effects (e.g. out-of-plane mass transfer) are sometimes called 'two-and-a-half' dimensional. Beyond the fact that lower-dimensional models are much easier to constrain and test, systems with significant three-dimensional variation can sometimes be understood as a succession of two-dimensional slices, if lateral mass fluxes are small relative to fluxes in the direction of the slices. Nonetheless, there are many cases, such as basins with multiple interacting source points, or combined lateral and transverse flow, or strongly three-dimensional subsidence, in which a fully three-dimensional model becomes necessary.

Rule-based vs. deductive models

Traditionally, quantitative models are developed by starting with fundamental governing equations, defining boundary conditions, making suitable simplifications and then solving the equations. The skill of the modeller is mostly in applying the right boundary conditions, making insightful approximations and finding the best solution method. As systems become complex, it becomes increasingly difficult to break them down into elementary objects and continua to which Newton's laws can be applied exactly. Typically the first recourse when things get too complicated is to substitute semi-empirical relations for fundamental equations. Trivial and, by now, standard examples in sedimentary-systems modelling include using drag coefficients and empirical sediment-flux laws, the latter in effect substituting for a momentum equation for sediment. A more extreme way of dealing with dynamical complexity is to construct a model from 'rules' that may range from attempts to get the gist of fundamental laws in simplified form, through attempts to abstract dynamics that are below the resolution of the model, to purely intuitive guesses. Models based on this approach are called *rule-based*. We will use *deductive* for models using the more traditional approach.

The rule-based approach to modelling is well developed in ecology, where a deductive approach starting from first principles is generally not practical (Starfield *et al.*, 1989, 1993; Starfield, 1990). In the physical sciences, the distinction between rule-based and deductive models is related to a more fundamental debate now going on between the so-called 'reductionist' viewpoint and an alternative that one could term the 'synthesist' viewpoint. The difference becomes important in the context of complex systems. Reductionism refers to taking the system apart into its fundamental components and constructing a model from the bottom up. The synthesist viewpoint (Tipper, 1992; Goldenfeld & Kadanoff, 1999; Werner, 1999) is that the dynamics of complex systems may occur on many levels (time or space scales), and that the dynamics at higher levels may be more or less independent of that at lower levels. Werner (1999) has emphasized the likelihood that higher-level dynamics is controlled by only a few crucial processes at the level below, with the implication that the modeller must focus on only those crucial processes. In this viewpoint, attempting to extract the dynamics at high levels from comprehensive modelling of everything going on at lower levels is at best clumsy and inefficient, and at worst entirely misleading – like analysing the creation of *La Bohéme* as a neurochemistry problem.

The synthesist viewpoint can be used as a basis for a rule-based approach to modelling complex sedimentary systems. The 'rules' are then the modeller's synthesis of the important lower-level dynamics. Models of sedimentary systems that are synthesist in spirit include Chase (1992), Werner & Hallet (1993), Werner & Fink (1993), Murray & Paola (1994) and Werner & Kocurek (1997). In current practice, however, rule-based models are also used simply because they are convenient, or for biological processes that cannot readily be modelled from first principles (e.g. carbonate precipitation). We will return to this subject later.

Coupled vs. uncoupled models

In general, *coupling* refers to connecting different components of a model system so that they can interact, via shared terms in governing equations, special 'coupling conditions' and/or shared boundaries ('internal boundaries'). The key issue in multicomponent models is the strength and direction of the feedbacks allowed. Modellers speak of *decoupling* components if they are either independent of one another or one is driven passively by the others. For instance, a fully coupled tectonics–sedimentation model accounts both for tectonic controls on sedimentation and for the effects of sediment transfer on tectonics (e.g. loading-induced subsidence, uplift forced by erosion). A decoupled (and very naïve) model would be one in which sedimentation was passively driven by tectonics, or they had nothing to do with one another. Since our focus here is on models of sedimentation, coupling will usually refer to joining single-environment models together to obtain a model of a whole sedimentary system. For instance, a complete model of the continental-margin system would require coupling models for the fluvial (erosional and depositional), continental shelf, continental slope and deep-marine environments. There is obviously some trade-off between the number of coupled components and the level of detail in each component. It is also easy to overlook the fact that the dynamics of a coupled model can be as sensitive to the conditions and solution techniques applied at the junctions between the component parts ('internal boundaries') as to the behaviour of the parts themselves (Swenson *et al.*, in press). Even a model composed of linear parts becomes nonlinear if the coupling between any of the parts is nonlinear. This can lead to surprisingly complex behaviour even from relatively simple models.

Analytical vs. simulation models

An *analytical model* means one designed primarily for analysing general aspects or behaviours of a system. It is typically intended to be an abstract, minimalist representation, retaining only those aspects of the dynamics that the modeller thinks are essential to the problem at hand. Analytical models are typically not intended to be used to reproduce particular cases. Rather they are aimed at illustrating fundamental principles or revealing general behaviours. Analytical models involve a minimal number of free parameters and are thus relatively testable, but they often involve high levels of simplification. *Simulation models* are intended to reproduce more of the details of specific cases. By design they are more 'realistic' than analytical models. Simulation models are generally much larger computationally than analytical models, and their flexibility usually comes at the cost of including more parameters that have to be set by the user, and so can be adjusted to fit particular cases. This makes simulation models harder to test. Their large size and number of parameters also makes it awkward to explore their parameter space comprehensively. The distinction between analysis as a search for general effects and simulation as an attempt to reproduce particular cases can as well be applied to the use of models as to models themselves. In this sense, simulation can be viewed as the modelling expression of what Sloss (1962) called the 'Twenhofelian' viewpoint, in which the emphasis is on understanding the unique character of each different basin. In contrast, analytical modelling would represent the search for 'order and system' that Sloss advocated as the key to moving stratigraphy out of its descriptive (adolescent) phase. A broader and more accurate view is that basins have both unique aspects that give them their 'personality', and general features and processes that they share more or less widely. A modeller might see a basin's 'personality' as the net effect of all the parameters and boundary and initial conditions that must be specified in a general process model in order to reproduce a particular case.

Probabilistic vs. deterministic models

A *probabilistic model* is one that includes random variables, i.e. variables that are represented by probability–density functions. A *deterministic model* does not. In most probabilistic models, the random variables are handled by using random-number generators, which means the model must be computational. (It is also possible to model probability–density functions via the calculus of random variables, but this has not been done in sedimentary-systems modelling.) Nearly all current models for basin filling are deterministic, at least at the largest scales. Random processes have been used extensively at the bed scale (which we will not cover here) and, in fluvial systems, at the mesoscale, as we will discuss below. Deterministic models can produce apparently stochastic behaviour if the system is sufficiently complex, a phenomenon often referred to as 'chaos'. Chaos is implicitly present in stochastic models, since random-number generators produce their output using deterministic computations.

THEORETICAL MODELS OF BASIN FILLING: A GUIDED TOUR

Geometric and carbonate models

In geometric models, the sediment surface has a prescribed geometry. The simplest geometry is a horizontal plane at sea-level; we term these 'rigid-lid' models. Since surface dynamics has been removed from the problem, rigid-lid models capture only the effects of subsidence in governing stratal geometry. The basic equations governing subsidence are by now a standard part of geodynamics (Turcotte & Schubert, 1982; Allen & Allen, 1990; Angevine *et al.*, 1990; Slingerland *et al.*, 1994). There is an excellent summary of the stratigraphic results of the initial rigid-lid models in Allen & Allen (1990); we will mention only a few highlights here. The McKenzie (1978) thermal subsidence model was one-dimensional, so the only stratigraphic variable was sediment thickness vs. time. Early modifications to the McKenzie model focused on including the related effects of differential extension (Royden & Keen, 1980), lateral heat flux (Steckler, 1985; Sawyer *et al.*, 1987) and finite rifting time (Jarvis & McKenzie, 1980; Cochran, 1983), and were aimed mainly at improving estimation of the subsidence history, particularly the partitioning of subsidence between the rift and thermal stages, and at modelling uplift adjacent to the rift. All of these early thermal models used a local isostatic balance. Local isostasy throughout deposition of the sediment column allowed for the development of relatively straightforward computational procedures for accounting for the role of sediment loading and compaction in amplifying subsidence. Removal of these effects along with water-depth variation allows one to estimate tectonic subsidence relative to eustatic sea-level through time, a procedure known as 'backstripping' (Angevine *et al.*, 1990; Gildner, 1990). More recent backstripping techniques also allow for flexural effects (Steckler *et al.*, 1993). Backstripping in two dimensions allows one to reconstruct approximate palaeotopography from present-day depositional geometry (Steckler *et al.*, 1999).

Two-dimensional rigid-lid models

More interesting stratal relations can be investigated by using two-dimensional subsidence models, particularly when flexural effects are included. One of the most important stratigraphic results of including flexural effects centres on the temporal evolution of flexural rigidity. The issue is important to stratigraphic modelling because the effects of variable rigidity could easily be misinterpreted as being caused by sea-level changes. Watts *et al.* (1982) pointed out that oceanic lithosphere tends to stiffen (increase in flexural rigidity) as it cools. For passive margins, this increase in rigidity causes the flexure at the margin to widen with time, producing stratigraphic onlap (Fig. 5a; Watts *et al.*, 1982). The opposite effect – stratal offlapping of the basin margin – can be produced by viscoelastic behaviour of the lithosphere (Fig. 5b; Beaumont, 1978). Viscoelasticity is a manifestation of the

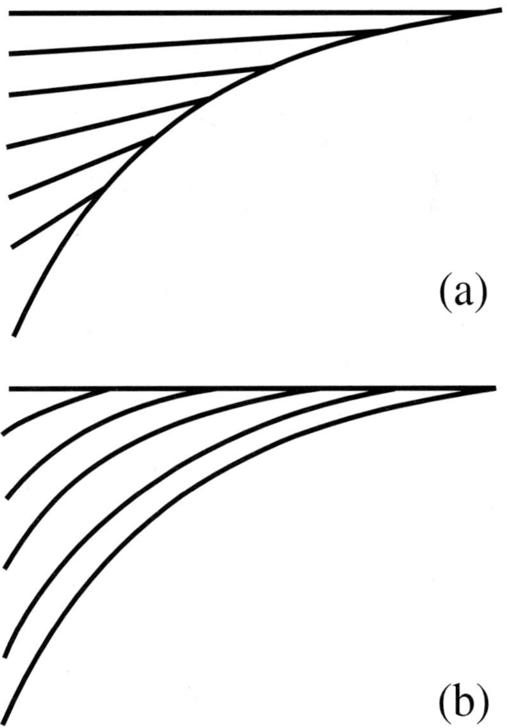

Fig. 5. Sketch showing the stratigraphic effects of two scenarios for time change in flexural rigidity at a basin margin. (a) Onlap associated with increasing rigidity due to cooling; (b) offlap associated with decreasing rigidity due to long-term viscoelastic relaxation.

familiar idea that rocks are elastic on short timescales and viscous on long timescales. Elasticity on seismic timescales and flow of the asthenosphere on timescales of the order of 10^4 years are well known aspects of viscoelasticity; what is in dispute is the upper limit of timescale on which viscous relaxation can be important. Karner *et al.* (1983), McNutt *et al.* (1988) and Kruse & Royden (1994) have presented evidence for long-term maintenance of flexural rigidity in basins and mountain ranges. Kusznir & Karner (1985) studied the behaviour of a multilayer model of the lithosphere based on measured material properties and concluded that most viscous relaxation takes place within 200 000 years of loading. Relaxation is slowest for thermally old (thick) lithosphere, but for that case also produces the smallest relative change in rigidity. There is now direct evidence for this multilayer model in the form of two scales of folding beneath Tibet (Jin *et al.*, 1994). On the whole there seems to be little evidence for significant viscoelastic behaviour in sedimentary basins on timescales of longer than a few hundred thousand years.

From a stratigraphic viewpoint, rigid-lid modelling is most appropriate for looking at stratal patterns generated by differential subsidence and rotation. It is still used for this (Hardy & Poblet, 1994; Mountney & Westbrook, 1997; ter Voorde *et al.*, 1997). Nonetheless, to anyone who has spent a career thinking about sedimentary processes, rigid-lid models seem extremely crude. Their utility in stratigraphy reflects the fact that the vertical scales associated with subsidence are many times greater than those of variation in the sediment surface generated by depositional processes. Apart from the continental slope and restricted subenvironments like alluvial fans and the edges of carbonate platforms, sedimentary slopes typically do not exceed a few parts per thousand. This means that over a basin length of the order of a hundred kilometres, depositional slopes produce elevation changes of the order of a few hundred metres. This is comparable with the range of eustatic sea-level variation, to the range of vertical flexing associated with intraplate stress variation (Cloetingh *et al.*, 1985; Cloetingh, 1986), and to the thickness of many individual formations. For models that attempt to resolve these features, it is inconsistent not to include surface variation, at least geometrically. Nonetheless, scales of a few hundred metres are small compared with the total subsidence of major sedimentary basins. The success of rigid-lid models highlights the dominant role of subsidence in governing stratigraphic patterns over long timescales. The rigid-lid approach also introduced the idea that these patterns might be used to measure the rheology of the lithosphere.

The next step is to model the sedimentary surface by imposing fixed geometric constraints on it, such as slopes or deposit lengths. This approach produces what are called 'geometric models'. The slope typically depends on the environment, with submarine slopes steeper than subaerial ones. A classical geometric model is the Pitman (1978) analysis of shoreline response to eustatic sea-level variation, which was discussed above. In this case, both subaerial and submarine parts of the coastal plain were assigned a single slope, and a sedimentation rate of zero was assumed for the shoreline. This forced the fluvial section to be always erosional. These simplifications allowed Pitman to derive a first-order linear differential equation for the movement of the shoreline. As discussed earlier, Pitman focused on the behaviour of this equation for slow eustatic cycles, i.e. those whose period is long compared with a tectonic time constant that appears as part of the solution to the governing equation.

Pitman's (1978) theory is also a good example of an analytical basin model. It was not intended to provide a detailed simulation of a particular stratigraphic section, but rather used judicious approximation to find a simplified equation whose behaviour could be readily analysed and used to obtain general results. The main basic prediction, that shoreline should track rate of sea-level rise rather than sea-level over long timescales, is straightforward and testable.

A wide range of geometric models have been developed subsequently. Two broad themes stand out. One is that geometric models are the main arena in which carbonate stratigraphy has been coupled with clastic sedimentation. The other is that geometric models are the ones most closely tied to sequence stratigraphy (Van Wagoner *et al.*, 1990; Weimer & Posamentier, 1994; Emery & Myers, 1996). This is not surprising: the dissection of stratigraphic sections into unconformity-bounded sequences is essentially a geometric process, and the conceptual models behind sequence stratigraphy are geometric, following on from Sloss (1962). In many cases, only quantitative implementation separates conceptual models from geometric computer models. The connection with sequence stratigraphy has led to the widespread use of geometric models to look at the effects of eustatic variation on the stratigraphic record. Ross (1991) gives a nice summary of the linked development of sequence stratigraphy and quantitative models.

Carbonate shelf models

Carbonate production is modelled using a rule-based approach, and at present it is difficult to imagine how a process so dominated by biological effects could be tackled any other way. The common feature of current models is that carbonate production rate depends primarily on depth, and very strongly (Fig. 6). The ways in which this is implemented illustrate the range of styles in rule-based modelling. The most flexible but least analytical is to specify a general form for the accumulation curve but to let the user set all the internal break points in the curve (Demicco & Spencer, 1989; Demicco, 1998). Since light attenuation is the basic cause of the depth dependence of carbonate production rate, more analytical forms can be obtained by relating carbonate sedimentation to known functions for light attenuation in the ocean (Bice, 1991; Bosscher & Schlager 1992, 1993; Bosscher & Southam, 1992; Galewsky, 1998) as well as other factors like turbidity and temperature (Lerche *et al.*, 1987) and rate of sea-level rise (Watney *et al.*, 1991). The user must still supply parameters, but their physical meaning is clearer and their values less arbitrary.

The depth control of carbonate sedimentation makes it possible to model carbonate sedimentation in one dimension; the fact that many shallow-water platform sequences are laterally continuous but show rapid vertical variation makes it natural to do so. Vertical shallowing-upward sequences modelled numerically vary in thickness from a few metres (e.g. Goldhammer *et al.*, 1987; Goldhammer & Harris, 1989) to a few tens of metres (Watney *et al.*, 1991). One-dimensional carbonate modelling has sought to explain these sequences in terms of superposition of cyclical eustatic forcing and depth-dependent sedimentation (Goldhammer *et al.*, 1987, 1993; Spencer & Demicco, 1989; Watney *et al.*, 1991). The dependence of model results on the form of production function assumed is illustrated by the work of Cisne *et al.* (1984), who assumed that in shallow water production rate increases with water depth. This allows for a steady-state solution in which water depth is proportional to the rate of relative sea-level rise, a result reminiscent of Schwarzacher's (1966) steady-state solution and Pitman's (1978) model of shoreline

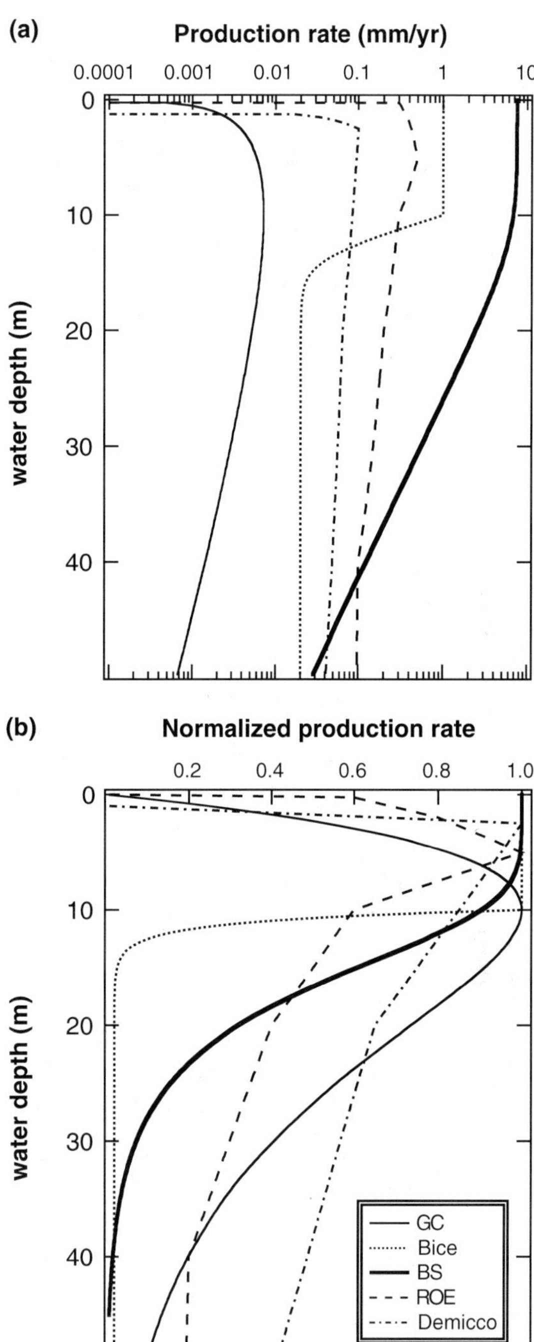

Fig. 6. Carbonate production functions used by various authors. (a) Sedimention rate in dimensional units, to show range of rates used, and (b) rate normalized to the maximum rate to show variation in shape. GC – Gildner & Cisne (1990), Bice – Bice (1991), BS – Bosscher & Schlager (1992), ROE – Read *et al.* (1991), Demicco – Demicco (1998). The low rates represent what Gildner & Cisne (1990) termed 'epeiric sea' conditions, the high rates 'reef' conditions.

position. A regime of positive depth dependence is included by Demicco & Spencer (1989), Bosence & Waltham (1990), Read *et al.* (1991), Goldhammer *et al.* (1993) and Demicco (1998). These and the more general model of Gildner & Cisne (1990) have a strongly decreasing production function below a depth of maximum production (Fig. 6). Depths of maximum production suggested by the above authors are typically in the range of a few metres. The increasing-production regime is absent from the semi-empirical reef model of Bosscher & Schlager (1992), which was adopted by Bice (1988) and Galewsky (1998).

Carbonate platforms also show significant lateral variation, particularly near their margins. Two-dimensional models have accounted for such features by including geometric treatment of sediment redistribution by physical transport. The strong depth dependence of carbonate production tends to produce steep-edged platforms (Fig. 7; Aigner *et al.*, 1989; Strobel *et al.*, 1989; Read *et al.*, 1991; Bosscher & Southam, 1992). Weakening the depth dependence of carbonate production, and/or strong offshore transport by physical processes, reduces this 'sharpening' tendency, leading to ramp rather than platform geometries (Aurell *et al.*, 1995, 1998). Progradation of carbonate platforms by physical transport produces clinoforms whose shape is related to the interplay of eustasy, transport and subsidence (Bice, 1988, 1991; Spencer & Demicco, 1989; Lawrence *et al.*, 1990; Bosence *et al.*, 1994; Galewsky, 1998). It seems that simple models with plausible parameter values can reproduce observed stratigraphic sections (Demicco *et al.*, 1991; Harris, 1991; Bosence *et al.*, 1994) and seismic patterns (Aigner *et al.*, 1989; Shuster & Aigner, 1994). It is also clear that the solutions are nonunique (Demicco *et al.*, 1991; Bosence *et al.*, 1994; Aurell *et al.*, 1995, 1998). A systematic approach along the lines of that described by Heller *et al.* (1993) for clastic systems would be a way of constraining this nonuniqueness. A characteristic feature of carbonate platforms related to carbonate production dynamics is the possibility of migration autocycles (Ginsburg, 1971) in which migration into deepening water leads to drowning (Schlager, 1981) and initiation of a new clinoform. This can be imposed via a 'drowning depth' (Demicco, 1998) but an analytical production function that decreases strongly with depth allows the autocycles to occur spontaneously (Galewsky, 1998).

Clastic and mixed systems

Geometric modelling of clastic systems started off in two dimensions (Pitman, 1978), following the original framework set out by Sloss (1962). Cant (1989) proposed a one-dimensional geometric model for clastic sequence stratigraphy, but incorrectly concluded from it that the causes of sequence boundaries cannot be distinguished using the stratigraphic record. Collier *et al.* (1990) have shown that a geometric shoreline model only slightly more complicated than Pitman's can reproduce sequences comparable to well-dated Quaternary sequences where the forcing is known. Comprehensive, two-dimensional geometric models for clastic sequence stratigraphy have been described by Jervey (1988) and Ross *et al.* (1995). The Ross *et al.* model focused on partitioning mud and sand among fluvial, shelf and turbidite systems and stands out as the first to separate the shoreline from the shelf–slope break. Both papers show how a geometric model based on sea-level, sediment supply and subsidence can be used to reproduce patterns in both lithostratigraphy (Fig. 8) and time–space diagrams. Jervey's approach was adopted in the SEDPAK model (Kendall *et al.*, 1991a, b), whose workings are explained in Strobel *et al.* (1989). SEDPAK is noteworthy for including both clastic and carbonate sedimentation. Like Jervey's and many other sequence-stratigraphic models, it is biased towards sediments rather than subsidence, although SEDPAK's treatment of subsidence goes as far as including basic flexural effects. In the sediment column it includes compaction along with deposition in fluvial, clastic-shelf, carbonate-shelf, continental-slope and deep-marine environments. The carbonate section in particular is fairly elaborate. Because of its many geometric parameters, SEDPAK requires a good deal of input from the user, and is highly tunable. Models like this are very hard to test, but in a comparison with seismic stratigraphy of the Great Bahama Bank (Eberli *et al.*, 1994), SEDPAK reproduces major features of the observed section using the Haq *et al.* (1987) sea-level curve and reasonable parameter values (Fig. 9a). SEDPAK is a simulation rather than an analytical model. It is, in effect, a very flexible tool for converting ideas about depositional geometry and external forcing into stratigraphy. In this regard, it has been useful as a way of systemizing and refining conceptual sequence-stratigraphic models (Helland-Hansen *et al.*, 1988).

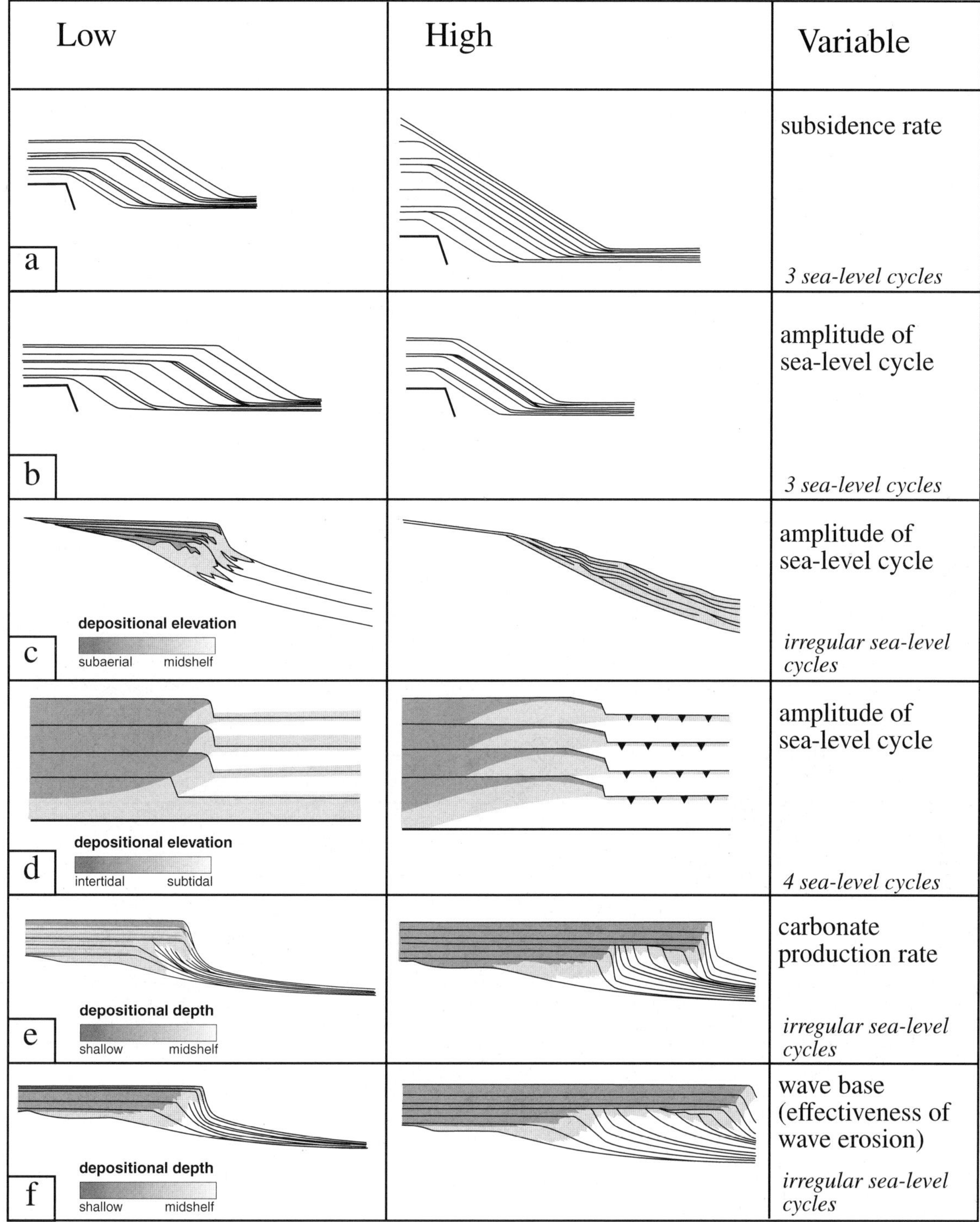

Fig. 7. A catalogue of carbonate stratal geometries produced from various models under various scenarios. In each case the variable being changed between the left image and the right image is given on the far right along with (in italics) the imposed forcing. Sea-level cycles are sinusoidal unless stated otherwise. The continuous lines are equally spaced time lines. Where an additional variable has been mapped onto the time lines it is shown in grey, with a relative scale bar below. The figures are redrawn and simplified from model results to give an idea of the stratigraphic pattern; readers should consult the original works for details. (a) Effect of changing subsidence rate (Bice, 1991); (b) effect of changing amplitude of sea-level fluctuation (Bice, 1991); (c) effect of changing amplitude of sea-level fluctuation (Read *et al.*, 1991); (d) effect of changing amplitude of sea-level fluctuation (Spencer & Demicco, 1989) – sawteeth indicate subaerial exposure; (e) effect of changing carbonate production rate (Bosence *et al.*, 1994); (f) effect of changing wave base (Bosence *et al.*, 1994).

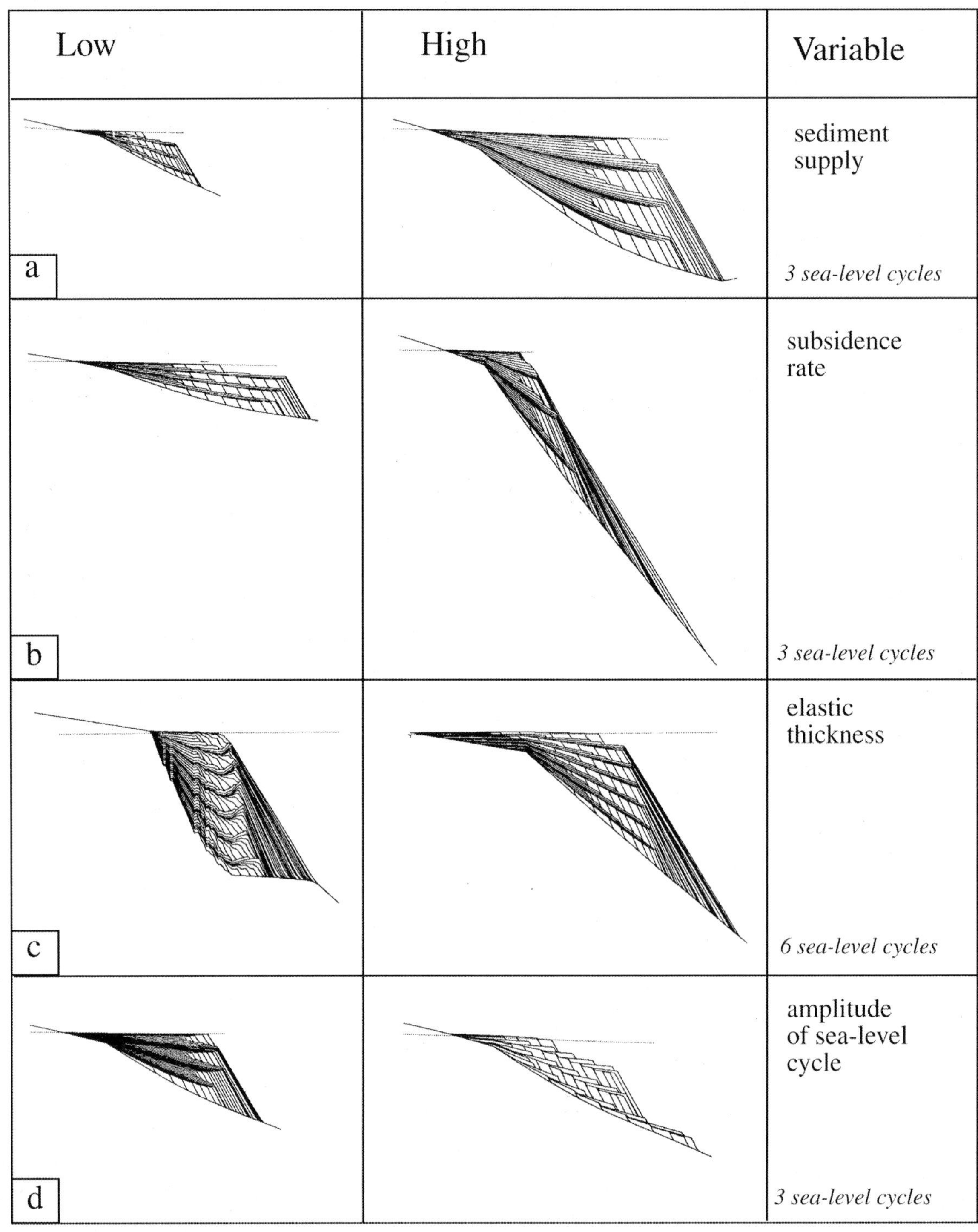

Perlmutter *et al.* (1998) provide a good example of using SEDPAK to formalize and sharpen a conceptual model. The idea is that climatically driven eustatic cycles are inherently linked to various other changes, especially in sediment supply (Perlmutter & Matthews, 1990). The latter may vary in phase relative to sea-level over 180°. The SEDPAK application allows one to visualize the effects of changing the phase of sediment supply relative to sea-level (Fig. 8g). As would be expected, the main effect is to change the partitioning of deposition between fluvial and submarine, with the maximum fluvial deposition occurring when the supply maximum coincides

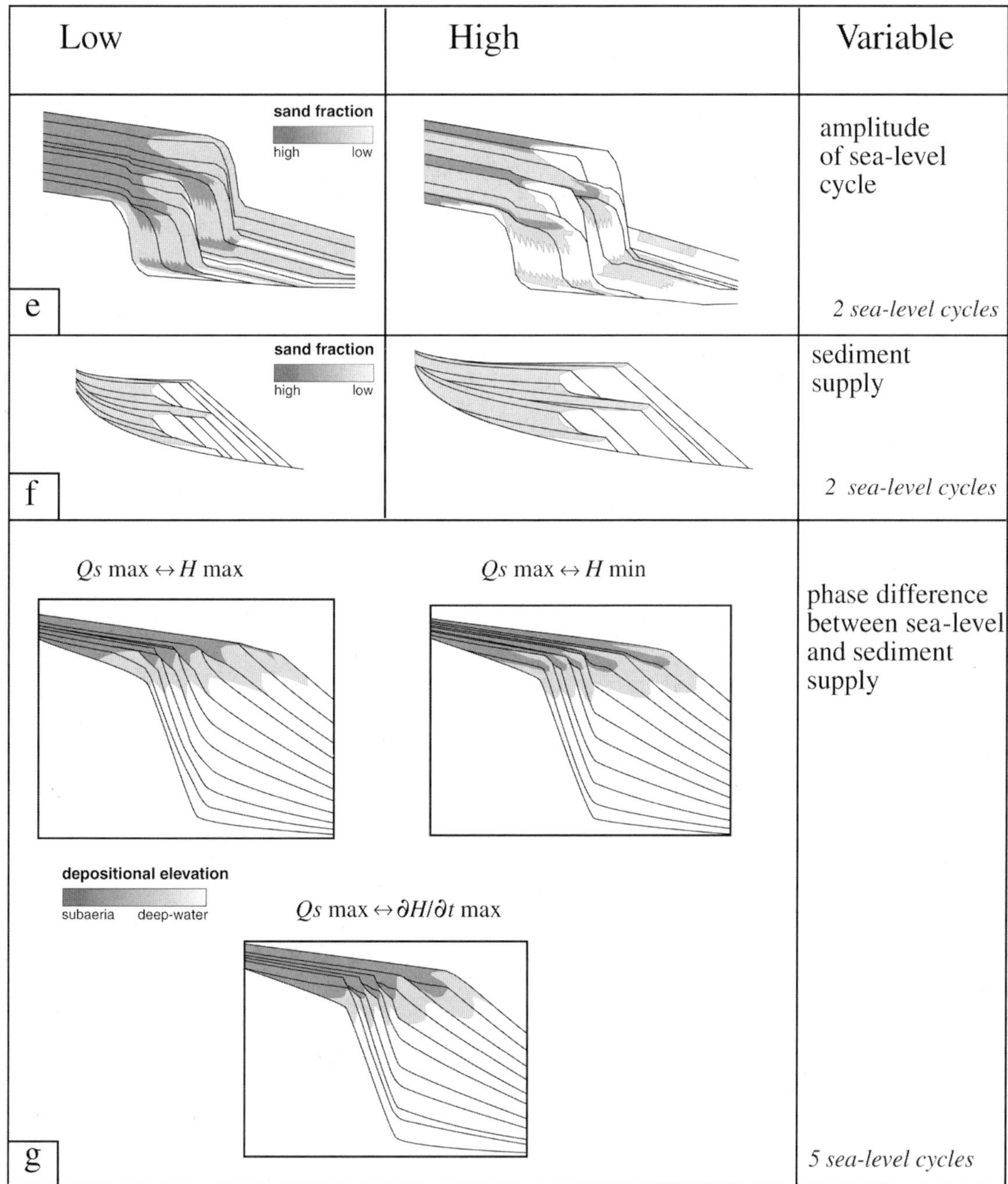

Fig. 8. A catalogue of clastic stratal geometries produced using geometric surface models under various scenarios. Layout, disclaimer and explanation are the same as for Fig. 7. (a) Effect of changing sediment supply; (b) effect of changing subsidence rate; (c) effect of changing elastic thickness; (d) effect of changing amplitude of sea-level cycle (a–d from Steckler *et al.*, 1993); (e) effect of changing amplitude of sea-level cycle (Ross *et al.*, 1995); (f) effect of changing sediment supply (Jervey, 1988); (g) effect of changing relative phase of sediment-supply and sea-level cycles (Perlmutter *et al.*, 1998).

with the maximum rate of sea-level rise. A more recent, and more adventurous, SEDPAK application is an attempt (Sen *et al.*, 1999) to date sedimentary sequences by pattern matching with the model, along the lines of magnetic-reversal stratigraphy. The idea is to generate model stratigraphies from the Haq *et al.* (1987) eustatic sea-level curve over various time intervals, then to date the section by choosing the interval that gives the best match to the real stratigraphic pattern. This is potentially a useful approach but requires a good deal of faith: in the reliability of the sea-level curves, in the accuracy of the model and in the assumption that changes in other forcing factors, especially subsidence rate, can be neglected.

Cant (1991) and Nummedal *et al.* (1993) have presented geometric models of sequence development on clastic continental margins aimed at, among other things, resolving ravinement surfaces formed by shallow submarine erosion. Cant does this by using curved depositional surfaces. He also suggests changing the geometry for times when the surfaces represent 'disequilibrium' conditions. This is a step in the direction of dynamic models that also illustrates the limitations of geometric modelling. In Nummedal *et al.* (1993) the shoreface migrates via 'Bruun's rule' (Bruun, 1962), a mass-conserving geometric model of shoreline retreat. This rule, although developed for engineering rather than stratigraphic purposes, is nonetheless a particularly nice example of using a geometric approach to capture the main effects of very complex physical processes, in this case sediment redistribution in the surf and shallow-offshore zones.

The SEQUENCE model of Steckler (1999) is the latest in a series of quantitative sequence-stratigraphy models that evolved from the rigid-lid approach. They are still 'geophysical' in spirit, coupling a dynamic treatment of subsidence, including both isostatic and flexural effects, to a geometric model of the sediment surface. A geometric clinoform model was used in this way by Reynolds *et al.* (1991) to show that variation in rigidity could lead to different stratigraphic sequences for the same eustatic forcing. Reynolds *et al.* (1991) and Steckler *et al.* (1993) clarified the independent roles of sea-level, tectonics (subsidence) and sediment supply in producing sequences (Fig. 8a–d). Their model also generates phase shifts in the timing of sequence-boundary formation reminiscent of the original Pitman (1978) result, although in this case dependent on amplitude and cycle time of the eustatic fluctuations, and on sediment supply. Two important additions in SEQUENCE, the most recent development in this line (Steckler, 1999), are a provision allowing for separation between the shoreline and the shelf-slope break (Ross *et al.*, 1995), and explicit accounting for the time needed to reach (flexural) isostatic equilibrium. The latter results from the need to accommodate changes in loading by viscous flow on timescales of the order of 10^4 years (e.g. glacioeustatic sea-level fluctuation). This is a timescale on which the relaxation effects originally examined by Beaumont (1978) are important, and it introduces another mode for dependence of stratigraphic response on cycle time. Separation of shoreline and the main clinoform break leads to phase differences between the two, for instance progradation of the shelf-slope break while shoreline is transgressing (Steckler, 1999).

Another model that uses geometric sedimentation coupled to dynamic subsidence is the Shell model described by Lawrence *et al.* (1990). This model also treats both carbonate and clastic sedimentation. Inclusion of subsidence mechanics allows it to produce some of the dynamical features of the early 'rheological' rigid-lid models, but with the addition of a full suite of sedimentary environments. Lawrence *et al.* (1990) and Aigner *et al.* (1989, 1990) show that by tuning their model (i.e. adjusting parameter values to achieve a best fit to observations) they can reproduce a variety of observed stratigraphic sections (e.g. Fig. 9b). This approach is continued by Shuster & Aigner (1994), who also generate synthetic seismic sections from the model output. Lawrence (1994) describes a more analytical application of the same model in which stratigraphic sequences are produced solely by varying sediment supply, as opposed to the usual eustatic forcing. This case also nicely illustrates one of the key differences between geometric and dynamic models: a dynamic simulation of variable sediment supply would also change the fluvial slope as a function of sediment supply, potentially producing updip unconformities during intervals of low supply (Fig. 4; Paola *et al.*, 1992). Nonetheless, the Shell model is one of the most complete stratigraphic models that exists today. It is unfortunate that it is not available for general use.

It is clear that one can model a number of stratigraphic features using predefined surface geometries. Particularly nice summaries of the physical basis for doing so are provided by Ross *et al.* (1995) and Nummedal *et al.* (1993). Nonetheless, there is no avoiding the fact that the surface morphology of basins is determined dynamically by the interplay of transport processes and subsidence. One of the immediate consequences of modelling sediment-surface dynamics (so-called 'morphodynamics') is that in a system with steady forcing and subsidence, the surface topography evolves towards a steady state in which sedimentation everywhere balances subsidence. This condition is the depositional

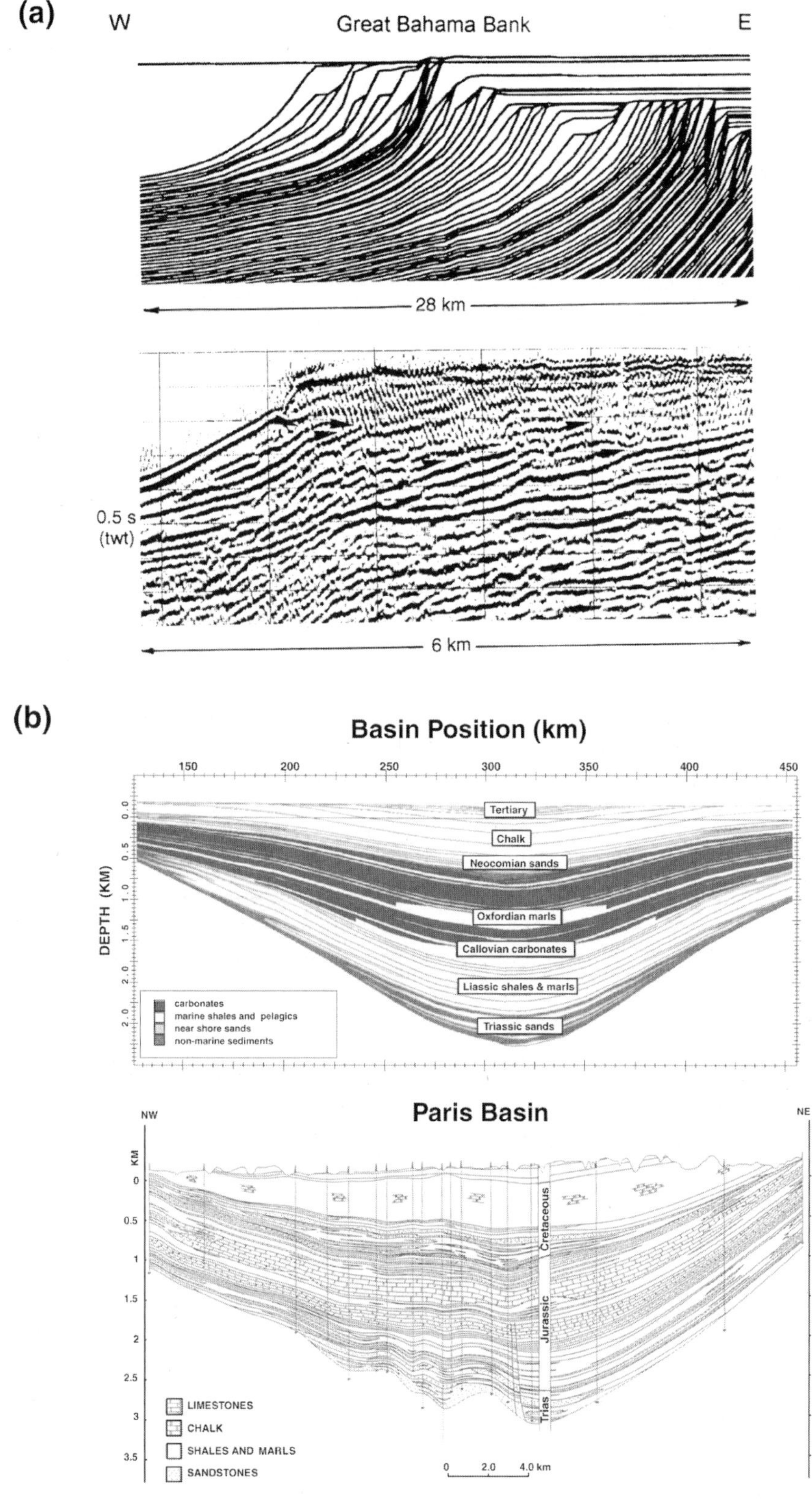

Fig. 9. Plausibility comparisons of geometric models. (a) SEDPAK output using the Haq *et al.* (1987) sea-level curve (upper) compared with seismic data (lower) from Great Bahama Bank (from Eberli *et al.*, 1994; © 1994, reprinted by permission of the American Association of Petroleum Geologists, whose permission is required for further use); (b) Shell model output (upper) using the Haq *et al.* (1987) sea-level curve and two iterations of tuning sediment parameters vs. stratigraphy (lower) from the Paris Basin (from Aigner *et al.*, 1989).

equivalent of the classical graded river. Let us call the timescale required to reach equilibrium T_{eq}; techniques for estimating it and numerical values will be discussed later. The point here is that dynamic variation in surface geometry becomes relatively less important as the period of forcing approaches and exceeds T_{eq} (strictly speaking this also depends on the amplitude of the forcing). Hence the formal justification for the geometric modelling approach is strongest for timescales greater than T_{eq}. As will be discussed below, in many cases higher-order (i.e. third order and above; Emery & Myers, 1996) eustatic oscillations occur on timescales that are below typical T_{eq} values, especially for large basins. This leads to potentially important effects associated with nonequilibrium response, which can be seen both theoretically (Jordan & Flemings, 1991; Swenson *et al.*, in press) and experimentally (Wood *et al.*, 1993, 1994; Koss *et al.*, 1994).

Dynamic models

Fluvial systems and the diffusion equation

Not surprisingly, the first dynamic stratigraphic models were for fluvial systems. The most commonly used equation has the form of the standard equation for the diffusion of heat or material and hence is usually called the diffusion equation:

$$\sigma + \frac{\partial \eta}{\partial t} = \frac{\partial \eta}{\partial x}\left(\nu \frac{\partial \eta}{\partial x}\right) \quad (1)$$

where σ is the subsidence rate, η the elevation of the sediment surface, t time, x downstream distance and ν the transport (diffusion) coefficient (Fig. 10). It is worth stressing that use of this equation does not mean that the sediment diffuses in a real physical sense. Rather, the combination of one-dimensional mass balance

$$\sigma + \frac{\partial \eta}{\partial t} = \frac{-1}{C_0}\frac{\partial q_s}{\partial x} \quad (2)$$

where q_s is the mean sediment flux per unit width of basin and C_0 the volume sediment concentration in the deposit, with a slope-dependent sediment flux

$$q_s = -\nu' \frac{\partial \eta}{\partial x} \quad (3)$$

where $\nu' = \nu C_0$ gives an equation that has the same form as that of true diffusion by random motion.

Various forms of the diffusion equation have long been used to model fluvial morphology in engineering and geomorphology (Adachi & Nakatoh, 1969; Begin *et al.*, 1981; Garde *et al.*, 1981; Jain, 1981; Soni, 1981; Gill, 1983; Jaramillo & Jain, 1984; Ribberink & van der Sande, 1985; Zhang & Kahawita, 1987). The first workers to couple diffusion and subsidence to generate stratigraphy were Flemings & Jordan (1989). Their work built on the flexural model of foreland-basin subsidence of Jordan (1981), and represents the first fully dynamic basin-filling model. Before we look at the results of this and subsequent work, it is worth reviewing the approximations involved in deriving the diffusion equation as it applies to fluvial systems, both to appreciate its limitations and to make clear that it can be derived from basic principles of channel flow and sediment transport and does not, as is sometimes claimed, come from unrelated models of hillslopes or deltas. We will follow the derivation of Paola *et al.* (1992). Consider a fluvial system such as that shown in Fig. 10. The flow is distributed in one or more channels, whose planform is unspecified. The important assumptions are as follows:

1 The flow can be represented at each downstream location x by an average depth, velocity, shear stress, etc. This ignores the effects of both spatial and temporal variation in these parameters. For time variability, one can run a model repeatedly, imposing individual floods according to some specified distribution (e.g. Mackey & Bridge, 1995). This works, but does not provide any direct way of relating properties of the flood hydrograph to overall transport properties of the river system. One means of handling internally generated variability is a statistical–mechanics approach, based on the Reynolds decomposition from turbulence theory (Paola, 1996; Paola *et al.*, 1999). Unfortunately, as is the case for turbulence, the decomposition adds new unknowns to the analysis, which must be constrained independently. Whatever approach is used, it is clear that there is a great need for data on spatially averaged sediment fluxes in relation to channel pattern, water supply, sediment properties and biota.

2 Within particular fluvial regimes, the dimensionless boundary shear stress is constant. This is a substantial generalization of the approach presented in Paola *et al.* (1992). The dimensionless (or Shields) stress is defined as

$$\tau_* = \tau_0/[\rho g(s-1)D] \quad (4)$$

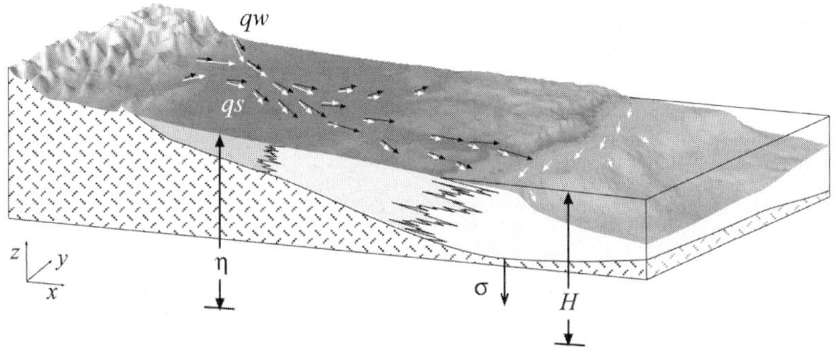

Fig. 10. Definition sketch for basin variables.

where τ_0 is the bottom shear stress, ρ the fluid density, s the sediment-specific gravity and D the median grain size. The set of basic equations for mean quantities in a fluvial system (e.g. velocity, depth) is underdetermined by one equation. The equation needed to close the system would impose a physical control on channel width. Setting τ_* constant (or, more generally, relating it to other variables in the system) supplies the additional equation and so is a way of indirectly setting the width. The idea is that limits on shear stress can be formulated directly in terms of bank erosion: if the stress is too high, bank erosion widens the channel and brings it back to its limiting value (Parker, 1978a,b). A more general stress condition would involve a dynamic equilibrium in which local bank erosion would balance local bank deposition on average.

In any case, Paola & Seal (1995) showed that τ_* estimated from mean depth, slope and grain size in coarse-grained braided rivers (data from Church & Rood, 1983) is indeed constant at about 1·4 times the critical value for initiating sediment movement (0·05) as long as the stream is not degrading and is not confined by cohesive banks or bedrock (Fig. 11). More surprisingly, Dade & Friend (1998) and Parker *et al.* (1998a) have presented data suggesting that for alluvial sand-bed streams τ_* is also constant, with a mean value between 1 and 2 (Fig. 11). The $\tau_* = \text{constant}$ approximation is rejected by Robinson & Slingerland (1998), who did not find the correlation between shear stress and grain size that would be expected if τ_* were constant. In part, this is because they applied a condition that is restricted to gravel rivers (Parker, 1978b; Paola *et al.*, 1992) to sand-bed ones (Parker, 1978a) as well. Nonetheless, their work nicely illustrates the ongoing debate about what controls river width, as well as the abrupt jump in τ_* at the gravel–sand transition in several rivers. It is worth stressing that the $\tau_* = \text{constant}$ condition in sand-bed rivers is not predicted by any existing theory.

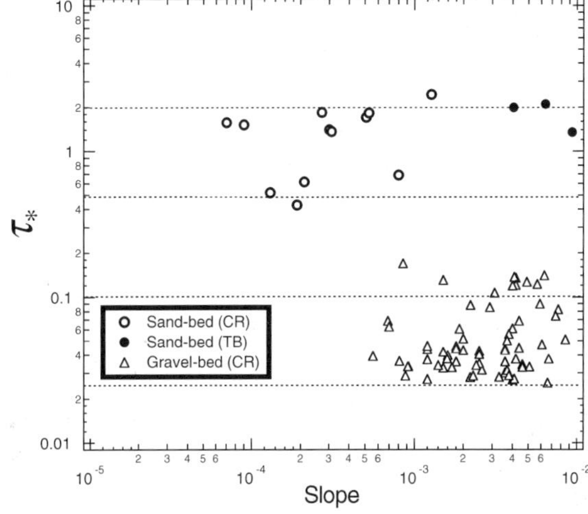

Fig. 11. Dimensionless shear stress (τ_*) for gravel-bed and sand-bed streams, suggesting two distinct regimes for fluvial diffusion models. Open points (CR) are from the data set of Church & Rood (1983). Points for both stream types were filtered as follows: characteristic flow condition, 2-year flood (preferred) or bankfull flood; channel type, meandering or higher; bank type, cohesive or higher; process code, not paved, degrading, or nonalluvial. The sand-bed data are for rivers with median size <1 mm, the gravel-bed data from rivers with median size >20 mm and a dimensionless stress high enough (>0·025) to indicate at least some level of modern channel activity. Filled points (TB) are from a mine tailings basin (Paola *et al.*, 1999). Dashed lines define a factor of 2 variation about stress values of 1 and 0·05 for the sand-bed and gravel-bed cases, respectively.

3 The momentum equation for the fluid can be reduced to a balance between bed shear stress and depth–slope product: $\tau_0 = \rho g S h$ where S is the mean topographic slope and h the mean depth.

This equation can be found in most introductory sedimentology texts, but it is not always clear under what conditions it applies. Basically, the following two conditions must be satisfied: $U/gST << 1$ and $H/SL << 1$ where U, H, S, T and L are representative scales for velocity, depth, slope, time variation and length variation, respectively (there is a more complete discussion in Paola, 1996). The first condition is satisfied for nearly any realistic meteorological flood. In the second condition, H/S serves as a scale for the distance over which spatial (convective) accelerations are important. The most common manifestation of these acceleration terms is a backwater profile, so the ratio H/S may be termed the 'backwater length'. Note that the product HS scales the bed shear stress. If for a given stress H is large and S small, the backwater length is large, and vice versa. Evidently the depth–slope product as an estimate of shear stress works best for shallow, high-slope streams and worst for deep, low-gradient streams. For problems in which satisfying $H/SL << 1$ is overly restrictive, one must keep the backwater term (e.g. Snow & Slingerland, 1987; Slingerland *et al.*, 1994). The cost of this is greater computational complexity. In particular, the system of governing equations no longer reduces to a single morphodynamic equation for the sediment surface.

4 The boundary shear stress can be related to the local mean velocity via a constant drag coefficient c_f. This is not strictly correct, because the drag coefficient depends on the ratio of depth to local bed roughness. The effect of varying roughness is most likely to be important for dune-covered beds (Smith & McLean, 1977). On the whole, compared with the first three assumptions, constant drag coefficient is a relatively minor simplification. Leaving in the weak depth dependence makes the system slightly nonlinear (Parker *et al.*, 1998a).

With these four assumptions, it is easy to show that the morphodynamic equation has the diffusional form given above. The transport (diffusion) coefficient is given by:

$$\nu = \left(\frac{q_{s*}}{\langle\tau_*\rangle^{3/2}}\right)\frac{c_f^{\ 1/2}\langle q_w\rangle}{(s-1)C_0} \tag{5}$$

where q_{s^*} is a dimensionless unit sediment flux determined by τ_* and a standard sediment-transport formula, and $\langle q_{\rm w}\rangle$ is the average unit water discharge. An additional correction term can be incorporated in the coefficient to account for transport fluctuations (Paola *et al.*, 1999). The linear dependence of the transport coefficient on discharge means that the diffusion approximation is equivalent to setting the sediment flux proportional to the unit stream power. Apart from revealing the structure of the transport coefficient, deriving the diffusion equation from first principles also shows how, once the topography has been found, it can be used to back-calculate the depth, velocity, etc. Fluvial diffusion basically works like this: decreasing the total sediment flux $Q_{\rm s}$ decreases the total stream width because of assumption 2. Since the velocity is kept constant by assumptions 2 and 4, this means that the depth increases to conserve water. The depth increase, in turn, forces the slope to decrease to maintain the constant shear stress. Hence decreasing $Q_{\rm s}$ forces the channel system to reduce its slope and form narrower and deeper channels, all changes that in nature would tend to shift the system from braiding to meandering.

The two most important objections to the above analysis are to assumptions 2 and 3. The limitations of 3 were described above and become important chiefly for small length scales and/or large, low-gradient rivers. With regard to 2, the main alternative to imposing a stress condition is to set the width directly. This is the approach advocated by Snow & Slingerland (1987), Slingerland *et al.* (1994) and Robinson & Slingerland (1998). There is considerable empirical support for a relation of the form

$$b \propto Q_{\rm w}^{\ 1/2} \tag{6}$$

where b is the width and $Q_{\rm w}$ the water discharge. The problem is that Eq. (6) has never been explained mechanistically. The use of empirical equations whose physical basis is unknown seems particularly risky for stratigraphic analysis, which may involve systems different from the modern ones on which such equations are based. However, if the stress-closure approach described above is correct, then Eq. (6) would require that sediment and water delivery systems organize themselves such that effective sediment flux $Q_{\rm s}$ varies as $Q_{\rm s} \propto Q_{\rm w}^{\ 1/2}$ within the gravel- and sand-dominated regimes. The lack of an explanation for how this self-organization might come about is a major weakness of the constant-τ_* approach.

It is important to note that the diffusion equation as formulated above is independent of grain size. In the diffusional context, it is tempting but incorrect to reason that the downstream slope decrease forced by deposition is directly responsible for downstream fining, on the grounds that the slope decrease reduces flow competence. Grain size comes into the picture only via τ_*, which as discussed above is some 25 times larger for sand-bed than for gravel-bed streams but seems to be roughly constant within each of these regimes. This implies that (1) within either regime downstream fining does not change τ_* and thus does not strongly influence the transport coefficient, (2) the slope decrease in the diffusional solution is forced only by deposition and is independent of grain size, and (3) that the most important grain-size change in fluvial systems is the rapid transition from gravel to sand. This transition typically occurs over a distance that is short compared with the entire length of the river (Sambrook Smith & Ferguson, 1995; Robinson & Slingerland, 1998); our group has previously approximated it as a front (Paola, 1988, 1990; Paola *et al.*, 1992). Within the gravel and sand regimes, downstream grain-size change is controlled mainly by the size distribution of the sediment supply. The τ_* constant condition implies that, to first order, it might be possible to decouple downstream grain-size variation from hydraulics, within each of the two size regimes (gravel and sand). This approach, which was adopted by Paola & Seal (1995), would be a powerful simplification of present size-sorting models (Parker, 1991; Bridge & Bennett, 1992; van Niekerk *et al.*, 1992; Hoey & Ferguson, 1994; Cui *et al.*, 1996). These require knowledge of the width and hence are difficult to apply on geological timescales, on which the width is a dependent variable.

An alternative treatment of grain-size variation in fluvial systems is to assign individual transport coefficients to sediment types (Rivenæs, 1992, 1997) and then solve a set of coupled diffusion equations for each type simultaneously. This approach produces a smooth, continuous downstream increase in the fraction of the more mobile type (i.e. higher transport coefficient). It also predicts that the effective transport coefficient (i.e. the composite value resulting from interaction of the various sediment types) increases smoothly and continuously downstream. This is because the transport coefficient is modelled as being a property of the material, rather than the stream system, and thus the coefficient tracks the material that carries it. For gravel–sand sorting, this continuous downstream variation is directly contrary to observed well-defined gravel–sand transitions. Since the size dependence in the transport coefficient enters via τ_* (Eq. 5), continuous downstream change in the overall transport coefficient also implies continuous downstream change in τ_* and thus is directly contradicted by the data presented in Fig. 11. The multiple-diffusivity approach might work for sand–mud sorting, particularly for settings like the continental shelf where it can be argued that the transport process really is diffusional (discussed below). In most river systems, it seems more likely that overall sand–mud sorting is controlled by the interaction of channels and floodplains (e.g. Mackey & Bridge, 1995; Ross *et al.*, 1995). This process is not treated physically in any existing diffusional models.

Robinson & Slingerland (1998) have for the first time coupled a physically based sorting model (van Niekerk *et al.*, 1992) to basinal fluvial aggradation. The depositional model in this case is not diffusional, because it uses a direct width closure (point 2 above), and because it retains the full backwater term (point 3 above). This work nicely illustrates the roles of both the rate and the spatial distribution of subsidence in controlling fining rate for a known input size distribution, and it is particularly valuable for showing how a physically based model can be used to compute quantities (here sedimentation rate and channel width) that can be estimated independently from the stratigraphic record. This is a significant step towards more testable stratigraphic models.

Results from diffusion-based fluvial models

Diffusional fluvial models have primarily been applied to the stratigraphic evolution of foreland basins. These have been coupled models that built on the approach developed by Flemings & Jordan (1989). In their model, both erosion of the mountain belt and the associated basin filling occur via linear diffusion, with a much lower transport (diffusion) coefficient in the erosional area than in the depositional basin. This leads to steeper slopes in the erosional area. The basement deformation that controls uplift and subsidence is controlled by the classical flexure equation, and the advancing orogenic wedge is assumed to be critically tapered. Viscous relaxation is ignored, for the reasons explained earlier. The Flemings and Jordan model is analytical as defined here; it was used primarily to explore fundamental behaviours rather than to simulate specific situations. It remains one of the most widely cited papers in basin modelling, and for good reason: it presents a model that has just enough of the relevant physics to be useful, and to reveal important general trends, but not so much detail that it has to be recalibrated for every situation. The most important overall result is that the model evolves to a state in which the thrust belt and basin form a wave that propagates with constant form, the basin forming its leading edge. The system is thus at steady state in a coordinate system moving at the shortening rate. Sedimentation balances subsidence, and input of material into the orogenic wedge balances erosion. The timescale for evolution to this steady state would be a useful reference point for studies of foreland-basin dynamics, but apparently has not yet been computed. Some of the key results are summarized in Fig. 12(a–d). At steady state, increasing the erosion coefficient makes the deposit wider, not only because more sediment is produced but because the sediment load widens the basin flexurally. Increasing the depositional transport coefficient makes the basin go from underfilled to overfilled and eventually buries the forebulge. Increasing the flexural rigidity produces the expected effect of widening the basin; increasing the thrusting rate narrows it, leading to underfilling, because basin deposits quickly become incorporated into the thrust. Flemings and Jordan were also able to produce a partially constrained comparison between their steady-state model and the basement and surface topographies of the sub-Andean foreland.

Flemings & Jordan (1990) and Jordan & Flemings (1990) went on to look at time-dependent aspects of their foreland-basin model. Key results included: (1) significant changes in basin wavelength, which might be interpreted as changes in lithospheric rheology but rather are due to sediment redistribution after an episode of thrusting; (2) the formation of piggyback basins; and (3) strong time variation in stratigraphy resulting from erosion and sediment redistribution after a thrust episode. Sinclair *et al.* (1991) used a similar coupled model to show how thickening and slowing of a critically tapered thrust wedge could produce unconformities that deepen approaching the forebulge.

The uncoupled behaviour of a diffusional system with simple geometric subsidence was examined by Paola *et al.* (1992) with the aim of understanding general aspects of diffusional basin filling. The addition of subsidence to the diffusion equation allows for a steady state in which the surface topography is adjusted so that sedimentation rate balances subsidence rate, i.e. $\partial\eta/\partial t = 0$ in Eq. (1). The time T_{eq} required to reach this condition is of the order of L^2/ν, where L is the length of the basin. T_{eq} is the basin analogue of the 'graded time' proposed by Schumm & Lichty (1965). Basin response to external forcing depends on the timescale of the forcing relative to T_{eq} (Fig. 12f–h). This result can be seen as a generalization of the findings of Pitman (1978) and Angevine (1989), and indeed it is easily shown that Pitman's tectonic timescale is approximately equivalent to T_{eq} as defined above. The main implication is that different basins can respond to the same external forcing in qualitatively different ways. Computed T_{eq} values for typical basin conditions fall within timescales of geological interest, as shown in Fig. 13.

Results from several other models with diffusional components will be discussed below under 'Coupled models'. A computer code with a graphical interface and documentation that allows one to explore many of the ideas discussed above has been made freely and readily available by Flemings & Grotzinger (1996). Beyond the computer code, Flemings and Grotzinger's model of openness in research is an important contribution to the geological community in its own right.

Alluvial architecture

In principle, there is no reason to separate the modelling of what is termed 'alluvial architecture' from alluvial basin filling in general. However, the work of Leeder (1978) discussed above set off a whole line of research specifically aimed at modelling the distribution of channel-belt sand bodies, first in vertical panels and later in three-dimensional blocks. The problem does seem to invite analysis on its own terms: fluvial sand bodies are usually distinct from enclosing floodplain material, and modelling avulsion as a stochastic process seems reasonable based on the apparently random distribution of sand bodies in outcrop. Leeder's original model was followed closely by Allen (1978), who incorporated the effect of buried fluvial topography, while later workers added effects of compaction, lateral tilting and lateral variation in floodplain sedimentation rate (Allen, 1979; Bridge & Leeder, 1979; Alexander & Leeder, 1987; Leeder & Gawthorpe, 1987; Bridge & Mackey, 1993a,b). We will refer to this whole series of theoretical models as the 'Leeder–Allen–Bridge' (LAB) models. Shanley & McCabe (1991, 1993) have coupled LAB-type models to sequence stratigraphy.

Bryant *et al.* (1995) showed experimentally that the avulsion rate can be a strong function of sedimentation rate, and Heller & Paola (1996) evaluated the effect of this within the LAB framework. At present, the most fully elaborated member of the LAB series is the model of Mackey & Bridge (1995), who have provided the first analysis of channel-belt architecture in three dimensions. The model is semi-empirical and serves to highlight both the value of considering the third dimension and the number of hard-to-constrain variables that must be added to make this possible. Unlike some cases, in which large-scale models seem to overwhelm their creators, Mackey and Bridge were able to extract a relatively simple and testable implication from their model: they predict that progessive upstream shift in avulsion points should lead to an 'avulsion sequence' in which sand bodies thin and become more widely dispersed going upward.

It seems likely that eventually 'architecture models' *per se* will disappear, becoming merged into overall basin-filling models. Avulsion and other channel-belt processes would retain their place as the dominant mesoscale process, but would occur spontaneously as part of the dynamics of the fluvial system. This approach would allow a unified approach to modelling the interplay between alluvial architecture and external forcings such as base-level change and variation in subsidence and sediment supply.

Outstanding problems in fluvial modelling

There are still several important pieces missing from first-order fluvial basin models. Current models lack a physical model for floodplain development, and although there have been a number of detailed studies of floodplain deposition, these do not yet provide a framework for developing a large-scale morphodynamic floodplain model. This is a first-order deficiency, since without floodplains it is impossible to model accurately the trapping of fines within fluvial systems, the downstream sorting of sandy and finer material, or avulsion. Another important point has to do with the so-called 'wash load', traditionally defined either as material finer than a cut-off small grain size (e.g. 40 μm), or as material in transport that is not present in the bed. Paola *et al.* (1999) have suggested a new definition more appropriate for morphodynamics: that the wash load is that portion of the load that does not affect the average bed slope. At present no method is available to determine the wash load defined this way, but its first-order importance is clear: any morphodynamic treatment that relates slope to sediment flux could greatly overestimate the slope by including in the sediment flux material that is in effect getting a free ride. In addition, in this view of wash load, downstream reduction in shear stress associated with both deposition and downstream fining would cause progressive conversion of wash load into effective (i.e. slope-forming) load, adding what would look like a continuously distributed sediment source to the flow. Finally, another outstanding problem in modelling fluvial systems is taking account of the effects of fluctuations in local flow conditions, either due to floods or to simple flow variability within channel systems. This was discussed earlier in the context of diffusion; the methods considered so far cover only spatial and flood-related variability. This is a specific instance of the more general problem of how to abstract the essential features of dynamics at one timescale so that they may be incorporated appropriately to the dynamics at longer timescales. We will return to this subject in the Discussion.

The continental shelf

The development of stratigraphic-scale models of fluvial systems has been paralleled by similar efforts for the continental shelf. The first step was to apply diffusional models to the continental shelf. Jordan & Flemings (1991), Johnson & Beaumont (1995) and Rivenæs (1992, 1997) have used standard linear diffusion for the shelf, assigning a lower diffusivity to represent the observed generally higher depositional slopes on shelves. Kaufman *et al.* (1991) used a depth-dependent diffusivity to account for attenuation of wave-induced bottom velocities in deep water. This depth dependence produces a rollover in the topography that was interpreted as representing the shelf–slope break. In none of these studies was any attempt made to derive the morphodynamic equation from a detailed analysis of flow and sediment transport on continental shelves. Niedoroda *et al.* (1995) have proposed a morphodynamic model of sedimentation by shallow-submarine processes. Their idea is to sum the effects of numerous short-term transport events, such as storms or river floods. These events are individually advective, but their net effect can be thought of as diffusional in aggregate. In this case,

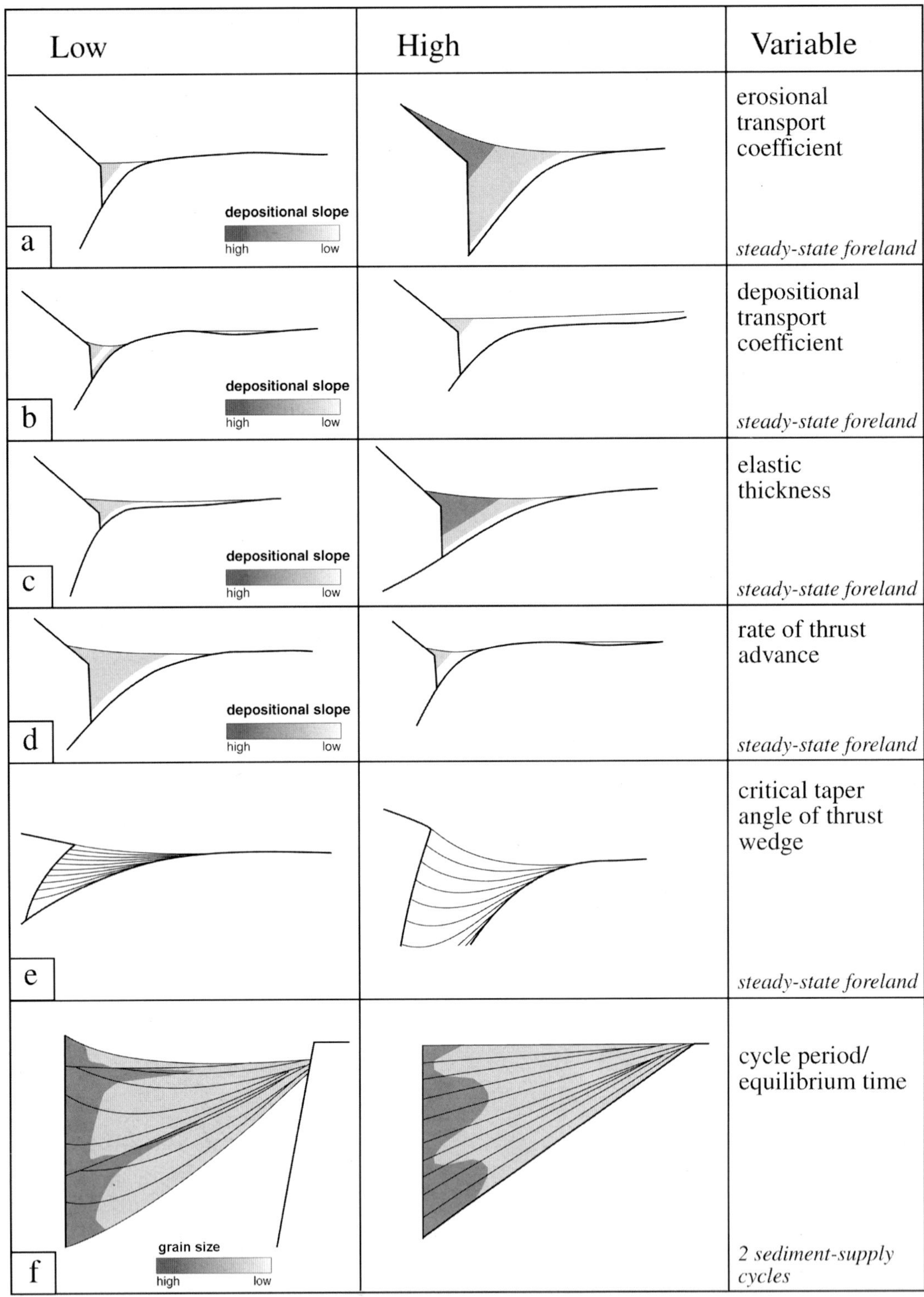

unlike the fluvial diffusion approximation, the transport process is actually diffusive in a physical sense. The model also includes a component of advective onshore transport in shallow water. The Niedoroda *et al.* approach to diffusion works best when the events are uncorrelated, which is reasonable for shelf transport. The effective diffusion coefficients turn out to be strong functions of position and are dependent on the input meteorological and

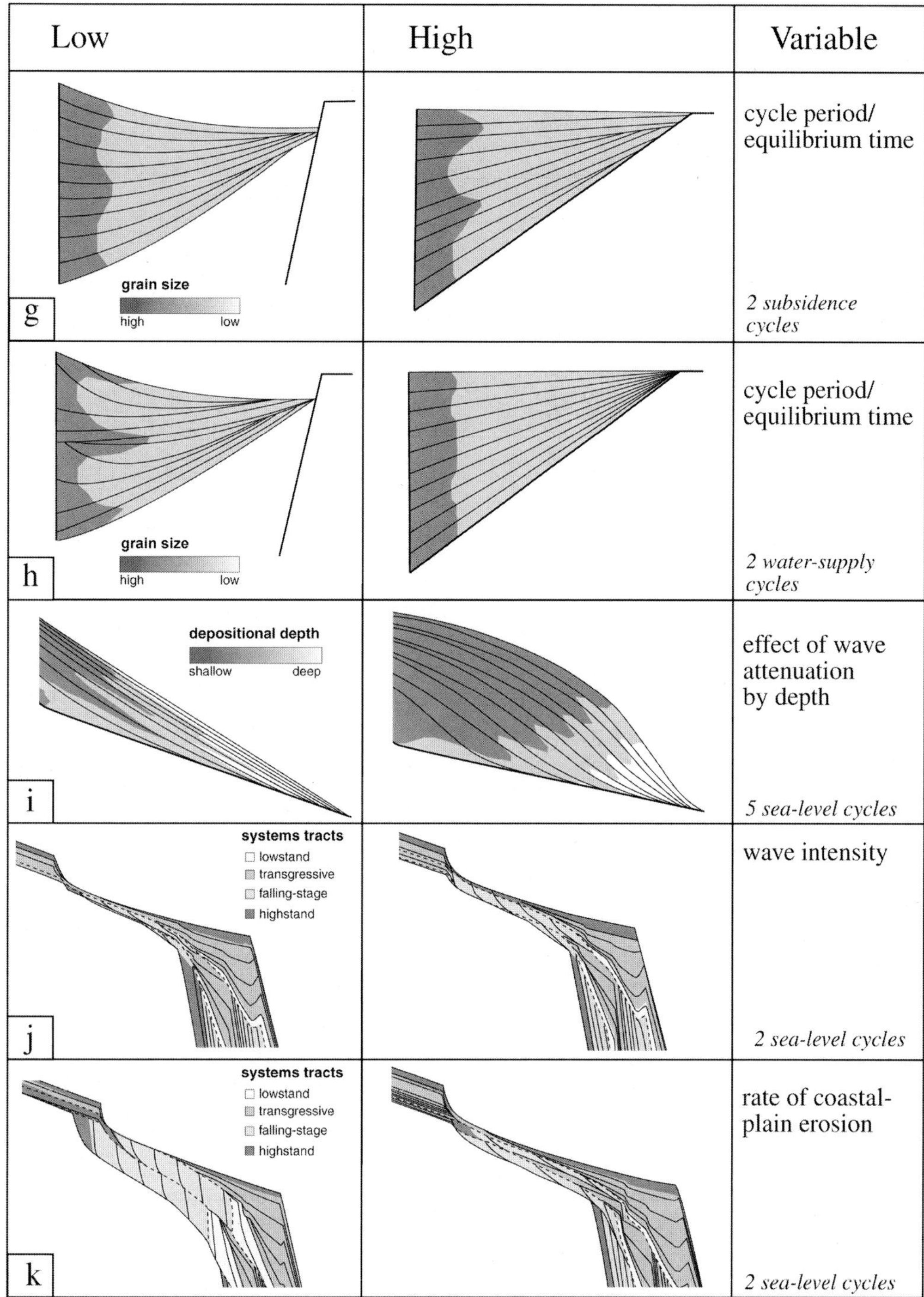

Fig. 12. A catalogue of clastic stratal geometries produced using diffusional surface models under various scenarios. Layout, disclaimer and explanation are the same as for Fig. 7. (a) Effect of changing erosional transport coefficient in thrust belt; (b) effect of changing depositional transport coefficient in foreland basin; (c) effect of changing elastic thickness; (d) effect of changing rate of thrust advance (a–d: Flemings & Jordan, 1989); (e) effect of changing critical taper angle of orogenic wedge (Sinclair *et al.*, 1991); (f) effect of changing period of imposed sediment-supply cycle relative to basin equilibrium time; (g) effect of changing period of imposed subsidence cycle relative to basin equilibrium time; (h) effect of changing period of imposed water-supply cycle relative to basin equilibrium time (f–h: Paola *et al.*, 1992); (i) effect of changing strength of depth dependence in depth-dependent diffusion (Kaufman *et al.*, 1991); (j) effect of changing wave strength in shelf advection–diffusion model; (k) effect of changing rate of coastal-plain erosion rate in shelf advection-diffusion model (j–k; Carey *et al.*, 1999).

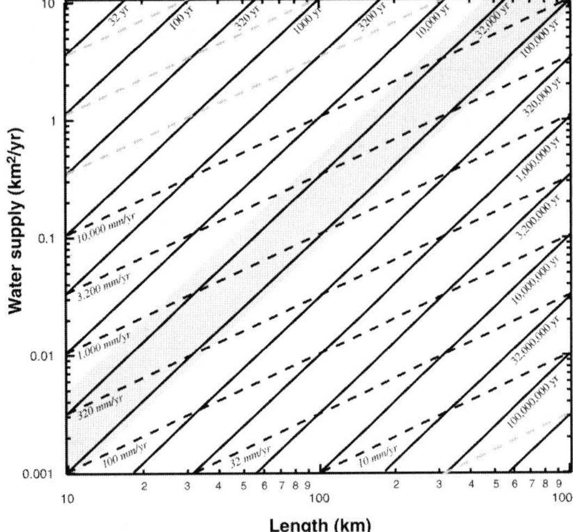

Fig. 13. Diffusional T_{eq} values (solid diagonal lines, plain labels) as a function of effective water supply and basin length. The effective water supply $\langle q_w \rangle$ is the product of rainfall rate and catchment length. The transport (diffusion) coefficient was calculated from Eq. (5) using $c_f = 0{\cdot}01$, $s = 1{\cdot}67$, $C_0 = 0{\cdot}6$, $\tau_* = 2$, and the sediment-transport formula of Engelund & Hansen (1967). The Milankovich frequency band is shown in light grey. The dashed diagonal lines show the rainfall values (italic labels) that would be required to produce the given values of $\langle q_w \rangle$ if the catchment length were equal to the basin length. Greyed rainfall lines represent implausible values.

hydrological event data used to compute them. In the present case, the greatest difficulty is that for most of the shelves for which abundant data are available (e.g. North American east coast), the present topography is substantially out of equilibrium, owing to the recent rapid rise in sea-level. The effect of the underlying topography on the computed effective diffusion coefficients is unclear but could be substantial. The Niedoroda *et al.* model has been coupled to a flexural subsidence model to produce a model of the coastal plain by Carey *et al.* (1999). The result is a semidynamic model: the shelf segment is treated dynamically, but the shoreline migrates via the geometric rule proposed by Bruun (1962) and the fluvial system is handled geometrically. One can understand dealing with the complexity of the nearshore region geometrically, but now that the physical basis of diffusion in river systems is clear, there is no reason to treat that side of the coastal plain geometrically while using a dynamic model for the submarine side. In any case, the Carey *et al.* model seems to produce reasonable shelf topography and stratigraphy but remains difficult to test, like most stratigraphic models.

A prominent feature of seismic sections of clastic shelves is prograding clinoforms. The simplest way to make a clinoform is by delta progradation (Kenyon & Turcotte, 1985; Syvitski & Daughney, 1992; Driscoll & Karner, 1999). However, the clinoform reconstructions of Steckler *et al.* (1999) suggest that the clinoform break is often in water depths of several tens of metres and hence is not a shoreline. An advective–diffusive submarine clinoform model was proposed by Pirmez *et al.* (1998). Here the advective component of the flow is a residual river current that drops sediment as it expands into deepening water, so the clinoform is still essentially deltaic. This produces the desired morphology, and is a step towards a more realistic shelf model in which waves and marine currents are the main agents that transport sediment. The advective–diffusive model of Niedoroda *et al.* (1995) can also produce clinoforms, but apparently only under conditions in which the rate of sea-level change is high relative to sediment supply (Carey *et al.*, 1999).

Since several shelf models appear to have adopted the diffusion equation with little or no dynamical justification, it is worth pointing out that there is no compelling *a priori* reason to choose diffusion as a first-order morphodynamic model. One can derive a diffusion equation either by invoking transport mechanics that is actually diffusive (Niedoroda *et al.*, 1995) or by showing that sediment flux is controlled by bottom slope, i.e. $q_s = f(S)$. The latter approach makes sense for systems like rivers (or turbidity currents) that are gravity-driven. For cases like the continental shelf or large aeolian systems in which the flow is driven by externally imposed pressure gradients, there is no particular reason why the bottom slope should be the main control on sediment flux. For clastic shelves, the dominant role played by clinoforms (Pirmez *et al.*, 1998; Steckler *et al.*, 1999) suggests that a wave equation might be a more natural choice than a diffusion equation. The Pirmez *et al.* (1998) clinoform model is really fluviodeltaic rather than marine, but it is not hard to see how one might develop a comparable wave-current model for shelves by simplifying the dynamics considered by Niedoroda *et al.* (1995). Assume that the characteristic event that moves sediment across the shelf is storm-induced downwelling. The cross-shelf flow in this case is due to the wind stress applied to the surface. Ignoring waves for the moment and confining attention to the inner shelf, where the effects of stratification and the Coriolis force can be neglected, the bottom stress is proportional to the wind stress. If the latter is constant, so are the bottom stress and (approximately) the near-bottom current, independent of depth or slope. As in the Niedoroda *et al.* model, it is waves coupled to this downwelling flow that give rise to spatial stress variation, via the exponential dependence of bottom orbital velocity on water depth. This suggests that to first order the structure of the sediment flux on the shelf could be $q_s = f(h)$, rather than $q_s = f(S)$.

Since the water surface is nearly horizontal, variations in water depth h and bed elevation η are linked via $d\eta = -dh$. Thus with a depth-controlled flux q_s we have

$$\frac{\partial \eta}{\partial t} = \frac{1}{C_0}\left(\frac{dq_s}{dh}\right)\frac{\partial \eta}{\partial x} \tag{7}$$

This describes a wave that propagates in one direction, with constant form if dq_s/dh is constant. In contrast, diffusion always tends to spread and flatten topography. The wave speed c in Eq. (7) is given by $c = (1/C_0)dq_s/dh$, which would be negative on the shelf. Thus the wave propagates offshore. The equilibrium time T_{eq} for the wave case is given by $T_{eq} = L/c$ where L is the length of the shelf. On the shelf, q_s should decrease faster than linearly with h, in which case the shallower parts of the bed wave move faster than the deeper ones. This produces a kinematic instability in which the wave steepens until some independent slope-related failure criterion is exceeded, at which point a failure-dominated front forms. The same instability has been invoked in the case of bedforms by Middleton & Southard (1984) to explain why they have steep lee faces. In the shelf case, the wave is a clinoform and the front is its steep leading edge.

The question of wave-like (advective) vs. diffusive behaviour has been somewhat muddied because diffusional morphodynamics can also produce clinoforms (Jordan & Flemings, 1991; Kaufman *et al.*, 1991; Rivenæs, 1997; Carey *et al.*, 1999). For dual-diffusivity models (Jordan & Flemings, 1991; Rivenæs, 1997), the clinoform is produced by the change in transport coefficient at the shoreline and is thus basically deltaic. The origin of the wave-like behaviour in fully submarine diffusion is hidden in the spatial variability of the diffusion coefficient. Applying the product rule to (1) above, we have

$$\sigma + \frac{\partial \eta}{\partial t} = \frac{\partial \nu}{\partial x}\left(\frac{\partial \eta}{\partial x}\right) + \nu\left(\frac{\partial^2 \eta}{\partial x^2}\right) \tag{8}$$

The first term on the right is a wave term similar to that in Eq. (7), and the second is the diffusive term. The wave term is zero if the diffusion coefficient is constant. If not, the overall behaviour of the equation is determined by which term dominates. In most cases the two terms are of the same order of magnitude, and in no case can the wave term dominate everywhere because it is not the highest-order term in the equation. The diffusive term tends to disperse wave-like topography, so the wave-like behaviour associated with spatial variation in a diffusional transport coefficient is significantly different from that produced by a wave equation. Hence, there really are two different schools of thought on shelf morphodynamics, and they should be distinguishable based on observations of shelf strata. On the one hand, there is the 'clinoform' view proposed by Pirmez *et al.* (1998), in which wave propagation dominates shelf morphodynamics. On the other hand, there is the 'diffusive' view of Niedoroda *et al.* (1995), in which clinoform development requires special circumstances (high rate of change of sea-level relative to sediment supply; Carey *et al.*, 1999). If the 'clinoform' viewpoint is correct, the wave term should be treated separately and not appear as a byproduct of spatially variable diffusion. One could retain a diffusive component but it would be uncoupled from the clinoform term. If the 'diffusive' viewpoint is correct, then the presence of clinoforms in shelf strata, which can be readily detected in seismic sections, is not a generic event but rather a byproduct of dominance of the wave-like part of Eq. (8) owing to particular conditions.

A detailed model of tidally driven flow in an epeiric sea (a shelf system interior to a continent, possibly lacking an associated continental slope and rise) was developed by Ericksen & Slingerland (1990). The approach was to adapt an existing numerical code to the boundary conditions appropriate to the Cretaceous interior seaway of west-central North America. The goal was not to produce a morphodynamic model but rather to see how effective tidal action could have been given the known palaeogeography of the interior seaway. This approach is ideally suited to a problem such as this one where attention is focused on a specific question (how strong might tidal currents have been in the seaway?). The same approach was used later to constrain the timing of uplifts that would have restricted entry of the tidal wave to a narrow Miocene seaway in southern Europe (Martel *et al.*, 1994). Apart from its importance to understanding the target deposits, this work is a reminder that there are many problems for which specialists have already developed sophisticated computational algorithms that could be adapted to stratigraphic purposes.

Outstanding problems in shelf modelling

The outstanding problem of continental-shelf research from a stratigraphic perspective is lack of a range of dynamical examples. Where for rivers there are hundreds of case studies showing how they respond to a wide spectrum of changes and inputs, for shelves we have only the single 'case' of the Holocene rapid rise in sea-level and, on shorter timescales, a handful of well-documented storms and storm beds. Morphodynamic insight on shelves will have to come from physical–oceanographic models, physical experiments, and careful analysis of the rock record – the past as the key to the present.

Excellent summaries of the state of understanding of transport dynamics on continental shelves are provided in Cacchione & Drake (1990) and Nittrouer & Wright (1994). It should be evident from the above discussion that it is not clear even what general form a first-order morphodynamic equation for siliciclastic continental shelves would take. Likewise, there does not seem to be general agreement within the physical–oceanography community about the basic processes that transport sand across shelves, even at a descriptive level. Is there a 'unit event' (presumably a representative storm, analogous to a characteristic flood in rivers) that, if iterated, could account for the first-order topography of a constructional shelf? Or is shelf sedimentation *inherently* dependent on variability of transport events, as implied in the Niedoroda *et al.* (1995) approach? Much of the uncertainty about these questions is due to the lack of data on sediment movement and associated flow dynamics for a large, sandy shelf the overall morphology of which is clearly of modern origin. Indeed, it is not even clear that such a shelf exists in the world today, though the Bering Sea, Arafura Sea and the shelf west of Taiwan seem to be possibilities. The Eel River site currently being studied in the STRATAFORM project (Nittrouer, 1999) provides an example with substantial Holocene deposition, but it is predominantly mud. Thus we have very little data on which to base a stratigraphic model of shelf sand transport and deposition. It is also worth noting that no first-order morphodynamic model has yet been proposed for tidally dominated shelves, although tidal effects were the main focus of the work on epeiric seas by Ericksen & Slingerland (1990).

Compared with the lack of a general consensus as to even the general form of the equations for overall shelf morphodynamics, the details that are missing seem less critical at this point. In almost every respect, our understanding of wave-influenced transport systems lags behind that of current-dominated systems. Physically based models for wave–current interaction and associated sediment transport, which must form the basis for any theory of sedimentation on shelves, did not even appear until the late 1970s, and these models have still not been comprehensively tested (Cacchione & Drake, 1990; Nittrouer & Wright, 1994). There is also considerable basic work to be done on wave-related bedforms, particularly for combined wave–current flows, and full-spectrum and directionally variable waves. Mud-dominated shelves are even less well understood, particularly the dynamics of so-called 'fluid muds' that may be able to transport large quantities of fine material near the bottom without complete suspension. There is also still much to be done on organism–sediment interactions in the shallow-marine realm.

Coupled models

Earlier, we considered models in which eroding compressional uplifts were coupled to an adjacent foreland basin via the diffusion equation. Here we look at models in which the primary emphasis is on the stratigraphic effects of coupling itself.

Humphrey & Heller (1995) studied the behaviour of a simplified coupled system of an eroding uplift and an associated depositional fluvial system. A unique aspect of their work is that the erosion was modelled as a first-order (wave) rather than a second-order (diffusion) process. The location of the erosion–deposition boundary was assumed to be pinned by a fault. The coupling of the two leads to a kind of

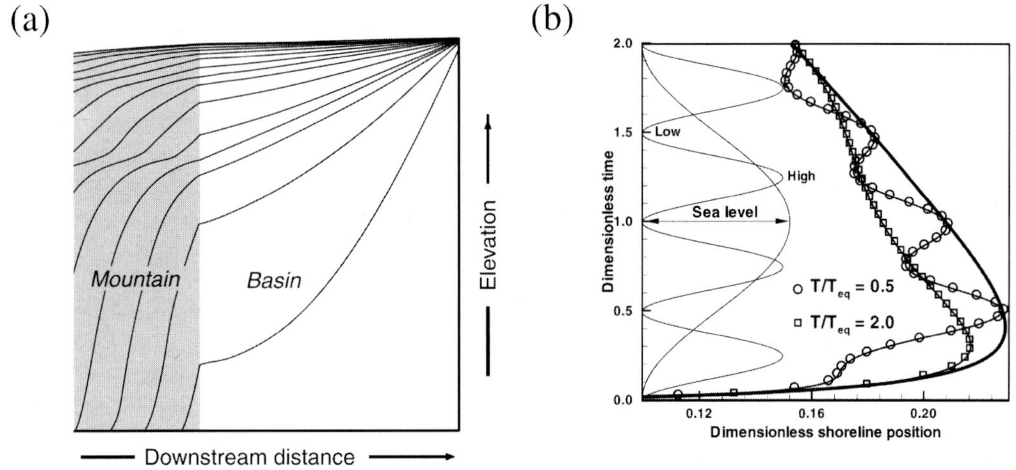

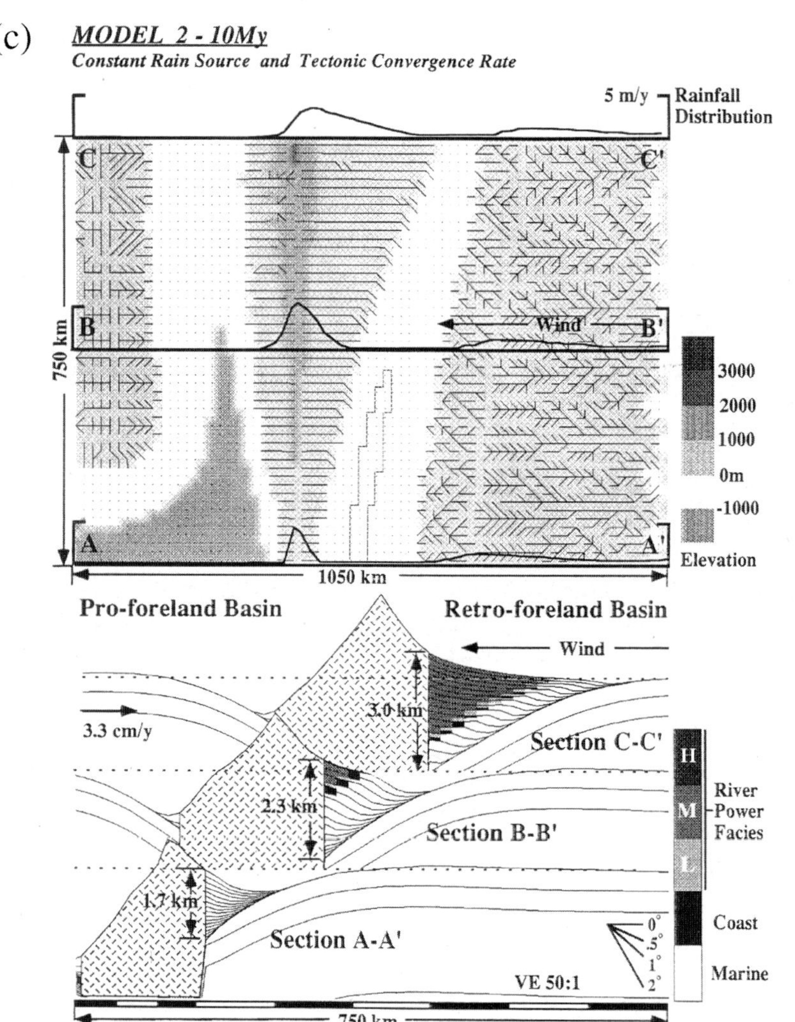

Fig. 14. Effects produced by coupled nonlinear models. (a) 'Ringing' cycles of deposition produced by a single, abrupt base-level rise (Humphrey & Heller, 1995); (b) shoreline autoretreat (heavy line) with superimposed shoreline migration cycles for low- (squares) and high- (circles) frequency sea-level variation with back-tilted linear subsidence (Swenson *et al.*, in press); (c) three-dimensional thrust-belt and foreland-basin topography produced by a coupled geodynamics/sediment-transport/rainfall model (from Johnson & Beaumont, 1995).

'ringing' in which the system responds to a step change in base level through a set of damped waves of aggradation (Fig. 14a).

A coupled model for the fluvial system and continental shelf was proposed by Jordan & Flemings (1991). The idea was to treat both the fluvial system and the shelf via linear diffusion, but with a lower transport coefficient on the shelf. As mentioned above, this model is essentially deltaic, and no detailed explanation of the coupling condition at the shoreline is given. The model produces phase shifts between shoreline position and sea-level reminiscent of those found by Pitman (1978). The results suggest that phase shifting becomes more important with increase in the ratio of cycle time to a characteristic time analogous to the T_{eq} defined above, but the paper presents only a set of case studies, from which the general form of the phase-shift function cannot be deduced. A more complete treatment of the shoreline coupling problem is presented by Swenson *et al.* (in press), who show that for a deltaic system with angle-of-repose foresets and a linear-diffusive fluvial system, shoreline migration is formally analogous to migration of a melting front in a two-phase (liquid–solid) medium heated from one end. The shoreline (phase transition) is a migrating internal boundary whose location is determined as part of the solution by specifying explicit coupling conditions at the boundary. The Swenson *et al.* model shows only minor phase shifting of shoreline relative to sea-level (Fig. 14b). There is no indication of a reduction of amplitude of shoreline response for long-period fluctuations, as was predicted by Angevine (1989). Another important result is that the model has no steady state; the shoreline tends to drift landward ('autoretreat') owing to lengthening of the submarine surface. This work highlights two key aspects of modelling coupled systems: (1) the coupling conditions are as important as the equations used to model the separate parts of the system, and (2) a system that couples linear parts with nonlinear coupling conditions can be strongly nonlinear.

The Dalhousie geodynamics group has developed the most complete model yet proposed for a foreland-basin system (Beaumont *et al.*, 1992; Kooi & Beaumont, 1994; Johnson & Beaumont, 1995). The latest version (Fig. 14c) is three-dimensional and couples erosion, deposition in fluvial and submarine systems, orographic rainfall and a geodynamic model of a convergent orogen. Hillslope transport is handled by assuming local linear diffusion. For channellized flow, it is assumed that at each point there is an equilibrium sediment flux proportional to the local stream power, i.e. equivalent to the diffusional approximation given above. Erosion is modelled as being detachment-limited in the sense that the system approaches its equilibrium sediment flux according to a first-order reaction law, i.e. exponentially over distance. The distance scale is a measure of the erodibility of the substrate. For erosion of unconsolidated material, or deposition, the distance scale tends to zero, making the model formally equivalent to diffusion as described above, but following channel axes. Submarine deposition is again handled with ordinary diffusion. In effect, the surface-process model is a composite of a detachment-limited drainage basin model, of which a number have been proposed (Willgoose *et al.*, 1991; Chase, 1992; Howard, 1994; Tucker & Slingerland, 1994), and diffusional fluvial and submarine depositional systems. The latter approach is similar to that proposed by Jordan & Flemings (1991). Apart from being the first to couple erosion to fluvial and submarine deposition in a three-dimensional model, the Dalhousie group's approach embodies two major advances: (1) it treats spatial variation in effective fluvial transport coefficient in a natural way, by letting water discharge accumulate and disperse over the three-dimensional grid; and (2) water is introduced by distributing an incident vapour flux over the terrain according to elevation and wind direction, thereby allowing for orographic precipitation effects. Having evolved from earlier geophysical subsidence models, this model also treats uplift and subsidence dynamically, using a simple elastic rheology. It is the first fully dynamic three-dimensional model of a foreland basin system. Including orographic effects highlights the importance of coupling climate and tectonics: the asymmetry between the basins on the leading and trailing edges of the orogen can be either reinforced or reduced depending on the relative wind direction.

The Stanford model

One of the most ambitious basin modelling programs to date is a three-dimensional model called SEDSIM that was developed at Stanford University. The basic algorithms and a first version of the code are presented in Tetzlaff & Harbaugh (1989). The approach is based on an adaptation of the marker-in-cell method originally developed for the study of complex, unsteady flows. The method tracks fluid units through a fixed computational grid; erosion and deposition occur as the units accelerate and decelerate owing to interactions with topography and each other (via pressure variations). Although it is very computationally intensive, this method has the advantage of allowing the flow to follow topography in a very natural way. This feature allows for development of remarkably realistic-looking patterns of flow and topography (Fig. 15). In addition to being fully dynamic and three-dimensional, the Stanford model is highly coupled, including modules for wave-influenced transport and turbidity currents along with compaction and hydrocarbon genesis and migration. Unfortunately, with the exceptions of the early version presented in Tetzlaff & Harbaugh (1989) and the nearshore model presented in Martinez & Harbaugh (1993), only general sketches of the workings of SEDSIM have been made available in the literature (Tetzlaff, 1990; Lee & Harbaugh, 1992; Martinez, 1992; Wendebourg & Ulmer, 1992; Harbaugh, 1993). In particular, no well-constrained tests of the accuracy of the model, with respect either to fluid mechanics or morphodynamics, appear to have been carried out. As such it remains an intriguing 'black box' on the basin modelling landscape.

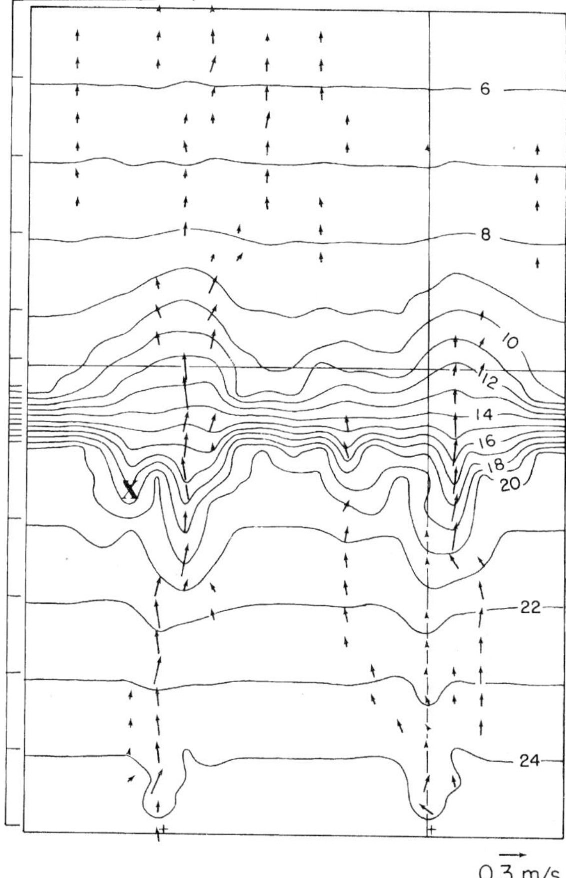

Fig. 15. Example of SEDSIM output (from Tetzlaff & Harbaugh, 1989) for development of alluvial fans and erosional valleys from an initial topographic step. Topography shown by contour lines in metres, flow by vector arrows.

The Penn State models

A wide-ranging summary of techniques for basin modelling is presented in Slingerland *et al.* (1994). Unlike SEDSIM, this is not a single, integrated model, but rather a collection of techniques for modelling specific subsystems. These include erosion and sediment supply, river flow in channels of variable width, size-selective sediment transport as bedload and in suspension, deltaic sedimentation, shelf transport by tides and storms, and compaction and subsurface flow. Aspects of the fluvial and shallow-marine parts of this collection have been discussed in previous sections. The collected work is noteworthy for including excellent introductory discussions of the derivation and physical meaning of the equations, a good deal of field data to go with the analytical material and all of the necessary computer code. Slingerland *et al.* (1994) is a toolbox, rather than a black box. It is an excellent summary of process-based basin modelling, with emphasis on doing a complete and careful job on individual components of the system.

The INSTAAR models

A comprehensive basin model called SEDFLUX has been developed at the Institute of Arctic and Alpine Research (INSTAAR), University of Colorado (Syvitski & Daughney, 1992; Syvitski & Alcott, 1993; Syvitski & Andrews, 1994; Syvitski & Alcott, 1995; Syvitski *et al.*, 1998a,b). SEDFLUX is aimed at simulating the transport and delivery of sediment on continental margins over timescales of up to tens of thousands of years. It includes the effects of eustatic variation, floods, storms and other relevant environmental factors (climate trends, random catastrophic events), using relatively short (daily to yearly) time steps. SEDFLUX combines several individual process–response models into one coupled model, distributing a multisized sediment load over a continental margin (Fig. 16). Processes currently included are river mouth dynamics, buoyant surface plumes, hyperpycnal flows, ocean storms, slope instabilities, turbidity currents and debris flows. The model includes compaction and load-induced isostatic subsidence; externally forced subsidence and other tectonic processes (faults, uplift) are predefined by the user. The modelled stratigraphy has a typical vertical resolution of 0·01–0·25 m, and a typical horizontal resolution of between 1 and 100 m.

At the moment, SEDFLUX is two-dimensional and marine. The INSTAAR group has a larger plan to develop a comprehensive three-dimensional model, at present called EARTHWORKS, that would integrate the fluvial system (erosional and depositional) as well as hydrology, thus allowing for the modelling of climatic effects. Such a model would be to sedimentary geology what global climate models are to atmospheric science. One of the best features of the INSTAAR group's approach is that it has been entirely open: thorough explanations of the algorithms, computer code, model output and the results of numerous tests of the component pieces against field data have been made available throughout development of the model series.

Unconventional models

The only explicitly stochastic basin-filling models proposed so far were introduced by Tipper (1991, 1992) and developed further by Pelletier & Turcotte (1997). Both are essentially diffusion models, but instead of invoking diffusion as a mean behaviour (i.e. Eq. 1), they model the morphodynamics as a random walk in two dimensions that behaves diffusionally in aggregate. These models are able to produce plausible-looking surface topography (Tipper, 1991) and power-law distributions of sedimentation rate as a function of averaging time (Pelletier & Turcotte, 1997).

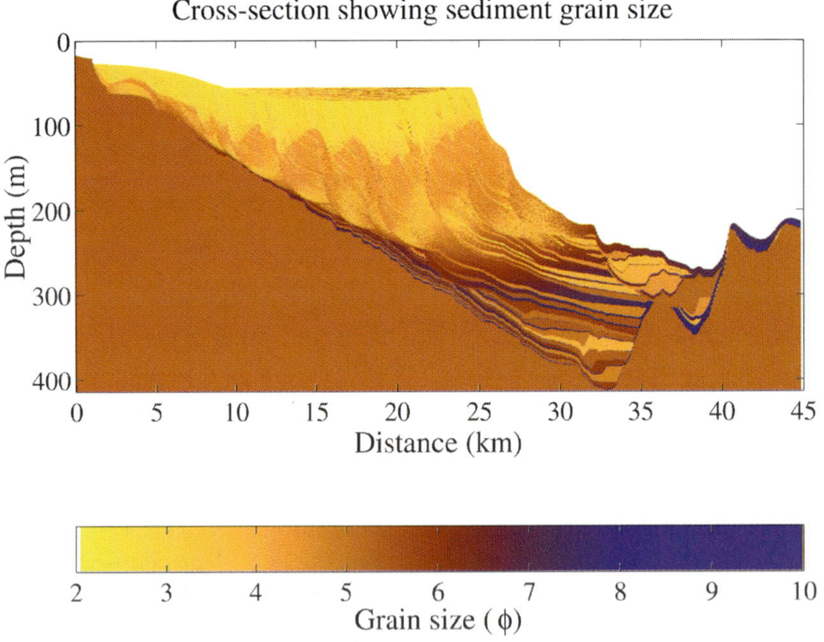

Fig. 16. Deltaic deposition as modelled with the INSTAAR model (references given in the text). The section is parallel to flow and the colours indicate grain size. The example is from a 12 000 year (daily) simulation for Knight Inlet, British Columbia. It includes the retreat of the glaciers from the ocean (about 10 kyr BP), an early to mid-Holocene warm period (to 6 kyr BP), slow cooling to the Little Ice Age (a few hundred years before present), rapid warming to the present, and then a hypothetical slower warming for 2 kyr into the future. The grain size shows the influence of a variety of processes. Hemipelagic drape from surface plumes can be seen across the entire basin. Debris flows (layers of uniform grain size) and graded turbidites occur in the deeper portions of the sub-basins, downslope from failure scarps. Delta front progradation of coarse particles from the dumping of the bedload seaward of the river mouth characterizes the upper portion of the deposits. Changes in sea-level caused by eustatic and isostatic influences of changing ice volume affect the position of the river mouth, forming delta plain and lagoonal deposits during relative sea-level rises.

FUZZIM (Norldund, 1999) is a rule-based, three-dimensional basin-filling model using fuzzy logic that has been developed and made freely available by Nordlund (1996). 'Fuzzy' logic replaces what would in traditional logic be binary choices by continuous variables. Thus 'false' (0) and 'true' (1) are replaced by a real number between 0 and 1 that expresses the degree of truth. In the basin model, this approach is implemented by defining state variables (e.g. 'far_from_source') that are true in varying degrees. Since FUZZIM is entirely rule-based, it is best thought of as a way of implementing conceptual ideas rather than a true model. Nonetheless, FUZZIM is fast, and accessible to people who are unfamiliar with numerical analysis. If nothing else, for these reasons it has great potential for introducing students to modelling. On a research level, it is not clear what is gained by replacing readily solvable equations with fuzzy-logic rules. On the other hand, there are aspects of depositional systems, such as those involving biota, where semi-empirical rules are a good starting point.

Model testing

Here 'testing' means a genuine attempt to falsify a model, as opposed to fitting it to a particular data set by varying its parameters. A true test requires that model predictions be compared with data that are independent of those used to constrain the model parameters. This rather exacting standard is very difficult to attain in historical geology, even for analytical models with relatively few parameters.

Geometric models

Because their depositional geometry must be specified by the user, geometric models are difficult to test. Hence, as described above, most of the effort in geometric modelling has gone to showing that the models can be calibrated to produce results that match real stratigraphy. One of the few geometric models that can in principle be tested in the strict sense is the original one of Pitman (1978). Although the main model prediction, phase-shifted long-term shoreline response, is relatively straightforward, it has been surprisingly difficult to test. Miller *et al.* (1993) report evidence of phase shifting of sequence boundaries relative to eustatic sea-level, but did not check quantitatively whether their cycles fit the temporal criterion for application of the model (Angevine, 1989).

Diffusional models

Again, despite their relative simplicity, the results of diffusion-based stratigraphic models have proven difficult to test. That many preserved basins show broad upward consistency in the locations of major facies boundaries (e.g. the gravel–sand transition) indicates that overall the idea of an average steady-state balance between subsidence and sedimentation is reasonable, but this is hardly a test. Flemings & Jordan (1989) showed that their coupled diffusion model could correctly reproduce the present-day surface and basement topography of the Andean foreland using reasonable parameter values. Heller & Paola (1992) provided some examples in which model-based inferences were consistent with previous interpretations of basin history based on classical geological evidence. Sinclair *et al.* (1991) modelled the development of a major unconformity with their diffusional foreland-basin model, and were able to show good agreement between computed and observed time–space diagrams for the unconformity. Although the quantitative details could not be constrained geologically, there was at least independent evidence for the nature and timing of the thickening event associated with the unconformity. This sort of constrained comparison between predicted and observed stratigraphic patterns is an important step towards real quantitative testing. On the whole, though, applications of diffusional modelling to the stratigraphic record amount to a series of arguments for plausibility, rather than true tests.

An unusual opportunity to test a diffusion-based basin model under highly constrained conditions was presented to our group several years ago. The problem involved a fluvial fan in a mine-tailings basin. The area of the basin was fixed, and accommodation was provided by raising the perimeter walls of the basin. Hence the system functioned as if it were in a basin with steady, uniform subsidence. The basin was supplied with water and sediment at a constant rate, and it developed a self-similar topography in which the sedimentation rate was everywhere constant (Fig. 17), as predicted by steady-state diffusional theory. Since all the inputs to the system were known and constant, it was an ideal case study for testing diffusion-based models. The engineering solution described by Parker

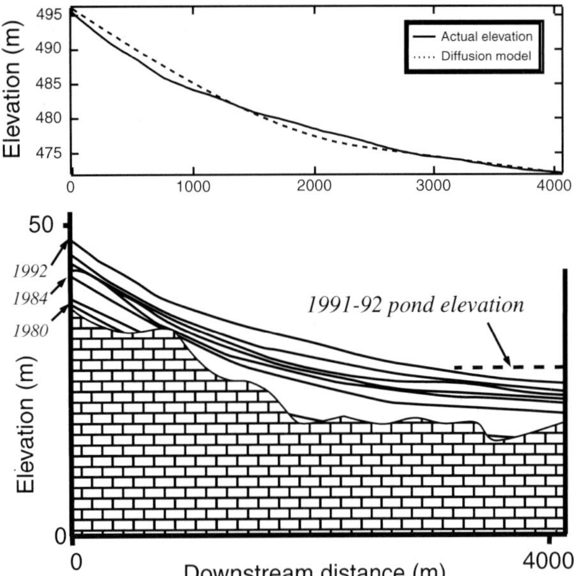

Fig. 17. Steady-state depositional profile developed in a mine-tailings basin (lower) and calculated profile from a diffusional solution (upper) using a transport coefficient computed from independent flow data (Paola *et al.*, 1999).

et al. (1998a,b) for channellized flow on the fan is, in effect, a linear diffusion equation in radial coordinates. It was fitted to the observed topography by adjusting a single parameter related to the nonlinear effect of zones of concentrated flow (e.g. channel confluences) in enhancing sediment transport. Later Paola *et al.* (1999) showed that this parameter could be estimated with reasonable accuracy from the observed distributions of flow velocity, depth and grain size (Fig. 17). Hence for this case at least, simple diffusion could be used to predict steady-state basinal topography on the basis of independently constrained input parameters.

PHYSICAL MODELS OF BASIN FILLING: A LOOK THROUGH THE DOOR

In contrast to all this theoretical modelling, very little experimental work has been done on basin filling. Most of the early experimental work in sedimentary geology focused on understanding outcrop-scale sedimentary structures. While there is still a good deal of work to be done in this area, in recent years experimental work has taken the same general direction as theory and shifted to larger-scale problems. There are now active programmes in experimental stratigraphy at Utrecht University, the University of Leeds and the University of Minnesota. Results from the latter two will be discussed below; reports of the Utrecht group's work can be found online at http://www.geo.uu.nl/Research/Sedimentology/Staff/gpostma/gpostma.html.

The main motivation for experimentation in stratigraphy is the same as in all other branches of science: it allows us to generate results under precisely known, repeatable conditions. This is especially valuable in the study of sedimentary basins, for which ancient examples usually involve boundary conditions that cannot be reconstructed precisely. Stratigraphic experiments have a particular advantage in compressing time. Much of the incompleteness of the stratigraphic record represents time when the sedimentary systems are inactive. In an experiment, the system is active all the time; for instance, in a fluvial experiment it is as if the rivers were constantly in flood. This represents a major compression of time. Continuing with the river example, if a typical natural river is activated by a flood that occurs every 1 or 2 years and lasts for a few days, the time compression from this effect alone in an experiment is more than two orders of magnitude. The price is that one loses effects such as plant growth and soil development that take place during the 'inactive' periods.

Previous work

One of the longest-running experimental programmes aimed at large-scale sediment processes is that at Colorado State University. A comprehensive summary of many of the main results is given by Schumm *et al.* (1987). Most of the Colorado State group's work has been on fluvial systems and is geomorphic rather than stratigraphic. A noteworthy exception is the work of Wood *et al.* (1993, 1994) and Koss *et al.* (1994), who studied the behaviour of fluvial systems and associated shelves in response to varying base level. Although the deposits were not analysed, the work was nonetheless aimed directly at sequence stratigraphy. Key results were: (1) shelf angle plays a major role in determining the total sediment volume derived from incision during relative fall, and hence the fraction of (presumably finer) shelf-derived material in the final deposit; (2) there are significant lags in delivery of fluvial sediment to offshore deltas associated with the finite time required for incision to propagate upstream; and (3) fluvial response to base-level fluctuation shows strong upstream decay. Points (2) and (3) also nicely illustrate the limitations of geometric theoretical models, which cannot capture this kind of wave-propagation effect, when applied to relatively high-frequency sea-level changes.

Alluvial architecture is another area in which stratigraphic experimentation is just beginning. The first experimental study directly aimed at architecture was that of Bryant *et al.* (1995), who in an unscaled experiment showed avulsion frequency to be a strongly increasing function of sedimentation rate, contradicting one of the key assumptions of the early LAB models. A much more detailed study is that of Ashworth *et al.* (1999), who modelled the internal architecture of a gravel-bed braided river (Ashburton River, New Zealand) using a 1:50 Froude-scale model (Fig. 18). The coarse-grained nature of the Ashburton system makes it an ideal target for physical modelling, and the experiments replicate both the observed distribution of subfacies in the field and the response of the Ashburton to changes in sedimentation rate. This response is a loud silence: no change in facies architecture was detectable for changes in sedimentation rate of a factor of two (laboratory) and nearly 10 (field).

There have been two major obstacles to experimentation as a tool in stratigraphy. The first is that until recently there has been no means of incorporating the effects of spatially variable subsidence into physical experiments. A solution to this is presented below. The second is that physical modelling of processes that in nature occur over large time and space scales seems to present formidable scaling problems. Since experimentation in stratigraphy is in its infancy, it is worth examining scaling issues before continuing any further.

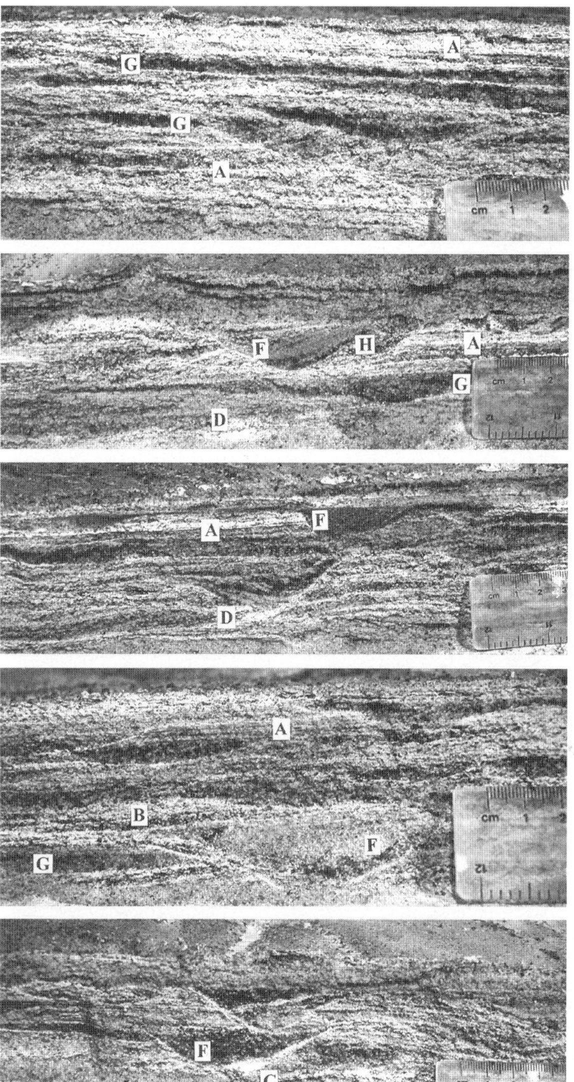

Fig. 18. Cross-sections showing depositional niches in a physical model of braided-river architecture (see Ashworth *et al.*, 1999, for details). Depositional niches identified in the flume sections are classified according to grain size for fine-grained (niches A–D) and coarser grained (niches F–H) sediments A: bar-top fines; B: bar margins; C: channel fills; D: erosional remnants; F: coarse-grained channel fills; G: bar cores; H: bar margins.

Scaling and the use of experimental results in stratigraphy

Although there is a substantial literature on the construction of dynamical scale models (Yalin, 1971; Peakall *et al.*, 1996, and References therein), in practice it is rarely possible to construct an exact scale model of any system involving fluid flow and sediment transport. For all practical purposes, water is kinematically the least viscous fluid available, which means that experimental models nearly always have significantly lower Reynolds numbers than their prototype systems. In loose-boundary experiments, it is difficult to scale down the ratio of particle size to system size because granular materials become cohesive as particle size decreases. There are a number of ways around these problems, but the key point is that scaling effects are by no means unique to stratigraphic experiments. The theoretical impossibility of creating exact scale models has not prevented engineers from solving countless design and management problems in rivers, estuaries and coastal areas with the aid of models. Beyond a certain amount of judgement and experience, the key to interpretation of the model results is that the experiments must be done in conjunction with physically based theory.

The dynamics of a system as complex as a sedimentary basin has many facets. At this point, a complete set of dimensionless numbers for basins, covering the whole range of relevant time and space scales, is not available. For a simplified basin model, Swenson *et al.* (in press) provide a set of dimensionless numbers that govern model response and stratal geometry. As the equations governing basin systems become better known, so will the relevant dimensionless numbers. Among these, some will be numbers that can readily be reproduced at laboratory scales, and some will not. The latter will tell us what aspects of the system we are missing in our experiments. At a more general level, basin processes and effects can be divided into three broad classes: (1) those that are scale-independent, (2) those that are present but distorted at laboratory scales and (3) those that cannot be represented at laboratory scales. At present, many of the first-order predictions of theoretical models are scale-independent. This is true for all geometric models (e.g. the shoreline response predicted by Pitman, 1978), which are scale-independent by definition. Neither are the existing dynamical models fundamentally scale-dependent, although some of them include effects (e.g. deposition of fine-grained sediments or floodplain formation) that are difficult to represent at small scales. In these cases, if a model is really robust, the modifications needed to apply it to a laboratory experiment should be dictated in a straightforward way by the model itself.

Experiments and theory are fundamentally interrelated in basin analysis, just as they are in engineering. Theoretical models that can successfully predict experimental results will have passed an important first hurdle. Although success in predicting experimental results does not guarantee that a model can be trusted in the field, it is very difficult to imagine trusting a model in the field if it cannot predict the results of a simplified, well-constrained experiment without a lot of tuning. Once a model has been tested, it becomes a vehicle for examining the effects of processes that cannot be scaled down to the experiments, for evaluating the effects of scaling distortions, and for examining new scenarios and suggesting further work. On the other hand, if experiments are done without a decent theoretical framework, one has no recourse but the naïve and often misleading approach of treating the experimental system as a miniature analogue.

One can also think of an experiment simply as being itself a model (Hooke, 1968). In this sense, it is similar to a theoretical model in that it involves simplified versions of the fundamental processes. The weakness of experiments is that they often involve parameters or conditions that are obviously different from natural systems. The strength, and the weakness, of theoretical models is that their parameters can be set to whatever values we want. If a model has enough of these 'knobs', it can be tuned to make its results look superficially like reality, whether it has actually captured reality or not. If we had complete confidence that the representations of natural processes embodied in our theoretical models were correct and complete, there would be no need for experiments. In my opinion, such confidence would be misplaced. Even a scaled-down water–sediment system, free to find its own transport paths and mechanisms, and to shape its own topography, is wonderfully complex. Seen as models (and not as miniatures), the advantage of experiments is that they are real, albeit simplified, physical systems. Their internal dynamics is not programmed, and they offer the possibility of new and unexpected behaviours. It is then necessary to determine, by analysis and/or field observation, if these are scale-related artefacts or not. Although experimental models of sedimentary systems are indeed simplifications, it is worth stressing that at present there is no theoretical model capable of capturing all of their dynamics.

An example experiment

Theoretical and philosophical arguments aside, the most direct way to appreciate the possibilities that experimentation has to offer in stratigraphy is to see the results of an actual experiment. Over the past 3 years the University of Minnesota group has been working to build an experimental basin with a programmable subsiding floor (Fig. 19). The subsidence mechanism is based on a honeycomb of independent hexagonal cells. The honeycomb forms the floor of an open basin. At the beginning of an experiment, the basin is filled with dry, well-sorted gravel. The top of the gravel is covered with a thin rubber membrane. Subsidence is induced by withdrawing gravel from the bottoms of the hexagonal cells. Each hexagon forms the top of a cone that tapers into an elbow pipe. Subsidence is induced by firing a

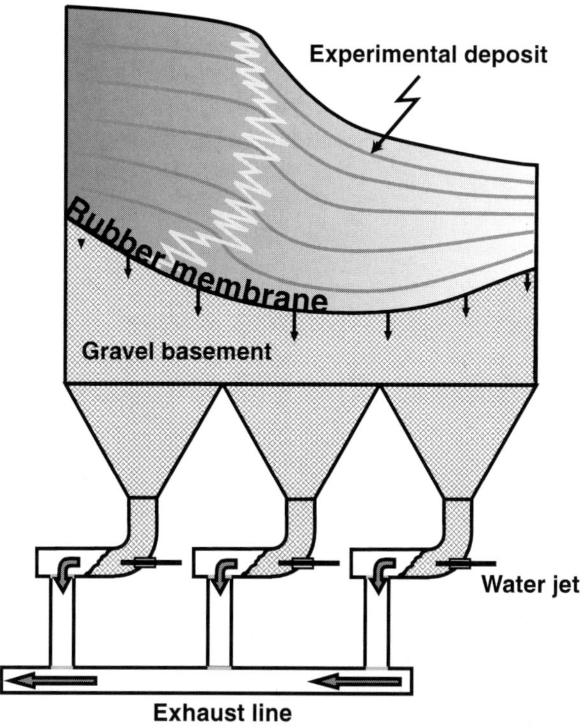

Fig. 19. Schematic diagram of the mechanism used to produce subsidence in the experimental stratigraphy facility discussed in the text. The subsidence cells are 0·4 m wide.

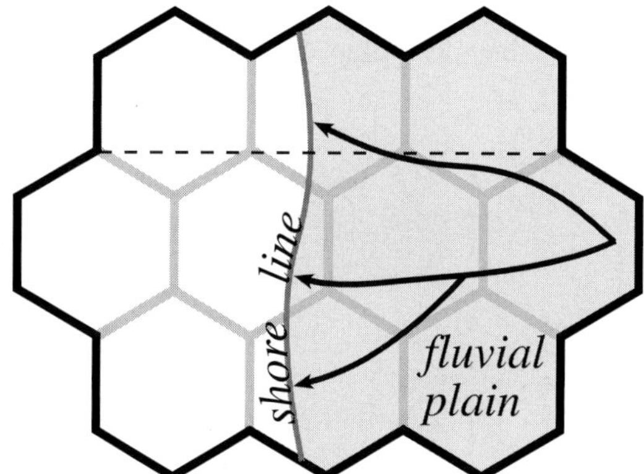

Fig. 20. Plan view of the base-level experiment described in the text. The dashed line indicates the section shown in Fig. 21.

pulse of high-pressure water into the gravel in the elbow. A small volume of gravel is knocked out of the elbow and into an exhaust line. Each pulse produces about 120 μm of subsidence. The subsidence is spatially continuous because the cells are separated only at floor level. Firing a single cell, for instance, produces a smooth bowl-shaped subsidence pattern that extends over the six adjoining cells. The full-size basin is 11·4 m long, 5·8 m wide and has about 1·3 m of usable accommodation. Further details of the experimental system can be found at http://www.umn.edu/safl/X_strat.htm

Stratigraphic signature of base-level cycles

This experiment was carried out in a small subsiding-floor basin that is 1·3 m long and 1 m wide, with 10 subsidence cells (Fig. 20). The experiment was designed to study the effects of slow and rapid falls in base level on shoreline position and the associated stratigraphy. Water and a 50:50 mixture (by volume) of quartz and coal sand were fed from a single source point. Base level was controlled from the opposite side of the basin. Subsidence was induced in a bowl-shaped pattern with a maximum in the centre of the basin. The rates of subsidence and water and sediment discharge were maintained constant throughout the run. The sediment discharge was set so that the initial shoreline position would be about in the centre of the basin.

First, the basin was run at equilibrium for 10 hours to stabilize the shoreline. Then the first, slow cycle of base-level fall and rise was begun. The time history of base level is shown in Fig. 21. The slow cycle took a total of 30 hours and had an amplitude of 0·1 m. The rate of fall was never greater than the maximum subsidence rate. The delta-front sediments were thus never exposed. The cycle time was longer than the equilibrium time for the basin, as defined in Paola *et al.* (1992). After base level had returned to its equilibrium position, the system was left undisturbed for 6 hours to re-equilibrate. It is clear from the deposit that during this period a series of growth faults developed near the shoreline, and that they produced space that accommodated a significant fraction of the total sedimentation. There was no visible surficial evidence of the faults during the experiment, however. After the equilibration period a rapid cycle of base-level fall and rise was begun. The amplitude of this cycle was the same as in the slow case, but the cycle time was reduced to 2 hours. Base-level fall caused exposure of the delta front and development of a major channel incision. Exposure of the delta front also led to 10–20 mm of visible gravitational collapse of the front, which occurred along the pre-existing growth faults. During the rise, all the

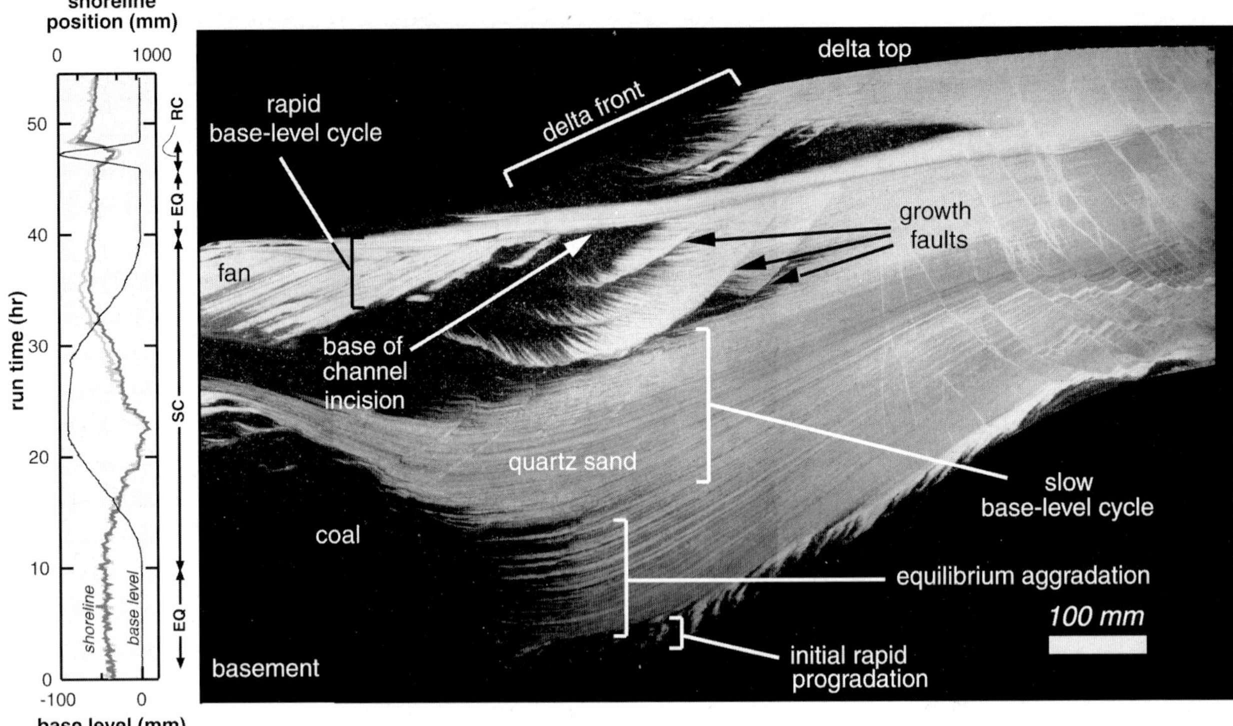

Fig. 21. Labelled photographic panel of the deposit produced in the prototype basin run. Transport was from right to left. Light tones are quartz sand, dark tones crushed coal. Base-level history is shown in the panel at left (black line), with the slow and rapid base-level cycles (SC and RC, respectively) and periods of equilibrium (EQ) marked. All other variables were held constant. Shoreline positions measured on longitudinal transects within and outside of the rapid-cycle incision are also shown at left (light and dark grey lines respectively).

sediment supplied to the system went to fill the incision, and as a result the shoreline substantially overshot its initial position at the end of the rise. The run ended with a regression as the shoreline migrated back toward the basin centre, during which time a new set of growth faults began to form.

The stratigraphic results of the experiment are shown in Fig. 21. The first, slow, base-level fall and rise produced a stratigraphically symmetric regression and transgression of the shoreline. This cycle produced only minor, localized incision, and the sand body produced was laterally extensive. On the other hand, the rapid base-level fall produced a narrow incised valley and an accompanying depositional lobe. Deposition across the entire system did not resume until the end of the cycle, when the incised valley was filled. Filling of the incised valley was accompanied by widening of the valley as sedimentation triggered active bank erosion. Outside of the valley, the subaerial part of the basin experienced the entire fall and rise cycle as a disconformity.

This experiment, despite its simplicity, produced a rich and complex stratigraphy. Other data from the experiment include a time-lapse video record of the surface pattern, and some 200 photomosaics of cross-sections parallel to the one in Fig. 21, spaced 2·5 mm apart. These have been assembled into a three-dimensional block diagram of the deposit, available for viewing at the web site given above. The scientific highlights of the run include: (1) the stratigraphic signature of slow base-level cycles is quite different from that for rapid cycles, as existing theory predicts; (2) the amplitude of response to the slow base-level cycle was larger than that for the rapid one, and neither showed strong phase-shifting of shoreline relative to base level; and (3) both base-level cycles, especially the rapid one, showed significant shoreline 'overshoots' at the end of the cycle, a feature not predicted by existing theory.

The use of stratigraphic experiments in model testing is illustrated in Fig. 22, which shows the measured base-level change and shoreline response in comparison with the theory of Swenson

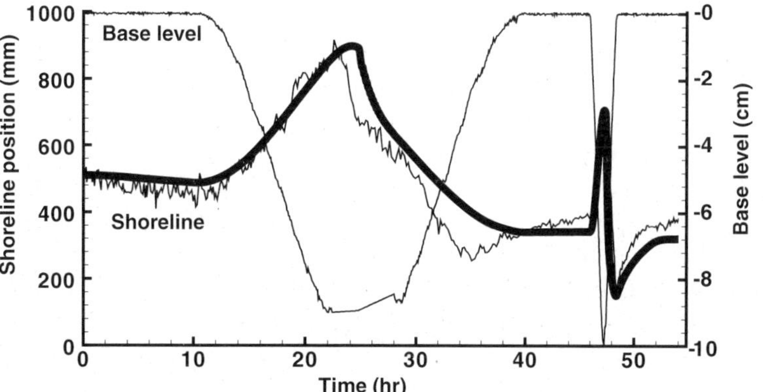

Fig. 22. Theoretical (heavy line) and experimental (light line) shoreline behaviour for two base-level cycles for the experiment described in the text. Theory from Swenson *et al.* (in press).

Table 1. Characteristic equations used in two-dimensional dynamic modelling and some of their basic properties. Symbols are defined in the text, except for ω and θ, which are growth coefficients with dimension time^{-1}

Characteristic equation	Environment(s)	Equilibrium condition	Equilibrium time (T_{eq})	Characteristic response	Stratigraphic signal
$\sigma + \frac{\partial \eta}{\partial t} = \omega(H - \eta)$	carbonate platform: shallow water (h< few m)	$\sigma = \omega(H - \eta)$	$\frac{1}{\omega}$	damped depth variation	stable palaeodepth
$\sigma + \frac{\partial \eta}{\partial t} = -\theta(H - \eta)$	carbonate platform: deep water (h> few m)	unstable	none	abrupt shallowing, drowning	shallowing-upward cycles; abrupt deepening
$\sigma + \frac{\partial \eta}{\partial t} = c\frac{\partial \eta}{\partial x}$	clastic shelf, erosional mountain belt	$\sigma = c\frac{\partial \eta}{\partial x}$	$\frac{L}{c}$	wave propagation	clinoforms
$\sigma + \frac{\partial \eta}{\partial t} = \frac{\partial}{\partial x}\left(\nu\frac{\partial \eta}{\partial x}\right)$	fluvial, clastic shelf, erosional mountain belt	$\sigma = \frac{\partial}{\partial x}\left(\nu\frac{\partial \eta}{\partial x}\right)$	$\frac{L^2}{\nu}$	spatially decaying	boundary-driven unconformities

et al. (in press). The only calibration involved estimating the effective transport coefficient from the supplied sediment flux and the measured fluvial slope. Thus the results shown constitute a true test, not a curve-fitting exercise.

One of the most useful aspects of experimental results such as those shown in Fig. 21 is that they provide a unique opportunity to apply one's interpretive techniques to a case for which the answer is known. We have done this exercise with a variety of interpreters: academic and industrial, experienced and inexperienced. The most interesting general result is not that almost no one gets it right – it is a very hard problem, after all – but that the error is nearly always on the side of overcomplicating the interpretation. We clearly have a lot to learn about how simple forcing can give rise to complex stratigraphy.

DISCUSSION

A synthesis of three basic model forms

If we consider the three main environments we have looked at in this review – carbonate shelf, clastic shelf and fluvial – we find that each has its own characteristic form of morphodynamic equation. These form a simple sequence if we express them in a common form, as given in Table 1. As discussed above, it is still not clear what is the best first-order morphodynamic equation for clastic shelves. I have chosen the wave equation simply because clinoforms are one of the things that make the shelf distinctive, and a wave equation is the most direct way to represent them. It is important to remember that the two clastic cases both make use of the general Exner (sediment-conservation) equation (Eq. 2). This equation applies as long as the deposited material is brought in (advected) from somewhere else. The carbonate equation does not conserve sediment because it refers to material produced within the water column. Physical transport of carbonates can be described with one of the clastic forms.

Each of these equations can be linearized to give a simple solution. The form of this solution can be interpreted as the mathematical expression of a characteristic depositional motif for each system, as illustrated in Fig. 23. The characteristic carbonate solution is exponential. If production depends positively on depth the solution tends to a stable equilibrium point where sedimentation (production) balances subsidence. If production depends negatively on depth an equilibrium point exists but is unstable, so the surface tends either to fill to sea-level or deepen rapidly if perturbed. The stratigraphic expressions of these two trends are shallowing-upward cycles and abrupt deepening, respectively. The characteristic clastic-shelf solution is an offshore-migrating wave, and the stratigraphic expression of this is clinoform deposition. The characteristic fluvial solution is an error function. The stratigraphic expression of this is response to forcing that tends to decay spatially away from the boundary where the forcing is applied.

Combining all three terms and extending them to two horizontal dimensions (x, y) we have the following generalized morphodynamic equation:

$$\sigma + \frac{\partial \eta}{\partial t} = p(H - \eta) + \left[c_x \frac{\partial \eta}{\partial x} + c_y \frac{\partial \eta}{\partial y}\right] + \left[\frac{\partial}{\partial x}\left(\frac{\nu \partial \eta}{\partial x}\right) + \frac{\partial}{\partial y}\left(\frac{\nu \partial \eta}{\partial y}\right)\right] \tag{9}$$

where p is a production function with units of (volume of sediment produced)/(bed area)/(time), and H is water level relative to some fixed datum. This is an advection–diffusion equation with one source term (the production rate p) and one sink term (the subsidence rate σ). So far p has been used only for carbonate production, but it could as well represent production of evaporites, biogenic silica or (if modelling a particular grain size) production or destruction of particles by aggregation–disaggregation processes. In the other two terms H does not appear explicitly, although the coefficients c and ν depend on at least the sign of $H-\eta$. They may depend independently on x and y as well. The second and third terms are both mass-conserving, and come from the Exner Eq. (2). The second term is a first-order wave equation (Eq. 7), also called an 'advective term', and describes the effect of dependence of sediment flux q_s on elevation η. The wave speed c has two vector components corresponding to the two horizonatal components of the sediment flux: $c_x = -(1/C_0)\mathrm{d}q_{sx}/\mathrm{d}\eta$ and $c_y = -(1/C_0)\mathrm{d}q_{sy}/\mathrm{d}\eta$. The third term is diffusive (Eq. 1), although if the diffusion coefficient ν depends on position, differentiation of this term produces additional wave-like terms, as discussed earlier. Overall, the ratio of the wave (second) term to the diffusive (third) term scales as $\mathbf{c}L/\nu$ where $\mathbf{c}$ is a characteristic value of the wave speed and L is a length scale. This number, which has the form of a Reynolds or Péclet number, is a measure of the importance of wave-like (advective) behaviour relative to diffusion. In the stratigraphic context it can be thought of as a ratio of clinoform to spatially decaying response styles.

The future of basin modelling

In the preceding sections, I have suggested some specific problems that deserve attention in the coming years. More generally, an entire area that has apparently been overlooked so far is large-scale quantitative modelling of aeolian systems. This could be done building on the conceptual framework provided by Kocurek & Havholm (1994). Other natural next steps involve better integration of various subsystems now being looked at separately. One of the grand challenges in this regard will be the continued development of fully coupled dynamic models for entire sedimentary systems, such as the continental margin. This will require much closer integration of carbonate and clastic modelling, which are often still pursued independently. There is also still plenty of work to be done on coupling erosional and depositional systems following the initial steps discussed earlier. Another integration problem is coupling models for basinal fluid flow and reaction (Harrison, 1990; Person *et al.*, 1996; Lee, 1997) to dynamic stratigraphic models. The Stanford SEDSIM group has, it is claimed, already accomplished some of this. Unfortunately, the development of SEDSIM has been closed, allowing for neither evaluation, testing nor use by other researchers. A much better approach is Syvitski *et al.*'s (1997) call for development of a 'community' stratigraphic model along the lines of community climate models. This approach is surely the way to develop the kind of large-scale, highly integrated models that will be needed to simulate sedimentary basins in detail.

As interesting as it will be to solve the many remaining technical problems in basin modelling, I hope we can do better than this. I hope we can also think of new and more creative ways to approach modelling itself. Here are a few thoughts on this.

Unlimited computing power

How does the advent of nearly unlimited computing power change our way of thinking about modelling in the earth sciences, or for that matter in science in general? Until the late twentieth century, simplification of problems was a necessary and universally accepted first step in modelling physical processes. One could argue that the most essential element of scientific skill was the ability to simplify insightfully, to extract the real essence of a problem. As computers become more powerful, simplification becomes less necessary. The obvious (and simple) response is that if we do not have to simplify, we should not do it. We all know that nature is complex, and so, according to this line of reasoning, our models of nature must be similarly complex. A model that includes more processes and/or does so in more detail is always better than a simpler model. The ideal model would, by extrapolation, include everything. I term this viewpoint 'literalism': the idea that theoretical models should always aim for the most exact, literal and detailed representation of natural systems that current knowledge and computing power allow. It is surprising how common this view of things is, so perhaps it is worth noting that the only entirely acceptable model in the literalist world would be an exact replica of the Earth itself (in an identical Solar System, of course).

In areas of science in which quantitative theory is part of the disciplinary tradition, the behaviour

(a) *carbonate production: stable*

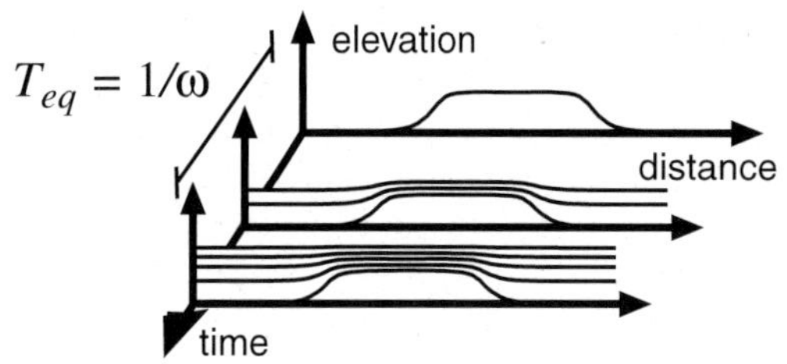

(b) *carbonate production: unstable*

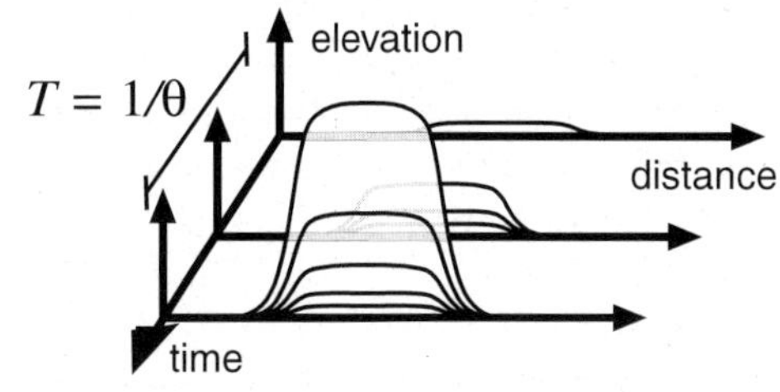

(c) *sediment transport: first-order wave*

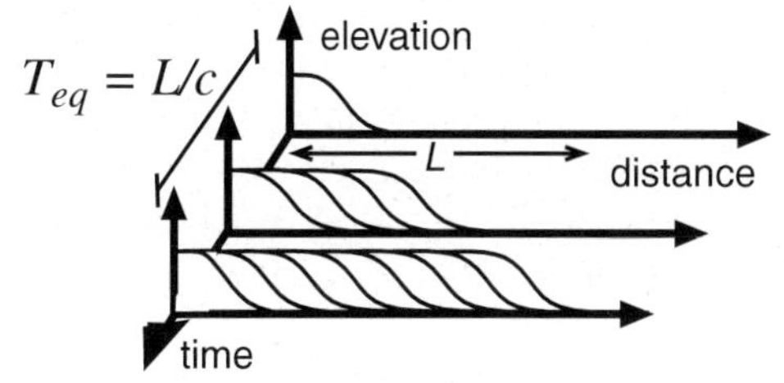

(d) *sediment transport: diffusive*

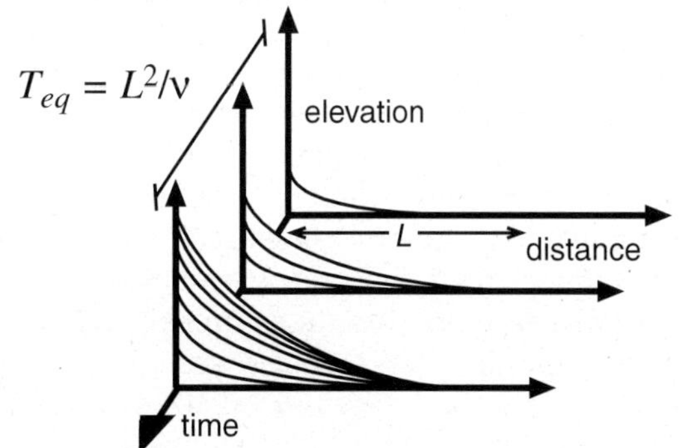

Fig. 23. Stratigraphic expressions of linearized solutions to the main equation forms used so far to model basin filling (Table 1). (a) Carbonate, shallow-water form (increasing production with depth). Stable equilibrium; the system responds to perturbation by returning to equilibrium on a timescale of the order ω^{-1}. (b) Carbonate, deep-water form (decreasing production with depth). Unstable equilibrium; the system responds to perturbation by exponential shallowing or deepening on a timescale of the order θ^{-1}. (c) Clastic shelf, first-order wave equation. Stable equilibrium; the system responds to perturbation by wave propagation on a timescale of the order L/c. (d) Fluvial system or clastic shelf, second-order linear diffusion equation. Stable equilibrium; the system responds to perturbation by a damped response on a timescale of the order L^2/ν. For perturbations with periods T significantly shorter than L^2/ν the response is damped over a distance of the order of $(T\nu)^{1/2}$.

and limitations of simplified models are well known. Large-scale computational models are developed not simply because they are feasible, or because of a generic belief that natural systems are complicated, but because it is known that they are required to solve a particular problem. There is no question that comprehensive stratigraphic models will be large, complex and very useful. But in sedimentary geology, it has happened that the conceptual breakthroughs that made analytical basin models possible have coincided with the manufacturing breakthroughs that made large-scale computational power available at low cost. Obviously, this is a good thing. However, it also means that it is now easy to create models that we ourselves do not understand, because we have skipped the initial 'simplistic' steps, or because we know more about programming than we do about analysis. There is a great temptation to program to the limits of our computers, rather than to the limits of our comprehension. Perhaps this is because, while it is easy to justify putting things into a model, it takes real understanding to justify leaving things out. It would be a shame if, in the pursuit of bigger and more comprehensive models, we were to lose sight of the equally valuable pursuit of the elegant, the insightful and the essential.

Model runs as experiments

Available computing power may be increasing exponentially, but the grasp of the human mind is not. As models become less comprehensible to their creators, running them becomes less like the final step in an analysis and more like an exploratory physical experiment. In what sense is a complicated model run analogous to an actual physical experiment? The criteria that a model must meet to be thought of in this way are: (1) the *governing equations* for the process must be completely known, and (2) the *representation* of those equations in the numerical model must have been thoroughly tested (e.g. by comparison with known solutions of the relevant equations), so that it is certain that the model output really represents the solution of the equations. Only if both conditions are satisfied is it safe to consider a computer simulation as equivalent to a physical experiment. At present, large-scale models of sedimentary basins do not pass criterion (1), because we do not know the complete governing equations for any sedimentary system. As discussed above, scaled-down physical experiments can provide a useful bridge between untested theories and hard-to-constrain field cases. Experimental testing is a crucial step in building confidence that our models correctly capture dynamics that is either scale-independent or scale-dependent in ways that we can account for. Nonetheless, experimental testing clearly is not enough. Ultimately, all geological models must be tested against natural systems. How can this be done?

Model testing

Time is to historical geology what distance is to cosmology. Models in both fields cannot be tested by direct experimentation on the objects under study because they are removed from us by impossibly large dimensions. Passage through both deep space and deep time progressively destroys information, but what reaches us from space lends itself far more readily to interpretation via physics than what time has left us. Ideally, we would look back into the past using only physical laws that we are certain are invariant. The fact that this approach is so hard to apply has to do both with the kind of information we have in geology and with the irreversible evolution of the Earth; it is why geology is not a branch of physics. Nonetheless, the more closely our models come directly from basic physical laws, the more confident we can be about applying them to systems where they cannot be tested directly.

After we have done our best to base our models on known physical laws and to test them experimentally, what remains must take the form of an integrated approach to narrowing the constraints on field cases as much as possible. We must make use of every available type of information, descriptive as well as quantitative. A crucial limitation in this regard is the lack of highly integrated basin-scale data sets. What could we learn if the best techniques we have, from all the relevant subdisciplines, were brought to bear in a concerted way on the best exposed and constrained basins in the world? The culture of geology suffers in this regard from the curse of easy access: we divide the world into little areas, and each of us works on 'our rocks', largely because we can. Integrated programmes, in which a number of researchers with different skills and viewpoints work together for something larger than any of them could achieve on their own, are much more common in fields like oceanography, where often there is no choice. One of the great challenges of the next few decades will be to develop in sedimentary geology a complementary culture of large-scale integrated field studies, without destroying the diversity and creative potential of our existing 'small-science' culture. This will involve learning from colleagues in fields like oceanography and geophysics, and co-operation with industrial colleagues as well.

Model design should also take field testing into account. So far, we have largely concentrated on making plausible-looking simulations and explaining general patterns. 'Testable design' would involve designing models to predict relations between aspects of preserved deposits that can be independently measured (e.g. Paola, 1990; Robinson & Slingerland, 1998), rather than just reproducing basin fills. In the fluvial realm, we have now been using diffusional models for years. As mentioned earlier, derivation of the equation from first principles allows us to predict downstream changes in variables like channel width and depth, but so far this opportunity has hardly been exploited. Little has been done to see how well one might constrain the transport (diffusion) coefficient from the stratigraphic record, based on careful analysis of grain size, sedimentary structures and other transport indicators in the deposits. Palaeohydraulic reconstructions of ancient rivers had a vogue in the 1970s and 1980s (e.g. Cotter, 1971; Baker, 1974; Ethridge & Schumm, 1978; Kochel & Baker, 1982; Gardner, 1983; Steer & Abbott, 1984) but have so far not been integrated into the fabric of sedimentary geology. Part of the problem has been a lack of context for the numbers: of what use is it to say that the discharge of a Cambrian river was $X\,\mathrm{m^3\,s^{-1}}$ and leave it at that? Now, however, quantitative basin models have given us a framework in which such numbers might be pieced together to help constrain transport coefficients and hence predictions of large-scale stratal geometry.

In this regard, an area that has received insufficient attention is careful analysis of what can actually be determined quantitatively from the stratigraphic record. Lessenger & Cross (1996) have argued that inversion of stratigraphy is possible, while Burton *et al.* (1987) and Burgess & Allen (1996) have stressed the nonuniqueness of the stratigraphic record. Heller *et al.* (1993) have outlined a method for constraining the space of possible forcing functions for a given stratigraphic pattern. It should be possible to compute the loss of information from ancient systems, and use this to design more testable models. In this regard, the gap between Quaternary geology and general sedimentology and stratigraphy still represents an underexploited opportunity. The review by Blum and Törnqvist (2000) shows the value of Quaternary studies in evaluating the stratigraphic effects of eustatic and climate change. Quaternary studies could be equally valuable for quantitative modelling; for instance, a systematic approach to model testing might involve starting with the shortest intervals and going back in time progressively. There also may be an analogy between the abstraction of dynamics from one level to the next in a multiple-timescale model and the loss of information as one goes deeper into the stratigraphic record.

Learning from climatology

There are some useful parallels between modelling climate and modelling sedimentary basins. The atmosphere and sedimentary basins are both complex, nonlinear systems; both are primarily physical but involve significant biological effects; and both involve dynamics over a wide range of timescales. In this analogy, sedimentary processes on human timescales, typically the domain of engineering and geomorphology, constitute the 'weather'. This points to a crucial aspect in which the analogy breaks down: weather modelling uses an agreed-upon set of basic equations, and weather models are tested many times every day. For all the major sedimentary environments, there are still fundamental questions about how to set up the governing flow and transport equations, and the database of case studies is much more limited. Another important way in which the analogy fails is that while we have only one atmosphere, we have many different basins. The general problem of modelling sedimentary basins is comparable with modelling not just the Earth's atmosphere but those of all other planets as well.

These differences make climate modelling easier than basin modelling. Because of this, and because it is topical, climate modelling is much more advanced than basin modelling. Thus we may be able to learn something from it. In particular, climate modellers have given considerable thought to dealing with multiple time and space scales. When we run a discretized model of a basin, we choose a time step $\mathrm{d}t$, and a grid size $\mathrm{d}x$. Processes with timescales less than $\mathrm{d}t$ and length scales less than $\mathrm{d}x$ are, by definition, below the resolution of the model. How do we extract the important dynamics of those processes so that their effects are correctly represented in our basin model?

The main way this is handled in climate modelling is through so-called 'parameterization', in which events or processes below the model resolution are represented as functions of variables that are resolved in the model (e.g. average temperature at a point). The most familiar example of this kind of parameterization is the representation of turbulent momentum flux via an eddy viscosity. In clastic basin modelling, the Exner sediment-continuity equation (Eq. 2) is the fundamental equation that is discretized and solved. Synthetic relations between sediment flux and, for instance, mean topographic slope (Eq. 3) are examples of parameterization. The transport coefficient ν is the parameter. In climate modelling, a good deal of effort has gone into calculating the parameter values by suitably averaging the underlying equations. In basin analysis, this process is only just beginning (Paola *et al.*, 1992, 1999; Niedoroda *et al.*, 1995; Paola, 1996).

The opposite problem, of resolving subgrid-scale local effects, has been approached via so-called 'nested' models (Fig. 24; Giorgi & Mearns, 1991). The idea is to have a series of grids, of increasingly fine resolution, one inside the other like a Russian doll. In the simplest kind of nested model, information flows downward, i.e. the solution at the large scale provides a set of boundary conditions for a more detailed model of processes within a particular grid cell. In the basin context, an overall diffusional model could, for instance, provide the boundary conditions for a detailed model of channel processes within a grid cell. However, information can also propagate upward in so-called 'adaptive multigrid' methods, in which the results of the fine-scale simulation can change parameter values at the larger scale (Barros & Lettenmaier, 1993).

Another useful idea that has been suggested in climate modelling is that the behaviour of a complex system over long timescales may be simpler and more predictable than its behaviour over short timescales. Mahlman (1998) has stressed the distinction between weather forecasting and climate modelling. Weather forecasting is a classical initial-value problem in a chaotic system and therefore cannot in principle be solved very far into the future. Modelling the effect of a secular change (e.g. a rise in CO_2 concentration) on climate is a boundary-value problem that in principle has a definite solution. (Here the climate represents a statistical steady state of the atmosphere.) Mahlman has proposed a particularly apt metaphor that also works for basin analysis: think of the atmosphere (basin) as a pinball machine. Forecasting the weather (evolution, for instance, of a single river channel belt) is like trying to forecast the precise path of a single ball, which is practically impossible. Modelling climate change (long-term basin response) is like trying to model the mean playing time over many balls, as a function of the tilt angle of the machine. This long-term problem is solvable in a way that the unpredictable short-term problem is not.

Emergent behaviour, self-organization and chaos

From an intellectual standpoint, the most far-reaching result to come out of climate research is the theory of complex systems, with its core result that even determinisitc systems can be fundamentally unpredictable (Lorenz, 1963). Unpredictable behaviour in deterministic

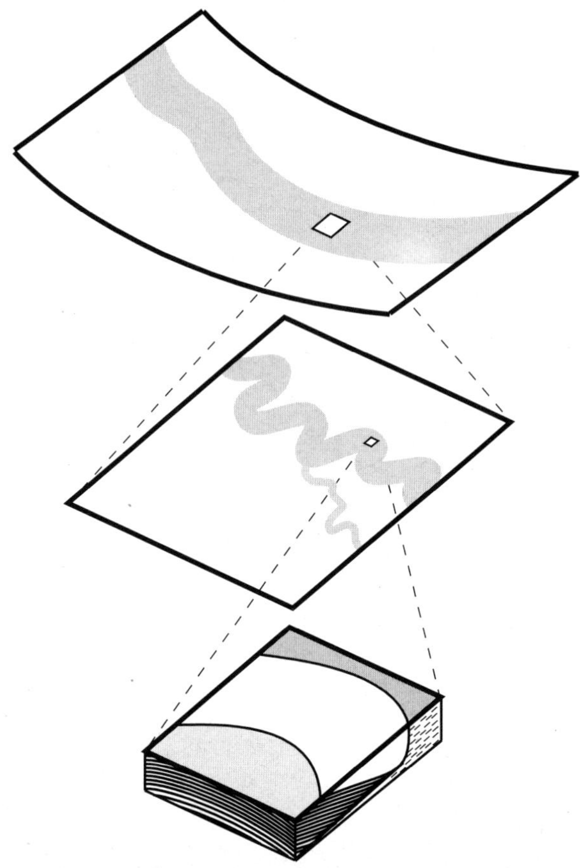

Fig. 24. How 'nested' numerical models might be used in stratigraphy. In the simplest scheme, the solution at each higher level provides boundary conditions for a more detailed solution at the lower level. In more complicated schemes, the lower-level solution can also feed information back up to higher levels.

systems was originally termed 'chaos', but this term has now become somewhat ambiguous. We will retain it here with the understanding that it carries no implications beyond the original definition. Although sedimentary basins are certainly complex systems, it is noteworthy how little impact complex-systems theory has had on modelling them. Slingerland (1990) and Nicolis & Nicolis (1991) have provided brief descriptions of complex-systems dynamics and chaos, but only a very broad outline of how these ideas might apply to sedimentary systems. At this point, most large-scale models for stratigraphy are of conventional (prechaos) form, evidently reflecting a belief on the part of their creators that the longest-period stratigraphic signals directly reflect external forcing. The Stanford SEDSIM model discussed earlier shows signs of unpredictability (Tetzlaff, 1990), but this has evidently not been analysed systematically. Even stochastic effects, certainly present in stratigraphic systems and themselves a form of chaos, have been incorporated explicitly in basin models only by Tipper (1991, 1992) and Pelletier & Turcotte (1997). If we step back and look at sedimentary processes as a whole, the general pattern that emerges is one of alternation between unpredictability and predictability as a function of scale. At the finest scale of interest, turbulence is the 'type example' of deterministic chaos. There are also signs of chaos at the level of bedform interactions (Rubin, 1992; Werner, 1995; Werner & Kocurek, 1997). Moving up in scale, sedimentary-system response to single events, like floods or storms, has been modelled using conventional predictive methods. This is reflected in engineering models of river or coastline behaviour over human timescales. In the fluvial case, this window of predictability occupies what might be called the 'in-channel' scale, i.e. the range of timescales on which the dynamics of a channel is determined primarily by local processes within that channel. In fluvial systems a new level of unpredictability arises for timescales on which interaction with other channels or the floodplain must be taken into account. Braiding defines the shortest of these timescales, nodal avulsion the longest. We might call this the 'multichannel' range of timescales. The unpredictability is reflected in the stochastic aspects of modelling avulsion (Leeder, 1978; Mackey & Bridge, 1995), and models of braiding that exhibit chaos (Murray & Paola, 1994; Murray & Paola, 1997) and power-law scaling (Sapozhnikov & Foufoula-Georgiou, 1996, 1997). Avulsion on deltas and submarine fans is presumably comparable, but beyond that it is not clear whether analogues of this mesoscale chaotic interval exist in other environments. In any case, at the largest scales, the view implicit in the modelling we have reviewed is that sedimentary systems are once again predictable.

Even within the apparently chaotic scale intervals, there seems to be little indication that the formal mathematical theory of chaos, with its emphasis on rigorous analysis of low-dimensional systems, is going to be of major relevance for basin analysis (for a possible counterexample see Gaffin & Maasch, 1991). But unpredictability is only one aspect of nonlinearity; other types of nonlinear behaviour are everywhere in basin modelling. Even apparently simple geometric models need not be linear, and coupled models involving linear components become nonlinear if any of the coupling conditions are nonlinear. Stratigraphic models that focus specifically on nonlinear effects include Gaffin & Maasch (1991), Humphrey & Heller (1995) and Swenson *et al.* (in press).

The idea of emergent behaviour is one of the more interesting and relevant developments in the theory of complex dynamical systems. 'Emergent' describes behaviours that seem to emerge from the interactions of the parts of a nonlinear system. Emergent behaviour occurs in nonlinear systems with many interacting parts, and cannot be derived directly from the equations governing the parts. A common type of emergent behaviour is formation of organized patterns through internal interactions; this is called 'self-organization'. Basins are obviously complex systems with many interacting parts, so it is reasonable to expect emergent behaviours in them. Knowing that emergence is possible and perhaps inevitable in this type of system raises the possibility that modelling them from the bottom up (i.e. by summing many small-scale or short-term events) might be inefficient, misleading, or as Werner (1999) has argued, fundamentally inconsistent. This is the basis for the 'synthesist' approach to basin modelling that was described earlier.

At the fine scale, self-organization models have been applied to bedforms and other geomorphic patterns (Werner & Fink, 1993; Werner & Hallet, 1993; Werner, 1995; Werner & Kocurek, 1997). All of these are highly organized patterns that clearly do develop spontaneously without external forcing. Likewise, in the model of stream braiding proposed by Murray & Paola (1994, 1997), the braid pattern emerges from a highly simplified model of the dynamics. In this realm there is still vigorous debate about whether the pattern formation requires an approach explicitly based on emergence, or can be better modelled with conventional 'reductionist' approaches.

There are aspects of emergence and self-organization in sedimentary systems that have not been identified as such. Autocyclicity (Beerbower, 1964) can be seen as an example of mesoscale self-organization in channellized systems. 'Autocyclic' has come to be used for internally generated stratigraphic cycles on length scales large enough that one might plausibly think they were externally forced. (Although it was used for good reason, the 'cycle' in 'autocycle' should not be taken too literally. Self-generated fluctuations need not be periodic, and may have very broad spectra. They are, however, a persistent, intrinsic feature of the system.) Typical autocyclic mechanisms include meander migration and cut-off, and channel-belt or lobe switching. In these cases, as in bedforms and bars, the transport system organizes itself into a highly structured spatial pattern that generates strong variations in sediment flux even with steady forcing. The spatial pattern leaves a corresponding signature in the stratigraphic record. If the essence of autocyclicity is internally generated variability in sedimentation owing to spatial pattern formation, then it encompasses a wide range of scales and forms, from bedforms through bars and small-scale channel switching, and up to large-scale lobe switching and avulsion. These can all be seen as aspects of self-organization in sedimentary systems.

The traditional question of how to tell autocyclic from allocyclic responses could be rephrased as that of distinguishing self-organized (internally generated) sedimentary patterns from externally forced ones. The better we understand the internally generated signals, the more accurately we can constrain the external forcing. Traditional wisdom says that autocycles are restricted in both spatial and temporal scale, and this is the main basis for identifying them. So at this point, it seems that self-organization, in channellized systems at least, is restricted to scales ranging from those of bedforms to the mesoscale of channel belts, lobes and avulsion. (The only hint so far of self-organization beyond the level of bedforms in the shallow-submarine or aeolian realms is the autocyclic drowning mechanism of Ginsburg (1971).) There is an element of circularity in our reasoning. What kinds of large-scale self-organized patterns might there be in sedimentary systems that we have not recognized because we have always assigned them external causes?

Another interesting aspect of self-organization that has received little attention so far is the interplay between self-organized and externally forced behaviours. One can see, however, the possibility that external forcing within the range of autocyclic timescales might either be completely masked in the autocyclic 'noise' or produce a very unintuitive response. Indeed, some aspects of what Schumm (1977) calls 'complex response' could be viewed as the superposition of self-generated and externally forced dynamics. Though 'complex response' is a bit vague, the basic concept of exploring system response to forcing on timescales comparable with those of self-generated variability is potentially a very fruitful area of research. It would bear revisiting with modern dynamical-systems concepts in mind.

Where are we now?

Is sedimentary geology a quantitative, analytical science, or a descriptive one? Is it model-driven or observation-driven? Inductive or deductive? Should we look at rocks or study mathematics? Of course, the answer is 'all of the above'. It is not clear whether basins are chaotic, but the state of modelling them certainly is. Within just the last 10 years we have seen publication of everything from the simplest of one-dimensional geometric models to three-dimensional, computationally intensive dynamical models, all aimed at explaining stratigraphy. This chaos is a good thing, and we should not try to escape from it too soon. Overall, we have learned two things:

1 It is not hard to invent plausible stratigraphic models.

2 It is very hard to test them.

This is not meant to be dismissive. Given how much we still do not understand about morphodynamics, the fact that we have been able to produce useful models at all is a sign that at least some of what we are interested in stratigraphically is not all that sensitive to the details of basin-filling processes. One of the common threads that has emerged from basin modelling using a variety of approaches is that even relatively simple forcing can produce wonderfully complex effects within basin systems. This should not be discouraging; rather, it is a motivation to redouble our efforts to be economical in invoking causes for observed stratigraphic patterns. We do not need a separate explanation for every observed stratigraphic feature. Rather what we need is a general framework for defining the space of possible forcings and parameters that could produce a given stratigraphic pattern, along the lines suggested by Heller *et al.* (1993) and Lessenger & Cross (1996). The model results reproduced in this review form an abbreviated catalogue of some of the stratigraphic patterns that modellers have seen. They can be used as a starting point in the search for the right combination of parameters to explain a given real sequence.

Another general result is that basin systems have intrinsic timescales, and that the nature of stratigraphic response to external forcing can depend strongly on the timescale of the forcing relative to these timescales. Some of the timescales that have been identified for depositional processes are given in Table 1. We have now formulated, and can estimate magnitudes for, timescales associated with erosion, deposition and isostatic relaxation. More generally, we also now have a number of theoretical cases in which systems with different timescales respond very differently to the same forcing (Angevine, 1989; Reynolds *et al.*, 1991; Paola *et al.*, 1992; Steckler *et al.*, 1993; Rivenæs, 1997). This is a possibility that must be borne in mind when trying to separate global from local effects by looking for similarity of response over a wide area. It would be useful to unify all these results in a common framework, based on fundamental dimensionless parameters (e.g. Swenson *et al.*, in press). The dependence of basin response on local characteristics should also be testable, for instance by well-constrained comparison of the response of basins with different timescales (e.g. different lengths) to the same forcing.

Overall, however, the main point that emerges from the modelling done to date is that even simple models readily surpass our ability to test them. As discussed earlier, the strategy will have to be to constrain the problem from as many directions as possible, using a combination of tactics: incorporation of the best fundamental understanding of processes; continuing the search for simple, testable behaviours within complex models; playing off different forms of stratigraphic information against one another; emphasizing testability as we develop models; critically comparing results from models using different approaches; pushing experimental tests to their limits; using our models to analyse both the uniqueness of model results and the limits placed on us by the stratigraphic record; and above all gathering comprehensive descriptive and quantitative data sets from the best field sites in the world.

Educating analytical sedimentologists

So far we have discussed what has been done in quantitative modelling, and what might be done in the future. In my view, however, the single most important problem facing us now is human, rather than scientific. We have to change the culture of the discipline. Geology has often served as a haven for students seeking to do science without mathematics, and within geology, sedimentology and stratigraphy have formed one of the more sheltered parts of the harbour. It will do little good to develop better models if most of our students see them as black boxes, or worse, are either terrified or awed by them. It is not that everyone in the field must be able to develop models themselves. On the contrary, it seems clear that inasmuch as the weak link in the chain is already not developing models but rather using and testing them, there will continue to be a much greater need for field researchers than for modellers in sedimentary geology. Rather, what is needed is researchers who are skilled in the field but at the same time understand what quantitative modelling is about: why and how people make approximations, why approaches to modelling can and must differ, and, above all, what the mathematics in the models means physically. Just as there is no substitute for experience in learning to work in the field, there is no substitute for experience in developing physical insight. And there is no shortcut: we need researchers who are good at both traditional, descriptive geology and quantitative, analytical geology. For the 'modal' sedimentary-geology student, it is not sophisticated computational skills or training in advanced calculus that is lacking, but rather the routine application of basic quantitative reasoning. This means things like estimating scales and rates for key processes, knowing the magnitudes of basic physical properties, and being able to estimate the relative importance of various processes in a particular setting. Understanding scales, rates and relative magnitudes is to quantitative science what recognizing quartz and feldspar is to field geology. Neither requires years of sophisticated training, but both require repetition until they become habitual.

In the 1960s, research on the origin, mechanics and sedimentary signature of bedforms led to a new appreciation of the value of fluid and sediment mechanics in sedimentary geology. Landmark publications in this regard were the works of Allen (1970) and Middleton & Southard (1984). The modelling of sedimentary basins and strata has provided another avenue of application of physical principles to sedimentary geology. To what extent has all of this worked its way into the basic fabric of our discipline?

Some 30 years after the initial 'physics scare' associated with bedforms and sedimentary structures, a set of basic principles from fluid and sediment mechanics now appears routinely in introductory sedimentology textbooks. Popular items include settling velocity and Stokes's Law, the Reynolds and Froude numbers, and the basic force balance for steady, uniform channel flow. This material is typically presented somewhere near the beginning of the book and then is largely ignored (though progress has been made in the recent texts of Allen (1997) and Leeder (1999)). There remains a striking contrast between the role of fluid and sediment physics in sedimentary geology and that of thermodynamics in igneous and metamorphic geology. In 'ig-met' texts the underlying thermodynamic principles are introduced and then applied repeatedly. Whereas in hard-rock petrology, thermodynamics permeates the discipline, in sedimentary geology, sediment mechanics still seems a little like taking vitamins: it is surely good for you, but most people cannot say exactly why. There are several reasons for this. In current practice, process-based interpretation is often applied in a piecemeal, descriptive way, to no apparent end beyond providing the interpreter with one more adjective. In addition, the quantitative material that is traditionally taught is often not the most important. For instance, a real appreciation of the implications of the sediment-continuity equation as the governing relation for physical sedimentation is far more useful than the details of sediment-transport formulae or even the definition of the Reynolds number. Finally, and most important, those of us in the 'quantitative school' simply have not provided a comprehensive interpretive framework that is sufficiently compelling to convince the larger community to learn a new way of thinking. At the moment it appears that the crossover of quantitative thinking into the mainstream may happen first in the deep-marine environment, where the fluid-mechanical approach to turbidites and debris flows is in a period of rapid acceleration (Kneller & Buckee, 2000). The physics of sediment-gravity flows is one level more subtle than that of rivers – perhaps just enough so that we cannot even understand deep-marine deposits qualitatively without the help of quantitative analysis. Wherever the practical utility of the quantitative approach is first demonstrated conclusively, it will spread from there to the rest of sedimentary geology. At the basin scale, the models discussed here have opened a door. Following through represents one of the great challenges for quantitative sedimentary geology in the next century.

Deciphering the language of sediments is one of the great enterprises of the Earth sciences. We try to base our interpretations as closely as possible on basic principles because only these do not change with time. We try to apply these principles quantitatively because quantitative models yield sharper predictions and can be more readily tested than qualitative ones. It is still sometimes said that the complexity of sedimentary systems places them beyond the reach of analysis, in a domain where the only recourse is endless description and cataloguing. Whether these objections arise from legitimate scepticism or from a simple distaste for mathematics, they will ultimately be overcome not by philosophical arguments but by the development of quantitative models that are simply too useful to resist. Nonetheless, the beginning of a new millennium is a time for reflection, as the papers in this special issue attest. Here in closing are some general thoughts:

1 At this point it is not known whether the inverse problems we are trying to solve in sedimentary geology have unique solutions. Even if they are nonunique, this does not mean they are incomprehensible. For problems without a unique solution, the role of analysis is to find the space of possible solutions. It may well be qualitative constraints that determine which theoretical possibility is correct.

2 The multivariate character of the stratigraphic record means that there are many sources of information to play off against one another.

3 There is no law of nature that says that 'complexity + complexity = greater complexity'. The sum of complicated processes may be simple, especially if the processes are uncorrelated. The possibility of simple behaviour in complex systems should never be dismissed until it has been shown to be absent by observation.

4 Quantitative modelling is a complement to, and an extension of, traditional conceptual modelling in sedimentary geology. A quantitative model can refine and add predictive power to a conceptual model, or show that it is physically impossible. But a correct quantitative model cannot be developed from an incorrect conceptual model.

5 Physical sedimentary geology deals with processes that obey known laws. There is a great deal of work to be done to find out how to apply these laws over the immense span of geological time. But there is no sign anywhere of processes, events or conditions that are beyond the reach of human physical insight.

ACKNOWLEDGEMENTS

It is a pleasure to thank Phil Allen, Phil Ashworth, Jim Best and Dan Bosence for thoughtful, comprehensive reviews that considerably improved this paper. I also thank Tom Hickson, Lincoln Pratson, John Swenson, Paul Heller, Gary Parker and Rudy Slingerland for helpful informal reviews, and Marcia McNutt, Gary Kocurek, Neil Lundberg, Shun Karato, Brad Murray and Chris Kendall for useful suggestions. James Syvitski and Eric Hutton produced Fig. 16 and Lincoln Pratson helped with Fig. 21. Finally, I am deeply grateful to Paul Heller, Gary Parker, David Mohrig and Giovanni Seminara for innumerable discussions that have clarified my thinking on many of the topics discussed in this paper. This work has been supported by US National Science Foundation Grants EAR 95-12472 and EAR 97-09217, and by the members of the St. Anthony Falls Experimental Stratigraphy consortium: ARCO Exploration and Production Technology Company, Conoco Inc., Exxon Production Research Company, Japan National Oil Corporation, Mobil Technology Company and Texaco Corporation.

REFERENCES

Adachi, S. and **Nakatoh, T.** (1969) Changes of top-set-bed in a silted reservoir. In: *13th Congress, International Association for Hydraulic Research*, **5**, 269–271.

Aigner, T., Brandenburg, A., van Vliet, A., Doyle, M., Lawrence, D. and **Westrich, J.** (1990) Stratigraphic modelling of epicontinental basins: two applications. *Sedim. Geol.*, **69**, 167–190.

Aigner, T., Doyle, M., Lawrence, D., Epting, M. and **van Vliet, A.** (1989) Quantitative modeling of carbonate platforms: some examples. In: *Controls on Carbonate Platform and Basin Development* (Ed. by P. D.Crevello, *et al.*), *Spec. Publ. Soc. Econ. Paleont. Miner.*, **44**, 323–338.

Alexander, J. and **Leeder, M.R.** (1987) Active tectonic control on alluvial architecture. In: *Recent Developments in Fluvial Sedimentology* (Ed. by F. G. Ethridge, R. M. Flores and M. D. Harvey), *Spec. Publ. Soc. Econ. Paleont. Miner.*, **39**, 243–252.

Allen, J.R.L. (1965) A review of the origin and characteristics of recent alluvial sediments. *Sedimentology*, **5**, 89–191.

Allen, J.R.L. (1970). *Physical Processes of Sedimentation.* George Allen and Unwin Ltd, London.

Allen, J.R.L. (1974) Studies in fluviatile sedimentation: implications of pedogenic carbonate units, Lower Old Red Sandstone, Anglo-Welsh outcrop. *Geol. J.*, **9**, 181–208.

Allen, J.R.L. (1978) Studies in fluviatile sedimentation: an exploratory quantitative model for the architecture of avulsion-controlled alluvial suites. *Sedim. Geol.*, **21**, 129–147.

Allen, J.R.L. (1979) Studies in fluviatile sedimentation: an elementary geometrical model for the connectedness of avulsion-related channel sand bodies. *Sedim. Geol.*, **24**, 253–267.

Allen, P.A. (1997) *Earth Surface Processes.* Blackwell Science, Oxford.

Allen, P.A. and **Allen, J.R.** (1990) *Basin Analysis: Principles and Applications.* Blackwell Scientific Publications, Oxford.

Angevine, C.L. (1989) Relationship of eustatic oscillations to regressions and transgressions on passive continental margins. In: *Origin and Evolution of Sedimentary Basins and Their Energy and Mineral Resources* (Ed. by R. A. Price), pp. 29–35. American Geophysical. Union, Washington, DC.

Angevine, C.L., Heller, P.L. and **Paola, C.** (Eds) (1990) *Quantitative Sedimentary Basin Modeling. American Association of Petroleum Geologists, Continuing Education Course Note Series* **32**.

Ashworth, P.J., Best, J.L., Peakall, J. and Lorsong, J.A. (1999) The influence of aggradation rate on braided alluvial architecture: field study and physical scale-modelling of the Ashburton River gravels, Canterbury Plains, New Zealand. In: *Current Research in Fluvial Sedimentology: Proceedings of the 6th International Conference on Fluvial Sedimentology* (Ed. by N. D. Smith and J. J. Rogers), *Spec. Publ. Int. Assoc. Sedimentol.*, **28**, 333–346.

Aurell, M., Bádenas, B., Bosence, D.W.J. and **Waltham, D.A.** (1998) Carbonate production and offshore transport on a Late Jurassic carbonate ramp (Kimmeridgian, Iberian Basin, NE Spain): evidence from outcrops and computer modelling. In: *Carbonate Ramps* (Ed. by V. P. Wright and T. P. Burchette), *Spec. Publ. Geol. Soc. London*, **149**, 137–161.

Aurell, M., Bosence, D. and **Waltham, D.** (1995) Carbonate ramp depositional systems from a late Jurassic epeiric platform (Iberian Basin, Spain): a combined computer modelling and outcrop analysis. *Sedimentology*, **42**, 75–94.

Baker, V.R. (1974) Paleohydraulic interpretation of Quaternary alluvium near Golden, Colorado. *Quaternary Res.*, **4**, 94–112.

Barros, A.P. and **Lettenmaier, D.P.** (1993) Multiscale aggregation and disaggregation of precipitation for regional hydroclimatological studies. In: *Macroscale Modelling of the Hydrosphere* (Ed. by W. B. Wilkinson), *IAHS Publ.*, **214**, 183–193.

Beaumont, C. (1978) The evolution of sedimentary basins on a viscoelastic lithosphere: theory and examples. *Geophys J. R. Astronom. Soc.*, **55**, 471–497.

Beaumont, C. (1981) Foreland basins. *Geophys J. R. Astronom. Soc.*, **65**, 291–329.

Beaumont, C., **Fullsack, P.** and **Hamilton, J.** (1992) Erosional control of active compressional orogens. In: *Thrust Tectonics* (Ed. by K. R. McClay), pp. 1–18. Chapman & Hall, London.

Beerbower, J.R. (1964) Cyclothem and cyclic depositional mechanisms in alluvial plain sedimentation. *Kansas Geol. Surv. Bull.*, **169**, 31–42.

Begin, Z.B., **Meyer, D.F.** and **Schumm, S.S.** (1981) Development of longitudinal profiles of alluvial channels in response to base-level lowering. *Earth Surf. Process. Landforms*, **6**, 49–68.

Bice, D. (1988) Synthetic stratigraphy of carbonate platform and basin systems. *Geology*, **16**, 703–706.

Bice, D.M. (1991) Computer simulation of carbonate platform and basin systems. In: *Sedimentary Modeling: Computer Simulations and Methods for Improved Parameter Definition* (Ed. by E. K. Franseen, W. L. Watney, C. G. S. C. Kendall and W. Ross), *Kansas Geol. Surv. Bull.*, **233**, 431–447.

Blum, M.D. and **Törnqvist, T.** (2000) Fluvial response to climate and sea-level change: a review and look forward. *Sedimentology*, **47** (Suppl. 1) 2–48.

Bosence, D. and **Waltham, D.** (1990) Computer modeling of the internal architecture of carbonate platforms. *Geology*, **18**, 26–30.

Bosence, D.W.J., **Pomar, L.**, **Waltham, D.A. and Lankester, T.H.G.** (1994) Computer modelling a Miocene carbonate platform, Mallorca, Spain. *AAPG Bull.*, **78**, 247–266.

Bosscher, H. and **Schlager, W.** (1992) Computer simulation of reef growth. *Sedimentology*, **39**, 503–512.

Bosscher, H. and **Schlager, W.** (1993) Accumulation rates of carbonate platforms. *J. Geol.*, **101**, 345–355.

Bosscher, H. and **Southam, J.** (1992) CARBPLAT – A computer model to simulate the development of carbonate platforms. *Geology*, **20**, 235–238.

Bridge, J.S. and **Bennett, S.J.** (1992) A model for the entrainment and transport of sediment grains of mixed sizes, shapes, and densities. *Water Resour. Res.*, **28**, 337–363.

Bridge, J.S. and **Leeder, M.R.** (1979) A simulation model of alluvial stratigraphy. *Sedimentology*, **26**, 617–644.

Bridge, J.S. and **Mackey, S.D.** (1993a) A revised alluvial stratigraphy model. In: *Alluvial Sedimentation* (Ed. by M. Marzo and C. Puigdefabregas), *Int. Assoc. Sedimentol. Spec. Publ.*, **17**, 319–336.

Bridge, J.S. and **Mackey, S.D.** (1993b) A theoretical study of fluvial sandstone body dimensions. In: *The Geological Modelling of Hydrocarbon Reservoirs and Outcrop Analogues* (Ed. by S. S. Flint and I. D.Bryant), *Int. Assoc. Sedimentol. Spec. Publ.*, **15**, 213–236.

Briggs, L.I. and **Pollack, H.N.** (1967) Digital model of evaporite sedimentation. *Science*, **155**, 453–456.

Bruun, P. (1962) Sea-level rise as a cause of shore erosion. *J. Waterways Harbors Division, ASCE*, **88**, 117–130.

Bryant, M., **Falk, P.** and **Paola, C.** (1995) Experimental study of avulsion frequency and rate of deposition. *Geology*, **23**, 365–368.

Burgess, P.M. and **Allen, P.A.** (1996) A forward-modelling analysis of the controls on sequence stratigraphic geometries. In: *Sequence Stratigraphy in British Geology* (Ed. by S. P. Hesselbo and D. N. Parkinson), *Spec. Publ. Geol. Soc. London*, **103**, 9–24.

Burton, R., **Kendall, C.G.S.C.** and **Lerche, I.** (1987) Out of our depth: On the impossibility of fathoming eustasy from the stratigraphic record. *Earth-Sci. Rev.*, **24**, 237–277.

Cacchione, D.A. and **Drake, D.E.** (1990) Shelf sediment transport: An overview with applications to the Northern California continental shelf. In: *The Sea*, Vol. 9: *Ocean Engineering Science* (Ed. by B. LeMehaute and D. M. Hanes), pp. 729–773. John Wiley and Sons, Inc., New York.

Cant, D.J. (1989) Simple equations of sedimentation: applications to sequence stratigraphy. *Basin Res.*, **2**, 73–81.

Cant, D.J. (1991) Geometric modelling of facies migration: theoretical development of facies successions and local unconformities. *Basin Res.*, **3**, 51–62.

Carey, J.S., **Swift, D.J.P.**, **Steckler, M.**, **Reed, C.W.** and **Niedoroda, A.** (1999) High resolution sequence stratigraphic modelling: effects of sedimentation processes. In: *Numerical Experiments in Stratigraphy* (Ed. by J. Harbaugh, L. Watney, G. Rankey, R. Slingerland, R. Goldstein and E. Franseen), *Spec. Publ. SEPM*, **62**, 151–164.

Chase, C.G. (1992) Fluvial landsculpting and the fractal dimension of topography. *Geomorphology*, **5**, 39–57.

Church, M. and **Rood, K.** (1983) *Catalogue of Alluvial River Channel Regime Data.* University of British Columbia, Department of Geography, Vancouver, BC.

Cisne, J.L., **Gildner, R.F.** and **Rabe, B.D.** (1984) Epeiric sedimentation and sea-level: synthetic ecostratigraphy. *Lethaia*, **17**, 267–288.

Cloetingh, S. (1986) Intraplate stresses: a new tectonic mechanism for fluctuations of relative sea-level. *Geology*, **14**, 617–620.

Cloetingh, S., **McQueen, H.** and **Lambeck, K.** (1985) On a tectonic mechanism for regional sealevel variations. *Earth Planet. Sci. Lett*, **75**, 157–166.

Cochran, J.R. (1983) Effects of finite rifting times on the development of sedimentary basins. *Earth Planet. Sci. Lett*, **66**, 289–302.

Collier, R.E.L., **Leeder, M.R.** and **Maynard, J.R.** (1990) Transgressions and regressions: a model for the influence of tectonic subsidence, deposition and eustasy, with application to Quaternary and Carboniferous examples. *Geol. Mag.*, **127**, 117–128.

Cotter, E. (1971) Paleoflow characteristics of a late Cretaceous river in Utah from analysis of sedimentary structures in the Ferron sandstone. *J. Sedim. Petrol.*, **41**, 129–138.

Cross, T.A. (Ed.) (1990) *Quantitative Dynamic Stratigraphy.* Prentice Hall, Englewood Cliffs, New Jersey.

Cui, Y., **Parker, G.** and **Paola, C.** (1996) Numerical simuation of aggradation and downstream fining. *J. Hydraulic Res.*, **34**, 185–204.

Dade, W.B. and **Friend, P.F.** (1998) Grain size, sediment-transport regime and channel slope in alluvial rivers. *J. Geol.*, **106**, 661–675.

Demicco, R.V. (1998) CYCOPATH 2D – a two-dimensional, forward model of cyclic sedimentation on carbonate platforms. *Computers & Geosciences*, **24**, 405–423.

Demicco, R.V. and **Spencer, R.J.** (1989) MAPS – a BASIC program to model accumulation of platform sediments. *Computers & Geosciences*, **15**, 95–105.

Demicco, R.V., **Spencer, R.J.**, **Waters, B.B.** and **Cloyd, K.C.** (1991) Two dimensional computer models of a Cambrian carbonate shelf deposit. In: *Sedimentary Modeling: Computer Simulations and Methods for Improved Parameter Definition* (Ed. by E. K. Franseen, W. L. Watney, C. G. S. C. Kendall and W. Ross), *Kansas Geol. Surv. Bull.*, **233**, 463–472.

Dietz, R.S. (1963) Collapsing continental rises: an actualistic concept of geosynclines and mountain building. *J. Geol.*, **71**, 314–333.

Driscoll, N.W. and **Karner, G.D.** (1999) Three-dimensional quantitative modeling of clinoform development. *Mar Geol*, **154**, 383–398.

Eberli, G.P., Kendall, C.G.S.C., Moore, P., Whittle, G.L. and **Cannon, R.** (1994) Testing a seismic interpretation of Great Bahama Bank with a computer simulation. *AAPG Bull.*, **78**, 981–1004.

Emery, D. and **Myers, K.** (Eds) (1996) *Sequence Stratigraphy*. Blackwell Science, Oxford.

Engelund, F. and **Hansen, E.** (1967) *A Monograph on Sediment Transport in Alluvial Streams*. Teknisk Forlag, Copenhagen.

Ericksen, M.C. and **Slingerland, R.** (1990) Numerical simulations of tidal and wind-driven circulation in the Cretaceous Interior Seaway of North America. *Geol. Soc. Am. Bull.*, **102**, 1499–1516.

Ethridge, F.G. and **Schumm, S.A.** (1978) Reconstructing paleochannel morphologic and flow characteristics: methodology, limitations, and assessment. In: *Fluvial Sedimentology* (Ed. by A. D. Miall), *Can. Soc. Petrol. Geol. Mem.*, **5**, 703–721.

Flemings, P.B. and **Grotzinger, J.P.** (1996) STRATA: Freeware for Solving Classic Stratigraphic Problems. *GSA Today*, **6**, 1–7.

Flemings, P.B. and **Jordan, T.E.** (1989) A synthetic stratigraphic model of foreland basin development. *J. Geophys. Res.*, **94**, 3851–3866.

Flemings, P.B. and **Jordan, T.E.** (1990) Stratigraphic modeling of foreland basins: Interpreting thrust deformation and lithosphere rheology. *Geology*, **18**, 430–434.

Franseen, E.K., Watney, W.L., Kendall, C.G.S.C. and **Ross, W.** (Eds) (1991) *Sedimentary Modeling: Computer Simulations and Methods for Improved Parameter Definition. Kansas Geol. Surv. Bull.*, **233**.

Gaffin, S.R. and **Maasch, K.A.** (1991) Anomalous cyclicity in climate and stratigraphy and modeling nonlinear oscillations. *J. Geophys. Res.*, **96**, 6701–6711.

Galewsky, J. (1998) The dynamics of foreland basin carbonate platforms: tectonic and eustatic controls. *Basin Res.*, **10**, 409–416.

Garde, R.J., Ranga Raju, K.G. and **Mehta, P.J.** (1981) Bed level variation in aggrading alluvial streams. *International Association for Hydraulic Research Congress*, **19**, 247–253.

Gardner, T.W. (1983) Paleohydrology and paleomorphology of a Carboniferous, meandering, fluvial sandstone. *J. Sediment. Petrol.*, **53**, 991–1005.

Gildner, R.F. (1990) STRIPPER: An interactive backstripping, decompaction and geohistory program. In: *Quantitative Dynamic Stratigraphy* (Ed. by T. A. Cross), pp. 165–180. Prentice Hall, Englewood Cliffs, NJ.

Gildner, R.F. and **Cisne, J.L.** (1990) Quantitative modeling of carbonate stratigraphy and water-depth history using depth-dependent sediment accumulation function. In: *Quantitative Dynamic Stratigraphy* (Ed. by T. A. Cross), pp. 417–432. Prentice Hall, Englewood Cliffs, New Jersey.

Gill, M.A. (1983) Diffusion model for degrading channels. *J. Hydraulic Res.*, **21**, 369–378.

Ginsburg, R.N. (1971) Landward movement of carbonate mud: new model for regressive cycles in carbonates. *Am. Assoc. Petrol. Geol. Bull.*, **55**, 340.

Giorgi, F. and **Mearns, L.O.** (1991) Approaches to the simulation of regional climate change: a review. *Rev. Geophysics*, **29**, 191–216.

Goldenfeld, N. and **Kadanoff, L.P.** (1999) Simple lessons from complexity. *Science*, **284**, 87–89.

Goldhammer, R.K., Dunn, P.A. and **Hardie, L.A.** (1987) High-frequency glacio-eustatic sealevel oscillations with Milankovitch characteristics recorded in Middle Triassic platform carbonates in northern Italy. *Am. J. Sci.*, **287**, 853–892.

Goldhammer, R.K. and **Harris, M.T.** (1989) Eustatic controls on the stratigraphy and geometry of the Latemar buildup (Middle Triassic), the Dolomites of northern Italy. In: *Controls on Carbonate Platform and Basin Development, Spec. Publ. Soc. Econ. Paleont. Miner.*, **44**, 323–338.

Goldhammer, R.K., Lehmann, P.J. and **Dunn, P.A.** (1993) The origin of high-frequency platform carbonate cycles and third-order sequences (Lower Ordovician El Paso Group, West Texas): constraints from outcrop data and stratigraphic modeling. *J. Sedim. Petrol.*, **63**, 318–359.

Haq, B.U., Hardenbol, J. and **Vail, P.R.** (1987) Chronology of fluctuating sea-levels since the Triassic (250 million years ago to present). *Science*, **235**, 1156–1167.

Harbaugh, J., Watney, L., Rankey, G., Slingerland, R., Goldstein, R. and **Franseen, E.** (Eds) (1999) *Numerical Experiments in Stratigraphy. Spec. Publ. SEPM*, **62**.

Harbaugh, J.W. (1993) Simulating geologic processes that create sedimentary basins: incorporation of a carbonate-ecology model. In: *Computerized Basin Analysis: the Prognosis of Energy and Mineral Resources* (Ed. by J. Harff and D. F. Merriam), pp. 11–23. Plenum Press, New York.

Harbaugh, J.W. and **Bonham-Carter, G.** (1970) *Computer Simulation in Geology*. John Wiley & Sons, New York.

Harbaugh, J. W. (1966) Mathematical simulation of marine sedimentation. *Kansas Geol. Surv. Computer Contrib.*, **1**.

Hardy, S. and **Poblet, J.** (1994) Geometric and numerical model of progressive limb rotation in detachment folds. *Geology*, **22**, 371–374.

Harris, M.T. (1991) Carbonate progradation geometry and significance of sediment accumulation patterns and syndepositional relief: Comparison of modeling results with outcrop data. In: *Sedimentary Modeling: Computer Simulations and Methods for Improved Parameter Definition* (Ed. by E. K. Franseen, W. L. Watney, C. G. S. C. Kendall and W. Ross), *Kansas Geol. Surv. Bull.*, **233**, 509–517.

Harrison, W.J. (1990) Modeling fluid/rock interactions in sedimentary basins. In: *Quantitative Dynamic Stratigraphy* (Ed. by T. A. Cross), pp. 195–231. Prentice Hall, Englewood Cliffs, New Jersey.

Haxby, W.F., Turcotte, D.L. and **Bird, J.M.** (1976) Thermal and mechanical evolution of the Michigan Basin. *Tectonophysics*, **36**, 57–75.

Helland-Hansen, W., Kendall, C.G.S.C., Lerche, I. and **Nakayama, K.** (1988) A simulation of continental basin margin sedimentation in response to crustal movements, eustatic sea-level change, and sediment accumulation rates. *Mathemat. Geol*, **20**, 777–802.

Heller, P.L., Burns, B. and **Marzo, M.** (1993) Stratigraphic solution sets for determining the roles of sediment supply, subsidence, and sea-level on transgressions and regressions. *Geology*, **21**, 747–750.

Heller, P.L. and **Paola, C.** (1992) The large-scale dynamics of grain-size variation in alluvial basins, 2: application to syntectonic conglomerate. *Basin Res.*, **4**, 91–102.

Heller, P.L. and **Paola, C.** (1996) Downstream changes in alluvial architecture: an exploration of controls on channel-stacking patterns. *J. Sedim. Res.*, **66**, 297–306.

Hoey, T.B. and **Ferguson, R.** (1994) Numerical simulation of downstream fining by selective transport in gravel bed rivers: Model development and illustration. *Water Resour. Res.*, **30,** 2251–2260.

Hooke, R.L. (1968) Model geology: prototype and laboratory streams: discussion. *Geol. Soc. Am. Bull.*, **79,** 391–394.

Howard, A.D. (1994) A detachment-limited model of drainage basin evolution. *Water Resour. Res.*, **39,** 2261–2285.

Humphrey, N.F. and **Heller, P.L.** (1995) Natural oscillations in coupled geomorphic systems: an alternative origin for cyclic sedimentation. *Geology*, **23,** 499–502.

Jain, S.C. (1981) River bed aggradation due to overloading. *J. Hydraulics Division ASCE*, **107,** 120–124.

Jaramillo, W.F. and **Jain, S.C.** (1984) Aggradation and degradation of alluvial-channel beds. *J. Hydraulic Engineering*, **110,** 1072–1085.

Jarvis, G.T. and **McKenzie, D.P.** (1980) Sedimentary basin formation with finite extension rates. *Earth Planet. Sci. Lett*, **48,** 42–52.

Jervey, M.T. (1988) Quantitative geological modeling of siliciclastic rock sequences and their seismic expression. In: *Sea-Level Changes—an Integrated Approach* (Ed. by C. K. Wilgus, B. S. Hastings, H. W. Posamentier, J. Van Wagoner, C. A. Ross and C. G. S. C. Kendall), *Spec. Publ. Soc. Econ. Paleont. Miner.*, **42,** 47–69.

Jin, Y., McNutt, M.K. and **Zhu, Y.** (1994) Evidence from gravity and topography data for folding of Tibet. *Nature*, **371,** 669–674.

Johnson, D.D. and **Beaumont, C.** (1995) Preliminary results from a planform kinematic model of orogen evolution, surface processes and the development of clastic foreland basin stratigraphy. In: *Stratigraphic Evolution of Foreland Basins* (Ed. by S. L. Dorobek and G. M. Ross), *Spec. Publ. Soc. Econ. Paleont. Miner.*, **52,** 3–24.

Jordan, T.E. (1981) Thrust loads and foreland basin evolution. *Am. Assoc. Petrol. Geol. Bull.*, **65,** 2506–2520.

Jordan, T.E. and **Flemings, P.B.** (1990) From geodynamic models to basin fill – a stratigraphic perspective. In: *Quantitative Dynamic Stratigraphy* (Ed. by T. A. Cross), pp. 149–163. Prentice Hall, Englewood Cliffs, New Jersey.

Jordan, T.E. and **Flemings, P.B.** (1991) Large-scale stratigraphic architecture, eustatic variation, and unsteady tectonism: a theoretical evaluation. *J. Geophys. Res.*, **96,** 6681–6699.

Karner, G.D., Steckler, M.S. and **Thorne, J.A.** (1983) Long-term thermo-mechanical properties of the lithosphere. *Nature*, **304,** 250–252.

Kaufman, P., Grotzinger, J.P. and **McCormick, D.S.** (1991) Depth-dependent diffusion algorithm for simulation of sedimentation in shallow marine depositional systems. In: *Sedimentary Modeling: Computer Simulations and Methods for Improved Parameter Definition* (Ed. by E. K. Franseen, W. L. Watney, C. G. S. C. Kendall and W. Ross), *Kansas Geol. Surv. Bull.*, **233,** 489–508.

Kendall, C.G.S.C., Moore, P., Strobel, J., Cannon, R., Perlmutter, M., Bezdek, J. and **Biswas, G.** (1991a) Simulation of the sedimentary fill of basins. In: *Sedimentary Modeling: Computer Simulations and Methods for Improved Parameter Definition* (Ed. by E. K. Franseen, W. L. Watney, C. G. S. C. Kendall and W. Ross), *Kansas Geol. Surv. Bull.*, **233,** 9–30.

Kendall, C.G.S.C., Strobel, J., Cannon, R., Bezdek, J. and **Biswas, G.** (1991b) The simulation of the sedimentary fill of basins. *J. Geophys. Res.*, **96,** 6911–6929.

Kenyon, P.M. and **Turcotte, D.L.** (1985) Morphology of a delta prograding by bulk sediment transport. *Geol Soc. Am. Bull.*, **96,** 1457–1465.

Kneller, B. and **Buckee, C.** (2000) The structure and fluid mechanics of turbidity currents: a review of some recent studies and their geological implications. *Sedimentology*, **47** (Suppl. 1), 63–94.

Kochel, R.C. and **Baker, V.R.** (1982) Paleoflood hydrology. *Science*, **215,** 353–361.

Kocurek, G. and **Havholm, K.G.** (1994) Eolian sequence stratigraphy—a conceptual framework. In: *Siliciclastic Sequence Stratigraphy: Recent Developments and Applications* (Ed. by P. Weimer and H. Posamentier), *AAPG Mem.*, **58,** 393–409.

Kooi, H. and **Beaumont, C.** (1994) Escarpment evolution on high-elevation rifted margins: Insights derived from a surface-processes model that combines diffusion, advection, and reaction. *J. Geophys. Res.*, **99,** 12191–12209.

Koss, J.E., Ethridge, F.G. and **Schumm, S.A.** (1994) An experimental study of the effects of base-level change on fluvial, coastal plain and shelf systems. *J. Sedim. Res.*, **B64,** 90–98.

Krumbein, W.C. and **Pettijohn, F.J.** (1938) *Manual of Sedimentary Petrography.* Appleton-Century-Crofts, Inc., New York.

Kruse, S.E. and Royden, L.H. (1994) Bending and unbending of an elastic lithosphere: the Cenozoic history of the Apennine and Dinaride foredeep basins. *Tectonics*, **13,** 278–302.

Kusznir, N. and **Karner, G.** (1985) Dependence of the flexural rigidity of the continental lithosphere on rheology and temperature. *Nature*, **316,** 138–142.

Lawrence, D.T. (1994) Evaluation of eustasy, subsidence, and sediment input as controls on depositional sequence geometries and the synchroneity of sequence boundaries. In: *Siliciclastic Sequence Stratigraphy: Recent Developments and Applications* (Ed. by P. Weimer and H. Posamentier), *AAPG Mem.*, **58,** 337–367.

Lawrence, D.T., Doyle, M. and **Aigner, T.** (1990) Stratigraphic simulation of sedimentary basins: Concepts and calibration. *AAPG Bull.*, **74,** 273–295.

Lee, M.-K. (1997) Predicting diagenetic effects of groundwater flow in sedimentary basins: a modeling approach with examples. In: *Basin-Wide Diagenetic Patterns: Integrated Petrologic, Geochemical, and Hydrologic Considerations* (Ed. by I. P. Montañez, J. M. Gregg and K. L. Shelton), *Spec. Publ. Soc. Econ. Paleont. Miner.*, **57,** 3–14.

Lee, Y.-H. and **Harbaugh, J.W.** (1992) Stanford's SEDSIM project: dynamic three-dimensional simulation of geologic processes that affect clastic sediments. In: *Computer Graphics in Geology* (Ed. by R. Pflug and J. W. Harbaugh), pp. 113–127. Springer-Verlag, Berlin.

Leeder, M.R. (1978) A quantitative stratigraphic model for alluvium, with special reference to channel deposit density and interconnectedness. In: *Fluvial Sedimentology* (Ed. by A. D. Miall), *Can. Soc. Petrol. Geol. Mem.*, **5,** 587–596.

Leeder, M. (1999) *Sedimentology and Sedimentary Basins: from Turbulence to Tectonics.* Blackwell Science, Oxford.

Leeder, M.R. and **Gawthorpe, R.L.** (1987) Sedimentary models for extensional tilt block/half-graben basins. *Spec. Publ. Geol. Soc. London*, **28,** 139–152.

Lerche, I., Dromgoole, E., Kendall, C.G.S.C., Walter, L.M. and **Scaturo, D.** (1987) Geometry of carbonate bodies: a quantitative investigation of factors influencing their evolution. *Carbonates Evaporites*, **2,** 15–42.

Lessenger, M.A. and **Cross, T.A.** (1996) An inverse stratigraphic simulation model – Is stratigraphic inversion possible? *Energy Exploration & Exploitation*, **14**, 627–637.

Lorenz, E.N. (1963) Deterministic nonperiodic flow. *J. Atmospheric Sci.*, **20**, 130–141.

Mackey, S.D. and **Bridge, J.S.** (1995) Three-dimensional model of alluvial stratigraphy: theory and application. *J. Sedim. Res.*, **B65**, 7–31.

Mahlman, J.D. (1998) Science and nonscience concerning human-caused climate warming. *Annu. Rev. Energy Environ.*, **23**, 83–106.

Martel, A.T., **Allen, P.A.** and **Slingerland, R.** (1994) Use of tidal-circulation modeling in paleogeographical studies: an example from the Tertiary of the Alpine perimeter. *Geology*, **22**, 925–928.

Martinez, P.A. (1992) Three-dimensional simulation of littoral transport. In: *Computer Graphics in Geology* (Ed. by R. Pflug and J. W. Harbaugh), pp. 129–142. Springer-Verlag, Berlin.

Martinez, P.A. and **Harbaugh, J.W.** (1993) *Simulating Nearshore Environments*. Pergamon Press, Oxford.

McKenzie, D.P. (1978) Some remarks on the development of sedimentary basins. *Earth Planet. Sci. Lett.*, **40**, 25–32.

McNutt, M.K., **Diament, M.** and **Kogan, M.G.** (1988) Variations of elastic plate thickness at continental thrust belts. *J. Geophys. Res.*, **93**, 8825–8838.

Meckel, L.D. (1967) Origin of Pottsville conglomerates (Pennsylvanian) in the central Appalachians. *Geol Soc. Am. Bull.*, **78**, 223–258.

Middleton, G.V. and **Southard, J.B.** (1984) *Mechanics of Sediment Movement. SEPM Eastern Section Short Course*, **3**.

Miller, K.G., **Thompson, P.R.** and **Kent, D.V.** (1993) Integrated Late Eocene – Oligocene stratigraphy of the Alabama coastal plain: correlation of hiatuses and stratal surfaces to glacioeustatic lowerings. *Paleoceanography*, **8**, 313–331.

Mountney, N.P. and **Westbrook, G.K.** (1997) Quantitative analysis of Miocene to Recent forearc basin evolution along the Colombian convergent margin. *Basin Res.*, **9**, 177–196.

Murray, A.B. and **Paola, C.** (1994) A cellular model of braided rivers. *Nature*, **371**, 54–57.

Murray, A.B. and **Paola, C.** (1997) Properties of a cellular braided stream model. *Earth Surf. Process. Landforms*, **22**, 1001–1025.

Nicolis, G. and **Nicolis, C.** (1991) Nonlinear dynamic systems in the geosciences. In: *Sedimentary Modeling: Computer Simulations* and Methods for Improved Parameter Definition (Ed. by E. K. Franseen, W. L. Watney, C. G. S. C. Kendall and W. Ross), *Kansas Geol. Surv. Bull.*, **233**, 33–42.

Niedoroda, A.M., **Reed, C.W.**, **Swift, D.J.P.**, **Arata, H.** and **Hoyanagi, K.** (1995) Modeling shore-normal large-scale coastal evolution. *Mar. Geol.*, **126**, 181–199.

van Niekerk, A., **Vogel, K.R.**, **Slingerland, R.L.** and **Bridge, J.S.** (1992) Routing of heterogeneous sediments over movable bed: model development. *J. Hydraulic Engineering*, **118**, 246–262.

Nittrouer, C.A. (1999) STRATAFORM: overview of its design and synthesis of its results. *Mar. Geol.*, **154**, 3–12.

Nittrouer, C.A. and **Wright, L.D.** (1994) Transport of particles across continental shelves. *Rev. Geophysics*, **32**, 85–113.

Nordlund, U. (1999) FUZZIM: forward stratigraphic modeling made simple. *Comput. Geosci.*, **25**, 449–456.

Nordlund, U. (1996) Formalizing geological knowledge—with an example of modeling stratigraphy using fuzzy logic. *J. Sedim. Res.*, **66**, 689–698.

Nummedal, D., **Riley, G.W.** and **Templet, P.L.** (1993) High-resolution sequence architecture: a chronostratigraphic model based on equilibrium profile studies. In: *Sequence Stratigraphy and Facies Associations* (Ed. by H. W. Posamentier, C. P. Summerhayes, B. U. Haq and G. P. Allen), *Spec. Publ. Int. Assoc. Sedimentol.*, **18**, 55–68.

Paola, C. (1988) Subsidence and gravel transport in alluvial basins. In: *New Perspectives in Basin Analysis* (Ed. by K. L. Kleinspehn and C. Paola), pp. 231–243. Springer-Verlag, New York.

Paola, C. (1990) A simple basin-filling model for coarse-grained alluvial systems. In: *Quantitative Dynamic Stratigraphy* (Ed. by T. A. Cross), pp. 363–374. Prentice Hall, Englewood Cliffs, New Jersey.

Paola, C. (1996) Incoherent structure: turbulence as a metaphor for stream braiding. In: *Coherent Flow Structures in Open Channels* (Ed. by P. J. Ashworth, S. J. Bennett, J. L. Best and S. J. McLelland), pp. 705–723. John Wiley & Sons, Chichester.

Paola, C., **Heller, P.L.** and **Angevine, C.L.** (1992) The large-scale dynamics of grain-size variation in alluvial basins, 1: Theory. *Basin Res.*, **4**, 73–90.

Paola, C., **Parker, G.**, **Mohrig, D.C.** and **Whipple, K.X.** (1999) The influence of transport fluctuations on spatially averaged topography on a sandy, braided fluvial fan. In: *Numerical Experiments in Stratigraphy: Recent Advances in Stratigraphic and Sedimentologic Computer Simulations* (Ed. by J. Harbaugh, L. Watney, G. Rankey, R. Slingerland, R. Goldstein and E. Franseen), *Spec. Publ. Soc. Econ. Paleont. Miner.*, **62**, 211–218.

Paola, C. and **Seal, R.** (1995) Grain-size patchiness as a cause of selective deposition and downstream fining. *Water Resour. Res.*, **31**, 1395–1407.

Parker, G. (1978a) Self-formed straight rivers with equilibrium banks and mobile bed. Part 1. The sand-silt river. *J. Fluid Mechanics*, **89**, 109–125.

Parker, G. (1978b) Self-formed straight rivers with equilibrium banks and mobile bed. Part 2. The gravel river. *J. Fluid Mechanics*, **89**, 127–146.

Parker, G. (1991) Selective sorting and abrasion of river gravel. I: theory. *J. Hydraulic Engineering*, **117**, 131–149.

Parker, G., **Paola, C.**, **Whipple, K.X.** and **Mohrig, D.C.** (1998a) Alluvial fans formed by channelized fluvial and sheet flow. I: theory. *J. Hydraulic Engineering*, **124**, 985–995.

Parker, G., **Paola, C.**, **Whipple, K.X.**, **Mohrig, D.C.**, **Toro-Escobar, C.M.**, **Halverson, M.** and **Skoglund, T.W.** (1998b) Alluvial fans formed by channelized fluvial and sheet flow. II: application. *J. Hydraulic Engineering*, **124**, 996–1004.

Peakall, J., **Ashworth, P.** and **Best, J.** (1996) Physical modelling in fluvial geomorphology: Principles, applications and unresolved issues. In: *The Scientific Nature of Geomorphology* (Ed. by B. L. Rhoads and C. E. Thorn), pp. 221–253. John Wiley & Sons, Chichester.

Pelletier, B.R. (1958) Pocono paleocurrents in Pennsylvania and Maryland. *Geol. Soc. Am. Bull.*, **69**, 1033–1064.

Pelletier, J.D. and **Turcotte, D.L.** (1997) Synthetic stratigraphy with a stochastic diffusion model of sedimentation. *J. Sedim. Res.*, **67**, 1060–1067.

Perlmutter, M.A. and **Matthews, M.D.** (1990) Global cyclostratigraphy — a model. In: *Quantitative Dynamic Stratigraphy* (Ed. by T. A. Cross), pp. 233–260. Prentice Hall, Englewood Cliffs, New Jersey.

Perlmutter, M.A., **Radovich, B.J.**, **Matthews, M.D.** and **Kendall, C.G.S.C.** (1998) The impact of high-frequency sedimentation cycles on stratigraphic interpretation. In: *Sequence Stratigraphy — Concepts and Applications* (Ed.

by F. M. Gradstein, K. O. Sandvik and N. J. Milton), *Norwegian Petrol. Soc. Spec. Publ.*, **8**, 141–170.

Person, M., **Raffensperger, J.P.**, **Ge, S.** and **Garven, G.** (1996) Basin-scale hydrogeologic modeling. *Rev. Geophysics*, **34**, 61–87.

Pettijohn, F.J. (1957) *Sedimentary Rocks.* Harper, New York.

Pettijohn, F.J. and **Potter, P.E.** (1964) *Atlas and Glossary of Primary Sedimentary Structures.* Springer-Verlag, Berlin.

Pettijohn, F.J., **Potter, P.E.** and **Siever, R.** (1973) *Sand and Sandstone.* Springer-Verlag, New York.

Pirmez, C., **Pratson, L.F.** and **Steckler, M.S.** (1998) Clinoform development by advection-diffusion of suspended sediment: modeling and comparison with natural systems. *J. Geophys. Res.*, **103**, 24141–24157.

Pitman, W.C. (1978) Relationship between eustacy and stratigraphic sequences on passive margins. *Geol. Soc. Am. Bull.*, **89**, 1389–1403.

Potter, P.E. and **Pettijohn, F.J.** (1977) *Paleocurrents and Basin Analysis.* Springer-Verlag, Berlin.

Read, J.F., **Osleger, D.** and **Elrick, M.** (1991) Two-dimensional modeling of carbonate ramp sequences and component cycles. In: *Sedimentary Modeling: Computer Simulations and Methods for Improved Parameter Definition* (Ed. by E. K. Franseen, W. L. Watney, C. G. S. C. Kendall and W. Ross), *Kansas Geol. Surv. Bull.*, **233**, 473–488.

Reynolds, D.J., **Steckler, M.S.** and **Coakley, B.J.** (1991) The role of the sediment load in sequence stratigraphy: the influence of flexural isostasy and compaction. *J. Geophys. Res.*, **96**, 6931–6949.

Ribberink, J.S. and **van der Sande, J.T.M.** (1985) Aggradation in rivers due to overloading – analytical approaches. *J. Hydraulic Res.*, **23**, 273–284.

Rivenæs, J.C. (1992) Application of a dual-lithology, depth-dependent diffusion equation in stratigraphic simulation. *Basin Res.*, **4**, 133–146.

Rivenæs, J.C. (1997) Impact of sediment transport efficiency on large-scale sequence architecture: results from stratigraphic computer simulation. *Basin Res.*, **9**, 91–105.

Robinson, R.A.J. and **Slingerland, R.L.** (1998) Origin of fluvial grain-size trends in a foreland basin: the Pocono Formation on the the central Appalachian Basin. *J. Sedim. Res.*, **68**, 473–486.

Ross, W.C. (1991) Cyclic stratigraphy, sequence stratigraphy, and stratigraphic modeling from 1964 to 1989: Twenty-five years of progress?. In: *Sedimentary Modeling: Computer Simulations and Methods for Improved Parameter Definition* (Ed. by E. K. Franseen, W. L. Watney, C. G. S. C. Kendall and W. Ross), *Kansas Geol. Surv. Bull.*, **233**, 3–8.

Ross, W.C., **Watts, D.E.** and **May, J.A.** (1995) Insights from stratigraphic modelling: Mud-limited versus sand-limited depositional systems. *AAPG Bull.*, **79**, 231–258.

Royden, L. and **Keen, C.E.** (1980) Rifting processes and thermal evolution of the continental margin of eastern Canada determined from subsidence curves. *Earth Planet. Sci. Lett.*, **51**, 343–361.

Rubin, D.M. (1992) Use of forecasting signatures to help distinguish periodicity, randomness, and chaos in ripples and other spatial patterns. *Chaos*, **2**, 525–535.

Sambrook Smith, G.H. and **Ferguson, R.I.** (1995) The gravel-sand transition along river channels. *J. Sedim. Res.*, **A65**, 423–430.

Sapozhnikov, V.B. and **Foufoula-Georgiou, E.** (1996) Self-affinity in braided rivers. *Water Resour. Res.*, **32**, 1429–1439.

Sapozhnikov, V.B. and **Foufoula-Georgiou, E.** (1997) Experimental evidence of dynamic scaling and indications of self-organized criticality in braided rivers. *Water Resour. Res.*, **33**, 1983–1991.

Sawyer, D.S., **Hsui, A.T.** and **Toksoz, M.N.** (1987) Extension, subsidence, and thermal evolution of the Los Angeles Basin — a two-dimensional model. *Tectonophysics*, **133**, 15–32.

Schlager, W. (1981) The paradox of drowned reefs and carbonate platforms. *Geol. Soc. Am. Bull.*, **92**, 197–211.

Schlee, J. (1957) Upland gravels of southern Maryland. *Geol. Soc. Am. Bull.*, **68**, 1371–1430.

Schumm, S.A. (1977) *The Fluvial System.* John Wiley, New York.

Schumm, S.A. and **Lichty, R.W.** (1965) Time, space, and causality in geomorphology. *Am. J. Sci.*, **263**, 110–119.

Schumm, S.A., **Mosley, M.P.** and **Weaver, W.E.** (1987) *Experimental Fluvial Geomorphology.* John Wiley, New York.

Schwarzacher, W. (1966) Sedimentation in a subsiding basin. *Nature*, **5043**, 1349–1350.

Sen, A., **Kendall, C.G.S.C.** and **Levine, P.** (1999) Combining a computer simulation and eustatic events to date seismic sequence boundaries: a case study of the Neogene of the Bahamas. *Sediment. Geol.*, **125**, 47–59.

Shanley, K.W. and **McCabe, P.J.** (1991) Predicting facies architecture through sequence stratigraphy—An example from the Kaiparowits Plateau, Utah. *Geology*, **19**, 742–745.

Shanley, K.W. and **McCabe, P.J.** (1993) Alluvial architecture in a sequence stratigraphic framework: a case history from the Upper Cretaceous of southern Utah, USA. In: *The Geological Modelling of Hydrocarbon Reservoirs and Outcrop Analogues* (Ed. by S. S. Flint and I. D. Bryant), *Spec. Publ. Int. Assoc. Sedimentol.*, **15**, 21–56.

Shuster, M.W. and **Aigner, T.** (1994) Two-dimensional synthetic seismic and log cross sections from stratigraphic forward models. *AAPG Bull.*, **78**, 409–431.

Sinclair, H.D., **Coakley, B.J.**, **Allen, P.A.** and **Watts, A.B.** (1991) Simulation of foreland basin stratigraphy using a diffusion model of mountain belt uplift and erosion: an example from the Central Alps, Switzerland. *Tectonics*, **10**, 599–620.

Sleep, N.H. (1971) Thermal effects of the formation of Atlantic continental margins by continental break-up. *Geophys. J. Royal Astronom. Soc.*, **45**, 325–350.

Slingerland, R. (1990) Predictability and chaos in quantitative dynamic stratigraphy. In: *Quantitative Dynamic Stratigraphy* (Ed. by T. A. Cross), pp. 45–53. Prentice Hall, Englewood Cliffs, New Jersey.

Slingerland, R., **Harbaugh, J.W.** and **Furlong, K.** (1994) *Simulating Clastic Sedimentary Basins.* PTR Prentice Hall, Englewood Cliffs, New Jersey.

Sloss, L.L. (1962) Stratigraphic models in exploration. *J. Sedim. Petrol.*, **32**, 415–422.

Sloss, L.L. (1991) The tectonic factor in sea-level change: a countervailing view. *J. Geophys. Res.*, **96**, 6609–6617.

Smith, J.D. and **McLean, S.R.** (1977) Spatially averaged flow over a wavy surface. *J. Geophys. Res.*, **82**, 1735–1746.

Snow, R.S. and **Slingerland, R.L.** (1987) Mathematical modeling of graded river profiles. *J. Geol.*, **95**, 15–33.

Soni, J.P. (1981) Laboratory study of aggradation in alluvial channels. *J. Hydrol.*, **49**, 87–106.

Spencer, R.J. and **Demicco, R.V.** (1989) Computer models of carbonate platform cycles driven by subsidence and eustasy. *Geology*, **17**, 165–168.

Starfield, A.M. (1990) Qualitative, rule-based modeling. *Bioscience*, **40**, 601–604.

Starfield, A.M., Cumming, D.H.M., Taylor, R.D. and **Quadling, M.S.** (1993) A frame-based paradigm for dynamic ecosystem models. *AI Applications*, **7**, 1–13.

Starfield, A.M., Farm, B.P. and **Taylor, R.H.** (1989) A rule-based ecological model for the management of an estuarine lake. *Ecol. Modelling*, **46**, 107–119.

Steckler, M.S. (1999) High resolution sequence stratigraphic modeling: 1. The interplay of sedimentation, erosion and subsidence. In: *Numerical Experiments in Stratigraphy* (Ed. by J. Harbaugh, L. Watney, G. Rankey, R. Slingerland, R. Goldstein and E. Franseen), *Spec. Publ. Soc. Econ. Paleont. Miner.*, **62**, 139–150.

Steckler, M.S. (1985) Uplift and extension at the Gulf of Suez: indications of induced mantle convection. *Nature*, **317**, 135–139.

Steckler, M.S., Mountain, G.S., Miller, K.G. and **Christie-Blick, N.** (1999) Reconstruction of Tertiary progradation and clinoform development on the New Jersey passive margin by 2-D backstripping. *Mar. Geol.*, **154**, 399–420.

Steckler, M.S., Reynolds, D.J., Coakley, B.J., Swift, B.A. and **Jarrard, R.** (1993) Modelling passive margin sequence stratigraphy. In: *Sequence Stratigraphy and Facies Associations* (Ed. by H. W. Posamentier, C. P. Summerhayes, B. U. Haq and G. P. Allen), *Spec. Publ. Int. Assoc. Sedimentol.*, **18**, 19–41.

Steer, B.L. and **Abbott, P.L.** (1984) Paleohydrology of the Eocene Ballena Gravels, San Diego County, Caliornia. *Sedim. Geol.*, **38**, 181–216.

Sternberg, H. (1875) Untersuchungen über Längen und Querprofil geschiebeführnde Flüsse. *Z Bauwesen*, **25**, 483–506.

Strobel, J., Cannon, R., Kendall, C.G.S.C., Biswas, G. and **Bezdek, J.** (1989) Interactive (SEDPAK) simulation of clastic and carbonate sediments in shelf to basin settings. *Computers & Geosciences*, **15**, 1279–1290.

Swenson, J.B., Voller, V., Paola, C. and **Parker, G.** (in press) Modeling the stratigraphic response of fan-deltas to base-level forcing as a generalized Stefan problem. *European J. Appl. Mechanics*, in press.

Syvitski, J.P.M. and **Alcott, J.M.** (1993) GRAIN2: Predictions of particle size seaward of river mouths. *Computers & Geosciences*, **19**, 399–446.

Syvitski, J.P.M. and **Alcott, J.M.** (1995) RIVER3: Simulation of river discharge and sediment transport. *Computers & Geosciences*, **21**, 89–151.

Syvitski, J.P.M. and **Andrews, J.T.** (1994) Climate Change: Numerical modelling of sedimentation and coastal processes, Eastern Canadian Arctic. *Arctic Alpine Res.*, **26**, 199–212.

Syvitski, J.P.M. and **Daughney, S.** (1992) DELTA2: Delta progradation and basin filling. *Computers & Geosciences*, **18**, 839–897.

Syvitski, J.P.M., Morehead, M. and **Nicholson, M.** (1998a) HYDROTREND: a climate-driven hydrologic-transport model for predicting discharge and sediment to lakes or oceans. *Computers & Geosciences*, **24**, 51–68.

Syvitski, J.P.M., Nicholson, M., Skene, K. and **Morehead, M.D.** (1998b) PLUME1.1: Deposition of sediment from a fluvial plume. *Computers & Geosciences*, **24**, 159–171.

Syvitski, J.P.M., Pratson, L. and **Morehead, M.D.** (1997) EARTHWORKS: a large spatial scale numerical model to study the flux of sediment to ocean basins and reworking of deposits over various time scales. *EOS Suppl.*, **78**, F258.

ter Voorde, M., Ravnås, R., Færseth, R. and **Cloetingh, S.** (1997) Tectonic modelling of the Middle Jurassic synrift stratigraphy in the Oseberg-Brage area, northern Viking Graben. *Basin Res.*, **9**, 133–150.

Tetzlaff, D.M. (1990) Limits to the predictive ability of dynamic models that simulate clastic sedimentation. In: *Quantitative Dynamic Stratigraphy* (Ed. by T. A. Cross), pp. 55–65. Prentice Hall, Englewood Cliffs, New Jersey.

Tetzlaff, D.M. and **Harbaugh, J.W.** (1989) *Simulating Clastic Sedimentation.* Van Nostrand Reinhold, New York.

Tipper, J. (1991) Modelling the fill of sedimentary basins. *Exploration Geophys.*, **22**, 397–400.

Tipper, J.C. (1992) Landforms developing and basins filling: three-dimensional simulation of erosion, sediment transport, and deposition. In: *Computer Graphics in Geology. Lecture Notes in Earth Sciences, Vol. 41.* (Ed. by R. Pflug and J. W. Harbaugh), pp. 155–170. Springer-Verlag, Berlin.

Tucker, G.E. and **Slingerland, R.L.** (1994) Erosional dynamics, flexural isostasy, and long-lived escarpments: a numerical modeling study. *J. Geophys. Res.*, **99**, 12229–12243.

Turcotte, D.L. and **Schubert, G.** (1982) *Geodynamics, Applications of Continuum Physics to Geological Problems.* John Wiley & Sons, New York.

Van Wagoner, J.C., Mitchum, R.M., Campion, K.M. and **Rahmanian, V.D.** (Eds.) (1990) Siliciclastic Sequence Stratigraphy in Well Logs, Cores, and Outcrops: Concepts for High-Resolution Correlation of Time and Facies. *AAPG Methods in Exploration Series*, **7**.

Waltham, D. (1992) Mathematical modelling of sedimentary basin processes. *Mar. Petrol. Geol.*, **9**, 265–273.

Watney, W.L., Wong, J.-C. and **French,J.A. Jr** (1991) Computer simulation of Upper Pennsylvanian (Missourian) carbonate-dominated cycles in western Kansas. In: *Sedimentary Modeling: Computer Simulations and Methods for Improved Parameter Definition* (Ed. by E. K. Franseen, W. L. Watney, C. G. S. C. Kendall and W. Ross), *Kansas Geol. Surv. Bull.*, **233**, 415–430.

Watts, A.B., Karner, G.D. and **Steckler, M.S.** (1982) Lithospheric flexure and the evolution of sedimentary basins. *Phil. Trans. Royal Soc. London*, **305A**, 249–281.

Weimer, P. and **Posamentier, H.** (Eds) (1994) *Siliciclastic Sequence Stratigraphy: Recent Developments and Applications. AAPG Memoir*, **58**.

Wendebourg, J. and **Ulmer, J.W.D.** (1992) Modelling compaction and isostatic compensation in SEDSIM for basin analysis and subsurface fluid flow. In: *Computer Graphics in Geology* (Ed. by R. Pflug and J. W. Harbaugh), pp. 143–153. Springer-Verlag, Berlin.

Werner, B.T. (1995) Sand dune attractors. *Geology*, **23**, 1111–1114.

Werner, B.T. (1999) Complexity in natural landform patterns. *Science*, **284**, 102–104.

Werner, B.T. and **Fink, T.M.** (1993) Beach cusps as self-organized patterns. *Science*, **260**, 968–971.

Werner, B.T. and **Hallet, B.** (1993) Numerical simulation of self-organized stripes. *Nature*, **361**, 142–145.

Werner, B.T. and **Kocurek, G.** (1997) Bed-form dynamics: does the tail wag the dog? *Geology*, **25**, 771–774.

Willgoose, G., Bras, R.L. and **Rodriguez-Iturbe, I.** (1991) A coupled channel network growth and hillslope evolution model. I. Theory. *Water Resour. Res.*, **27**, 1671–1684.

Wood, L.J., Ethridge, F.G. and **Schumm, S.A.** (1993) The effects of rate of base-level fluctuation on coastal-plain, shelf and slope depositional systems; an experimental approach. In: *Sequence Stratigraphy and Facies Associations* (Ed. by H. W. Posamentier, C. P. Summerhayes, B. U. Haq and G. P. Allen), *Spec. Publ. Int. Assoc. Sedimentol.*, **18**, 43–53.

Wood, L.J., Ethridge, F.G. and **Schumm, S.A.** (1994) An experimental study of the influence of subaqueous shelf angles on coastal plain and shelf deposits. In: *Siliciclastic Sequence Stratigraphy: Recent Developments and Applications* (Ed. by P. Weimer and H. Posamentier), *AAPG Mem.*, **58,** 381–391.

Yalin, M.S. (1971) *Theory of Hydraulic Models.* MacMillan, London.

Yeakel, L.S. (1962) Tuscarora, Juniata and Bald Eagle paleocurrents and paleogeography in the central Appalachians. *Geol Soc. Am. Bull.*, **73,** 1515–1540.

Zhang, H. and **Kahawita, R.** (1987) Nonlinear model for aggradation in alluvial channels. *J. Hydraulic Engineering*, **113,** 353–369.

NOMENCLATURE

Symbol	Meaning	Dimensions (**L** length, **M** mass, **T** time)
b	stream width	**L**
c	wave (phase) speed	$\mathbf{LT^{-1}}$
c_f	drag coefficient $\tau_0/(\pi u^2)$	
C_0	volumetic sediment concentration in the bed	
D	grain diameter	**L**
f	is a function of	from context
$f_D(D)$	grain diameter probability density	$\mathbf{L^{-1}}$
g	gravitational acceleration	$\mathbf{LT^{-2}}$
h	flow depth	**L**
H	base level elevation relative to a fixed datum; representative depth	**L**
L	basin length; length scale	**L**
p	*in situ* sediment production weight	$\mathbf{LT^{-1}}$
q_s	sediment flux (discharge) per unit width	$\mathbf{L^2T^{-1}}$
q_{s*}	dimensionless sediment flux $q_s/[g^{1/2}(s-1)^{1/2}D^{3/2}]$	0
q_w	unit water flux (discharge) per unit width	$\mathbf{L^2T^{-1}}$
Q_s	volumetric sediment flux (discharge)	$\mathbf{L^3T^{-1}}$
Q_w	volumetric water flux (discharge)	$\mathbf{L^3T^{-1}}$
s	sediment specific gravity	0
S	downstream bed slope	
t	time	**T**
T	representative time; cycle period	**T**
T_{eq}	time required for system to reach equilibrium	**T**
u	flow velocity	$\mathbf{LT^{-1}}$
U	representative velocity	$\mathbf{LT^{-1}}$
x	horizontal distance along the direction of mean sediment flux	**L**
$\mathbf{x}$	two dimensional location (x, y)	(**L**, **L**)
y	horizontal distance across the direction of mean sediment flux	**L**
η	sediment surface elevation relative to a fixed datum	**L**
θ	deep-water carbonate production parameter	$\mathbf{T^{-1}}$
ν	transport coefficient	$\mathbf{L^2T^{-1}}$
ν'	modified transport coefficient. νC_0	$\mathbf{L^2T^{-1}}$
ρ	fluid density	$\mathbf{ML^{-3}}$
σ	subsidence rate	$\mathbf{LT^{-1}}$
τ_0	boundary shear stress	$\mathbf{ML^{-1}T^{-2}}$
τ_*	dimensionless boundary shear stress $\tau_0/[\rho g(s-1)D]$	
ϕ	logarithmic grain size measure, $-\log_2(D)$, D in mm	
ω	shallow-water carbonate production parameter	$\mathbf{T^{-1}}$
Modifiers		
0	value supplied at edge of depositional basin; bed value	
$\langle\rangle$	value averaged per unit basin width	

Sedimentology (2000) **47** (Suppl. 1), 179–214

Microbial carbonates: the geological record of calcified bacterial–algal mats and biofilms

ROBERT RIDING
Department of Earth Sciences, Cardiff University, Park Place, PO Box 914, Cardiff CF10 3YE, UK (E-mail: riding@cardiff.ac.uk)

ABSTRACT

Deposits produced by microbial growth and metabolism have been important components of carbonate sediments since the Archaean. Geologically best known in seas and lakes, microbial carbonates are also important at the present day in fluviatile, spring, cave and soil environments. The principal organisms involved are bacteria, particularly cyanobacteria, small algae and fungi, that participate in the growth of microbial biofilms and mats. Grain-trapping is locally important, but the key process is precipitation, producing reefal accumulations of calcified microbes and enhancing mat accretion and preservation. Various metabolic processes, such as photosynthetic uptake of CO_2 and/or HCO_3^- by cyanobacteria, and ammonification, denitrification and sulphate reduction by other bacteria, can increase alkalinity and stimulate carbonate precipitation. Extracellular polymeric substances, widely produced by microbes for attachment and protection, are important in providing nucleation sites and facilitating sediment trapping.

Microbial carbonate microfabrics are heterogeneous. They commonly incorporate trapped particles and *in situ* algae and invertebrates, and crystals form around bacterial cells, but the main component is dense, clotted or peloidal micrite resulting from calcification of bacterial cells, sheaths and biofilm, and from phytoplankton-stimulated whiting nucleation. Interpretation of these texturally convergent and often inscrutable fabrics is a challenge. Conspicuous accumulations are large domes and columns with laminated (stromatolite), clotted (thrombolite) and other macrofabrics, which may be either agglutinated or mainly composed of calcified or spar-encrusted microbes. Stromatolite lamination appears to be primary, but clotted thrombolite fabrics can be primary or secondary. Microbial precipitation also contributes to hot-spring travertine, cold-spring mound, calcrete, cave crust and coated grain deposits, as well as influencing carbonate cementation, recrystallization and replacement. Microbial carbonate is biologically stimulated but also requires favourable saturation state in ambient water, and thus relies uniquely on a combination of biotic and abiotic factors. This overriding environmental control is seen at the present day by the localization of microbial carbonates in calcareous streams and springs and in shallow tropical seas, and in the past by temporal variation in abundance of marine microbial carbonates. Patterns of cyanobacterial calcification and microbial dome formation through time appear to reflect fluctuations in seawater chemistry.

Stromatolites appeared at ~3450 Ma and were generally diverse and abundant from 2800 to 1000 Ma. Inception of a Proterozoic decline, variously identified at 2000, 1000 and 675 Ma, has been attributed to eukaryote competition and/or reduced lithification. Thrombolites and dendrolites mainly formed by calcified cyanobacteria became important early in the Palaeozoic, and reappeared in the Late Devonian. Microbial carbonates retained importance through much of the Mesozoic, became scarcer in marine environments in the Cenozoic, but locally re-emerged as large agglutinated domes, possibly reflecting increased algal involvement, and thick micritic reef crusts in the late Neogene. Famous modern examples at Shark Bay and Lee Stocking Island are composite coarse agglutinated domes and columns with complex bacterial–algal mats occurring in environments that are both stressed and current-swept: products of mat evolution, ecological refugia, sites of enhanced early lithification or all three?

Keywords Bacteria, biofilm, calcification, mat, microbe.

INTRODUCTION

Microbes are widespread and often abundant in Earth-surface environments (Brock *et al.*, 1994), and appear to have been so from early in Earth history (Schopf, 1983). They can live on virtually all wet sedimentary surfaces, at all depths in lakes and seas, and in sediment to depths of kilometres. Microbes interact widely with other organisms, both competitively and cooperatively, play major roles in chemical cycles that influence atmosphere–hydrosphere composition (Fenchel & Finlay, 1995; Banfield & Nealson, 1997; Ehrlich, 1998), and are extensively involved in the production, accumulation and diagenesis of sediment (Nealson, 1997; Riding & Awramik, 2000). The effects of microbial mediation on precipitation and localization of carbonate sediments are noticeable in a wide variety of depositional settings. These include marine reefs, tidal flats, lacustrine whitings, fluviatile tufas, hot-spring travertines and cave crusts, as well as subaerial calcretes and other coatings, grains and matrices within sediment. Microbial carbonates form a huge area for research. My aims here are to outline key aspects of these sediments and, in particular, to focus on some of the challenges that these studies encounter. I will necessarily be selective.

Microbes, most simply defined as all microscopic organisms, are generally considered to encompass bacteria (including cyanobacteria), fungi, small algae and protozoans (Fig. 1; Brock *et al.*, 1994). With such inclusivity, the boundaries of microbial carbonate studies are not well defined, or at least are not narrow (see Pia, 1926; Camoin, 1999). Many studies of microbial carbonates focus mainly on bacteria in general, and cyanobacteria in particular, and emphasize their ability to precipitate as well as to localize sediment. In contrast, in the case of small algae and protozoans, only their nonskeletal, e.g. trapping and binding, roles are usually included in microbial carbonate research. Small calcified organisms, such as coccolithophore algae and foraminiferal protozoans, are technically microbes, but deposits in which both can be abundant, such as the Cretaceous Chalk of north-west Europe, are not normally regarded as microbial carbonates. The reason is probably that coccoliths and tests are distinctive discrete skeletal components. In contrast, many, although not all, microbial carbonates consist of rather less clearly defined nonskeletal precipitates, together with trapped grains. The focus of microbial carbonate studies is mainly calcified biofilms and microbial mats, and their geological record.

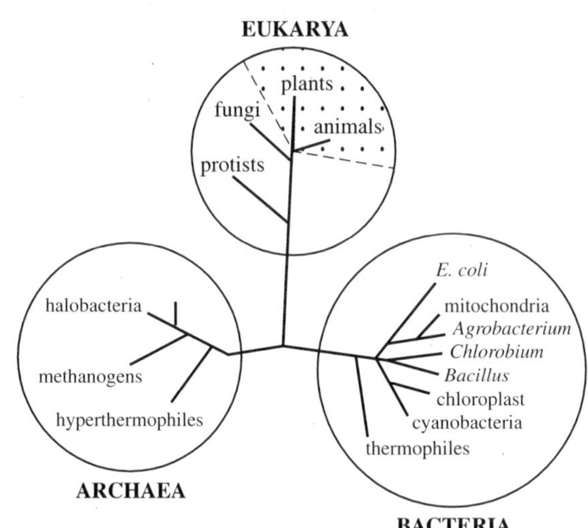

Fig. 1. Cloverleaf of life: microbial diversity (bacteria, algae, protozoans and fungi) in a molecular phylogeny perspective based on 16S rRNA gene-sequences (after Nealson, 1997, fig. 1). Microbes occupy all except the shaded area. Archaea and Bacteria are two major groupings of bacteria. Archaea include all known methane-producing bacteria (methanogens), and most bacteria living in very salt-rich (>10% NaCl, halophiles) and very hot (>80°C, hyperthermophiles) conditions (see Brock *et al.*, 1994). Bacteria include cyanobacteria and many other types. In contrast to bacteria, the eukaryote cell of the Eukarya contains membrane-bound cell organelles. Protists encompass most algae and protozoans.

MICROBES

Bacteria

Bacteria are the key organisms in the formation of microbial carbonates. In practice, the terms bacteria and prokaryotes are synonymous: prokaryotes lack discrete nuclei and other membrane-bound cell organelles; all other organisms are eukaryotes (Eukarya). Bacteria exhibit immense diversity (Brock *et al.*, 1994). Their phylogenetic study and classification has been revolutionized by genetic similarity comparisons, particularly based on analysis of ribosomal ribonucleic acid (rRNA) gene sequences. Although gene transfer complicates these interpretations (see Wiegel & Adams, 1998), a major result has been recognition of two distinct bacterial groups (Fig. 1), archaebacteria and eubacteria (Woese, 1987), now, respectively, termed Archaea and Bacteria (see Brock *et al.*, 1994). These names can cause some terminological confusion; here I refer in general to both groups as bacteria.

Cyanobacteria (previously variously termed blue–green algae, blue–green bacteria, cyanophytes) are very important sedimentologically. These relatively large and somewhat alga-like eubacteria are essentially aerobic phototrophs – using sunlight as energy. In shallow-water oxygenated environments they can thrive in the water column and at the sediment–water interface. Many other bacteria are anaerobic heterotrophs and can occupy dark anaerobic conditions such as sediment pore surfaces (Nealson, 1997). They mineralize (decompose) organic material to inorganic components by redox processes using electron acceptors in order of their energy-yield. This results in progressive depletion in ambient O_2, NO_3^-, SO_4^{2-} and CO_2 (see Fenchel & Finlay, 1995, p. 173). The catabolic (breakdown, usually energy-yielding) processes involved, denitrification, sulphate reduction and methanogenesis, occur successively with increasing depth in sediments. Sulphate reduction is especially important in marine environments, owing to the high sulphate concentration of seawater. As the supply of organic material is reduced, bacterial activity in sediments declines and is often slight at depths below 10–20 cm (Jørgensen, 1977; Fenchel & Finlay, 1995, p. 195).

Algae

Small algae in modern microbial carbonates (see Wallner, 1935; Walter, 1972, pp. 88–89; Golubic, 1976) are usually associated with microbial mats, where they trap and stabilize sediment. They include brown, green and red algae that are mainly relatively large filamentous forms (Scoffin, 1970; Riding *et al.*, 1991a). In contrast, diatoms are small unicellular forms, but they attach by mucilaginous stalks, secrete mucilage and locally are sufficiently abundant to be important sediment stabilizers in marine (Neumann *et al.*, 1970; Awramik & Riding, 1988) and nonmarine (Winsborough & Golubic, 1987; Winsborough, 1999) carbonate environments. In association with microbial carbonates, small calcified algae are prominent in fresh water (Wallner, 1934; Golubic, 1976), but can also be important in the sea (e.g. Dravis, 1983). Fungi become important in subaerial impermanently wetted settings, especially in soils, and are symbiotically associated with algae, and cyanobacteria, in lichens (Klappa, 1979; Jones & Wilson, 1986).

Extracellular polymeric substances

Extracellular polymeric substances (EPS), widely produced by microbes (Decho, 1990), are of central importance in the formation of microbial carbonates. EPS accumulates outside cells to form a protective and adhesive matrix that attaches microbes to substrates (Costerton *et al.*, 1978), provides physical and chemical protection, and can also aid nutrient absorption (Christensen & Characklis, 1990). Bacteria, cyanobacteria and diatoms can all secrete copious amounts of EPS. Sedimentologically important EPS exist in a variety of forms. On sediment surfaces, these range from well-defined protective envelopes surrounding cells, as in cyanobacterial sheaths, to masses that anchor and provide a favourable medium for microbial populations and communities on a scale from biofilms (100s of nanometres) to microbial mats (1000s of nanometres). In addition, microbes can form blooms (dense concentrations of phytoplanktic cells) and floccules, as in marine and lake snow (millimetric EPS-based organic aggregates suspended in the water column). These are typically complex communities which can also be sites of precipitation, as in whiting formation. Exopolymers directly promote the accretion and preservation of the biofilms and microbial mats that they help to constitute by favouring both sediment trapping and mineral precipitation.

Biofilms

Biofilms are generally submillimetric veneers of bacterial populations and communities in an EPS matrix, attached to substrates in aquatic environments. The EPS that they secrete provide stability, facilitate metabolic interaction and act as a diffusion barrier and adsorbent (Lawrence *et al.*, 1994), advantages that have drawn comparisons with eukaryote tissue (Costerton *et al.*, 1995, p. 713). Biofilms can exhibit structural heterogeneity (Wolfaardt *et al.*, 1994, fig. 4; Costerton *et al.*, 1995, fig. 8; Decho, 2000). Biofilm architecture (Lawrence *et al.*, 1991) may include a 'regular array of channels and aggregates', on a scale of 20–40 μm. The aggregates of cells form irregular three-dimensional networks separated by canal systems maintained by water currents and acting as conduits for nutrients, oxygen and waste materials. These are common in pure-culture biofilms under low-flow conditions, but mixed-species biofilms show less regular architecture. The difference could reflect nutrient-limitation in pure-culture biofilms, and niche-exploitation in mixed-species biofilms (Costerton *et al.*, 1995, pp. 719–720). Biofilms are usually only tens or hundreds of micrometres in thickness and their components can only readily be observed by electron microscopy. Biofilm sedimentology is complex (Jones, 1995); they may trap very small grains, but can be calcified by bacterial processes (Pedley, 1992; Perry, 1999).

Microbial mats

Mats are complex structures that form the accreting surface of most stromatolites and other large-scale benthic microbial carbonates. A modern living mat commonly has a thickness of millimetres, and relatively large microbes, including filamentous cyanobacteria, diatoms and filamentous algae, can create intertwined turf-like communities in which the main individual components can commonly be seen with a hand-lens. Mats readily trap micritic sediment, and where they are thick they may also trap sand or even coarser grains (Fig. 2). In addition to stability and protection, microbes require energy and nutrients. In mats these activities can be closely interrelated, involving complex syntrophic communities in which photosynthesis in the upper mat is balanced by decomposition below (Brock *et al.*, 1994, p. 636). There are steep gradients in both light and oxygen. These fluctuate diurnally (see Fenchel & Finlay, 1995, pp. 28, 181–186) and result in well-defined microbial stratification. Typically, aerobic phototrophs (cyanobacteria) are near the surface, anoxygenic phototrophs below, followed by chemo-organotrophs that require neither oxygen nor light (Revsbech *et al.*, 1983; Jørgensen *et al.*, 1983; Pierson *et al.*, 1987). The activities of these microbes determine mat calcification and preservation (Canfield & Raiswell, 1991; and see 'Other bacterial calcification' below). Algae add a further complexity to the bacterial surface community.

Mats in harsher environments such as hot springs and salinas are simpler and less disturbed by bioturbation than those on subtidal marine shelves. However, the details of mat complexity are often rapidly lost as the organisms die, although the organic framework persists (Défarge *et al.*, 1996). In most microbial carbonates, calcified cyanobacteria are the best-preserved microbes.

Fig. 2. Complex mat, trapping ooid–peloid sand (light coloured grains), from the surface of a subtidal dome in the tidal channel between Lee Stocking Island and Norman's Pond Cay, Exuma, The Bahamas. The mat is dominated by microbes (cyanobacteria, green algae and diatoms), but also supports a prominent canopy of hydrozoans and macroalgae (see Riding *et al.*, 1991a). Width of view ~5 cm.

PROCESSES

Many bacteria obtain nutrients and energy by degrading minerals and other compounds, and these activities are facilitated when bacteria adhere to sediment and rock surfaces. The mineral dissolution that results is a key aspect of weathering, and releases products that can be precipitated (Ehrlich, 1998). These intimate relationships between microbes and sediments make microbes of global importance in the biogeochemical cycling of elements. They are also responsible for the two main processes that create microbial carbonates: grain trapping and mineral precipitation.

Grain trapping

The size, motility, orientation and mutual relationships of microbes determine their ability to trap and stabilize sediment (see Gebelein, 1969; Monty, 1976; Golubic *et al.*, 1999) and EPS production is also important (Golubic, 1976). *Stabilization* is well seen on tidal and lacustrine flats where microbial growth occurs during pauses in sedimentation (Noffke, 1998). Firm mats form fibrous or leathery coverings and are further strengthened by desiccative early lithification. Flooding of the flats ruckles and folds mats, forming petee structures (Gavish *et al.*, 1985, p. 192). As a result, storm layers of coarse sediment alternate with mats, some of which are folded (Shinn *et al.*, 1969; Kendall & Skipwith, 1969).

Whereas microbial overgrowth stabilizes previously deposited sediment, *trapping* localizes sediment passing across mats (Black, 1933). This can involve fine to coarse, and siliciclastic as well as carbonate grains. The grade of particles trapped (Fig. 2) depends on ambient sediment supply and on mat structure (Ginsburg *et al.*, 1954). The process of trapping can involve simple physical blockage (*baffling*) of grain movement, and also adhesion on, and in, EPS. Both processes are facilitated where mats have irregular surface topography, such as that formed by relatively large microbes with abundant erect filaments, and also mats with abundant soft EPS, which may trap coarse sediment if it is available, as in some Shark Bay and Lee Stocking Island domes. In contrast, smooth mats or films with little surface topography trap only very fine grains or none at all. Trapping, the localization of hitherto mobile grains on mat surfaces, goes hand-in-hand with the dual process of *binding*: organic overgrowth of trapped grains by cells, filaments and biofilms, which incorporates them into the mat, as in the classic diagram of Gebelein (1969, fig. 14). Grains thus become the substrate for further surficial microbial growth. The mat surface rises, and accretion has occurred.

Precipitation

Although trapping was given prominence by work in modern intertidal environments (Black, 1933; Ginsburg *et al.*, 1954), and is occasionally still emphasized as the key process in the formation of stromatolites (e.g. Ehrlich, 1998), early lithification is also crucial for the formation of microbial carbonates (Walcott, 1914; Logan, 1961; Ginsburg, 1991). Conservation of accreted trapped sediment requires early lithification to strengthen the deposit, to allow microbial sediment to keep pace with or outpace surrounding sediment accumulation, and to maintain domical and columnar mats. Nonetheless, the importance of precipitation for microbial carbonates goes well beyond cementation of trapped material. Dendrolites, skeletal stromatolites and tufa stromatolites, for example, are fundamentally precipitated deposits dominated by calcified microbes, probably cyanobacteria. In addition, and significantly, other bacteria that degrade organic matter below the mat surface create conditions that result in precipitation.

Cyanobacterial calcification

Cyanobacterial calcification appears to result from creation of alkalinity gradients in mucilaginous sheaths, associated with photosynthetic uptake of CO_2 and/or HCO_3^- uptake that raises alkalinity (see Pentecost & Riding, 1986; Merz, 1992; Merz-Preiß, 1999). Preliminary assumptions made by Pentecost & Riding (1986) still appear to be valid: calcification is intimately related to the cyanobacterial mucopolysaccharide sheath, is specific (and see Défarge *et al.*, 1994a) but not obligate, and is dependent on suitable environmental conditions. Sheath impregnation appears to occur in environments where calcification is less rapid, such as lakes and pools, whereas sheath encrustation occurs in fast-flowing rapidly precipitating streams, although this requires further study (see Merz-Preiß & Riding, 1999). Although cyanobacterial calcification appears mainly to occur during life, the converse can also apply, as in desiccated *Entophysalis* in Shark Bay (Golubic & Hofmann, 1976). At the

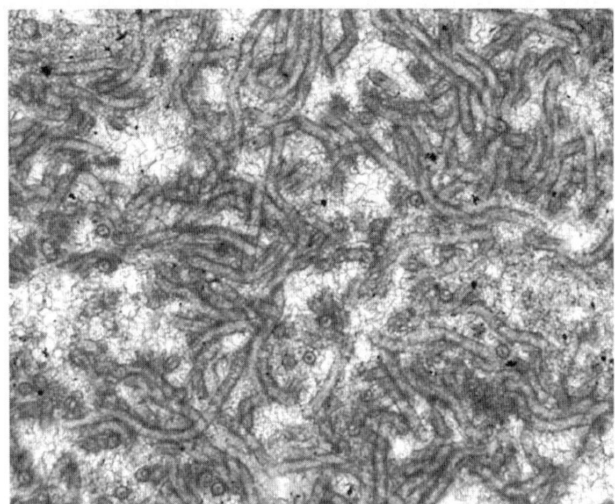

Fig. 3. *Girvanella*, a common calcified cyanobacterium in Palaeozoic–Mesozoic shallow-marine environments, preserved by sheath impregnation (low mid-Ordovician, Lunnan, northern Tarim Basin, China). Width of view = 1·5 mm.

present day, intense and widespread cyanobacterial calcification appears to be essentially a freshwater phenomenon and is scarcely known in modern subtidal environments, although it has been widespread at times in the past (Fig. 3). None the less, it occurs in present-day brackish marine environments (Rasmussen *et al.*, 1993).

EPS calcification

Microbially produced EPS can favour $CaCO_3$ precipitation by providing diffusion-limited sites that create alkalinity gradients in response to metabolic processes, and by attracting and binding calcium ions to negatively charged sites (Pentecost, 1985). However, EPS can also inhibit precipitation, depending on environmental conditions (see Arp *et al.*, 1999a). At high alkalinities, as in alkaline lakes, calcification appears not to be directly associated with living cyanobacteria because photosynthesis in these conditions will not create a sufficient alkalinity gradient in the sheath; instead, calcification preferentially occurs on surfaces poor in EPS (Arp *et al.*, 1999b).

Other bacterial calcification

A wide range of bacterial processes, in addition to photosynthesis in cyanobacteria, can lead to HCO_3^- concentration and raised alkalinity favouring $CaCO_3$ precipitation (see von Knorre & Krumbein, 2000). These include ammonification, denitrification, sulphate reduction and anaerobic sulphide oxidation (see Krumbein, 1979a; Visscher *et al.*, 1992; Castanier *et al.*, 1999). These processes are localized within decaying mats as heterotrophic bacteria degrade organic matter, including cyanobacteria (Bartley, 1996), and they lead to fine-grained precipitation, typically micritic or slightly coarser (Chafetz & Buczynski, 1992; Reitner, 1993; Défarge *et al.*, 1994b; Reitner *et al.*, 1995). These processes are of fundamental importance in the lithification of microbial mats (Krumbein *et al.*, 1977; Krumbein, 1978, 1979a, 1983, p. 495; Chafetz & Buczynski, 1992; Chafetz, 1994). The general result, in microbial carbonates containing calcified cyanobacteria, would be to create a micritic matrix around the calcified sheath material. In microbial carbonates primarily formed by agglutination of coarse sediment on and in mats, the fine-grained matrix created by synsedimentary bacterial precipitation will superficially resemble trapped fines, or micritic cement. Although the importance of these processes in the synsedimentary lithification of microbial carbonates is evident, linking particular products to specific bacterial processes remains an elusive goal. Défarge *et al.* (1996) observed diverse precipitates within stromatolitic mats. These probably nucleate at various sites: around cells, on organic and grain surfaces, and within EPS. The specific site of precipitation appears to influence the mineralogy of the precipitate. Buczynski & Chafetz (1991) found calcite, rather than aragonite, precipitation in more viscous laboratory media, and also in viscous mucilage in natural mats (Chafetz & Buczynski, 1992).

Organomineralization

Bacterially induced precipitation within mats is clearly important and complex. However, precipitation of fine-grained $CaCO_3$ in association with nonliving organic macromolecules (organomineralization; Trichet & Défarge, 1995), independent of biological activity, may also be a sedimentologically significant process (see Reitner *et al.*, 1995; Neuweiler *et al.*, 1999) that overlaps in environment, timing and product with bacterial micrite. Discrimination between these various closely associated fine-grained precipitates in ancient microbial carbonates and mud mounds is a major challenge for future research.

Whiting precipitation

Bacteria have long been linked to the precipitation of water-column carbonates such as whitings, relatively small (generally < 20 µm) $CaCO_3$ crystals (Fig. 4) in lakes and shallow seas (e.g. Bavendamm, 1932). Recently, both lacustrine and marine whitings have been specifically linked to picoplanktonic cyanobacteria (Thompson & Ferris, 1990; Robbins & Blackwelder, 1992). It remains likely, however, that algal and cyanobacterial blooms in general, together with environmental conditions that may vary seasonally, combine to stimulate precipitation through localized removal of CO_2 and HCO_3^-. If whitings are indeed water-column precipitates then their volume would be of considerable significance, potentially accounting several times over for all late Holocene bank-top lime mud on Great Bahama Bank, for example (Robbins & Evans, 1997). However, doubts continue about the nature of marine whitings (Morse & He, 1993; Thompson, 2000).

RECOGNITION

A central problem for the study of microbial carbonates concerns recognition of both the organisms and the processes involved. Many ancient microbial carbonates do not, or at least do not readily appear to, preserve clear evidence of the organisms responsible for their formation. This immediately raises the question, 'then how do we know that they are microbial'. This is not a new problem. Many early workers thought that the deposits that would come to be named stromatolites were made by simple animals. Hence, Hall (1883) named Cambrian examples, from near Saratoga Springs in New York State, *Cryptozoon* and Gürich (1906) called Early Carboniferous examples in Belgium spongiostromids, which he regarded as a new group of protozoan-like organisms that created distinctive microfabrics composed of rounded aggregates traversed by minute canals. When Kalkowsky (1908) proposed the term stromatolite he was convinced that they (and ooids) were organic, but he could not prove it, and he was not sure which organisms were responsible, although he suspected that they were simply organized plants.

Fig. 4. Calcite whiting crystal, 8 µm long, from sediment trap at −15 m, north basin Lake Lugano, Italy, July 1992 (sample provided by A. Hofmann; see Hofmann, 1996).

The essential answers were of course provided by modern analogues. The first organisms to be confidently linked to stromatolites in this way were cyanobacteria, when Walcott (1914) compared modern freshwater cyanobacterial tufa crusts and oncoids with Proterozoic stromatolites of the Belt rocks in Montana. Bradley (1929) did the same for Eocene stromatolites in the Green River Formation of Wyoming. Cyanobacteria were further implicated as key organisms in stromatolite formation by the work of Black (1933) and Ginsburg (1955) on tidal flats in Andros Island and Florida, respectively (see Hoffman, 1973, p. 180). Thus, Logan's (1961) attribution of the Shark Bay domes to cyanobacterial trapping, while at the same time terming them *Cryptozoon*, had the resonance of 75 years research.

Yet, in spite of this, the external appearance, internal fabrics and geochemical signatures of ancient examples of microbial carbonates are usually insufficient to provide unequivocal evidence of their origins, and it is often difficult to relate organism, process and product at both micro- and macrofabric levels. The question of demonstrating biogenicity resurfaces whenever the interpretation of a deposit is sufficiently important that really firm evidence of origin is demanded. A prime example is recent discussions about earliest Archaean stromatolites in the ~3450 Ma Towers Formation (Warrawoona Group) of Western Australia (Lowe, 1980). These occur as cones, domes and layers in chertified limestones and dolostones and appear to have formed in shallow, often evaporative, basins in volcanic greenstone belts (Walter, 1994). Biogenicity problems that afflict most microbial carbonates assume critical importance in these very old deposits. Arguments have been marshalled both against (Lowe, 1994, 1995) and for (Buick *et al.*, 1995) the biological origin of these and other stromatolite-like structures older than 3200 Ma. They remain widely regarded as probably microbial in origin (e.g. Schopf, 1994; Hofmann *et al.*, 2000; Hofmann, 1999) but, as often as not, definitive evidence is scarce. Such difficulties in establishing biogenicity have had a strong influence on how terms (e.g. see 'Stromatolite', below) should be defined. Internal fabrics are a prime source of evidence concerning possible origins.

FABRICS AND COMPONENTS

Components and fabrics are the key to understanding the organisms and processes involved in microbial carbonates (Monty, 1973), but since many of them form at the scale of the microbes themselves, their correspondingly small size complicates their interpretation. Trapped coarse grains, calcified cyanobacteria and subordinate skeletal encrusting eukaryotes are usually the most readily recognizable components. However, many microbial carbonates are dominated by micritic fabrics whose apparent uniformity masks a range of possible origins. Although a good deal of effort has been put into describing microfabrics and the finer details of mesofabrics in ancient stromatolites, these have proved hard to categorize and even harder to interpret (see Vologdin, 1962; Bertrand-Sarfati, 1976; Komar, 1989; Knoll & Semikhatov, 1998), except where calcified microbes are present (e.g. Bertrand-Sarfati *et al.*, 1994). It is possible to distinguish a variety of principal components and tentatively to suggest origins for them (Table 1), but comprehensively relating processes to fabrics and components remains a challenging research goal.

Micrite

(i) Dense micrite. Possible microbial origins:

(I) *Calcified bacterial cells.* Dead and dying bacteria can become calcified during lysis (Krumbein, 1979a) giving rise to discrete rounded bodies, small varieties (< 1 μm) of which (Maurin & Noël, 1977) have been attributed to nanobacteria (Folk, 1993), although similar objects can form abiotically (Kirkland *et al.*, 1999) and there is continuing dispute concerning nanobacteria (Abbott, 1999).

(II) *Whitings.* These planktonic precipitates (see 'Processes', above) include micrite as well as larger crystals (Fig. 4). They have long been attributed to bacteria, and recently specifically to cyanobacterial picoplankton (e.g. *Synechococcus*, Thompson & Ferris, 1990; Thompson, 2000).

(III) *Calcified biofilm.* Bacteria preferentially adhere to grain and other surfaces for stability, and create biofilm communities that are augmented by other microbes. Consequently, some constructive micrite envelopes (Fig. 5) probably

Table 1. Common components of microbial carbonates and their possible microbial origins.

Component	Possible origins
Micrite	
dense micrite	calcified cells, microbial whitings, calcified biofilm
clotted fabric	EPS calcification and trapped micrite
calcified sheath	impregnated cyanobacterial and other sheaths
peloid	calcified bacterial aggregates
Microspar & spar	extent of microbial origin uncertain; common: in freshwater tufas on peloids in travertine shrubs on cyanobacterial filaments as crystal aggregates on bacterial cells
Allochthonous grains	trapped by microbial mats
Pores	interstices, fenestrae and growth cavities, in microbial reefs

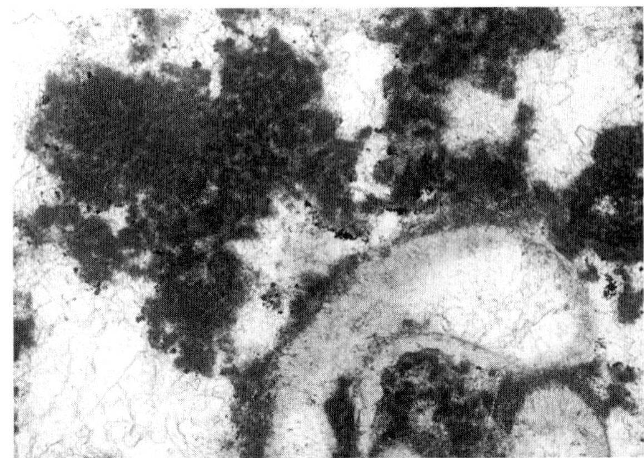

Fig. 5. Peloidal micrite veneers and clumps associated with a vermiform gastropod in skeletal stromatolite-thrombolite. Lower Carboniferous, Liddesdale, southern Scotland. Width of view, 2 mm.

represent biofilm calcification (see Kobluk & Risk, 1977; Perry, 1999; Webb *et al.*, 1999) .

(ii) Clotted fabrics. This microfabric, typically formed by an irregular sponge-like network of micrite, is widespread in stromatolites (and thrombolites, Pratt & James, 1989; but should not be confused with the *macroscopic* fabric of thrombolites). It has been referred to as spongiostrome (Gürich, 1906) and 'grumeleuse' (clotted; Cayeux, 1935, p. 271) fabric, and can grade to dense micrite (Figs 6 and 7). Vermiform fabric (Walter, 1972 – see Bertrand-Sarfati, 1976, p. 255; Pratt, 1982a) is similar. Diffusely clotted micrite appears to represent EPS calcification, although the precise origins and specific processes involved are usually not clear; it possibly also incorporates trapped micrite.

(iii) Calcified sheaths. Calcification of microbial external polysaccharide protective sheaths yields calcified fossils that are relatively readily recognizable. Many of them appear to be impregnated cyanobacterial sheaths, e.g. *Angusticellularia*, *Cayeuxia*, *Girvanella* (Fig. 3) and *Ortonella* (Fig. 7). However, the affinities of a host of other fossils, such as *Archaeolithoporella*, *Rothpletzella* and *Shamovella* (= *Tubiphytes*), which have at times been regarded as microbial, are more uncertain (Flügel, 1981; Riding, 1991a; Senowbari-Daryan & Flügel, 1993; Pratt, 1995).

Peloids

Granular micritic aggregates of uncertain origin (McKee & Gutschick, 1969) are a common constituent of modern tropical carbonate sediments (Illing, 1954, p. 24) and may exhibit a range of silt to sand sizes. Peloids that appear to be in-place precipitates are common in Holocene reefs (e.g. Lighty, 1985). These typically silt-size (20–60 µm) spherules of Mg-calcite have been regarded as cement fabrics (Macintyre, 1984, 1985), but also as calcified bacterial aggregates

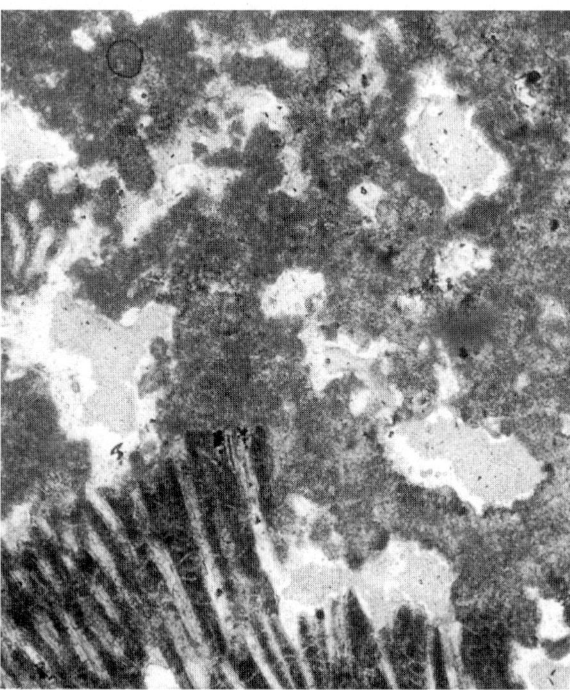

Fig. 6. Diffuse masses of clotted/spongiostrome micrite enclosing irregular uncemented pores and associated with calcified rivulariacean cyanobacteria (filaments to left and below) in a brackish-water bioherm. Pleistocene, Gulf of Corinth, Greece (see Richter *et al.*, 1979). Thin-section courtesy of Ernst Ott. Width of view, 1·2 mm.

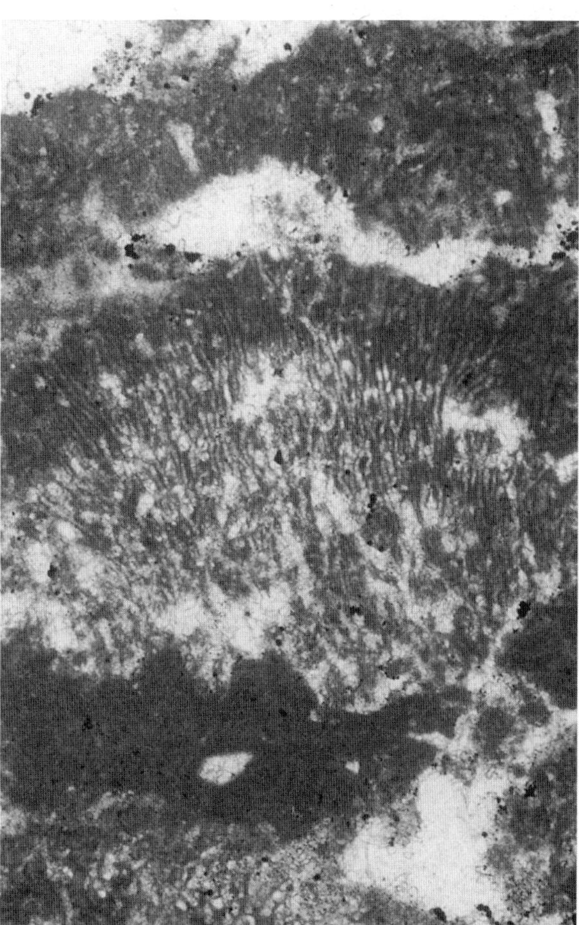

Fig. 7. Alternating laminae of dense micrite, clotted micrite, irregular fenestrae and the calcified cyanobacterium *Ortonella* (erect branching filaments in centre), in skeletal stromatolite–thrombolite. Lower Carboniferous, Liddesdale, southern Scotland. Width of view, 1·2 mm.

Table 2. Environmental distribution of microbial carbonates.

Environment	Type of deposit	Variety
Suspended		whitings in seas and lakes*
On sediment		
subaquatic	marine	domes, columns, reefs of stromatolite, thrombolite, etc.
		crusts in algal–metazoan reefs
		coated grains (oncoids, envelopes)*
	intertidal	stromatolites
	lacustrine–fluviatile	tufas and oncoids
subaerial		hot-spring travertine
		tufa mounds
		cave crusts
In sediment		
subaquatic		intergranular matrices, crusts, films, peloids
soil		calcretes, vadoids
'diagenetic'		precipitation and transformation

*Allochthonous deposits; others are mainly authochthonous.

rimmed by euhedral calcite crystals (Chafetz, 1986). Peloids that appear to be microbial precipitates are widespread and often very common in ancient reefs (e.g. Flügel & Steiger, 1981; Reid, 1987; Sun & Wright, 1989) and stromatolites (Fig. 5).

Microspar and spar

Fibrous, equant and dendritic spar precipitates are common as external crusts and rinds on organic tissue and grain and mineral surfaces in microbial carbonates. They are well seen in freshwater tufas (Irion & Müller, 1968), karst surfaces (Jones & Kahle, 1986; Jones, 1999), coating peloids (Chafetz, 1986) and contributing to travertine shrubs (Guo & Riding, 1994). The extent to which these crusts may be microbially induced is uncertain. They also form around cyanobacterial filaments, and on bacterial cells, where they can create distinctive crystal aggregates: spherulites, rosettes, crosses and sheaf- and dumbbell-shaped fascicles (see Krumbein *et al.*, 1977; Krumbein, 1979b; Buczynski & Chafetz, 1991; Guo & Riding, 1992a; Défarge *et al.*, 1996).

Pores and allochthonous grains

Diverse voids, ranging from tiny interstices to large growth cavities (Pratt, 1995), are associated with microbial carbonates. They include irregular fenestrae, such as microspar-filled spaces in clotted fabrics (Fig. 6), and larger tidal-flat vugs ('birdseyes'; Shinn, 1983a).

Trapped grains are important components of some microbial carbonates, such as agglutinated stromatolites. Agglutination is easy to recognize where the grains are silt- or sand-sized or larger, but trapped micrite can be difficult to distinguish from precipitated micrite in EPS. Micritized trapped grains can add to the fine-grained appearance of microbial carbonates (Hoffman, 1973, p. 183). Consequently, the relative importance of trapping and precipitation in micritic stromatolites can be unclear (Fairchild, 1991), although the clotted microfabrics resulting from biofilm calcification are more distinctive. Large grains in gravity-defying stromatolitic crusts, e.g. on overhangs (Riding *et al.*, 1991c), in otherwise micritic matrix demonstrate trapping, as does selective incorporation of grains that differ in size from those in adjacent interbeds (Riding *et al.*, 1991a, Fig. 3).

PRODUCTS

Introduction

Microbial carbonates can form extensive rock units, in reefs for example, in which they may be the principal component, or be relatively minor constituents of alga- or metazoan-dominated reefs and of allochthonous sediments. The details of these deposits that reflect organisms, processes and components are mainly microscopic, but more readily observed meso- and macroscale characters provide the main subdivisions. Microbial carbonates that form at the

Fig. 8. Detail of branching in the stromatolite *Tielingella*. Late Mesoproterozoic (~1100 Ma) Tieling Formation, near Jixian, northern China. Width of view, 15·5 cm.

aquatic sediment surface include (i) stromatolites, (ii) thrombolites, (iii) dendrolites and (iv) leiolites. These mainly form a variety of domes, columns, thick layers and crusts (Table 2). Tufas and travertines are subaquatic or subaerial, and are generally only partly microbial. Microbial coatings on allochthonous grains form oncoids or micrite envelopes. Other subaerial or subsurface deposits which may locally have a significant microbial component include cave crusts and calcretes. Microbial deposits that form as particles in suspension, but which accumulate on lake- and sea-floors, include whitings.

Major categories: domes, columns and crusts

Macrofabrics, usually readily distinguished in the field, are used to define major categories (Table 3) of microbial carbonate: stromatolite (laminated), thrombolite (clotted), dendrolite (dendritic) and leiolite (aphanitic). Process of origin is reflected in subcategories, such as agglutinated stromatolite and skeletal stromatolite.

Table 3. Principal categories of microbial carbonate deposits.

Major categories	Variety
Stromatolite	Skeletal
	Agglutinated
	Fine-grained
	Tufa
	Terrestrial
Thrombolite	Calcified microbial
	Coarse agglutinated
	Arborescent
	Tufa
	Postdepositional – bioturbation, enhancement, secondary creation
Dendrolite	
Leiolite	

Stromatolite

Definition

The term stromatolite was created by Kalkowsky (1908) as stromatolith (Greek *stromat*, to spread out, Latin *stroma*, bed covering; Greek *lithos*, stone) for lacustrine examples in the Lower Buntsandstein (lowest Triassic) in northern Germany (see Paul & Peryt, 1985). Some workers restrict the term to structures, such as domes and columns, with primary relief, but stromatolites may be originally flat deposits. Stromatolites are also not necessarily carbonate; there are siliceous (Walter *et al.*, 1972) and evaporite (Rouchy & Monty, 1981; Gerdes *et al.*, 1985; Renaut, 1993) examples (and see Hoffman, 1973, p. 346). Carbonate–siliciclastic domes occur (Martín *et al.*, 1993; Bertrand-Sarfati, 1994), and essentially flat siliciclastic examples are widespread on modern temperate tidal flats (e.g. Cameron *et al.*, 1985), where microbial mats may influence the formation and preservation of matground and other structures (Noffke, 1998; Pflüger, 1999). Many stromatolites, unlike dendrolites and thrombolites, lack calcified microbes (the main exception being skeletal stromatolites).

Despite being an old and widely used term, there is currently no generally accepted definition of stromatolite. Kalkowsky (1908) evidently regarded stromatolites as laminated, as the name indicates (Fig. 8), and microbial, but his long and intricate paper failed to specify a single clear definition. By the 1970s, research had revealed that: (i) not all microbial deposits are laminated, and (ii) it can be difficult to demonstrate biogenicity in ancient examples. Attempts to deal with these two complications resulted in new definitions that were as different as the problems that they attempted to resolve. In order to encompass nonlaminated deposits, one definition enlarged the term stromatolite into a broad 'genetic' term that encompassed all microbial deposits irrespective of macrofabric (Awramik & Margulis, 1974). In contrast, another definition avoided the problem of demonstrating biogenicity by being 'descriptive', and allowing stromatolite to apply to abiogenic as well as biogenic laminated deposits (Semikhatov *et al.*, 1979).

Although endorsed in the seminal 1970s publication on stromatolites by Walter (1976), the genetic definition did not discriminate between microbial deposits with markedly differing macrofabrics. As the term thrombolite (Aitken, 1967) for macroscopically clotted microbial carbonates gained acceptance, it became clear that a broad term, similar to that suggested by Awramik & Margulis (1974), was needed when referring to laminated stromatolites and clotted thrombolites as a whole, but that stromatolite would not do: it would be too confusing to try to use stromatolite simultaneously as both a specific term for laminated structures and as an umbrella term for microbial deposits in general. As a result, Burne & Moore (1987) took the Awramik & Margulis (1974) definition and applied it to a new, all encompassing, term: microbialite. This not only permitted the terms stromatolite (laminated) and thrombolite (clotted) to coexist, but subsequently fostered further nomenclatorial innovation to accommodate other macrofabrics: dendrolite for dendritic (Riding, 1991b) and leiolite for aphanitic (Braga *et al.*, 1995) forms.

However, the term stromatolite remained untouched by these clarifications. Whereas thrombolite, dendrolite and leiolite, as well as microbialite, each had single uncontested definitions, by the late 1970s the term stromatolite had three definitions, which variously viewed it as being essentially:

1 laminated and microbial (Kalkowsky, 1908);
2 just microbial (Awramik & Margulis, 1974);
3 just laminated (Semikhatov *et al.*, 1979).

Setting aside Awramik & Margulis's (1974) definition in favour of microbialite still leaves the choice between Kalkowsky's (1908) genetic definition and Semikhatov *et al.*'s (1979) descriptive definition unresolved. Semikhatov *et al.*'s (1979) definition of stromatolite has three deficiencies. (i) Its descriptive rationale conflicts with the genetic one whose usage is now well established for other microbial deposits. The terms dendrolite, leiolite and microbialite are all genetic because they are defined as microbial. (ii) It negates the essential microbial nature of stromatolites; no present-day deposits are termed stromatolites if they are not microbial. (iii) It would include laminated domes as diverse as mineral deposits, diagenetic concretions, travertine crusts and speleothem, as well as microbial deposits, within the term stromatolite. Use of dual definitions of stromatolite, descriptive for ancient and genetic for present-day examples (Ginsburg, 1991, p. 27), would not reduce these deficiencies, and could increase confusion.

Kalkowsky (1908) was convinced that stromatolites are organic, even though he could not prove it and was unsure which organisms were responsible (although he suspected simply organized plants). Consequently, his definition requires having to demonstrate biogenicity. This is not a problem in stromatolites that preserve the organisms responsible for their formation, such as skeletal stromatolites, but it can be difficult in other types. However, interpretation of microbial carbonate fabrics, including microfabrics, is improving and can provide the ability to achieve the level of confidence required to apply Kalkowsky's (1908) definition. There will, of course, be examples in which preservation will be inadequate, and in these cases uncertainty can be indicated by 'possible, probable' prefixes. This is no different from the difficulties involved in recognizing other fossils; the fact that we cannot always confidently recognize fossils in all geological situations does not prevent us usefully defining them. But it remains necessary to realize the difficulties and to apply biogenic criteria rigorously.

It could be, therefore, that not only the oldest but also the best definition of stromatolite is that of Kalkowsky (1908) which can simply be restated: '*a stromatolite is a laminated benthic microbial deposit*' (Riding, 1991b, 2000). This adds stromatolite to the suite of microbial deposits distinguished according to their differing macrofabrics, stromatolite, thrombolite, dendrolite and leiolite, with microbialite as the overarching term.

Varieties

Carbonate stromatolites vary considerably in origin and components, and also in quality of lamination. Episodic, even accretion promotes layering, including lamination (Table 4; Braga *et al.*, 1995), and may be related to seasonal growth, periodic sedimentation or both. Lamination (Gerdes *et al.*, 1991) is a key characteristic of stromatolites, and distinct and continuous lamination is typical of most Precambrian stromatolites (Walter, 1972, p. 11). It is disrupted by desiccation in intertidal to supratidal environments, and by bioturbation. Layering is thicker, and may be crude, where components are large and/or erect, as in skeletal stromatolites and dendrolites. In agglutinated stromatolites, crude lamination may be a direct result of coarse sediment supply (Walter, 1972, p. 6, fig. 4), but also suggests irregular and uneven accretion (Braga *et al.*, 1995).

Skeletal stromatolite. Stromatolites 'in which the organisms responsible for its formation are commonly preserved as calcified fossils' (Riding, 1977; also porostromate stromatolite, Monty, 1981). It is arguable whether the term skeletal is appropriate for microbial fossils (Golubic & Campbell, 1981), but it conveys the relative distinctiveness and rigidity of these calcified microbes, mainly cyanobacteria. The resulting frames also contain biofilms, trapped particles and early cements. Skeletal oncoids are widespread in Palaeozoic and Mesozoic marine environments, and in modern calcareous streams. Skeletal stromatolites appear less common, possibly because the erect filaments that usually form them disrupt lamination and thereby create dendrolitic fabrics.

Table 4. Major dome and column categories of microbial carbonates comparing macrofabrics and accretion patterns.

Term	Macrofabric	Accretion
Stromatolite	laminate	episodic
Thrombolite	clotted	irregular
Dendrolite	dendritic	irregular
Leiolite	aphanitic	irregular

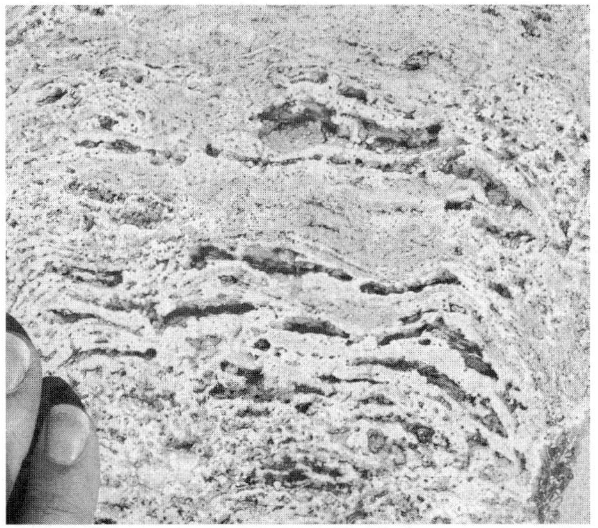

Fig. 9. Detail of vertical section of a Shark Bay microbial column from Hamelin Pool, Western Australia, showing coarse and irregular layering. Slab supplied by R. V. Burne.

Agglutinated stromatolite. 'Produced by the trapping/binding of particulate sediment' (Riding, 1991b). They may incorporate grains of any size from micrite to gravel, but are most readily recognized where the grains are silt-sized or coarser (Fig. 2). Increased coarseness appears to correlate with cruder layering, and coarse-grained microbial domes and columns are often at least partly thrombolitic, as in the famous Shark Bay (Fig. 9) and Lee Stocking Island examples. Ancient examples of coarse-grained domes are still remarkably rare (see Riding *et al.*, 1991b; Martín *et al.*, 1993; Braga *et al.*, 1995).

Fine-grained stromatolite. Interpretation of fine-grained stromatolites is a challenge. Are they agglutinated or precipitated (Fairchild, 1991), and which microbes were involved (Riding & Sharma, 1998)? These questions are compounded, especially in the Proterozoic, by the formation of a variety of distinctive morphotypes, such as digitate forms and large inverted cones, that lack direct modern analogues. Laminar intertidal stromatolites are also commonly at least partly fine grained. They occur in a variety of modern tropical environments, from the humid shores of Andros Island, where they pass laterally towards freshwater marsh with calcified cyanobacteria (Monty, 1967; Monty & Hardie, 1976), to evaporative sabkhas (Shinn, 1983b), and may be largely agglutinated. However, Neogene examples of fine-grained stromatolitic crusts and layers in coral and algal reefs could be largely precipitated (Riding *et al.*, 1991c; Montaggioni & Camoin, 1993; Reitner, 1993; Braga *et al.*, 1996), and offer both a reminder of the continuing significance of microbial carbonates in tropical reefs and a key to the past.

Tufa stromatolite. Freshwater fluviatile tufas locally exhibit oncoids (Fritsch, 1959) and thick laminated crusts formed by calcified cyanobacteria (Pentecost, 1978; Pedley, 1990) in which calcification appears largely to be by encrustation of the sheath (Merz-Preiβ & Riding, 1999). These tufa stromatolites are therefore mainly produced by precipitation of minerals on, rather than in, organic tissue (Riding, 1991b). This distinguishes them from skeletal stromatolites. However, both the specific site of precipitation and the quality of lamination differ according to environment and organisms (see 'Tufa mounds', below), and impregnation appears more common in lacustrine cyanobacteria. Tufa can include calcified algae such as *Cladophora*, *Gongrosira*, *Oocardium* and *Vaucheria* (see Wallner, 1934; Golubic, 1976) that coarsen the fabric. Locally, large tufa bioherms constructed by green algae (probably *Cladophora*) form in association with stromatolitic crusts, as in the Miocene Ries Crater Lake of southern Germany (Wolff & Füchtbauer, 1976; Arp, 1995). In lakes, tufa crusts with significant microbial (particularly cyanobacterial, diatom and green algal) components can form complex metre-scale domical mounds, for example those in Green Lake, New York (Eggleston & Dean, 1976). Cyanobacterial and associated tufa can record environmental and climatic information in stable carbon and oxygen isotope values (Andrews *et al.*, 1997).

Terrestrial stromatolite. Laminar calcretes formed by microbial activity have been termed lichen (Klappa, 1979), terrestrial (Wright, 1989) and subaerial stromatolites (Riding, 1991b). Although efforts have been made to discriminate between laminar calcrete and stromatolite (Read, 1976), and no doubt processes such as evaporation contribute significantly to $CaCO_3$ precipitation in calcretes, the local importance of microbes in calcrete formation (e.g. Krumbein, 1968) justifies the view that some calcretes may be regarded as microbial deposits (Wright, 1989; and see 'Microbial calcretes', below).

Thrombolite

Definition

Thrombolite (Greek: *thrombos*, clot; *lithos*, stone), defined on the basis of Cambro-Ordovician examples of the southern Canadian Rocky Mountains (Aitken, 1967; also see Aitken, 1966), has a macroscopically clotted fabric (not to be confused with clotted *microfabric*). Thrombolite clots, more-or-less discrete rounded to irregular patches that differ in colour and/or texture from intervening areas, create a blotchy, generally unlayered, mesoscale fabric. Clots range from diffuse to distinct, and from primary to secondary. Thus, clottedness, as a character, is broader and less readily defined than lamination. As a result, although the term thrombolite has become almost as well known as stromatolite, there is some confusion surrounding it, deriving from the variety of fabrics and origins it encompasses.

Varieties

Some thrombolites, like stromatolites, are primary products of microbial calcification or agglutination. They appear to be essentially subtidal and form columns, domes, layers and thick crusts (e.g. Pratt & James, 1982; Riding *et al.*, 1991b; Armella, 1994; Kennard, 1994), but are characteristically domical metre-scale doughnut-like masses that have been termed calyptrae (Luchinina, 1973; Rowland & Gangloff, 1988; kalyptrae). Thrombolites may prove to be locally important in the Proterozoic (Aitken & Narbonne, 1989), but currently are best known to become conspicuous near the Proterozoic–Cambrian boundary (e.g. Schmitt & Monninger, 1977), and were important throughout much of the Cambrian and Early Ordovician (e.g. Armella, 1994; Kennard, 1994). The view that thrombolites could be bioturbated stromatolites, and that Late Neoproterozoic – Early Palaeozoic thrombolite formation represented a response by stromatolites to metazoan disturbance (Walter & Heys, 1985), is at odds with the evident skeletal nature of many Early Palaeozoic thrombolites (see Kennard & James, 1986; Pratt & James, 1982; Walter, 1994), although some Neogene agglutinated thrombolites do appear bioturbated. Reports of post-Ordovician thrombolites are sporadic (see Kennard & James, 1986), although they were locally important in the Jurassic (Leinfelder *et al.*, 1993; Leinfelder & Schmid, 1999). Feldmann & McKenzie (1998) suggested that ancient thrombolites may represent eukaryotic structures. This could well apply to Late Miocene agglutinated thrombolites (Riding *et al.*, 1991b), but is unsupported for Early Palaeozoic examples dominated by *Epiphyton–Renalcis–Angusticellularia* group calcified microbes.

Calcified microbial thrombolites. Thrombolite fabric can be produced by microbial calcification. Examples in the Lower Palaeozoic (Kennard & James, 1986) typically display well-defined clots (Fig. 10) and were probably formed by cyanobacteria, as were dendrolites.

In oblique or transverse sections, dendrolite fabric appears thrombolitic. Nonetheless, many Early Palaeozoic thrombolites are not so evidently formed by calcified microbes (see Kennard, 1994), and it remains to be seen just how these examples, and also the 'type' thrombolites of Aitken (1967), some examples of which show diffuse poorly defined clotted fabrics, have formed. Present-day examples associated with sponge-tissue degradation (Reitner, 1993; Reitner *et al.*, 1995) should shed some light on the complex processes whereby bacterial calcification produces thrombolitic and other fabrics.

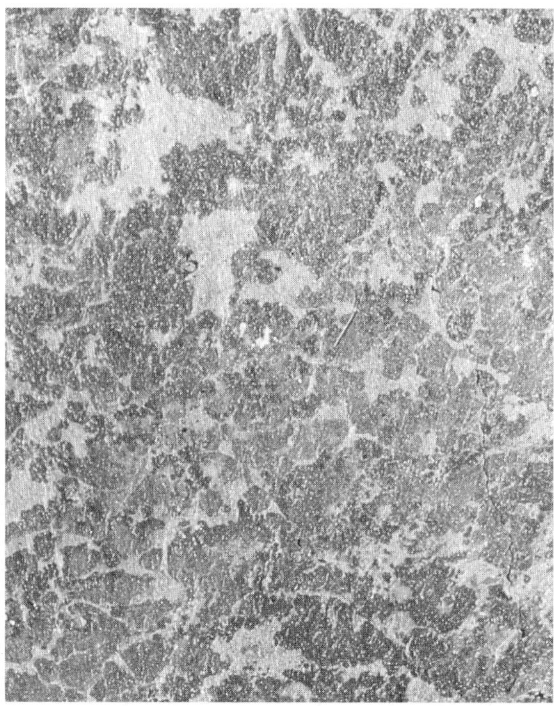

Fig. 10. Thrombolite formed by probable calcified cyanobacteria including *Tarthinia*. Lower Cambrian Lie de Vin Series, near Tiout, Morocco (see Latham & Riding, 1990). Width of view, 5 cm.

Coarse agglutinated thrombolites are best documented in the Late Miocene of SE Spain (Fig. 11; Martín *et al.*, 1993; Braga *et al.*, 1995; Feldmann & McKenzie, 1997) and show similarities to present-day Shark Bay and Lee Stocking Island domes (Riding *et al.*, 1991b). They incorporate sand- and even gravel-sized oolitic and bioclastic sediment, and commonly exhibit complex internal variation that includes stromatolitic, crudely layered blotchy, or relatively structureless fabrics. Blotchy agglutinated fabrics could be due to the coarseness of the sediment (Walter, 1972, p. 6), uneven accretion (Braga *et al.*, 1995) or other factors, including bioturbation. It has long been recognized that Shark Bay columns are often not well laminated (Fig. 9; Logan, 1961), although they are widely termed stromatolites. However, coarse agglutinated Neogene domes evidently differ from Early Palaeozoic thrombolites that are dominantly calcified microbial in structure, and do not usually display such relatively well-defined clots (but see Feldmann & McKenzie, 1998). Further work may recognize the blotchy appearance of Neogene examples as a distinct fabric type, but at present they can be regarded as a variety of thrombolite.

Arborescent 'thrombolites'. Palaeozoic thrombolites are commonly associated with decimetric dendritic fabrics (larger than dendrolite fabric which is typically centimetric) that have also been termed thrombolite (e.g. Schmitt & Mon-

Fig. 11. Coarse-grained, mainly oolitic, agglutinated composite leiolite–stromatolite–thrombolite domes. Messinian (Upper Miocene), Joyazo, Almería, Spain (see Riding *et al.*, 1991b). The dome in the left foreground is ~1 m wide.

Fig. 12. Dendrolite with dark centimetric bush-like fabrics formed by the probable calcified cyanobacteria *Gordonophyton* and *Tarthinia*. Mid-Cambrian Zhangxia Formation near Jinan, Shandong, China. Width of view = 14 cm.

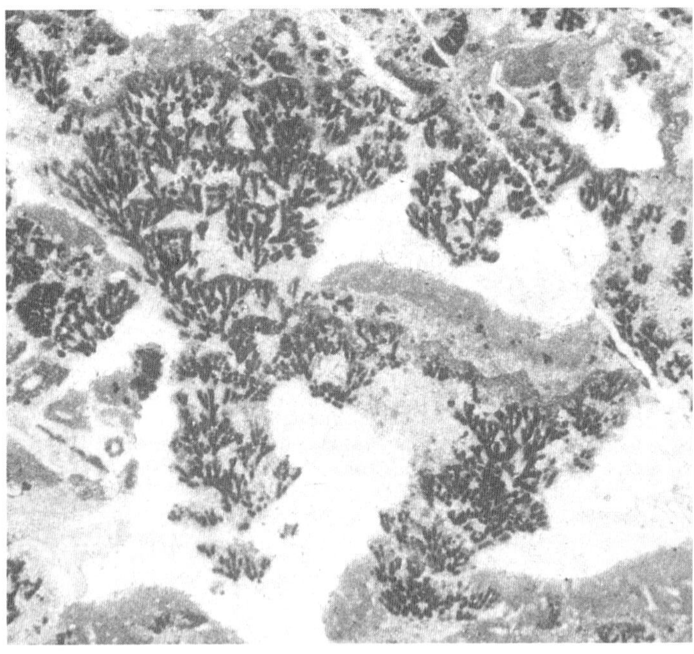

Fig. 13. Photomicrograph of dendrolite skeletal framework mainly formed by the probable calcified cyanobacteria *Gordonophyton*. Mid-Cambrian Zhangxia Formation near Jinan, Shandong, China. Plane polarized light. Width of view = 8·5 mm.

ninger, 1977; Pratt & James, 1982; Armella, 1994; type 1 of Kennard, 1994). Similar fabrics occur in the Jurassic (Leinfelder *et al.*, 1993). These show a transitional relationship with thrombolites, and also possibly with dendrolites.

Tufa thrombolites (new term). Just as intense $CaCO_3$ encrustation on organic surfaces generates distinctive tufa stromatolite fabrics, clotted fabrics can form in freshwater lakes and streams (Moore & Burne, 1994). At Kelly Lake, British Columbia, Ferris *et al.* (1997) suggested that thrombolitic fabrics reflect a greater degree of calcification than in stromatolites.

Post-depositional thrombolites. Clotted macrofabrics can be syndepositionally produced, diagenetically enhanced or diagenetically created. This adds a dimension to thrombolite studies that does not appear to apply to stromatolites. Some of the blotchy fabrics in Neogene agglutinated thrombolites resemble bioturbation patterns, and clots in skeletal thrombolites are commonly diagenetically enhanced by partial recrystallization and/or replacement, including selective dolomitization of matrix between clots (Fig. 10). Clotted fabrics in carbonates appear in some cases to develop secondarily, as in some Late Proterozoic 'thrombolites' (e.g. Mesoproterozoic Wumishan Fm, near Beijing, China).

Dendrolite

Definition

Dendrolite (Greek *dendron*, tree; *lithos*, stone; Riding, 1991b, p. 34) has a macroscopic centimetric bush-like fabric (Fig. 12), typically produced by calcified microbes. Dendrolites, like stromatolites and thrombolites, form large domes

Fig. 14. Tufa pinnacles formed by precipitation localized by calcium-rich spring-water entering an alkaline lake subaqueously. Fall in lake-level has exposed the pinnacles, which are each approximately 0·5–1 m diameter. Mono Lake, California.

and columns, and also extensive reefal masses that may be organized into large ovoid doughnut-like calyptrae. Unlike stromatolites and thrombolites, they are only known to form by microbial calcification, and not by agglutination of particles.

Varieties and distribution

Like thrombolites, dendrolites in hand-specimen consist of areas differing in colour and/or texture from intervening areas. But whereas in thrombolites these areas are rounded to irregular, in dendrolites they are bushy or tree-like, with vertically erect or radial orientation. Dendrolites lack lamination, but are commonly layered, reflecting irregularity in accretion. This results in coarse but well-defined layering in domes and columns. Regular accretion produces massive deposits as in some Early Cambrian reefs. The constituent dendroids range from small shrub-like forms to elongate radial fingers that grade towards arborescent forms that have been regarded as thrombolites (see 'Arborescent thrombolites', above). In oblique or transverse section, dendrolites resemble thrombolites.

Although 'dendriform' stromatolites (Aitken, 1989) and *Angusticellularia*-like calcified fossils (Turner *et al.*, 1993) are known from the Neoproterozoic, the earliest documented dendrolites remain Early Cambrian. They are best known from rocks of Cambrian to Early Ordovician and Late Devonian age, where they are built by calcified microbes belonging to one or more of the *Epiphyton*, *Renalcis* and *Angusticellularia* groups (Fig. 13; Riding, 1991b). In these time intervals, dendrolites were commonly major reef builders, either on their own or in association with sponges, such as archaeocyaths and *Archaeoscyphia* in the Cambrian – Early Ordovician and stromatoporoids in the Devonian (e.g. Mountjoy & Riding, 1981).

Leiolite

Definition

Leiolite (Greek *leios*, uniform or smooth, *lithos*, stone) has a relatively structureless, aphanitic, macrofabric lacking clear lamination, clots or dendritic fabrics. The term is founded on examples in the Late Miocene of SE Spain (Braga *et al.*, 1995, p. 347) where leiolites form large domes in association with stromatolites and thrombolites (Fig. 11). Leiolite fabric may reflect irregular accretion (Table 4) and/or compositional homogeneity, as in well-sorted oolite sand (Braga *et al.*, 1995) and in micritic reefal crusts, for example in Jurassic sponge reefs (Leinfelder *et al.*, 1993; Dupraz & Strasser, 1999; Leinfelder & Schmid, 1999).

The temptation to give microbial crusts and domes a well-known label, such as stromatolite or thrombolite, may have obscured the fact that many of these deposits lack the macrofabrics implied by such names. 'Stromatolitic' crusts are sometimes only weakly laminated (e.g. Riding *et al.*, 1991c, fig. 9b), and well-known modern domes and columns, such as at Shark Bay and Lee Stocking Island, include portions that appear leiolitic.

Other categories

Large homogeneous concentrations, exemplified by stromatolite, thrombolite and similar domes and masses, are not the only significant deposits of microbial carbonates, although they are perhaps the most noticeable. Many other microbial carbonates (Table 2) are commonly, although not always, intimately associated with carbonates of other origins. They can be locally important in spring, soil and cave deposits, as coated grains in a variety of depositional environments, in benthic sediments accumulated from water-column precipitation and in intergranular deposits, including cements. All deserve far fuller treatment than I can attempt here.

Spring microbial carbonates

Hot-spring travertine. As hot groundwater charged with calcium and bicarbonate in solution reaches the surface, pressure reduction, cooling and evaporation promote evasion of CO_2. Precipitation is localized around the vents, and lateral facies change is rapid as precipitated crusts give way to pool and marsh deposits (Guo & Riding, 1999). These travertines are exemplified by Mammoth Hot Springs (Wyoming), Rapolano Terme (Tuscany), and Pamukkale (western Turkey). Close to vents they appear mainly abiogenic, although microbial carbonates locally form close to springs (Guo & Riding, 1992b) and as crusts on $CaCO_3$ crystal layers (Guo & Riding, 1992a). The suggestion that small shrub-like precipitates in travertine pools and marshes are microbial (Chafetz & Folk, 1984) has some support from petrographic (Guo & Riding, 1994) and isotope studies (Guo *et al.*, 1996), but see also Pentecost (1990), Jones & Renaut (1995) and Chafetz & Guidry (1999).

Tufa mounds, pinnacles and towers. Spectacular carbonate towers form where calcium-rich spring water emerges subaqueously in alkaline lakes, triggering precipitation of $CaCO_3$ minerals. Well-known examples include Pyramid Lake, Nevada (Benson, 1994), Searles Basin, California (Scholl, 1960), Lake Van, Turkey (Kempe *et al.*, 1991), and Mono Lake, California (Fig. 14). Microbial activity has been suggested as a major factor in the precipitation of alkaline lake spring tufas (e.g. Scholl & Taft, 1964; Kempe *et al.*, 1991), but it seems likely that physicochemical processes based on mixing of spring and lake waters should be the principal factor localizing precipitation, although microbes are likely to have a significant role in determining microfabric. However, early diagenetic changes, such as ikaiite to calcite transformation, can seasonally destroy primary fabrics in cold lakes (Shearman *et al.*, 1989).

Microbial calcretes

In terrestrial stromatolites, cyanobacteria, fungi and lichens can be important influences on precipitation, mineralogy and fabrics, including the formation of micritic and microscopic dendritic, needle and spherulitic fabrics (Jones & Kahle, 1985, 1986; Jones & Wilson, 1986; Wright, 1986; Jones & Pemberton, 1987; Wright *et al.*, 1988; Verrecchia & Verrecchia, 1994; Verrecchia *et al.*, 1995). Larger vadoid nodules also form and can be confused with subaqueous oncoids (see Peryt, 1983a).

Microbial cave surface deposits

Cave crusts can support microbial populations that variously influence precipitation, mineralogy, solution, alteration and fabrics of carbonate deposits (Jones, 1995, 1999). An example is moonmilk (Mondmilch), a soft paste-like or powdery deposit in which microbes occur (Thrailkill, 1976; Gradzinski *et al.*, 1997) and from which calcifying bacteria have been isolated (Mason-Williams, 1959).

Microbially coated grains

Oncoids (Onkoid, from Greek *onkos*, nodule, Heim, 1916, p. 566) are unattached spherical stromatolites (oncolites, originally Pia, 1927, p. 37) that were widespread in ancient marine environments until at least the Jurassic or Cretaceous (Peryt, 1983b). Modern marine examples are scarce, but heavily calcified cyanobacterial oncoids are common in unpolluted calcareous lakes and streams. Some micrite envelopes on grains (Bathurst, 1966; creating cortoids, Flügel, 1982) may be microbial coatings.

Microbial intergranular and 'diagenetic' deposits

The influence and importance of microbes in a wide range of early diagenetic precipitation and transformation involving carbonate minerals, including cementation, concretion formation, dolomite formation, recrystallization and replacement, has long been realized. Many details remain to be elucidated (see, e.g. Berner, 1971a,b; Irwin *et al.*, 1977; Jones, 1985; Compton, 1988; Middleburg *et al.*, 1990; Vasconcelos *et al.*, 1995; De Craen *et al.*, 1999; Wright, 1999; Machel & Foght, 2000)

Microbial whiting deposits

These micrite to silt-grade precipitates are best known, as benthic deposits, from lacustrine varve-like layers, where organic-, silica- (diatom frustules) and carbonate-laminae alternate, reflecting seasonal variations in surface water productivity and precipitation (e.g. Kelts & Hsu, 1978). The carbonate laminae incorporate calcified algae (*Phacotus*) but the dominant fraction is made up of calcite rhombs (Fig. 4) and other aggregates, mainly less than ~20 μm in size, that have often been suggested to have a microbial origin (see 'Processes, Whiting precipitation').

TIME TRENDS

It is to be expected that over 3500 Myr microbial carbonates will have changed, as microbes and other organisms evolved and Earth-surface environments altered (Fig. 15). The classic view is that microbial carbonates, largely in the form of stromatolites, were widespread and abundant during much of the Precambrian, but that Phanerozoic competition from metazoans and other eukaryotes progressively relegated them to environmentally protected refuges, such as intertidal flats in general, and hypersaline Shark Bay in particular. How true is this, what changes can we perceive in the character, abundance and distribution of microbial carbonates in space and time, and to what extent can controlling factors be disentangled?

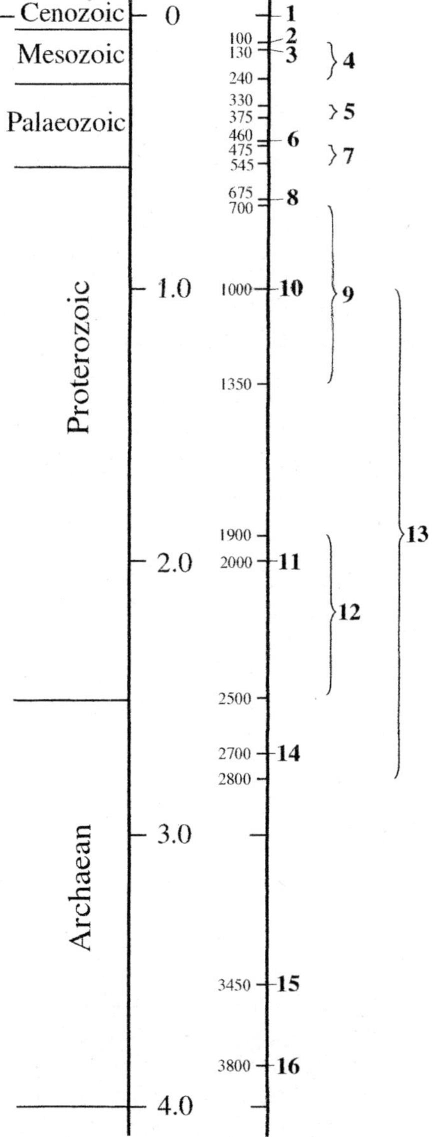

Fig. 15. Key developments in the history of microbial carbonates. Cyanobacterial calcification episodes (CCEs): Riding (1994b). Time-scale: Harland *et al.* (1990), except for base of Cambrian (Bowring *et al.*, 1993). **1**: 0–6 Ma, Late Neogene micritic reef crusts and coarse domes-columns. **2**: 100 Ma, Late Mesozoic decline of marine microbial carbonates. **3**: 130 Ma, first diatoms (Tappan, 1980). **4**: 100–240 Ma, Mid-Triassic to Early Cretaceous CCE. **5**: 330–375 Ma, Late Devonian to Early Carboniferous CCE. **6**: 460 Ma, stromatolite decline (Fischer, 1965). **7**: 475–545 Ma, Cambrian – Early Ordovician CCE, abundant thrombolite–dendrolite. **8**: 675 Ma, stromatolite decline (Awramik, 1971). **9**: 700–1350 Ma, Late Proterozoic CCE. **10**: 1000 Ma, stromatolite decline (Walter & Heys, 1985). **11**: 2000 Ma, stromatolite decline (Grotzinger, 1990). **12**: 1900–2500 Ma, Early Proterozoic CCE. **13**: 1000–2800 Ma, Late Archaean – Proterozoic stromatolite acme. **14**: 2700 Ma, definite cyanobacteria, probable algae (Brocks *et al.*, 1999). **15**: 3450 Ma, first stromatolites (Lowe, 1980), probable cyanobacteria (Awramik *et al.*, 1983). **16**: 3800 Ma, isotope evidence of microbes (Schidlowski, 1988).

Archaean–Proterozoic

Origination

Carbon-isotope values in the Isua Group of southern Greenland are evidence for the existence of microbes at 3800 Ma (Schidlowski, 1988), but currently the oldest tangible fossils are bacteria-like filaments comparable in size to cyanobacteria, in cherts in the ~3450-Ma Towers Formation of the Warrawoona Group of Western Australia (Awramik *et al.*, 1983). Stromatolites are present in the same unit (Lowe, 1980); thus, neatly, the oldest microbes and oldest microbial carbonates co-occur. However, doubts have been raised about the biogenicity not only of these Warrawoona stromatolites but of all stromatolite-like structures older than 3200 Ma (Lowe, 1994). The balance of opinion continues to favour their being microbial (Buick *et al.*, 1995; Hofmann *et al.*, 1999; Hofmann, 2000), but this debate is a reminder of the need for objective biogenicity criteria for ancient stromatolites (see Buick *et al.*, 1981; Walter, 1994, pp. 272–274).

Reports of Archaean stromatolites increased from 11 in 1980 (Walter, 1983) to 32 in the latest compilation (Hofmann, 2000). Nonetheless, their significance continues to outweigh their number, because they record all that is currently known about microbial carbonates during a critical 1000 Myr of early Earth history. The oldest examples are domes, cones and layers in chertified limestones and dolostones. Their typical sedimentary settings, ephemeral often evaporative basins on small cratons subjected to prolonged volcanism, might be a reason for their scarcity (Walter, 1994). Similarly, marked increases in stromatolite abundance, size and diversity at about 2800 Ma (Walter, 1994; Hofmann, 2000) could reflect increased craton size and more stable conditions.

Fig. 16. Microdigitate stromatolites less than 1 cm in height in partly silicified dolomite. Mid-Mesoproterozoic (~1300 Ma) lowermost Wumishan Formation, near Jixian, northern China. Width of view = 13 cm.

Acme

Twenty or more stromatolite morphotypes have been recorded from the late Archaean (Hofmann, 2000). The scale and diversity of this development, for example in the Cambellrand part of the Transvaal Supergroup of South Africa (Beukes, 1987), sets the scene for the Early Proterozoic. The immense period from 2800 Ma to the end of the Mesoproterozoic (1000 Ma) was the Golden Age of stromatolites. Diverse morphotypes, ranging from stratiform and microdigitate stromatolites (Fig. 16) on tidal flats, through shallow subtidal domes and branching columns (Fig. 8), to deep-water cones, participated extensively in the creation of carbonate platforms and the first extensive reefs (Hoffman, 1974; Grotzinger, 1989; Fig. 17). Sea-floor carbonate precipitates complicate interpretation of Archaean and Early Proterozoic stromatolites (Grotzinger, 1994; Grotzinger & Knoll, 1999; Sumner, 2000). An example is the microdigitate forms with distinctive radial fibrous microfabric that first became abundant at about 2700 Ma (Hofmann, 2000). They have been given Linnaean-style names, such as *Asperia* and *Pseudogymnosolen* (Liang *et al.*, 1985), but have also been termed tidal-flat tufas (Grotzinger & Read, 1983), and there has been debate about their biogenicity (in favour: Grey & Thorne, 1985; against: Hofmann & Jackson, 1987).

Fig. 17. Detail of a bioherm composed of *Baicalia*-type stromatolites. Bhander Limestone, possibly ~1000 Ma, near Satna, Madhya Pradesh, India.

Proterozoic stromatolite decline

It could be argued that it is less remarkable that stromatolites eventually declined, than that they prospered for so long. Nonetheless, probably the most evident and remarked upon aspect of microbial carbonate history is the decline of stromatolites from their 2800–1000 Ma acme. Treatment of this topic resembles that of the Decline and Fall of the Roman Empire: was it inevitable, was it quick, when did it start, were the causes simple or complex? Discussion of possible reasons clearly necessitates establishment of the essential facts; but this is easier said than done. Fischer (1965, pp. 1208–1209) and Cloud & Semikhatov (1969) noted stromatolite decline, but they recognized it from the mid-Ordovician onwards. This raises the question whether Proterozoic decline simply continued into the Early Palaeozoic (there was certainly dramatic decline from the Late Cambrian to Late Ordovician), or whether there was significant Early Palaeozoic recovery, not only of microbial carbonates in general, owing to the spectacular rise of dendrolites and thrombolites, but of stromatolites too. There is some evidence to support this latter view.

Proterozoic stromatolite decline was first clearly recognized from examination of stromatolite taxonomic data. Precambrian stromatolites provided rare potential for the subdivision and correlation of sedimentary sequences otherwise lacking in macrofossils. Their abundance and, in particular, their variety of shape fostered taxonomic studies with a practical aim: Proterozoic biostratigraphy. This approach, pioneered on sequences in the northern Urals and Siberia (see Cloud & Semikhatov, 1969; Semikhatov, 1980), was applied widely (see Bertrand-Sarfati & Walter, 1981) but raised many questions. Are stromatolite taxa consistently defined? Does stromatolite shape reflect organisms (Fenton & Fenton, 1937, p. 1941), environment (Logan *et al.*, 1964), abiogenic accretion patterns (Grotzinger & Rothman, 1996; Grotzinger & Knoll, 1999) or combinations of these (Serebryakov, 1976)? Which is more significant: microfabric, macrofabric or external shape? Does the stratigraphy work and, if so, is it because the microbes evolved or because the environments changed with time, or both? It has been suggested that external morphology mainly reflects environmental controls whilst microfabric reflects organic influences (see Semikhatov & Raaben, 1999), but the details of these questions have still largely to be resolved.

Nonetheless, stromatolite taxonomy proliferated numerous quasi-Linnaean form-genera and -species which, if the uncertainties are overlooked, represent an impressive data set. Plots of stromatolite taxa revealed a dramatic latest Proterozoic decline, commencing at ~675 Ma, that coincided with the current perceptions of metazoan evolutionary diversification (Awramik, 1971). Subsequent re-examination of these data suggested that decline commenced earlier, near 1000 Ma, although it remained sharpest later in the Neoproterozoic (Walter & Heys, 1985). However, qualitative abundance data suggest a much earlier commencement of decline near 2000 Ma (Grotzinger, 1990, fig. 6; Grotzinger & Knoll, 1999, p. 345, fig. 11). Thus, the bare bones of these various assessments suggest that stromatolite decline commenced either at 2000, 1000 or 675 Ma. Despite these disparities, and the continuing uncertainties about the significance of the taxonomic data, there is widespread agreement that stromatolite abundance and morphological diversity was much reduced after ~1000 Ma (Meso-Neoproterozoic boundary), and that microdigitate and coniform varieties in particular became far less common (Walter *et al.*, 1992, p. 259).

Phanerozoic

Despite possible Neoproterozoic (Aitken & Narbonne, 1989) and Palaeoproterozoic (Grotzinger & Hoffman, 1983) antecedents, thrombolites appeared dramatically near the base of the Cambrian, prompting initial suggestions that they might represent burrowed stromatolites, but more closely reflecting the development of abundant calcified microbes, particularly the *Epiphyton*, *Angusticellularia* (= *Angulocellularia*) and *Renalcis* groups (Riding, 1991b). During the Cambrian and Early Ordovician, thrombolites and dendrolites participated extensively in reef formation, either in association with metazoans such as archaeocyaths or alone (Copper, 1974). Stromatolites also remained important, and may have been relatively more abundant during the Late Cambrian (Zhuravlev, 1996), but they contrast with typical Proterozoic stromatolites: the distinctive coniform and microdigitate forms had disappeared, branched stromatolites were scarce and columnar forms often show thrombolitic interiors (Kennard, 1994, fig. 3). The Early Palaeozoic thus marked both a recovery for microbial carbonates and an alteration in their character.

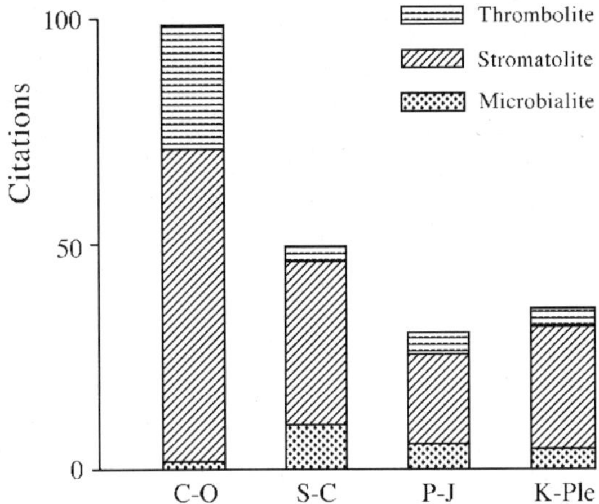

Fig. 18. Numbers of GEOREF citations (August 1999) of 'microbialite', 'stromatolite' and 'thrombolite' for four Phanerozoic intervals: Cambro-Ordovician 106 Myr, Silurian–Carboniferous 149 Myr, Permian–Jurassic 144 Myr, Cretaceous–Pleistocene 146 Myr (Harland *et al.*, 1990). These rough data suggest rapid initial decline followed by relatively minor changes in abundance of microbial carbonates during the remainder of the Phanerozoic.

The mid-Ordovician appearance of bryozoans, tabulate corals, stromatoporoid sponges and other skeletal metazoans was a major change for reef-building (Fagerstrom, 1987, p. 343). However, stromatolites remained both important and conspicuous in reefs, as in those of the Chazy of Quebec and Vermont (Pitcher, 1964), and detailed studies of younger reefs throughout much of the remainder of the Palaeozoic and Mesozoic usually reveal the continuing importance of microbial carbonates up until the present day, as perceptively noted by Pratt (1982a,b; see also Fagerstrom, 1987; Geldsetzer *et al.*, 1988; Monty *et al.*, 1995; Webb, 1996). This can be briefly demonstrated from just a few examples, including older literature: Silurian of England (Scoffin, 1971) and Alaska (Soja, 1994); Devonian of Western Australia (Wray, 1967); Carboniferous of Japan (Ota, 1969); Permian Capitan reef complex (Babcock, 1977); Triassic of the Dolomites (Fois & Gaetani, 1984); Jurassic of southern Germany (Gwinner, 1976); Cretaceous of Arizona (Scott, 1981). However, microbial carbonates appear to have been scarcer in the Palaeogene (see Webb, 1996, pp. 955–956). Miocene revival is largely focused on the Mediterranean area, which could represent unusual conditions. Nonetheless, sites in SE Spain present instructive examples of coarse-grained stromatolites, thrombolites, leiolites (Fig. 11) and also micritic reefal crusts (Riding *et al.*, 1991b,c; Martín *et al.*, 1993; Braga *et al.*, 1995, 1996).

It is tempting therefore to suggest that the stromatolite decline from the mid-Ordovician onwards, noted by Fischer (1965), actually marks the time when microbial carbonates became subsumed within (or themselves subsumed!) algal–metazoan reefs. Plots of references to microbial carbonates, albeit based on crude data, support the view that, following abrupt decline in the Early Palaeozoic, microbial carbonates exhibit no marked changes in abundance through the remainder of the Phanerozoic (Fig. 18). Nonetheless, a notable feature within this broad pattern is temporal variation in calcified microbe abundance (see 'Cyanobacterial calcification events'). Modern marine carbonate environments continue to host a diversity of microbial deposits (Browne *et al.*, 2000). These include stromatolite–thrombolite reefal crusts and matrices (Montaggioni & Camoin, 1993; Reitner, 1993; Webb *et al.*, 1998). Coarse-grained domes, such as at Shark Bay and Lee Stocking Island, may reflect changes in mat communities (see 'Microbial evolution', and 'Modern analogues' below).

Cyanobacterial calcification events

Supersaturation with respect to carbonate minerals varies environmentally in natural waters and directly affects microbial carbonate sedimentation, as can be seen by comparing tufa stromatolite formation in freshwater, at one extreme, with the absence of shallow-marine domical stromatolites from cold marine waters at the other. It has been suggested that variation in abundance of marine calcified cyanobacteria through geological time could reflect temporal variation in the saturation state of seawater (Riding, 1991a) and that cyanobacterial calcification events (CCEs; Riding, 1992) also broadly coincide with periods of unusual abundance of ooids and marine cements. These environmentally controlled calcification events (Riding, 1993) are especially marked in the Cambrian to Early Ordovician, Late Devonian and Permian–Triassic. If CCEs represent periods of elevated carbonate saturation, they may correspond with one or more of the following: high global temperature (which enhances precipitation rate); low sea-level and low skeletal abundance (which increase availability of calcium and bicarbonate); and development of alkalinity pumps from stratified basins

Fig. 19. Large molar tooth-shaped microbial column, partly buried by ooid sand dunes at a depth of 10 m in a channel between Lee Stocking Island and Norman's Pond Cay, Bahamas. Strong tidal currents supply sediment, deter competitors and may facilitate early lithification. Column top and sides are colonized by a diverse mat community (see Fig. 2).

(see Riding, 1993). Phanerozoic CCEs have been suggested to reflect saturation states >0·8 (Merz-Preiß & Riding, 1999; and see Kempe & Kazmierczak, 1994). Instead of simple unidirectional decline, the temporal distribution of microbial carbonates appears patterned in response to fluctuations in seawater chemistry.

DISCUSSION

The Precambrian enigma

Episodic abundances shown by marine calcified cyanobacteria during the Phanerozoic (Fig. 15) are only weakly matched in the Proterozoic (Riding, 1994b), in which calcified cyanobacteria are unexpectedly scarce. Although a variety of microbes could have been involved in building Precambrian stromatolites, it has been widely assumed that cyanobacteria were the principal ones. With locally intense synsedimentary carbonate precipitation (Grotzinger, 1989), Proterozoic carbonates would be expected to incorporate abundant calcified cyanobacterial microfossils. Their actual scarcity has been termed the 'Precambrian enigma' (Riding, 1994b). A possible explanation is that cyanobacteria were less important than has been generally assumed in stromatolite formation in the early mid-Proterozoic (Riding, 1994b, p. 432; Riding & Sharma, 1998). Reduction in calcification potential (Grotzinger, 1990), and environmental impact such as increasing oxygen levels on noncyanobacterial bacteria, together may have contributed to Proterozoic stromatolite decline (Riding, 1997). However, additional factors, such as nucleation problems (Knoll *et al.*, 1993) or phosphate levels (Merz-Preiß, 2000), could well have been important.

Competition

Grazing, or competition for habitats, by eukaryotes has long been an attractive explanation for the perceived Precambrian abundance and relative Phanerozoic scarcity of stromatolites. One of the possible causes advanced by Fischer (1965, pp. 1208–1209) to account for stromatolite decline from the mid-Ordovician onwards was competition with algae, but although Monty (1973) supported algal competition as a factor, attention dramatically switched to invertebrates following Garrett's (1970) suggestion that gastropod grazing was responsible for the apparent restriction of well-developed microbial mats to intertidal environments on the Bahama Banks. This view of metazoan competition was greatly strengthened when Awramik (1971) related the sharp late Proterozoic reduction in stromatolite diversity to metazoan appearance and rapid diversification at, and just prior to, the Neoproterozoic–Cambrian boundary. The realization that microbial carbonates continue in importance during the Phanerozoic (Pratt, 1982b) has damaged this hypothesis less than might have been expected, possibly because Phanerozoic reefal microbial carbonates usually appear quite different from Precambrian stromatolites, and also because the occurrence of the Shark Bay columns in an environment that is at least seasonally hypersaline is an ever-present reminder of the concept of stromatolite survival in stressed refugia.

Competition is difficult to test. It has even been suggested (Vologdin, 1962; Monty, 1973, p. 603) that microbial competition with metazoans delayed metazoan radiation, and therefore that microbial decline in the Neoproterozoic may have favoured metazoan emergence. In the same vein, in Early Cambrian reefs where dendrolites are often volumetrically dominant, it has been suggested that archaeocyath sponges were outcompeted by calcified microbes (see Riding, 1997).

Refugia

A concept direcfly derived from that of metazoan competition, is that when and where metazoans are scarce, stromatolites will become more abundant. It is consistent with the relative abundance of marine stromatolites on desiccated tidal flats, and with the presence of *Cryptozoon*-like columns in Shark Bay where hypersalinity seasonally reduces potential grazers (Logan, 1961, p. 520). It is not necessarily weakened by the existence of even larger columns in normal-salinity waters at Lee Stocking Island (Dill *et al.*, 1986) because these can be regarded as current-stressed (Fig. 19). The concept has also been invoked to account for the unusual abundance of large microbial carbonate domes in the Western Mediterranean about the time of the Late Miocene Messinian Salinity Crisis (Martín & Braga, 1994). It can similarly be applied to intervals during the Phanerozoic when metazoans were temporarily reduced by mass extinction. This view of stromatolites as 'disaster forms' during the first 4–5 Myr of the Early Triassic, immediately following the end-Permian event, was suggested by Schubert & Bottjer (1992; but see Soja, 1994). However, it could also be that some of the conditions connected to extinction events, such as raised temperatures and/or the saturation state of seawater with respect to $CaCO_3$ minerals, actually favour microbial carbonate formation irrespective of competitive interactions (Riding, 1997).

Importance of lithification

A discrete hypothesis that can stand in opposition to that of eukaryote competition to account for the distribution of microbial carbonates in time and space is the microbial calcification model, emphasizing both the importance of calcification in the formation of microbial carbonates and the environmental factors that significantly control it. Recognition of the importance of lithification for microbial carbonates is not new. Walcott (1914), Logan (1961, p. 520), Fischer (1965), Monty (1973, p. 614; 1977, pp. 23–24), Serebryakov & Semikhatov (1974) and Gebelein (1976, p. 505), and more recently Ginsburg (1991, p. 32) and Reid & Browne (1991), have all in various ways stressed that microbial carbonate formation depends on early lithification as well as on processes such as grain trapping. Without calcification, microbial carbonates can neither form nor be maintained and preserved.

The microbial calcification model proposes that microbial calcification is essentially micritic, metabolically induced, but environmentally dependent, calcification of extracellular polymeric substances (EPS), as biofilm or sheath mucilage. This process produces both clotted (spongiostrome) fabrics and calcified (porostromate) fossils, accounts for most microbial carbonates and is postulated to be the principal factor, after microbial growth, controlling their distribution and abundance in time and space. This has a number of implications:

1 Agglutinated micritic stromatolites are geologically scarce, and are perhaps most widely seen in Cenozoic intertidal environments.

2 The majority of microbial micrite is neither particulate grains nor external cement, but represents initial internal, micritic calcification of organic tissue (biofilm and sheath) that is promoted by metabolic activity and environmentally facilitated.

3 Micritic impregnation results in the formation of both spongiostrome and porostromate fabrics.

4 The scarcity of recognizable calcified cyanobacterial sheath in clotted microbial micrites suggests that spongiostrome fabrics, including those in Precambrian stromatolites, are formed by calcification of EPS biofilms created mainly by noncyanobacterial bacteria.

5 Steep-sided micritic stromatolites in the Precambrian indicate early lithification, interpreted as biocalcification. If calcified cyanobacteria are scarce, it suggests that other bacteria were responsible for these deposits.

6 The history of microbial carbonates mainly reflects calcification potential, both biological and environmental.

7 Calcification outweighs metazoan competition in determining the abundance and distribution of microbial carbonates.

8 Trapping and binding (agglutination) of particulate sediment is a relatively minor process in the history of microbial carbonate formation.

If these inferences are substantially correct, then the temporal, and spatial, distribution of microbial carbonates should correspond to enhanced environmentally facilitated precipitation, such as ooids and early cements.

Microbial evolution

The significance of temporal changes in mat communities (Awramik *et al.*, 1976) for microbial carbonate history may have been underestimated,

Fig. 20. Shark Bay columns at Carbla Point, Hamelin Pool, Western Australia, stranded in the intertidal zone due to Holocene relative sea-level fall. The main columns developed subtidally but have acquired an intertidal caps (see Fig. 21). Columns are ~25 cm high. Photograph courtesy of S. M. Awramik.

probably because clear evidence of the impact of microbial evolution on carbonate deposition remains limited. The oldest microbial filaments at ~3450 Ma (Awramik *et al.*, 1983) are cyanobacterium-like. Oscillatoriacean-like filaments are present in the 2750-Ma Fortescue Group, also of Western Australia (Schopf & Walter, 1982). Biomarkers of similar age (2600–2700 Ma) confirm the presence of not only cyanobacteria but also eukaryotes, presumably algae, by this time (Brocks *et al.*, 1999). The earliest reported body-fossils of algae are 2100 Ma (Han & Runnegar, 1992). Although it seems reasonable to expect that algae should have been involved in Proterozoic stromatolites (Walter, 1972, pp. 90–91), there is still little information concerning occurrences of either algae or bacteria other than cyanobacteria in Proterozoic microbial carbonate communities (e.g. Walter *et al.*, 1992, p. 255; but see Riding & Sharma, 1998). At the same time, it is noteworthy that evidence for cyanobacteria is sparse in subtidal Proterozoic stromatolites (Grotzinger & Knoll, 1999, p. 349; but see Bartley, 1996). It has been assumed that stromatolite morphotypes reflect particular microbial communities (Fenton & Fenton, 1937, p. 1941) and suggested that microbial evolution is reflected by Proterozoic stromatolite fabrics and possibly morphotypes (Semikhatov *et al.*, 1979), although again there seems little firm evidence to support this (but see Awramik & Semikhatov, 1979; Knoll & Semikhatov, 1998). Nonetheless, the diversity of Proterozoic stromatolite shape and microstructure is striking and contrasts with the Early Palaeozoic.

Early Palaeozoic development of dendrolites and thrombolites may at least partly represent cyanobacterial evolution (Riding, 1994b). It has been suspected that microbial carbonates have progressively incorporated bacteria, cyanobacteria and algae (Riding, 1991b; Walter, 1994; Fig. 8), yet there is little specific evidence (but see Feldmann & McKenzie, 1997) of significant algal involvement in microbial dome formation until late in geological history. Diatoms became abundant during the Cretaceous. These relative new-

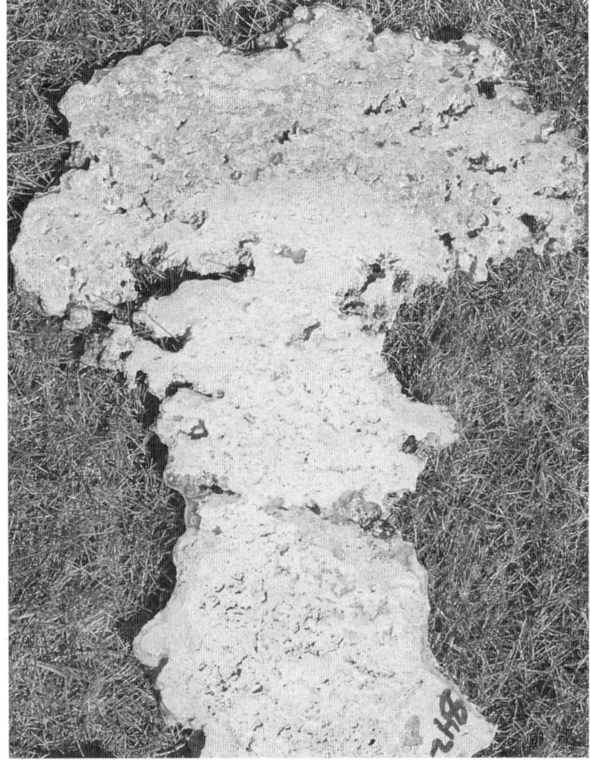

Fig. 21. Vertical section of a Shark Bay microbial column from Hamelin Pool, showing the lower narrow subtidal part and darker expanded intertidal cap. Cap is 25 cm wide. Slab courtesy of R. V. Burne.

comers may have transformed the trapping ability of some mat communities (Awramik & Riding, 1988). Coarse-grained agglutinated domes and columns are not restricted to the Cenozoic (Bertrand-Sarfati, 1994); but the best-known examples are all Neogene. These include carbonate and also carbonate–siliciclastic stromatolites, thrombolites and leiolites in the Late Miocene of SE Spain (Riding *et al.*, 1991b; Martín *et al.*, 1993) as well as Shark Bay and Lee Stocking Island examples. This relatively recent development of coarse-grained domes contrasts with the concept of long-term decline of microbial carbonates.

Modern analogues

Major developments, such as Walcott's (1914) recognition of the importance of cyanobacteria and lithification, Black's (1933) emphasis on sediment trapping, and Logan's (1961, p. 520) inference of the possible importance of metazoan predation, were all derived from examination of modern analogues. However, the diversity and long history of microbial carbonates permits modern analogues to distract as well as inform. After all, stromatolite research has been viewed as a succession of dogmas (Hoffman, 1973; Monty, 1977) largely arising from uncritical application of modern analogues.

Shark Bay and Lee Stocking Island

Modern marine microbial carbonates are well known in tidal flat settings, and are becoming increasingly evident in coral–algal reefs, but without doubt the most famous examples are restricted to seasonally hypersaline embayments of the Indian Ocean in Western Australia (Figs 20 and 21). It was tempting to regard these impressive modern microbial carbonate columns, at Hamelin Pool in Shark Bay, as analogues for Proterozoic examples (Logan, 1961), but it was soon recognized that this view is too simple (Hoffman, 1973, p. 188; Playford & Cockbain, 1976, pp. 408–409). Shark Bay columns contain much trapped coarse sediment, are often only crudely layered, and in places are thrombolitic or leiolitic. In contrast, many Proterozoic microbial carbonates are fine-grained and well-laminated stromatolites. A possible explanation for the coarse-grained nature of Shark Bay (and also Lee Stocking Island) subtidal columns is the floral composition of their mat communities. In addition to cyanobacteria, algae, and in particular diatoms and chlorophytes, are conspicuous mat-components in both Shark Bay and Lee Stocking Island (Fig. 2) columnar stromatolites (Awramik & Riding, 1988; Riding *et al.*, 1991a). These algae are generally larger than cyanobacteria and may be responsible for trapping the coarse sediment that is typical of these columns and that distinguishes them from most pre-Cenozoic stromatolites, although it must be remembered that this does not exclude the continuing importance of cyanobacteria in dome formation (Golubic & Browne, 1996).

Therefore, rather than being precise analogues for the Precambrian, these well-known modern examples of peritidal microbial columns instead possibly represent a relatively modern development in which algae have combined with prokaryotes to produce complex, faster growing and more effective sediment-accreting communities (Awramik & Riding, 1988). If this is correct, the question arises whether and where modern analogues for Precambrian stromatolites actually exist.

Temporal controls

The essential prerequisite for a good explanation of the historical development of microbial carbonates is confident recognition of their distributions. The emerging picture cannot yet be described as detailed or complete, but, based on the information to hand, can we reach a reasonable preliminary assessment?

The competition hypothesis, for long an influential concept, has encountered serious obstacles. Research on the older Proterozoic has pushed back the timing of stromatolite decline [although algae (Brocks *et al.*, 1999) and metazoans (Seilacher *et al.*, 1998) are also being reported earlier], but the Phanerozoic persistence, and success, of microbial carbonates in coexistence with invertebrates challenges the view that grazing and competition for space are major limiting factors. It has also never been made clear whether the anticipated effects of eukaryotes on microbial carbonates are attributed directly to grazing or to competition for substrate. On the other hand, was the dramatic Early Palaeozoic appearance of thrombolites and dendrolites in part a calcification response to provide protection? Yet potential early grazers in the Cambrian were small (Farmer, 1992) and, in any case, Archaean and Proterozoic stromatolites show all the signs of early lithification. Nonetheless, when we look again at the present day, it is striking that the most impressive examples of microbial carbonates are large domes and columns in, at least partly, stressed environments, whether by salinity or tidal currents, at Shark Bay and Lee Stocking Island. The less impressive, but volumetrically far more abundant, laminar stromatolites on tidal flats are also in environments that are severely stressed by desiccation and temperature fluctuation. Thus, although the initial strengths of the competition hypothesis have been weakened, it cannot yet be dispensed with.

The calcification model emphasizes the overriding importance of the environment in facilitating early lithification. It suggests that the supersaturation state of seawater waxed and waned, directly influencing microbial carbonate accretion and therefore abundance. The weaknesses of the competition hypothesis are strengths for the calcification model. Inception of decline in abundance of stromatolites relatively early in the Proterozoic could reflect changing water conditions that are also reflected in abiogenic precipitates (Grotzinger, 1990; Grotzinger & Knoll, 1999). The Phanerozoic development of microbial carbonates is patterned by fluctuations expressed in cyanobacterial calcification episodes (CCEs; Riding, 1992) that may also occur in the Proterozoic (Riding, 1994b), and which can speculatively be related to factors such as temperature (Riding, 1992) that influenced the saturation state with respect to $CaCO_3$ minerals (Merz-Preiß & Riding, 1999). Thus, early Proterozoic decline, Cambrian resurgence and Phanerozoic episodicity, which beleaguer the competition hypothesis, are accounted for by the calcification model. Not least, it offers a mechanism to account for significant Phanerozoic fluctuations in abundance of microbial carbonates. But how does the calcification model fare when confronted with the presence of Shark Bay and Lee Stocking Island columns in apparent stressed refugia? The concept of eukaryote competition as a determining factor (e.g. Gebelein, 1976; Walter & Heys, 1985) fits well with the present-day occurrence of marine microbial carbonates in hostile environments, including hypersaline lagoons, desiccated tidal flats and areas of highly mobile sediment (Awramik, 1971, p. 825). But these environments may also enhance carbonate precipitation. Shark Bay and Lee Stocking Island are, respectively, swept by waves and currents that can promote early cementation. On intertidal flats, evaporation has a similar effect. These could therefore not simply be refuges from metazoan interference, but environments of rapid early lithification, as are freshwater lakes and pools with heavily calcified tufa stromatolites (Riding, 1997).

On the other hand, as algal evolution provided new components for mat ecosystems, their trapping and calcification abilities may have altered. Have diatoms, originating during the Mesozoic, made a major change to sediment stabilization and accretion? Are the coarse-grained domes and columns of Shark Bay and Lee Stocking Island simply the expression of the improved abilities of such regenerated communities to accumulate sediment (Awramik & Riding, 1988)?

Experience generally shows us that one-sided views ultimately fail to account for complex situations. Surely, environmental effects, biological interactions and microbial evolution must all have importance. The range of possibilities and the time and space involved, not to mention the diversity of microbial carbonates, all counsel against hard and fast conclusions. Indeed, could everyone be right?: stromatolite abundance decline commenced around 2000 Ma (Grotzinger, 1990), diversity decline commenced around 1000 Ma (Walter & Heys, 1985) and was greatest after 675 Ma (Awramik, 1971, 1991), in response to a combination of factors; cyanobacterial evolution together with fluctuations in calcification (Riding, 1992, 1994b) and the effect of emergence of invertebrates and algae (Fischer, 1965; Garrett, 1970) strongly patterned the development of Phanerozoic microbial carbonates, in which incorporation of newly evolved microbes added yet another factor (Awramik & Riding, 1988). Thus, despite possible inhibition in marine environments owing to the relatively low saturation state of seawater, the present day reveals diverse microbial carbonates, coexisting in abundance with algae and invertebrates in reefs, colonizing stressed tidal flats, locally creating spectacular columns in bays and channels, and forming in freshwater and hot-spring environments where $CaCO_3$ precipitation is facilitated. Whereas competition can inhibit or modify mat growth, lithification can independently determine preservation and accretion. So we might contrast microbial carbonate development in high-competition/low-calcification environments with that in low-competition/high-calcification situations, and so on. How we test these ideas and attempt to disentangle overlapping controls such as competition, lithification and microbial evolution, while avoiding the cycle of dogmatism that has so long prevailed in microbial carbonates, is the continuing challenge, but we can be confident that the results will be more surprising than our preconceptions.

CONCLUDING REMARKS

Microbial carbonates have the longest geological history and most extensive facies distribution of all biogenic carbonates. For much of the Precambrian they are the most conspicuous evidence of life on Earth. Yet they rely on synsedimentary calcification that is not fully organically controlled. Biological factors, such as microbial evolution and competition with metazoans, must have had a significant influence, but ecological interactions in particular may have been over-emphasized in attempts to account for fluctuations in the abundance of microbial carbonates in marine environments during the past 1000 Myr. The role of environmental factors suggests that geological variations in microbial calcification could be a proxy for fluctuations in the supersaturation state of seawater with respect to carbonate minerals that have affected the accretion, abundance and preservation of these deposits in space and time.

There is continuing need to document the geological distribution of microbial carbonates. Present-day examples, such as domes and columns at Shark Bay and Lee Stocking Island, are not simply occupying stressed refugia, and these 'analogues' show clear differences from the much older microbial carbonates with which they have often been compared. It is curious that, in many respects, the Archaean and Proterozoic history of microbial carbonates should be better known, or at least appear more clear, than that of subsequent periods. A partial explanation must be the difficulty of recognizing and interpreting microbial carbonates amongst Phanerozoic skeletal fabrics. As large subaqueous domes and columns, microbial carbonates possess an evident and impressive geological record, but mixed with algal and invertebrate skeletons their significance can be obscured. Recognition of microbial carbonates will be aided by improved understanding of their microfabrics. These commonly fine-grained textures remain the key to the evolutionary history, and to the environmental and ecological interactions, of the microbes that contributed to their formation. The significance of variations in these fabrics, and recognition of the organisms and processes that have produced them, are key goals. They can be achieved by detailed comparative and interdisciplinary studies of present-day and ancient examples.

ACKNOWLEDGEMENTS

I am indebted to reviewers Hank Chafetz and Brian Pratt for stimulating and helpful comments. Ian Jarvis gave much thoughtful editorial guidance and encouragement throughout. Stanley M. Awramik, Robert V. Burne, Annette Hofmann and Ernst Ott generously provided material for Figs 4, 6, 9, 20 and 21.

REFERENCES

Abbott, A. (1999) Battle lines drawn between 'nanobacteria' researchers. *Nature*, **401**, 105.

Aitken, J.D. (1966) Middle Cambrian to Middle Ordovician cyclic sedimentation, southern Rocky Mountains of Alberta. *Bull. Can. Petrol. Geol.*, **14**, 405–441.

Aitken, J.D. (1967) Classification and environmental significance of cryptalgal limestones and dolomites, with illustrations from the Cambrian and Ordovician of southwestern Alberta. *J. Sedim. Petrol.*, **37**, 1163–1178.

Aitken, J.D. (1989) Giant 'algal' reefs, Middle/Upper Proterozoic Little Dal Group (>770, <1200 Ma), Mackenzie Mountains, N.W.T., Canada. In: *Reefs, Canada and Adjacent Area* (Ed. by H. H. J. Geldsetzer, N. P. James and G. E. Tebbutt), *Can. Soc. Petrol. Geol., Mem.*, **13**, 13–23.

Aitken, J.D. and **Narbonne, G.M.** (1989) Two occurrences of Precambrian thrombolites from the Mackenzie Mountains, northwestern Canada. *Palaios*, **4**, 384–388.

Andrews, J.E., Riding, R. and **Dennis, P.F.** (1997) The stable isotope record of environmental and climatic signals in modern terrestrial microbial carbonates from Europe. *Palaeogeog. Palaeoclimatol. Palaeoecol.*, **129**, 171–189.

Armella, C. (1994) Thromolitic-stromatolitic cycles of the Cambro–Ordovician boundary sequence, Precordillera Oriental Basin, western Argentina. In: *Phanerozoic Stromatolites II* (Ed. by J. Bertrand-Sarfati and C. Monty), pp. 421–441. Kluwer, Dordrecht.

Arp, G. (1995) Lacustrine bioherms, spring mounds and marginal carbonates of the Ries-impact-crater (Miocene, southern Germany). *Facies*, **33**, 35–90.

Arp, G., Hofmann, J. and **Reitner, J.** (1999a) Microbial fabric formation in spring mounds ('microbialites') of alkaline salt lakes in the Badain Jaran Sand Sea, PR China. *Palaios*, **13**, 581–592.

Arp, G., Reimer, A. and **Reitner, J.** (1999b) Calcification in cyanobacterial biofilms of alkaline salt lakes. *Eur. J. Phycol.*, in press.

Awramik, S.M. (1971) Precambrian columnar stromatolite diversity: reflection of metazoan appearance. *Science*, **174**, 825–827.

Awramik, S.M. (1991) Archaean and Proterozoic stromatolites. In: *Calcareous Algae and Stromatolites* (Ed. by R. Riding), pp. 289–304. Springer-Verlag, Heidelberg.

Awramik, S.M. and **Margulis, L.** (1974) *Stromatolite Newsletter*, **2**, 5.

Awramik, S.M. and **Riding, R.** (1988) Role of algal eukaryotes in subtidal columnar stromatolite formation. *Proc. Natl. Acad. Sci. USA*, **85**, 1327–1329.

Awramik, S.M., **Margulis, L.** and **Barghoorn, E.S.** (1976) Evolutionary processes in the formation of stromatolites. In: *Stromatolites* (Ed. by M. R. Walter), *Developments in Sedimentology*, **20**, pp. 149–162. Elsevier, Amsterdam.

Awramik, S.M. and **Semikhatov, M.A.** (1979) The relationship between morphology, microstructure, and microbiota in Gunflint stromatolites. *Can. J. Earth Sci.*, **16**, 484–495.

Awramik, S.M., **Schopf, J.W.** and **Walter, M.R.** (1983) Filamentous fossil bacteria from the Archean of Western Australia. *Precamb. Res.*, **20**, 357–374.

Babcock, J.A. (1977) Calcareous algae, organic boundstones, and the genesis of the Upper Capitan Limestone (Permian, Guadalupian), Guadalupe Mountains, west Texas and New Mexico. In: *Upper Guadalupian Facies, Permian Reef Complex, Guadalupe Mountains, New Mexico West Texas* (Ed. by M. E. Hileman and S. J. Mazzullo), *SEPM Permian Basin Section, Spec. Publ.*, **77-6**, 3–44.

Banfield, J.F. and **Nealson, K.H.** (Eds) (1997) Interactions between microbes and minerals. Mineralogical Society of America, Washington, DC, *Rev. Miner.*, **35**.

Bartley, J.K. (1996) Actualistic taphonomy of cyanobacteria: implications for the Precambrian fossil record. *Palaios*, **11**, 571–586.

Bathurst, R.G.C. (1966) Boring algae, micrite envelopes and lithification of molluscan biosparites. *J. Geol.*, **5**, 15–32.

Bavendamm, W. (1932) Die mikrobiologische Kalkfaellung in der tropischen See. *Arch. Mikrobiol.*, **3**, 205–276.

Benson, L. (1994) Carbonate deposition, Pyramid Lake Subbasin, Nevada; 1, Sequence of formation and elevational distribution of carbonate deposits (tufas). *Palaeogeog. Palaeoclimatol. Palaeoecol.*, **109**, 55–87.

Berner, R.A. (1971a) Bacterial processes effecting the precipitation of calcium carbonate in sediments. In: *Carbonate Cements* (Ed. by O. P. Bricker), *The Johns Hopkins University, Studies in Geology*, **19**, 247–251.

Berner, R.A. (1971b). *Principles of Chemical Sedimentology*. McGraw-Hill, New York.

Bertrand-Sarfati, J. (1976) An attempt to classify late Precambrian stromatolite microstructures. In: *Stromatolites* (Ed. by M. R. Walter), *Developments in Sedimentology*, **20**, pp. 251–259. Elsevier, Amsterdam.

Bertrand-Sarfati, J. (1994) Siliciclastic-carbonate stromatolite domes, in the Early Carboniferous of the Ajjers Basin (eastern Sahara, Algeria). In: *Phanerozoic Stromatolites II* (Ed. by J. Bertrand-Sarfati and C. Monty), pp. 395–419. Kluwer, Dordrecht.

Bertrand-Sarfati, J., **Freytet, P.** and **Plaziat, J.C.** (1994) Microstructures in Tertiary nonmarine stromatolites (France). Comparison with Proterozoic. In: *Phanerozoic Stromatolites II* (Ed. by J. Bertrand-Sarfati and C. Monty), pp. 155–191. Kluwer, Dordrecht.

Bertrand-Sarfati, J. and **Walter, M.R.** (1981) Stromatolite biostratigraphy. *Precamb. Res.*, **15**, 353–371.

Beukes, N.J. (1987) Facies relations, depositional environments and diagenesis in a major early Proterozoic stromatolitic carbonate platform to basinal sequence, Campbellrand Subgroup, Transvaal Supergroup, Southern Africa. *Sediment. Geol.*, **54**, 1–46.

Black, M. (1933) The algal sedimentation of Andros Island Bahamas. *Phil. Trans. Roy. Soc. London, Series B: Biol. Sci.*, **222**, 165–192.

Bowring, S.A., **Grotzinger, J.P.**, **Isachsen, C.E.**, **Knoll, A.H.**, **Pelechaty, S.M.** and **Kolosov, P.** (1993) Calibrating rates of Early Cambrian evolution. *Science*, **261**, 1293–1298.

Bradley, W.H. (1929) Algae reefs and oolites of the Green River Formation. *U.S. Geol. Surv. Prof. Pap.*, **154-G**, 203–223.

Braga, J.C., **Martín, J.M.** and **Riding, R.** (1995) Controls on microbial dome fabric development along a carbonate-siliciclastic shelf-basin transect, Miocene, S.E. Spain. *Palaios*, **10**, 347–361.

Braga, J.C., **Martín, J.M.** and **Riding, R.** (1996) Internal structure of segment reefs: *Halimeda* algal mounds in the Mediterranean Miocene. *Geology*, **24**, 35–38.

Brock, T.D., **Madigan, M.T.**, **Martinko, J.M.** and **Parker, J.** (1994) *Biology of Microorganisms*, 7th edn. Prentice Hall, New Jersey.

Brocks, J.J., **Logan, G.A.**, **Buick, R.** and **Summons, R.E.** (1999) Archean molecular fossils and the early rise of eukeryotes. *Science*, **285**, 1033–1036.

Browne, K.M., **Golubic, S.** and **Lee Seong-Joo** (2000) Shallow marine microbial carbonate deposits. In: *Microbial Sediments* (Ed. by R. Riding and S. M. Awramik), pp. 233–249. Springer-Verlag, Heidelberg.

Buczynski, C. and **Chafetz, H.S.** (1991) Habit of bacterially induced precipitates of calcium carbonate and the influence of medium viscosity on mineralogy. *J. Sedim. Petrol.*, **61**, 226–233.

Buick, R., **Dunlop, J.S.R.** and **Groves, D.I.** (1981) Stromatolite recognition in ancient rocks: an appraisal of irregularly laminated structures in an Early Archaean chert-barite unit from North Pole, Western Australia. *Alcheringa*, **5**, 161–181.

Buick, R., **Groves, D.I.** and **Dunlop, J.S.R.** (1995) Abiological origin of described stromatolites older than 3.2 Ga: comment and reply. Comment. *Geology*, **23**, 191.

Burne, R.V. and **Moore, L.S.** (1987) Microbialites: organosedimentary deposits of benthic microbial communities. *Palaios*, **2**, 241–254.

Cameron, B., **Cameron, D.** and **Jones, J.R.** (1985) Modern algal mats in intertidal and supratidal quartz sands, northeastern Massachusetts, U.S.A. In: *Biogenic Structures: Their Use in Interpreting Depositional Environments* (Ed. by H. A. Curran), *SEPM Spec. Publ.*, **35**, 211–223.

Camoin, G.F. (Ed.) (1999) Microbial mediation in carbonate diagenesis. *Sediment. Geol.*, **126**.

Canfield, D.E. and **Raiswell, R.** (1991) Carbonate precipitation and dissolution: its relevance to fossil preservation. In: *Taphonomy: Releasing the Data Locked in the Fossil Record* (Ed. by P. A. Allison and D. E. G.Briggs), pp. 411–453. Plenum, New York.

Castanier, S., **Le Métayer-Levrel, G.** and **Perthuisot, J.-P.** (2000) Bacterial roles in the precipitation of carbonate minerals. In: *Microbial Sediments* (Ed. by R. Riding and S. M. Awramik), pp. 32–39. Springer-Verlag, Heidelberg.

Cayeux, L. (1935) *Les Roches Sédimentaires De France;. Roches Carbonatées (Calcaires Et Dolomies). Fascicule 1.* Masson, Paris.

Chafetz, H.S. (1986) Marine peloids; a product of bacterially induced precipitation of calcite. *J. Sedim. Petrol.*, **56**, 812–817.

Chafetz, H.S. (1994) Bacterially induced precipitates of calcium carbonate and lithification of microbial mats. In: *Biostabilization of Sediments* (Ed. by W. E. Krumbein, D. M. Paterson and L. J. Stal), pp. 149–163. Bibliotheks und Informationssystem der Universität Oldenburg-Verlag, Oldenburg.

Chafetz, H.S. and **Buczynski, C.** (1992) Bacterially induced lithification of microbial mats. *Palaios*, **7**, 277–293.

Chafetz, H.S. and **Folk, R.L.** (1984) Travertines; depositional morphology and the bacterially constructed constituents. *J. Sedim. Petrol.*, **54**, 289–316.

Chafetz, H.S. and **Guidry, S.A.** (1999) Bacterial shrubs, crystal shrubs, and ray-crystal shrubs: bacterial vs. abiotic precipitation. *Sediment. Geol.*, **126**, 57–74.

Christensen, B.E. and **Characklis, W.G.** (1990) Physical and chemical properties of biofilms. In: *Biofilms* (Ed. by W. G. Characklis and K. C.Marshall), pp. 93–130. John Wiley & Sons, New York.

Cloud, P. and **Semikhatov, M.A.** (1969) Proterozoic stromatolite zonation. *Am. J. Sci.*, **267**, 1017–1061.

Compton, J.S. (1988) Degree of supersaturation and precipitation of organogenic dolomite. *Geology*, **16**, 318–321.

Copper, P. (1974) Structure and development of Early Paleozoic reefs. *Proc. Second Int. Coral Reef Symp.*, **1**, 365–386.

Costerton, J.W., **Geesey, G.G.** and **Cheng, K.J.** (1978) How bacteria stick. *Sci. Am.*, **238**, 86–95.

Costerton, J.W., **Lewandowski, Z.**, **Caldwell, D.E.**, **Korber, D.R.** and **Lappin-Scott, H.M.** (1995) Microbial biofims. *Ann. Rev. Microbiol.*, **49**, 711–745.

Decho, A.W. (1990) Microbial exopolymer secretions in ocean environments: their role(s) in food webs and marine processes. *Oceanogr. Mar. Biol. Ann. Rev.*, **28**, 73–153.

Decho, A.W. (2000) 'Exopolymer microdomains' as a structuring agent for heterogeneity within microbial biofilms. In: *Microbial Sediments* (Ed. by R. Riding and S. M. Awramik), pp. 9-15. Springer-Verlag, Heidelberg.

De Craen, M., **Swennen, R.**, **Keppens, E.M.**, **Macauley, C.I.** and **Kiriakoulakis, K.** (1999) Bacterially mediated formation of carbonate concretions in the Oligocene Boom Clay of northern Belgium. *J. Sedim. Res.*, **69**, 1098–1106.

Défarge, C., **Trichet, J.** and **Couté, A.** (1994a) On the appearance of cyanobacterial calcification in modern stromatolites. *Sediment. Geol.*, **94**, 11–19.

Défarge, C., **Trichet, J.**, **Maurin, A.** and **Hucher, M.** (1994b) Kopara in Polynesian atolls; early stages of formation of calcareous stromatolites. *Sediment. Geol.*, **89**, 9–23.

Défarge, C., **Trichet, J.**, **Jaunet, A.-M.**, **Robert, M.**, **Tribble, J.** and **Sansone, F.J.** (1996) Texture of microbial sediments revealed by cryo-scanning electron microscopy. *J. Sedim. Res.*, **66**, 935–947.

Dill, R.F., **Shinn, E.A.**, **Jones, A.T.**, **Kelly, K.** and **Steinen, R.P.** (1986) Giant subtidal stromatolites forming in normal salinity waters. *Nature*, **324**, 55–58.

Dravis, J.J. (1983) Hardened subtidal stromatolites, Bahamas. *Science*, **219**, 385–386.

Dupraz, C. and **Strasser, A.** (1999) Microbialites and microencrusters in shallow coral bioherms (Middle to Late Oxfordian, Swiss Jura Mountains). *Facies*, **40**, 101–130.

Eggleston, J.R. and **Dean, W.E.** (1976) Freshwater stromatolitic bioherms in Green Lake, New York. In: *Stromatolites* (Ed. by M. R. Walter), *Developments in Sedimentology*, **20**, pp. 479–488. Elsevier, Amsterdam.

Ehrlich, H.L. (1998) Geomicrobiology: its significance for geology. *Earth-Sci. Rev.*, **45**, 45–60.

Fagerstrom, J.A. (1987). *The Evolution of Reef Communities.* John Wiley & Sons, New York.

Fairchild, I.J. (1991) Origins of carbonate in Neoproterozoic stromatolites and the identification of modern analogues. *Precamb. Res.*, **53**, 281–299.

Farmer, J.D. (1992) Grazing and bioturbation in modern microbial mats. In: *The Proterozoic Biosphere; a Multidisciplinary Study* (Ed. by J. W. Schopf and C. Klein), pp. 295–297. Cambridge University Press, Cambridge.

Feldmann, M. and **McKenzie, J.A.** (1997) Messinian stromatolite–thrombolite associations, Santa Pola, SE Spain: an analogue for the Palaeozoic? *Sedimentology*, **44**, 893–914.

Feldmann, M. and **McKenzie, J.A.** (1998) Stromatolite–thrombolite associations in a modern environment, Lee Stocking Island, Bahamas. *Palaios*, **13**, 201–212.

Fenchel, T. and **Finlay, B.J.** (1995). *Ecology and Evolution in Anoxic Worlds. Oxford Series in Ecology and Evolution.* Oxford University Press, Oxford.

Fenton, C.L. and **Fenton, M.A.** (1937) Belt series of the north; stratigraphy, sedimentation, paleontology. *Geol. Soc. Am. Bull.*, **48**, 1873–1969.

Ferris, F.G., **Thompson, J.B.** and **Beveridge, T.J.** (1997) Modern freshwater microbialites from Kelly Lake, British Columbia, Canada. *Palaios*, **12**, 213–219.

Fischer, A.G. (1965) Fossils, early life, and atmospheric history. *Proc. Nat. Acad. Sci.*, **53**, 1205–1215.

Flügel, E. (1981) Paleoecology and facies of Upper Triassic reefs in the Northern Calcareous Alps. In: *European Fossil Reef Models* (Ed. by D. F. Toomey), *SEPM Spec. Publ.*, **30**, 291–359.

Flügel, E. (1982). *Microfacies Analysis of Limestones.* Springer-Verlag, Berlin.

Flügel, E. and **Steiger, T.** (1981) An upper Jurassic sponge-algal buildup from the northern Frankenalb, West Germany. In: *European Fossil Reef Models* (Ed. by D. F. Toomey), *SEPM Spec. Publ.*, **30**, 371–397.

Fois, E. and **Gaetani, M.** (1984) The recovery of reef-building communities and the role of cnidarians in carbonate sequences of the Middle Triassic (Anisian) in the Italian Dolomites. *Palaeontogr. Am.*, **54**, 191–200.

Folk, R.L. (1993) SEM imaging of bacteria and nannobacteria in carbonate sediments and rocks. *J. Sedim. Petrol.*, **63**, 990–1000.

Fritsch, F.E. (1959). *The Structure and Reproduction of the Algae, 2.* Cambridge University Press, Cambridge.

Garrett, P. (1970) Phanerozoic stromatolites: noncompetitive ecologic restriction by grazing and burrowing animals. *Science*, **169**, 171–173.

Gavish, E., **Krumbein, W.E.** and **Halevy, J.** (1985) Geomorphology, mineralogy and groundwater geochemistry as factors of the hydrodynamic system of the Gavish Sabkha. In: *Hypersaline Ecosystems; the Gavish Sabkha* (Ed. by G. M. Friedman and W. E. Krumbein). Springer-Verlag, Berlin. *Ecological Studies*, **53**, 186–217.

Gebelein, C.D. (1969) Distribution, morphology, and accretion rate of recent subtidal algal stromatolites, Bermuda. *J. Sedim. Petrol.*, **39**, 49–69.

Gebelein, C.D. (1976) The effects of the physical, chemical and biological evolution of the earth. In: *Stromatolites* (Ed. by M. R. Walter), *Developments in Sedimentology*, **20**, pp. 499–515. Elsevier, Amsterdam.

Geldsetzer, H.H.J., **James, N.P.** and **Tebbutt, G.E.** (Eds) (1988) Reefs, Canada and adjacent area, *Canadian Soc. Petrol. Geol., Mem.*, **13**.

Gerdes, G., **Krumbein, W.E.** and **Holtkamp, E.** (1985) Salinity and water activity related zonation of microbial communities and potential stromatolites of the Gavish Sabkha. In: *Hypersaline Ecosystems; the Gavish Sabkha* (Ed. by G. M.

Friedman and W.E.Krumbein). Springer-Verlag, Berlin. *Ecological Studies*, **53**, 238–266.

Gerdes, G., Krumbein, W.E. and **Reineck, H.E.** (1991) Biolaminations; ecological versus depositional dynamics. In: *Cycles and Events in Stratigraphy* (Ed. by G. Einsele, R. Werner and A. Seilacher), pp. 592–607. Springer-Verlag, Berlin.

Ginsburg, R.N. (1955) Recent stromatolitic sediments from south Florida. *J. Paleontol.*, **29**, 723–724.

Ginsburg, R.N. (1991) Controversies about stromatolites: vices and virtues. In: *Controversies in Modern Geology; Evolution of Geological Theories in Sedimentology, Earth History and Tectonics* (Ed. by D. W. Müller, J. A. McKenzie and H. Weissert), pp. 25–36. Academic Press, London.

Ginsburg, R.N., Isham, L.B., Bein, S.J. and **Kuperberg, J.** (1954) Laminated algal sediments of South Florida and their recognition in the fossil record. *Marine Laboratory, University of Miami, Coral Gables, Florida, Unpublished Report*, **54-20**.

Golubic, S. (1976) Organisms that build stromatolites. In: *Stromatolites* (Ed. by M. R.Walter), *Developments in Sedimentology*, **20**, pp. 113–126. Elsevier, Amsterdam.

Golubic, S. and **Browne, K.M.** (1996) *Schizothrix gebeleinii* sp. nov. builds subtidal stromatolites, Lee Stocking Island, Bahamas. *Algol. Studies*, **83**, 273–290.

Golubic, S. and **Campbell, S.E.** (1981) Biogenically formed aragonite concretions in marine *Rivularia*. In: *Phanerozoic Stromatolites; Case Histories* (Ed. by C. Monty), pp. 209–241. Springer, Berlin.

Golubic, S. and **Hofmann, H.J.** (1976) Comparison of Holocene and mid-Precambrian Entophysalidaceae (Cyanophyta) in stromatolitic algal mats; cell division and degradation. *J. Paleontol.*, **50**, 1074–1082.

Golubic, S., Lee Seong-Joo and **Browne, K.M.** (2000) Cyanobacteria: architects of sedimentary structures. In: *Microbial Sediments* (Ed. by R. Riding and S. M. Awramik), pp. 57–67. Springer-Verlag, Heidelberg.

Gradzinski, M., Szulc, J. and **Smyk, B.** (1997) Microbial agents of moonmilk calcification. In: *Proc. 12th Internat. Congr. Speleology* (Ed. by P. Y. Jeannin), **1**, pp. 275–278. International Union of Speleology.

Grey, K. and **Thorne, A.M.** (1985) Biostratigraphic significance of stromatolites in upward shallowing sequences of the early Proterozoic Duck Creek Dolomite, Western Australia. *Precamb. Res.*, **29**, 183–206.

Grotzinger, J.P. (1989) Facies and evolution of Precambrian carbonate depositional systems: emergence of the modern platform archetype. In: *Controls on Carbonate Platform and Basin Development* (Ed. by P. D. Crevello, J. L.Wilson, J. F. Sarg and J. F. Read), *SEPM Spec. Publ.*, **44**, 79–106.

Grotzinger, J.P. (1990) Geochemical model for Proterozoic stromatolite decline. *Am. J. Sci.*, **290-A**, 80–103.

Grotzinger, J.P. (1994) Trends in Precambrian carbonate sediments and their implications for understanding evolution. In: *Early Life on Earth* (Ed. by S. Bengtson), *Nobel Symposium*, **84**, pp. 245–258. Columbia University Press, New York.

Grotzinger, J.P. and **Hoffman, P.** (1983) Aspects of the Rocknest Formation, Asiak thrust-fold belt, Wopmay Orogen, District of Mackenzie. *Geol. Surv. Can Pap.*, **83-1B**, 83–92.

Grotzinger, J.P. and **Knoll, A.H.** (1999) Stromatolites in Precambrian carbonates: evolutionary mileposts or environmental dipsticks? *Annu. Rev. Earth Planet. Sci.*, **27**, 313–358.

Grotzinger, J.P. and **Read, J.F.** (1983) Evidence for primary aragonite precipitation, lower Proterozoic (1.9-Ga) Rocknest Dolomite, Wopmay Orogen, Northwest Canada. *Geology*, **11**, 710–713.

Grotzinger, J.P. and **Rothman, D.R.** (1996) An abiotic model for stromatolite morphogenesis. *Nature*, **383**, 423–425.

Guo, L. and **Riding, R.** (1992a) Aragonite laminae in hot water travertine crusts, Rapolano Terme, Italy. *Sedimentology*, **39**, 1067–1079.

Guo, L. and **Riding, R.** (1992b) Microbial micritic carbonates in uppermost Permian reefs, Sichuan Basin, southern China: some similarities with Recent travertines. *Sedimentology*, **39**, 37–53.

Guo, L. and **Riding, R.** (1994) Origin and diagenesis of Quaternary travertine shrub fabrics, Rapolano Terme, central Italy. *Sedimentology*, **41**, 499–520.

Guo, L. and **Riding, R.** (1999) Rapid facies changes in Holocene fissure ridge hot spring travertines, Rapolano Terme, Italy. *Sedimentology*, **46**, 1145–1158.

Guo, L., Andrews, J., Riding, R., Dennis, P. and **Dresser, Q.** (1996) Possible microbial effects on stable carbon isotopes in hot spring travertines. *J. Sedim. Res.*, **66**, 468–473.

Gürich, G. (1906) Les Spongiostromides du Viséen de la province de Namur. *Mus. Roy. Hist. Natur. Belg. Mém.*, **3.**, 23 pls.

Gwinner, M.P. (1976) Origin of the Upper Jurassic limestones of the Swabian Alb (southwestern Germany). *Contrib. Sedimentol.*, **5**.

Hall, J. (1883) *Cryptozoön*, n.g., *Cryptozoön proliferum*, n.sp. *New York State Museum of Natural History, 36th Annual Report of the Trustees*, plate 6.

Han, T.-M. and **Runnegar, B.** (1992) Megascopic eukaryotic algae from the 2.1 billion-year-old Negaunee Iron-Formation, Michigan. *Science*, **257**, 232–235.

Harland, W.B., Armstrong, R.L., Cox, A.V., Craig, L.E., Smith, A.G. and **Smith, D.G.** (1990). *A Geologic Time Scale 1989*. Cambridge University Press, Cambridge.

Heim, A. (1916) Monographie Churfürsten-Mattstock-Gruppe. 3. Lithogenesis. *Beitrage Geologische Karte Schweiz, NF*, **20**, 369–662.

Hoffman, P. (1973) Recent and ancient algal stromatolites: seventy years of pedagogic cross-pollination. In: *Evolving Concepts in Sedimentology* (Ed. by R. N. Ginsburg), pp. 178–191. The Johns Hopkins University Press, Baltimore.

Hoffman, P.F. (1974) Shallow and deep-water stromatolites in lower Proterozoic platform-to-basin facies change, Great Slave Lake, Canada. *Bull. Am. Assoc. Petrol. Geol.*, **58**, 856–867.

Hofmann, A. (1996) Caractéristiques géochimiques et processus de transport de la matière particulaire dans le bassin nord du lac de Lugano (Suisse, Italie). PhD Dissertation, University of Geneva, F-A. Forel Inst., *Terre et Environnement*, **8** (2 Volumes).

Hofmann, H.J. (2000) Archean stromatolites as microbial archives. In: *Microbial Sediments* (Ed. by R. Riding and S. M. Awramik), pp. 315–327. Springer-Verlag, Heidelberg, in press.

Hofmann, H.J., Grey, K., Hickman, A.H. and **Thorpe, R.I.** (1999) Origin of 3.45 Ga coniform stromatolites in Warrawoona Group, Western Australia. *Geol. Soc. Am. Bull.*, **111**, 1256–1262.

Hofmann, H.J. and **Jackson, J.D.** (1987) Proterozoic ministromatolites with radial fibrous fabric. *Sedimentology*, **34**, 963–971.

Illing, L.V. (1954) Bahaman calcareous sands. *Bull. Am. Assoc. Petrol. Geol.*, **38**, 1–95.

Irion, G. and **Müller, G.** (1968) Mineralogy, petrology and chemical composition of some calcareous tufa from the Schwäbische Alb, Germany. In: *Recent Developments in Carbonate Sedimentology in Central Europe* (Ed. by G. Müller and G. M. Friedman), pp. 157–171. Springer, Berlin.

Irwin, H., Curtis, C. and **Coleman, M.** (1977) Isotopic evidence for source of diagenetic carbonates formed during burial of organic-rich sediments. *Nature*, **269**, 209–213.

Jones, B. (1995) Processes associated with microbial biofilms in the twilight zone of caves; examples from the Cayman Islands. *J. Sedim. Res.*, **A65**, 552–560.

Jones, B. (2000) Microbial sediments in tropical karst terrains: a model based on the Cayman Islands. In: *Microbial Sediments* (Ed. by R. Riding and S. M. Awramik), pp. 171–178. Springer-Verlag, Heidelberg.

Jones, B. and **Kahle, C.F.** (1985) Lichen and algae: agents of biodiagenesis in karst breccia from Grand Cayman Island. *Bull. Can. Petrol. Geol.*, **33**, 446–461.

Jones, B. and **Kahle, C.F.** (1986) Dendritic calcite crystals formed by calcification of algal filaments in a vadose environment. *J. Sedim. Petrol.*, **56**, 217–227.

Jones, B. and **Pemberton, S.G.** (1987) The role of fungi in the diagenetic alteration of spar calcite. *Can. J. Earth Sci.*, **24**, 903–914.

Jones, B. and **Renaut, R.W.** (1995) Noncrystallographic calcite dendrites from hot-spring deposits at Lake Bogoria, Kenya. *J. Sedim. Petrol.*, **A65**, 154–169.

Jones, D. and **Wilson, M.J.** (1986) Biomineralization in crustose lichens. In: *Biomineralization in Lower Plants and Animals* (Ed. by B. S. C. Leadbeater and R. Riding), *The Systematics Association, Special Volume*, **30**, 91–105.

Jones, J.G. (1985) Microbes and microbial processes in sediments. In: *Geochemistry of Buried Sediments* (Ed. by G. Eglinton, C. D. Curtis, D. P. McKenzie and D. G. Murchison), *Phil. Trans. Roy. Soc. London, Series A*, **315**, 3–17.

Jørgensen, B.B. (1977) The sulfur cycle of a coastal marine sediment (Limfjorden, Denmark). *Limnol. Oceanogr.*, **22**, 814–832.

Jørgensen, B.B., Revsbech, N.P. and **Cohen, Y.** (1983) Photosynthesis and structure of benthic microbial mats: microelectrode and SEM studies of four cyanobacterial communities. *Limnol. Oceanogr.*, **28**, 1075–1093.

Kalkowsky, E. (1908) Oolith und Stromatolith im norddeutschen Buntsandstein. *Z. Deutsch. Geol. Gesellschaft*, **60**, 68–125. pls 4–11.

Kelts, K. and **Hsu, K.J.** (1978) Freshwater carbonate sedimentation. In: *Lakes; Chemistry, Geology, Physics* (Ed. by A. Lerman), pp. 295–323. Springer-Verlag, New York.

Kempe, S. and **Kazmierczak, J.** (1994) The role of alkalinity in the evolution of ocean chemistry, organization of living systems, and biocalcification processes. *Bull. Inst. Océanogr. Monaco, Num. Spéc.*, **13**, 61–117.

Kempe, S., Kazmierczak, J., Landmann, G., Konuk, T., Reimer, A. and **Lipp, A.** (1991) Largest known microbialites discovered in Lake Van, Turkey. *Nature*, **349**, 605–608.

Kendall, C.G.S.C., Skipwith, P.A. and **d'E.** (1969) Holocene shallow-water carbonate and evaporate sediments of Khor al Bazam, Abu Dhabi, southwest Persian Gulf. *Bull. Am. Assoc. Petrol. Geol.*, **53**, 841–869.

Kennard, J.M. (1994) Thrombolites and stromatolites within shale-carbonate cycles, Middle–Late Cambrian Shannon Formation, Amadeus Basin, central Australia. In: *Phanerozoic Stromatolites II* (Ed. by J. Bertrand-Sarfati and C. Monty), pp. 443–471. Kluwer, Dordrecht.

Kennard, J.M. and **James, N.P.** (1986) Thrombolites and stromatolites; two distinct types of microbial structures. *Palaios*, **1**, 492–503.

Kirkland, B.L., Lynch, F.L., Rahnis, M.A., Folk, R.L., Molineux, I.J. and **McLean, R.J.C.** (1999) Alternative origins for nannobacteria-like objects in calcite. *Geology*, **27**, 347–350.

Klappa, C.F. (1979) Lichen stromatolites; criterion for subaerial exposure and a mechanism for the formation of laminar calcretes (caliche). *J. Sedim. Petrol.*, **49**, 387–400.

Knoll, A.H. and **Semikhatov, M.A.** (1998) The genesis and time distribution of two distinctive Proterozoic stromatolite microstructures. *Palaios*, **13**, 408–422.

Knoll, A.H., Fairchild, I.J. and **Swett, K.** (1993) Calcified microbes in Neoproterozoic carbonates; implications for our understanding of the Proterozoic/ Cambrian transition. *Palaios*, **8**, 512–525.

von Knorre, H. and **Krumbein, W.E.** (2000) Bacterial calcification. In: *Microbial Sediments* (Ed. by R. Riding and S. M. Awramik), pp. 25–31. Springer-Verlag, Heidelberg.

Kobluk, D.R. and **Risk, M.J.** (1977) Calcification of exposed filaments of endolithic algae, micrite envelope formation and sediment production. *J. Sedim. Petrol.*, **47**, 517–528.

Komar, V.A. (1989) Classification of the microstructures of the Upper Precambrian stromatolites. *Himalayan Geol.*, **13**, 229–238.

Krumbein, W.E. (1968) Geomicrobiology and geochemistry of the 'Nari-limecrust' (Israel). In: *Recent Developments in Carbonate Sedimentology in Central Europe* (Ed. by G. Muller and G. M. Friedman), pp. 138–147. Springer-Verlag, New York.

Krumbein, W.E. (1978) Algal mats and their lithification. In: *Environmental Biogeohemistry and Geomicrobiology*, Vol. 1: *the Aquatic Environment* (Ed. by W. E. Krumbein), pp. 209–225. Proceedings of the Third International Symposium on Environmental Biogeohemistry, Wolfenbüttel; Ann Arbor Science, Ann Arbor.

Krumbein, W.E. (1979a) Calcification by bacteria and algae. In: *Biogeochemical Cycling of Mineral-Forming Elements* (Ed. by P. A. Trudinger and D. J. Swaine), pp. 47–68. Elsevier, Amsterdam.

Krumbein, W.E. (1979b) Photolithotrophic and chemoorganotrophic activity of bacteria and algae as related to beachrock formation and degradation (Gulf of Aquaba, Sinai). *Geomicrobiol. J.*, **1**, 139–203.

Krumbein, W.E. (1983) Stromatolites; the challenge of a term in space and time. *Precamb. Res.*, **20**, 493–531.

Krumbein, W.E., Cohen, Y. and **Shilo, M.** (1977) Solar Lake (Sinai). 4. Stromatolitic Cyanobacterial Mats. *Limnol. Oceanog.*, **22**, 635–656.

Latham, A. and **Riding, R.** (1990) Fossil evidence for the location of the Precambrian/Cambrian boundary in Morocco. *Nature*, **344**, 752–754.

Lawrence, J.R., Korber, D.R., Hoyle, B.D., Costerton, J.W. and **Caldwell, D.E.** (1991) Optical sectioning of microbial biofilms. *J. Bacteriol.*, **173**, 6558–6567.

Lawrence, J.R., Wolfaardt, G.M. and **Korber, D.R.** (1994) Determination of diffusion coefficients in biofilms by confocal laser microscopy. *Appl. Environ. Microbiol.*, **60**, 1166–1173.

Leinfelder, R.R., Nose, M., Schmid, D.U. and **Werner, W.**

(1993) Microbial crusts of the Late Jurassic: composition, palaeoecological significance and importance in reef construction. *Facies*, **29**, 195–230.

Leinfelder, R.R. and **Schmid, D.U.** (2000) Mesozoic reefal thrombolites and other microbolites. In: *Microbial Sediments* (Ed. by R. Riding and S. M. Awramik). pp. 289–294. Springer-Verlag, Heidelberg.

Liang, Y., Zhu, S., Zhang, L., Cao, R., Gao, Z. and **Bu, D.** (1985) Stromatolite assemblages of the late Precambrian in China. *Precamb. Res.*, **29**, 15–32.

Lighty, R.G. (1985) Preservation of internal reef porosity and diagenetic sealing of submerged early Holocene barrier reef, southeast Florida shelf. In: *Carbonate Cements* (Ed. by N. Schneidermann and P. M. Harris), *SEPM Spec. Publ.*, **36**, 123–151.

Logan, B.W. (1961) *Cryptozoon* and associated stromatolites from the Recent, Shark Bay, Western Australia. *J. Geol.*, **69**, 517–533.

Logan, B.W., Rezak, R. and **Ginsburg, R.N.** (1964) Classification and environmental significance of algal stromatolites. *J. Geol.*, **72**, 68–83.

Lowe, D.R. (1980) Stromatolites 3,400–3,500 Myr old from the Archean of Western Australia. *Nature*, **284**, 441–443.

Lowe, D.R. (1994) Abiological origin of described stromatolites older than 3.2 Ga. *Geology*, **22**, 387–390.

Lowe, D.R. (1995) Abiological origin of described stromatolites older than 3.2 Ga: comment and reply. Reply. *Geology*, **23**, 191–192.

Luchinina, V.A. (1973) Algal ecology and microstructure of algal biostromes of the Lower Cambrian of the middle Lena River. In: *Environment and Life in the Geological Past (Late Precambrian and Palaeozoic of Siberia).* Nauka, Novosibirsk, Institute of Geology and Geophysics, Siberian Branch, **169**, 69–71. In Russian.

Machel, H.G. and **Foght, J.** (2000) Products and depth limits of microbial activity in petroliferous subsurface settings. In: *Microbial Sediments* (Ed. by R. Riding and S. M. Awramik), pp. 105–120. Springer-Verlag, Heidelberg.

Macintyre, I.G. (1984) Extensive submarine lithification in a cave in the Belize Barrier Reef Platform. *J. Sedim. Petrol.*, **54**, 221–235.

Macintyre, I.G. (1985) Submarine cements – the peloidal question. In: *Carbonate Cements* (Ed. by N. Schneidermann and P. M. Harris), *SEPM Spec. Publ.*, **36**, 109–116.

Martín, J.M. and **Braga, J.C.** (1994) Messinian events in the Sorbas Basin in southeastern Spain and their implications in the Recent history of the Mediterranean. *Sediment. Geol.*, **90**, 257–268.

Martín, J.M., Braga, J.C. and **Riding, R.** (1993) Siliciclastic stromatolites and thrombolites, late Miocene, S.E. Spain. *J. Sedim. Petrol.*, **63**, 131–139.

Mason-Williams, A. (1959) The formation and deposition of moonmilk. *Trans. Cave Res. Group Great Britain*, **5**, 133–138.

Maurin, A.F. and **Noël, D.** (1977) A possible bacterial origin for Famennian micrites. In: *Fossil Algae; Recent Results and Developments* (Ed. by E. Flügel), pp. 136–142. Springer, Berlin.

McKee, E.D. and **Gutschick, R.C.** (1969) History of Redwall Limestone of northern Arizona. *Geol. Soc. Am. Mem.*, **114**.

Merz, M.U.E. (1992) The biology of carbonate precipitation by cyanobacteria. *Facies*, **26**, 81–102.

Merz-Preiß, M. (2000) Calcification in cyanobacteria. In: *Microbial Sediments* (Ed. by R. Riding and S. M. Awramik), pp. 50–56. Springer-Verlag, Heidelberg.

Merz-Preiß, M. and **Riding, R.** (1999) Cyanobacterial tufa calcification in freshwater streams: ambient environment, chemical thresholds and biological processes. *Sediment. Geol.*, **126**, 103–124.

Middleburg, J.J., de Lange, G.J. and **Kreulen, R.** (1990) Dolomite formation in anoxic sediments of Kau Bay, Indonesia. *Geology*, **18**, 399–402.

Montaggioni, L.F. and **Camoin, G.F.** (1993) Stromatolites associated with coralgal communities in Holocene high-energy reefs. *Geology*, **21**, 149–152.

Monty, C.L.V. (1967) Distribution and structure of Recent stromatolitic algal mats, eastern Andros Island, Bahamas. *Ann. Soc. Géol. Belg.*, **90**, 55–100.

Monty, C.L.V. (1973) Precambrian background and Phanerozoic history of stromatolitic communities, an overview. *Ann. Soc. Géol. Belg.*, **96**, 585–624.

Monty, C.L.V. (1977) Evolving concepts on the nature and the ecological significance of stromatolites. In: *Fossil Algae, Recent Results and Developments* (Ed. by E. Flügel), pp. 15–35. Springer-Verlag, Berlin.

Monty, C.L.V. (1976) The origin and development of cryptalgal fabrics. In: *Stromatolites* (Ed. by M. R. Walter), *Developments in Sedimentology*, **20**, pp. 193–249. Elsevier, Amsterdam.

Monty, C.L.V. (1981) Spongiostromate vs. porostromate stromatolites and oncolites. In: *Phanerozoic Stromatolites* (Ed. by C. Monty), pp. 1–4. Springer-Verlag, Berlin.

Monty, C.L.V., Bosence, D.W.J., Bridges, P.H. and **Pratt, B.R.** (Eds) (1995) *Carbonate Mud-Mounds, Their Origin and Evolution. Spec. Publ. Int. Assoc. Sedim.*, **23**.

Monty, C.L.V. and **Hardie, L.A.** (1976) The geological significance of the freshwater blue-green algal calcareous marsh. In: *Stromatolites* (Ed. by M. R. Walter), *Developments in Sedimentology*, **20**, pp. 447–477. Elsevier, Amsterdam.

Moore, L.S. and **Burne, R.V.** (1994) The modern thrombolites of Lake Clifton, Western Australia. In: *Phanerozoic Stromatolites II* (Ed. by J. Bertrand-Sarfati and C. Monty), pp. 3–29. Kluwer, Dordrecht.

Morse, J.W. and **He, S.** (1993) Influences of T, S and P_{CO_2} on the pseudo-homogeneous precipitation of $CaCO_3$ from seawater; implications for whiting formation. *Mar. Chem.*, **41**, 291–297.

Mountjoy, E.W. and **Riding, R.** (1981) Foreslope stromatoporoid-renalcid bioherm with evidence of early cementation, Devonian Ancient Wall reef complex, Rocky Mountains. *Sedimentology*, **28**, 299–319.

Nealson, K.H. (1997) Sediment bacteria: who's there, what they are doing, and what's new. *Ann. Rev. Earth Planet. Sci.*, **51**, 403–434.

Neumann, A.C., Gebelein, C.D. and **Scoffin, T.P.** (1970) The composition, structure, and erodability of subtidal mats, Abaco, Bahamas. *J. Sedim. Petrol.*, **40**, 274–297.

Neuweiler, F., Gautret, P., Thiel, V., Lange, R., Michaelis, W. and **Reitner, J.** (1999) Petrology of Lower Cretaceous carbonate mud mounds (Albian, N. Spain): insights into organomineralic deposits of the geological record. *Sedimentology*, **46**, 837–859.

Noffke, N. (1998) Multidirected ripple marks rising from biological and sedimentological processes in modern lower supratidal deposits (Mellum Island, southern North Sea). *Geology*, **26**, 879–882.

Ota, N. (1969) Reef deposits in the *Millerella* zone of the Akiyoshi limestone group. *Palaeont. Soc. Japan, Spec. Pap.*, **14**, 1–12.

Paul, J. and **Peryt, T.M.** (1985) Oolithe und Stromatolithen im Unteren Buntsandstein des Heeseberges bei Jerxheim, Kreis Wolfenbuettel. *Bericht Naturhist. Gesellschaft Hannover*, **128**, 175–186.

Pedley, H.M. (1990) Classification and environmental models of cool freshwater tufas. *Sediment. Geol.*, **68**, 143–154.

Pedley, H.M. (1992) Freshwater (phytoherm) reefs: the role of biofilms and their bearing on marine reef cementation. *Sediment. Geol.*, **79**, 255–274.

Pentecost, A. (1978) Blue-green algae and freshwater carbonate deposits. *Proc. Roy. Soc. London*, **B200**, 43–61.

Pentecost, A. (1985) Association of cyanobacteria with tufa deposits: identity, enumeration and nature of the sheath material revealed by histochemistry. *Geomicrobiol. J.*, **4**, 285–298.

Pentecost, A. (1990) The formation of travertine shrubs; Mammoth Hot Springs, Wyoming. *Geol. Mag.*, **127**, 159–168.

Pentecost, A. and **Riding, R.** (1986) Calcification in cyanobacteria. In: *Biomineralization in Lower Plants and Animals* (Ed. by B. S. C. Leadbeater and R. Riding), *Systemat. Assoc. Spec. Vol.*, **30**, 73–90.

Perry, C.T. (1999) Biofilm-related calcification, sediment trapping and constructive micrite envelopes: a criterion for the recognition of ancient grass-bed environments? *Sedimentology*, **46**, 33–45.

Peryt, T.M. (1983a) Vadoids. In: *Coated Grains* (Ed. by T. M. Peryt), pp. 437–449. Springer-Verlag, Berlin.

Peryt, T.M. (1983b) Oncoids: comment to recent developments. In: *Coated Grains* (Ed. by T. M. Peryt), pp. 273–275. Springer-Verlag, Berlin.

Pflüger, F. (1999) Matground structures and redox facies. *Palaios*, **14**, 25–39.

Pia, J. (1926). *Pflanzen Als Gesteinsbildner.* Gebrüder Borntraeger, Berlin.

Pia, J. (1927) Thallophyta. In: *Handbuch der Paläobotanik*, Vol. **1** (Ed. by M. Hirmer), pp. 31–136. Oldenbourg, Munich.

Pitcher, M. (1964) Evolution of Chazyan (Ordovician) reefs of eastern United States and Canada. *Bull. Can. Petrol. Geol.*, **12**, 632–691.

Pierson, B.K., **Oesterle, A.** and **Murphy, G.L.** (1987) Pigments, light penetration, and photosynthetic activity in the multilayered microbial mats of Great Sippewisset Salt Marsh, Massachusetts. *FEMS Microbiol. Ecol.*, **45**, 365–376.

Playford, P.E. and **Cockbain, A.E.** (1976) Modern algal stromatolites at Hamelin Pool, a hypersaline barred basin in Shark Bay, Western Australia. In: *Stromatolites* (Ed. by M. R. Walter), *Developments in Sedimentology*, **20**, pp. 389–411. Elsevier, Amsterdam.

Pratt, B.R. (1982a) Stromatolitic framework of carbonate mud-mounds. *J. Sedim. Petrol.*, **52**, 1203–1227.

Pratt, B.R. (1982b) Stromatolite decline – a reconsideration. *Geology*, **10**, 512–515.

Pratt, B.R. and **James, N.P.** (1982) Cryptalgal-metazoan bioherms of Early Ordovician age in the St. George Group, western Newfoundland. *Sedimentology*, **29**, 543–569.

Pratt, B.R. (1995) The origin, biota and evolution of deep-water mud-mounds. In: *Carbonate Mud-Mounds, Their Origin and Evolution* (Ed. by C. L. V. Monty, D. W. J. Bosence, P. H. Bridges and B. R. Pratt), *Spec. Publ. Int. Assoc. Sedim.*, **23**, 49–123.

Pratt, B.R. and **James, N.P.** (1989) Early Ordovician thrombolite reefs, St. George Group, western Newfoundland. In: *Reefs, Canada and Adjacent Area* (Ed. by H. H. J. Geldsetzer, N. P. James and G. E. Tebbutt), *Can. Soc. Petrol. Geol., Mem.*, **13**, 1231–1240.

Rasmussen, K.A., **Macintyre, I.G.** and **Prufert, L.** (1993) Modern stromatolite reefs fringing a brackish coastline, Chetumal Bay, Belize. *Geology*, **21**, 199–202.

Read, J.F. (1976) Calcretes and their distinction from stromatolites. In: *Stromatolites* (Ed. by M. R. Walter), *Developments in Sedimentology*, **20**, pp. 55–71. Elsevier, Amsterdam.

Reid, R.P. (1987) Nonskeletal peloidal precipitates in Upper Triassic reefs, Yukon Territory (Canada). *J. Sedim. Petrol.*, **57**, 893–900.

Reid, R.P. and **Browne, K.M.** (1991) Intertidal stromatolites in a fringing Holocene reef complex, Bahamas. *Geology*, **19**, 15–18.

Reitner, J. (1993) Modern cryptic microbial/metazoan facies from Lizard Island (Great Barrier Reef, Australia) – formation and concepts. *Facies*, **29**, 3–40.

Reitner, J., **Gautret, P.**, **Marin, F.** and **Neuweiler, F.** (1995) Automicrites in a modern marine microbialite. Formation model via organic matrices (Lizard Island, Great Barrier Reef, Australia). *Bull. Inst. Océanogr. Monaco, Num. Spéc.*, **14**, 237–263.

Renaut, R. (1993) Morphology, distribution, and preservation potential of microbial mats in the hydromagnesite-magnesite playas of the Cariboo Plateau, British Columbia, Canada. *Hydrobiologia*, **267**, 75–98.

Revsbech, N.P., **Jørgensen, B.B.**, **Blackburn, T.H.** and **Cohen, Y.** (1983) Microelectrode studies of the photosynthesis and O_2, H_2S and pH profiles of a microbial mat. *Limnol. Oceanog.*, **28**, 1062–1074.

Richter, D.K., **Herforth, A.** and **Ott, E.** (1979) Pleistozäne, brackische Blaugrünalgenriffe mit *Rivularia haematites* auf der Perachorahalbinsel bei Korinth (Griechenland). *N. Jb. Geol. Paläont. Abh.*, **159**, 14–40.

Riding, R. (1977) Skeletal stromatolites. In: *Fossil Algae, Recent Results and Developments* (Ed. by E. Flügel), pp. 57–60. Springer-Verlag, Berlin.

Riding, R. (1991a) Calcified cyanobacteria. In: *Calcareous Algae and Stromatolites* (Ed. by R. Riding), pp. 55–87. Springer-Verlag, Berlin.

Riding, R. (1991b) Classification of microbial carbonates. In: *Calcareous Algae and Stromatolites* (Ed. by R. Riding), pp. 21–51. Springer-Verlag, Berlin.

Riding, R. (1992) Temporal variation in calcification in marine cyanobacteria. *J. Geol. Soc. London*, **149**, 979–989.

Riding, R. (1993) Phanerozoic patterns of marine $CaCO_3$ precipitation. *Naturwissenschaften*, **80**, 513–516.

Riding, R. (1994) Evolution of algal and cyanobacterial calcification. In: *Early Life on Earth* (Ed. by S. Bengtson), *Nobel Symposium*, **84**, pp. 426–438. Columbia University Press, New York.

Riding, R. (1997) Stromatolite decline: a brief reassessment. In: *Biosedimentology of Microbial Buildups* (Ed. by F. Neuweiler, J. Reitner and C. Monty), IGCP Project 380, Proceedings of 2nd Meeting, Göttingen, Germany. *Facies*, **36**, 227–230.

Riding, R. (2000) The term stromatolite: towards an essential definition. *Lethaia*, in press.

Riding, R. and **Awramik, S.M.** (Eds) (2000). *Microbial Sediments.* Springer-Verlag, Heidelberg.

Riding, R. and **Sharma, M.** (1998) Late Palaeo-proterozoic (~1800–1600 Ma) stromatolites, Cuddapah Basin, southern India: cyanobacterial or other bacterial microfabrics? *Precamb. Res.*, **92**, 21–35.

Riding, R., Awramik, S.M., Winsborough, B.M., Griffin, K.M. and **Dill, R.F.** (1991a) Bahamian giant stromatolites: microbial composition of surface mats. *Geol. Mag.*, **128**, 227–234.

Riding, R., Braga, J.C. and **Martín, J.M.** (1991b) Oolite stromatolites and thrombolites, Miocene, Spain: analogues of Recent giant Bahamian examples. *Sediment. Geol.*, **71**, 121–127.

Riding, R., Martín, J.M. and **Braga, J.C.** (1991c) Coral stromatolite reef framework, Upper Miocene, Almería, Spain. *Sedimentology*, **38**, 799–818.

Robbins, L.L. and **Blackwelder, P.L.** (1992) Biochemical and ultrastructural evidence for the origin of whitings; a biologically induced calcium carbonate precipitation mechanism. *Geology*, **20**, 464–468.

Robbins, L.L. and **Evans, C.A.** (1997) Temporal and spatial distribution of whitings on Great Bahama Bank and a new lime mud budget. *Geology*, **25**, 947–950.

Rouchy, J.M. and **Monty, C.** (1981) Stromatolites and cryptalgal laminites associated with Messinian gypsum of Cyprus. In: *Phanerozoic Stromatolites; Case Histories* (Ed. by C. Monty), pp. 155–180. Springer, Berlin.

Rowland, S.M. and **Gangloff, R.A.** (1988) Structure and paleoecology of Lower Cambrian reefs. *Palaios*, **3**, 111–135.

Schidlowski, M. (1988) A 3,800-million-year isotopic record of life from carbon in sedimentary rocks. *Nature*, **333**, 313–318.

Schmitt, M. and **Monninger, W.** (1977) Stromatolites and thrombolites in Precambrian/Cambrian boundary beds of the Anti-Atlas, Morocco: preliminary results. In: *Fossil Algae, Recent Results and Developments* (Ed. by E. Flügel), pp. 80–85. Springer-Verlag, Heidelberg.

Scholl, D.W. (1960) Pleistocene algal pinnacles at Searles Lake, California. *J. Sedim. Petrol.*, **30**, 414–431.

Scholl, D.W. and **Taft, W.H.** (1964) Algae, contributors to the formation of calcareous tufa, Mono Lake, California. *J. Sedim. Petrol.*, **34**, 309–319.

Schopf, J.W. (Ed.) (1983) *Earth's Earliest Biosphere: its Origin and Evolution.* Princeton University Press, Princeton.

Schopf, J.W. (1994) The oldest known records of life early Archean stromatolites, microfossils, and organic matter. In: *Early Life on Earth* (Ed. by S. Bengtson), *Nobel Symposium*, **84**, pp. 193–206. Columbia University Press, New York.

Schopf, J.W. and **Walter, M.R.** (1982) Origin and early evolution of cyanobacteria: the geological evidence. In: *The Biology of Cyanobacteria* (Ed. by N. G. Carr and B. A. Whitton), pp. 543–564. Blackwell, London.

Schubert, J.K. and **Bottjer, D.J.** (1992) Early Triassic stromatolites as post-mass extinction disaster forms. *Geology*, **20**, 883–886.

Scoffin, T.P. (1970) The trapping and binding of subtidal carbonate sediments by marine vegetation in Bimini Lagoon, Bahamas. *J. Sedim. Petrol.*, **40**, 249–273.

Scoffin, T.P. (1971) The conditions of growth of the Wenlock reefs of Shropshire (England). *Sedimentology*, **17**, 173–219.

Scott, R.W. (1981) Biotic relations in Early Cretaceous coral-algal-rudistid reefs, Arizona. *J. Paleontol.*, **55**, 463–478.

Seilacher, A., Bose, P.K. and **Pflüger, F.** (1998) Triploblastic animals more than 1 billion years ago: trace fossil evidence from India. *Science*, **282**, 80–83.

Semikhatov, M.A. (1980) On the upper Precambrian stromatolite standard of north Eurasia. *Earth-Sci. Rev.*, **16**, 235–247.

Semikhatov, M.A., Gebelein, C.D., Cloud, P., Awramik, S.M. and **Benmore, W.C.** (1979) Stromatolite morphogenesis – progress and problems. *Can. J. Earth Sci.*, **16**, 992–1015.

Semikhatov, M.A. and **Raaben, M.E.** (2000) Proterozoic stromatolite taxonomy and biostratigraphy. In: *Microbial Sediments* (Ed. by R. Riding and S. M. Awramik), pp. 295–306. Springer-Verlag, Heidelberg.

Senowbari-Daryan, B. and **Flügel, E.** (1993) *Tubiphytes* Maslov, an enigmatic fossil: classification, fossil record and significance through time. Part I: discussion of Late Palaeozoic material. *Boll. Soc. Pal. Ital. Spec. Vol.* **1**, 353–382.

Serebryakov, S.N. (1976) Biotic and abiotic factors controlling the morphology of Riphean stromatolites. In: *Stromatolites* (Ed. by M. R. Walter), *Developments in Sedimentology*, **20**, pp. 321–336. Elsevier, Amsterdam.

Serebryakov, S.N. and **Semikhatov, M.A.** (1974) Riphean and Recent stromatolites: a comparison. *Am. J. Sci.*, **274**, 556–574.

Shearman, D.J., McGugan, A., Stein, C. and **Smith, A.J.** (1989) Ikaite, $CaCO_3.6H_2O$, precursor of the thinolites in the Quaternary tufas and tufa mounds of the Lahontan and Mono Lake basins, western United States. *Geol. Soc. Am. Bull.*, **101**, 913–917.

Shinn, E.A. (1983a) Birdseyes, fenestrae, shrinkage pores, and loferites; a reevaluation. *J. Sedim. Petrol.*, **53**, 619–628.

Shinn, E.A. (1983b) Tidal flat environment. In: *Carbonate Depositional Environments* (Ed. by P. A. Scholle, D. G. Bebout and C. H. Moore), *Am. Assoc. Petrol. Geol., Memoir*, **33**, 171–210.

Shinn, E.A., Lloyd, R.M. and **Ginsburg, R.N.** (1969) Anatomy of a modern carbonate tidal-flat, Andros island, Bahamas. *J. Sedim. Petrol.*, **39**, 1202–1228.

Soja, C.M. (1994) Significance of Silurian stromatolite-sphinctozoan reefs. *Geology*, **22**, 355–358.

Sumner, D.Y. (2000) Microbial versus environmental influences on the morphology of late Archean fenestrated microbialites. In: *Microbial Sediments* (Ed. by R. Riding and S. M. Awramik), pp. 307–314. Springer-Verlag, Heidelberg.

Sun, S.Q. and **Wright, V.P.** (1989) Peloidal fabrics in Upper Jurassic reefal limestones, Weald Basin, southern England. *Sediment. Geol*, **65**, 165–181.

Tappan, H. (1980) *The Paleobiology of Plant Protists.* W.H. Freeman, San Francisco.

Thompson, J.B. (2000) Microbial whitings. In: *Microbial Sediments* (Ed. by R. Riding and S. M. Awramik), pp. 250–260. Springer-Verlag, Heidelberg.

Thompson, J.B. and **Ferris, F.G.** (1990) Cyanobacterial precipitation of gypsum, calcite, and magnesite from natural alkaline lake water. *Geology*, **18**, 995–998.

Thrailkill, J. (1976) Speleothems. In: *Stromatolites* (Ed. by M. R. Walter), *Developments in Sedimentology*, **20**, pp. 73–86. Elsevier, Amsterdam.

Trichet, J. and **Défarge, C.** (1995) Non-biologically supported organomineralization. *Bull. Inst. Océanogr. Monaco, Num. Spéc.*, **14**, 203–236.

Turner, E.C., Narbonne, G.M. and **James, N.P.** (1993) Neoproterozoic reef microstructures from the Little Dal Group, northwestern Canada. *Geology*, **21**, 259–262.

Vasconcelos, C., McKenzie, J.A., Bernasconi, S., Grujic, D. and **Tien, A.J.** (1995) Microbial mediation as a possible mechanism for natural dolomite formation at low temperatures. *Nature*, **377**, 220–222.

Verrecchia, E.P. and **Verrecchia, K.E.** (1994) Needle-fiber calcite: a critical review and a proposed classification. *J. Sedim. Res.*, **A64**, 650–664.

Verrecchia, E.P., Freytet, P., Verrecchia, K.E. and **Dumont, J.L.** (1995) Spherulites in calcrete laminar crusts; biogenic $CaCO_3$ precipitation as a major contributor to crust formation. *J. Sedim. Res.*, **A65**, 690–700.

Visscher, P.T., Prins, R.A. and **van Gemerden, H.** (1992) Rates of sulfate reduction and thiosulfate consumption in a marine microbial mat. *FEMS Microbiol. Ecol.*, **86**, 283–294.

Vologdin, A.G. (1962). *The Oldest Algae of the USSR.* Academy of Sciences, Moscow. In Russian.

Walcott, C.D. (1914) Cambrian geology and paleontology III: Precambrian Algonkian algal flora. *Smithsonian Misc. Collection*, **64**, 77–156.

Wallner, J. (1934) Beitrag zur Kenntnis der Vaucheria-Tuffe. *Z. Bakteriol. Parasitkd.*, **90**, 150–193.

Wallner, J. (1935) Diatomeen als Kalkbildner. *Hedwigia*, **75**, 137–141.

Walter, M.R. (1972) Stromatolites and the biostratigraphy of the Australian Precambrian and Cambrian. *Spec. Pap. Palaeontol.*, **11**, 33 pls.

Walter, M.R. (1976) Introduction. In: *Stromatolites* (Ed. by M. R. Walter), *Developments in Sedimentology*, **20**, pp. 1–3. Elsevier, Amsterdam.

Walter, M.R. (1983) Archean stromatolites: evidence of Earth's earliest benthos. In: *Earth's Earliest Biosphere: its Origin and Evolution* (Ed. by J. W. Schopf), pp. 187–213. Princeton University Press, Princeton.

Walter, M.R. (1994) Stromatolites: the main geological source of information on the evolution of the early benthos. In: *Early Life on Earth* (Ed. by S. Bengtson), *Nobel Symposium*, **84**, pp. 270–286. Columbia University Press, New York.

Walter, M.R. and **Heys, G.R.** (1985) Links between the rise of the Metazoa and the decline of stromatolites. *Precamb. Res.*, **29**, 149–174.

Walter, M.R., Bauld, J. and **Brock, T.D.** (1972) Siliceous algal and bacterial stromatolites in hot spring and geyser effluents of Yellowstone National Park. *Science*, **178**, 402–405.

Walter, M.R., Grotzinger, J.P. and **Schopf, J.W.** (1992) Proterozoic stromatolites. In: *The Proterozoic Biosphere* (Ed. by J. W. Schopf and C. Klein), pp. 253–260. Cambridge University Press, Cambridge.

Webb, G.E. (1996) Was Phanerozoic reef history controlled by the distribution of non-enzymatically secreted reef carbonates (microbial carbonate and biologically induced cement)? *Sedimentology*, **43**, 947–971.

Webb, G.E., Baker, J.C. and **Jell, J.S.** (1998) Inferred syngenetic textural evolution in Holocene cryptic reefal microbialites, Heron Reef, Great Barrier Reef, Australia. *Geology*, **26**, 355–358.

Webb, G.E., Jell, J.S. and **Baker, J.C.** (1999) Cryptic intertidal microbialites in beachrock, Heron Island, Great Barrier Reef: implications for the origin of microcrystalline beachrock cement. *Sediment. Geol.*, **126**, 317–334.

Wiegel, J. and **Adams, M.W.W.** (Eds) (1998). *Thermophiles: the Keys to Molecular Evolution and the Origin of Life?* Taylor & Francis, London.

Winsborough, B.M. (2000) Diatoms and benthic microbial carbonates. In: *Microbial Sediments* (Ed. by R. Riding and S. M. Awramik), pp. 76–83. Springer-Verlag, Heidelberg.

Winsborough, B.M. and **Golubic, S.** (1987) The role of diatoms in stromatolite growth: two examples from modern freshwater settings. *J. Phycol.*, **23**, 195–201.

Woese, C.R. (1987) Bacterial evolution. *Microbiol. Rev.*, **51**, 221–271.

Wolfaardt, G.M., Lawrence, J.R., Robarts, R.D., Caldwell, S.J. and **Caldwell, D.E.** (1994) Multicellular organization in a degradative biofilm community. *Appl. Environ. Microbiol.*, **60**, 434–446.

Wolff, M. and **Füchtbauer, H.** (1976) Die karbonatishe Randfazies der tertiären Süβwasserseen des Nördlinger Ries und des Steinheimerbeckens. *Geol. Jahrb*, **D14**, 3–53.

Wray, J.L. (1967) Upper Devonian calcareous algae from the Canning Basin, Western Australia. *Colorado School Mines Prof. Contrib.*, **3**.

Wright, D.T. (1999) The role of sulphate-reducing bacteria and cyanobacteria in dolomite formation in distal ephemeral lakes of the Coorong region, South Australia. *Sediment. Geol.*, **126**, 147–157.

Wright, V.P. (1986) The role of fungal biomineralization in the formation of early Carboniferous soil fabrics. *Sedimentology*, **33**, 831–838.

Wright, V.P. (1989) Terrestrial stromatolites and laminar calcretes; a review. *Sediment. Geol.*, **65**, 1–13.

Wright, V.P., Platt, N.H. and **Wimbledon, W.A.** (1988) Biogenic laminar calcrete: evidence of calcified root-mat horizons in paleosols. *Sedimentology*, **35**, 603–620.

Zhuravlev, A. Yu. (1996) Reef ecosystem recovery after the Early Cambrian extinction. In: *Biotic Recovery from Mass Extinction Events* (Ed. by M. B. Hart), *Geol. Soc. London Spec. Publ.*, **102**, 79–96.

Sedimentology (2000) **47** (Suppl. 1), 215–238

Deposition and early alteration of evaporites

B. CHARLOTTE SCHREIBER* and MOHAMED EL TABAKH†
*Appalachian State University, Department of Geology, 195 Rankin Science Building, Boone, NC 28608, USA (E-mail: geologo@aol.com)
†Queens College (CUNY), Department of Earth and Environmental Sciences, 65–30 Kissena Boulevard, Flushing, NY 11367, USA (E-mail: eltabakh@hotmail.com)

ABSTRACT

The depositional settings for primary and early diagenetic evaporite deposits generally fall into three categories: marginal (mixed shallow-subaqueous and subaerial), shallow and deep subaqueous. These three environmental groupings hold for both marine and nonmarine settings, although the details of continental evaporites may be far more complex than in most marine-fed water bodies. The primary evaporite morphologies from many continental (playa), hypersaline marine and marine-marginal depositional settings are reasonably well understood, because of the numerous detailed studies of recent, Holocene and Cenozoic deposits that serve as models for sedimentary interpretation. The sedimentological features that develop in deeper water settings are inferred, based on examination of unaltered Cenozoic deposits. Each environmental setting develops characteristic depositional features and patterns, and one facies grades into the next. Because there may be significant physiochemical changes in water composition during deposition as well as sudden change(s) in both water depth and basinal circulation, description and interpretation of evaporative rocks should not be based on mineralogy alone, but on the distinct sedimentary characteristics of each part of a deposit. In cases where sediments still reflect their primary mineralogy and morphology, most of the environments can now be recognized; however, geochemical studies are commonly required to determine the source(s) of the original water. The determinative geochemical techniques that presently serve as a support to sedimentologic study include studies of fluid inclusions, bromine content of chloride salts, and the stable isotopes of strontium, sulphur, carbon and oxygen. Only after these and perhaps other chemical analyses are considered can the full depositional history of a region be reasonably unravelled.

INTRODUCTION

Evaporites include a wide range of chemical precipitates that form at the Earth's surface or in near-surface environments from brines concentrated by solar evaporation in restricted basins. This paper discusses the deposition and very early (synsedimentary) diagenesis of evaporative deposits. The best known of these deposits are the marine-marginal (sabkha: Butler *et al.*, 1982; Warren & Kendall, 1985) and shallow hypersaline environments (salina or shallow-water body: Schreiber *et al.*, 1976; Busson *et al.*, 1982; Ortí *et al.*, 1984; Geisler-Cussey, 1997; Rosell *et al.*, 1998). Much information concerning deeper environments has been inferred from Cenozoic and older deposits (Schreiber *et al.*, 1976; Schlager & Bolz, 1977; Lowenstein & Hardie, 1985). In the past few years, studies of entirely nonmarine evaporative sediments document some of these more complex assemblages (Busson & Schreiber, 1997; Niemi *et al.*, 1997). Only the settings and a general summary of nonmarine evaporites are presented in this paper, because of their vast complexity and diversity; however, morphologically, they are similar to marine-sourced deposits.

Geochemical studies greatly aid in the definition of the origin and history of evaporites. The geochemical analyses that are useful in determining the origins of many evaporites include studies of fluid inclusions, trace elements, bromine-to-chlorine ratios, $^{87}Sr/^{86}Sr$ ratios, sulphur and oxygen isotopes, and the oxygen and carbon isotope ratios in associated carbonates. These studies facilitate an understanding of the chemical evolution of the waters from which evaporites originate.

CLIMATIC RELATIONSHIPS

As water evaporates and solutes become more concentrated, the rate of evaporation slows, because of the increasing density and surface tension of the brine. Because very saline waters are denser than fresh water, a higher proportion of the infrared of incoming sunlight refracts back into the water and causes the internal temperature of the water to rise as high as 35–55°C. The high temperature permits evaporation to continue even in moderately humid regions. At high elevations (lower atmospheric pressure) or in very windy areas, the evaporation process accelerates. Additionally, halophylic bacteria commonly live in saline water (Colwell *et al.*, 1979; Sammy, 1985; Javor, 1989), colouring it pink or red. This causes the water temperature to rise 3–6°C above similar waters without these bacteria.

In regions where relative humidity is above 65%, halite may form, but is preserved with difficulty (Kinsman, 1976). When the humidity falls below 65%, halite forms and is preserved. Low relative humidity throughout the year (below 35%) is needed to continue evaporation beyond halite, to form and preserve potassic and magnesian salt precipitates. Such saline waters and the resulting salts are strongly hygroscopic and diurnal change in humidity is commonly enough to dissolve most potassium/magnesium salts that form (carnallite, $MgCl_2 \cdot KCl \cdot 6H_2O$ and sylvite, KCl being the most common). However, polyhalite, which is a syngenetic mineral, is somewhat more stable ($2CaSO_4 \cdot MgSO4 \cdot K2SO4 \cdot 2H_2O$; Pierre, 1983; Peryt & Pierre, 1994). Settings arid enough to precipitate potassic and magnesian salts are rare, occurring either at very high elevations and/or within orographic shadows developed in hot and arid regions. For details of evaporation mechanisms see Steinhorn and Assaf (1977) and Steinhorn (1997). At least one study (Braitsch, 1964) suggests that, in order for carnallite ($MgCl_2 \cdot KCl \cdot 6H_2O$) to precipitate, water temperatures must be considerably elevated, in the range 41–47°C (d'Ans, 1947; Casas *et al.*, 1992). Maximum salina temperatures observed in a number of studies of modern environments fall even above this range, as in the supratidal areas of the Arabian Gulf (Kinsman, 1969) and the solar lakes along the Gulf of Elat (Pierre, 1989).

FORMATIVE ENVIRONMENTS

Evaporite minerals may form where the rate of evaporation of a water body is greater than the total water inflow, leaving a concentrated mineral residue. There are three critical controlling factors in evaporative mineral formation and accumulation: initial ionic content (and ratios), temperature and relative humidity. Backreactions with already-formed sediments and with associated rock may also enter into the system (Harvie *et al.*, 1980). A fresh-water lake may become intensely saline as a result of evaporation and concentration, as well as a marine-fed water body; however, the time required to accumulate a given volume of evaporites from nonmarine water is usually much greater. Momenzadeh (1990) has suggested that hydrothermal rift waters may generate major evaporative water bodies (a volcanogenic model) with comparatively little marine input; the bulk of precipitating evaporites arising from hot springs and active hydrothermal sources, as in the lakes of the East African Rift zone. Generally, the major evaporative sequences appear to have largely formed in marine-fed or mixed-water environments, although massive evaporite deposits do exist in entirely nonmarine settings.

A further complication to understanding the formative setting of marine-sourced evaporative sequences is that the composition of sea water might have changed through geologic time. Stanley and Hardie (1998, 1999) suggest that marine compositional changes are directly tied to rates of sea-floor spreading, while Holland *et al.* (1996) propose that these same variations are the result of major changes in early diagenetic processes on land, affecting nonmarine input into oceans.

ENVIRONMENTS OF DEPOSITION

Observation of evaporite deposition in modern sabkhas, playa flats and shallow subaqueous evaporite deposits demonstrates that evaporites can form into thick accumulations of sediment in a number of different settings, all within arid hypersaline environments (outlined in Fig. 1). Massive accumulation of evaporites may take place in geologically short depositional intervals (>10 m thickness per thousand years; Schreiber & Hsü, 1980). Evaporites that form in deeper water from stratified hypersaline water bodies probably accumulate at slower rates (1·8 m per thousand years; Kirkland *et al.*, 1999), because the evaporation rate is slower as a result of a lower ratio of surface area to water volume.

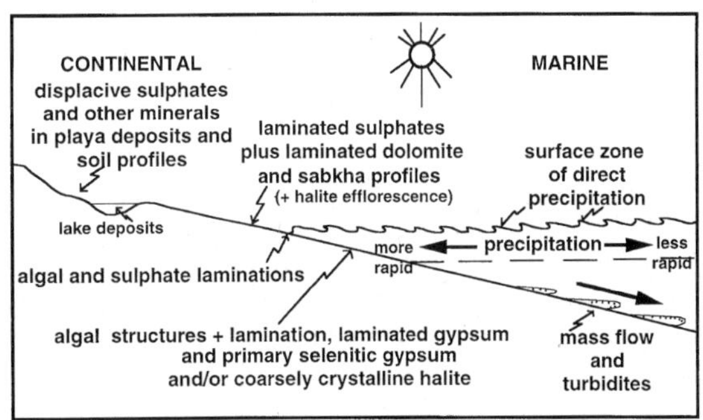

Fig. 1. Diagram illustrating the diverse environments of evaporite formation from subaerial (continental) to hypersaline deep basin (modified from Schreiber *et al.*, 1977).

Marine-marginal sabkhas

Marine-marginal sabkhas (arid region salt marshes) accumulate as part of outbuilding topset beds along a generally regressive shoreline, but some sabkhas may lie within the topmost beds of prograding sequences. The sabkha deposits of the Arabian Gulf are well studied in Iran, Kuwait and the Emirates (Shearman, 1963; Butler, 1970; Purser, 1973, 1985; Butler *et al.*, 1982; Kirkham, 1997). Similar deposits are also well documented from the coasts of the Nile Delta, Tunisia, Morocco and Australia, as presented by West (1979) in the modern sabkhas of the Mediterranean coast of Egypt; from South Australia (Warren, 1982); and North Africa (Perthuisot, 1975, 1980). Such shoreline deposits can prograde a thickness of about 1 m across 1 km of shoreline per thousand years (as clastic, carbonate or mixed carbonate plus clastic accumulations, all with evaporative sediments; Schreiber & Hsü, 1980). In this environment, the evaporites form within the mineralized zone of the soil profile of the sabkha, similarly to a caliche (an aridisol). Observations of deposition in modern marine sabkhas (Fig. 2A) show that such settings can form substantial accumulations of sediment, covering broad areas in geologically short intervals of time. Comparable accumulations also form in the mudflats at the margins of saline lakes.

Depositional profiles

The traditional sabkha profile, first mapped by Butler (1970) and then elaborated in many papers and summarized by Butler *et al.* (1982), (Fig. 2A) is more complex than the simple progradation processes originally envisioned. In the type section in Abu Dhabi, sabkha growth was apparently controlled by the existence of two older morphological features that also controlled sedimentation (Kirkham, 1997): (1) storm features (regressive), such as spits that built out between barrier islands or tombolos tied to the mainland; and (2) older Flandrian (transgressive) evaporitive sabkhas that were deposited 1–2 m higher than the present sabkha surface (Kirkham, 1997). It is possible that some of the thicker deposits reported in Butler *et al.* (1982) are the product of this superposition and amalgamation of two sabkha deposits of different ages (Fig. 2B).

As noted by Shearman (1978), the upper surface of sabkhas is a nearly level erosion or deflation surface controlled by the groundwater capillary zone, which governs the thickness of present-day sediment accumulation. The sabkha progrades seaward through the accumulation and stabilization of water- and wind-driven particles held together by the binding action of algal/bacterial mats and early cementation. This mostly aggradational process builds a soil profile within the sabkha that is dominated by interaction with saline, usually alkaline groundwater. Sporadically flooded areas accumulate behind what is left of partially eroded Flandrian beach ridges (see Fig. 2B). In these ponds, shallow subaqueous evaporite deposits form that are later incorporated into the general sabkha profile (see model in Warren & Kendall, 1985). Kirkham (1997) suggests that the ponded water is nonmarine, but Shearman points out that sabkhas are only 1–2 m above present sea level, i.e. well within the range of storm-driven marine water (a common occurrence), so that replenishment by marine waters is a regular feature of the setting.

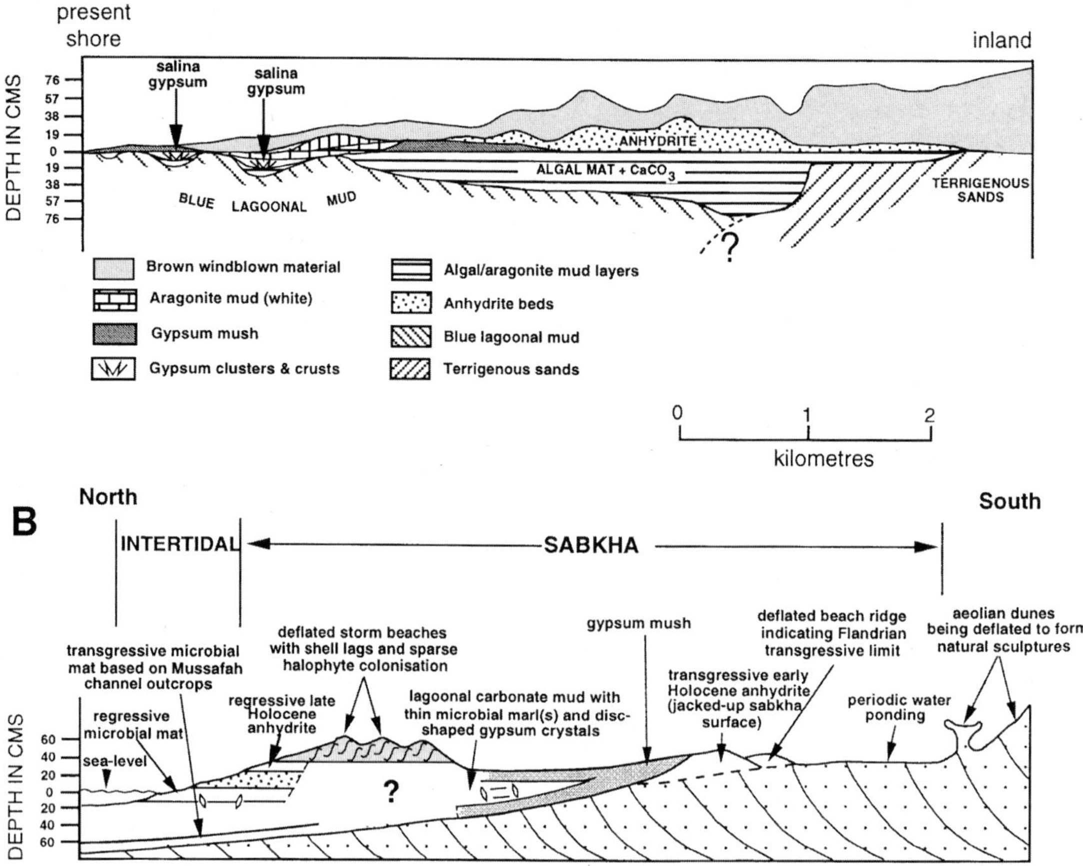

Fig. 2. Idealized cross-section views of the Abu Dhabi sabkha, illustrating different stratigraphic presentations: (A) as adapted from Butler *et al.* (1982); (B) as adapted from Kirkham (1997). The horizontal scale remains the same in A and B. The vertical scale in A is as given by Butler *et al.* (1982), while the vertical scale in B is approximate (not given by Kirkham, 1997).

Whether the simple Shearman–Butler sabkha model (see, for example, Shearman, 1978), or the more complex Kirkham (1997) model is employed, when a single-cycle sabkha is complete, it is composed of three basic components (Fig. 3): at the base, it has laminar, intertidal algal/bacterial mats with intercalated muds (commonly containing displacive, lenticular gypsums); in the middle (the capillary zone of the soil profile), it contains displacive nodular anhydrite/or gypsum with coalescent sulphate nodules and enterolithic sulphate layers in a matrix composed of carbonate, siliciclastic or mixed sands and muds; and, at the top, there is a truncation surface cut either by deflation or storm action. The total thickness of a typical sabkha deposit ranges from 30 cm to 1–2 m.

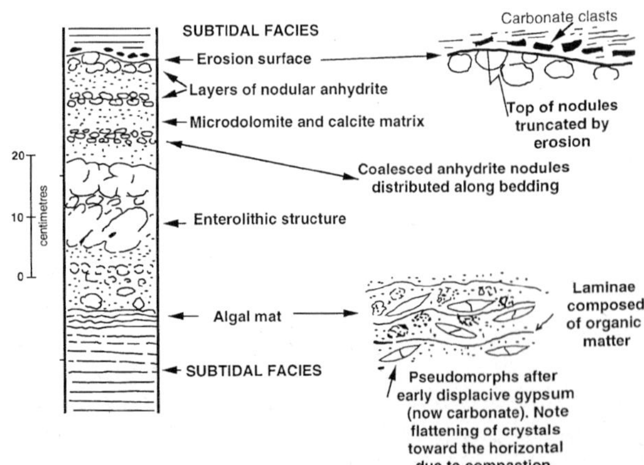

Fig. 3. Idealized sedimentary section of a sabkha. Adapted from Shearman (1978). This section was drawn from a core (Warlingham borehole).

Gypsum and anhydrite

Gypsum crystals form in ponds on the sabkha surface (prismatic and/or twinned swallow-tail form), along vertical desiccation cracks in the algal mats (elongate lenticular forms, aligned along the cracks), and as displacive subsurface crystals (lenticular and/or tabular) in the algal mats and in the algal peat deposits that lie below the depositional surface (Aref, 1998). Gypsum (usually in lenticular to tabular forms) forms within the vadose and capillary groundwater zones. It also forms in the phreatic zone (as lenticular forms). During arid periods, the pore water in the capillary zone becomes very saline. Most of the gypsum in this zone is then dehydrated to hemihydrate or anhydrite, losing much of its original crystal form and becoming irregular nodules and layers (Shearman, 1978). During periods of increased rainfall, much of the dehydrated sulphate is rehydrated to form gypsum, but this new microcrystalline gypsum retains the same 'dehydrated' format, as small polycrystalline nodules (gypsum crystals are 0·2–0·3 mm). The original crystal shapes of the primary gypsum are lost. These nodules may grow and enlarge as a result of repeated periods of hydration and dehydration accompanied by inflow of new sulphate-rich waters from adjacent shallow-marine environments. The earlier sulphates serve as nucleation sites for additional anhydrite or gypsum precipitation (Shearman & Fuller, 1969). Sulphate and halite also may be added to the groundwater as aerosols from the ocean (Eckardt & Spiro, 1999). Additional sulphate accumulation enlarges nodules that commonly merge and become continuous layers with a nodular mosaic structure (either as gypsum and/or anhydrite; Fig. 3).

Because regional climatic shifts take place slowly, over a span of hundreds or thousands of years, substantial layers of anhydrite may develop within the soil profile, with or without phases of (re)hydration. In some cycles, the sulphate is sparse, perhaps only present as occasional nodules. However, in other cases, a single cycle is composed almost entirely of nodules, coalescent nodular mosaics, and buckled, convoluted enterolithic layers (soft sediment deformation; Fig. 3). In areas where a great deal of sulphate is added as a result of progressive evaporation, the surface of the sabkha is jacked up (rises). The raised areas then become prone to further erosional truncation, losing the upper portion of the original depositional profile. Nodules of earlier sulphates may be reworked and deposited, mixed with fine clastics or bioclasts, with the onset of a new sabkha cycle. With the transgression of new marine water over the sabkha (still in restricted environments), well-bedded gypsum or anhydrite layers may form above the sequence.

Halite on the sabkha

The first well-documented study of a modern analogue for a sabkha salt deposit was by Shearman (1970) at Salina Ometepec (Southern California). Because that halite was actually formed under a salina, which dried sporadically, it will be discussed in detail under 'subaqueous deposits'. Similar but scattered lagoons and embayments of shallow-water halite are incorporated into a sabkha profile. Generally, these layers are only 5–20 cm thick, underlain by vertically aligned anhydrite nodules (relics of gypsum crystals), and pinch out laterally (perhaps as a result of early dissolution by groundwater).

Supratidal halite also develops on modern sabkhas from salt spray or the migration of saline groundwater to the surface of drying marginal salt flats of salinas. In these settings, halite commonly grows as delicate dendritic whiskers and crusts, a superficial efflorescence. Similar halite efflorescence develops on exposed playa surfaces on mudflats associated with saline lakes (see Smoot & Castens-Seidell, 1994). These fine salt structures are not readily preserved in the rock record, although the salt may accumulate and be preserved as wispy layers that fill concave desiccation polygons. Halite ooids or salt spheroids are occasionally found in these hollows (Castanier *et al.*, 1999) and form in shallow saline pans such as those in the southern Dead Sea (Weiler *et al.*, 1974) and on the marginal sabkhas of Lac Asal of Djibouti (Perthuisot *et al.*, 1993).

In the same sabkha settings, or from desiccating salinas or playas, displacive halite cubes may develop from groundwater within the upper portions of the underlying mud or sand (Gornitz & Schreiber, 1981). Groundwater is drawn by capillarity to the exposed sabkha surface, and precipitation of halite begins just at or above the groundwater table. Where extreme supersaturation exists, delicate frameworks of skeletal halite crystals grow displacively within soft tidal flat deposits (Southgate, 1982). In the Trucial Coast cores from the study by Butler *et al.* (1982), several layers contain skeletal displacive halite crystals. Where such crystals grow very quickly, they develop a marked skeletal aspect, having more enclosed matrix and rapid growth on their corners and edges, forming 'pagoda' halites. Where displacive halite growth is plentiful, the bedding of the host sediment becomes entirely disrupted and chaotic (Handford, 1982).

Sabkha summary

The depositional record formed within a sabkha setting is defined by the presence of a sedimentary profile developed adjacent to a water body in an arid setting. The matrix may be carbonate, mixed or siliciclastic, and can be emplaced by wind and/or water. The evaporites accumulate as part of a soil profile in the upper phreatic and vadose zones. Gypsum, anhydrite and halite are the most common evaporite minerals, although continental water influx may cause other minerals to form. Sabkha accumulations are commonly thin sequences (30 cm to 1–2 m), with each cycle topped by truncated wind or water-cut surfaces. Stacked, repeated sequences are not uncommon in the rock record; however, very thick, massive, regular beds of nodular sulphate do not represent a sabkha accumulation.

Shallow-water subaqueous evaporite deposits

It is somewhat easier to identify processes in artificial salinas than in a natural setting, because the ponds are subdivided and carefully controlled. The salina studies conducted at Salins-du-Midi (southern France: Busson *et al.*, 1982) and at the Salina Santa Pola (Alicante, Spain: Ortí *et al.*, 1984) form the core of modern understanding of subaqueous evaporite deposition (Fig. 4; reviewed in Geisler-Cussey, 1997). Their observations reveal the wide diversity in organic and carbonate accumulations as well as in gypsum and halite morphology, and were placed into a clear environmental framework by Rosell *et al.* (1998). Comparable diversity exists in nonmarine subenvironments (Geisler-Cussey, 1997). Lowenstein and Hardie (1985) discuss petrographic features that aid in the definition of both shallow and deeper water evaporite features.

Specific environments and biota

Much of what is known about shallow-water evaporites and associated biota comes from studies of modern 'salt' ponds. The first ponds into which seawater is drawn in a controlled, artificial salina, the 'evaporators', are commonly very broad, having many times the area of the gypsum and halite 'crystallizers'. A fairly diverse marine fauna lives in the first ponds, 35–36 $g\,L^{-1}$ salinity up to 55 $g\,L^{-1}$ (Total Dissolved Solids; TDS), but it becomes more restricted with rising salinity. Specially adapted fish, gastropods and pelecypods are common in this zone, often in quantities great enough to have commercial value. In the ponds with a salinity of >55 $g\,L^{-1}$, only a very restricted faunal grouping remains, often having a large population but with a low diversity. The floral population of these first ponds is fairly diverse with green and brown algae. Live diatoms persist up to salinities of 150 $g\,L^{-1}$, but with a very restricted population of only two species by 128 $g\,L^{-1}$ (Noel, 1984). Above salinities of 120 $g\,L^{-1}$, the population is largely composed of a restricted assemblage of cyanobacteria, with plentiful *Aphanotheca* sp. and *Dunaliella* sp. (Cornee, 1982, 1984). The bacteria live both in the water column and on the bottom of the ponds. They form rubbery mats on the bottom and grow in millimetre-scale laminae, building up substantial thicknesses. Occasional desiccation causes shrinkage polygons in the sediment surface, so the mats are not wholly continuous. Storm-generated 'roll-ups', folded and torn mats, and dismembered clots are common in the shallower ponds. Because the water is considerably more saline than seawater, there are no burrowing fauna and, other than desiccation cracks, the mats form fairly continuous beds of laminites. Carbonate, in the form of accumulations of seasonally controlled Cerithid gastropods, is common in some salinas, but most of it generally consists of micritic $CaCO_3$ coatings on bacterial filaments (Cornée, 1982, 1984).

Above 150 $g\,L^{-1}$ salinity, in the evaporator ponds, the biotal composition becomes almost entirely halophylic, salt-acclimatized bacteria (*Halobacterium*) that give the water a pink colour (in ponds of 325–350 $g\,L^{-1}$, the water is red). Brine shrimp (*Artemia* sp.) live in these ponds in large numbers and thrive on the halophylic bacteria.

Evaporative carbonates and organic matter

Carbonates formed under evaporative conditions in ponds having salinities of 120–150 $g\,L^{-1}$ are commonly similar in appearance to nonevaporative pond deposits (see above and Fig. 4). The major difference is that there are few or no fossils and little or no burrowing in the evaporative sediments. In thin section, they range from laminar to clotted, weakly bedded micrite. This

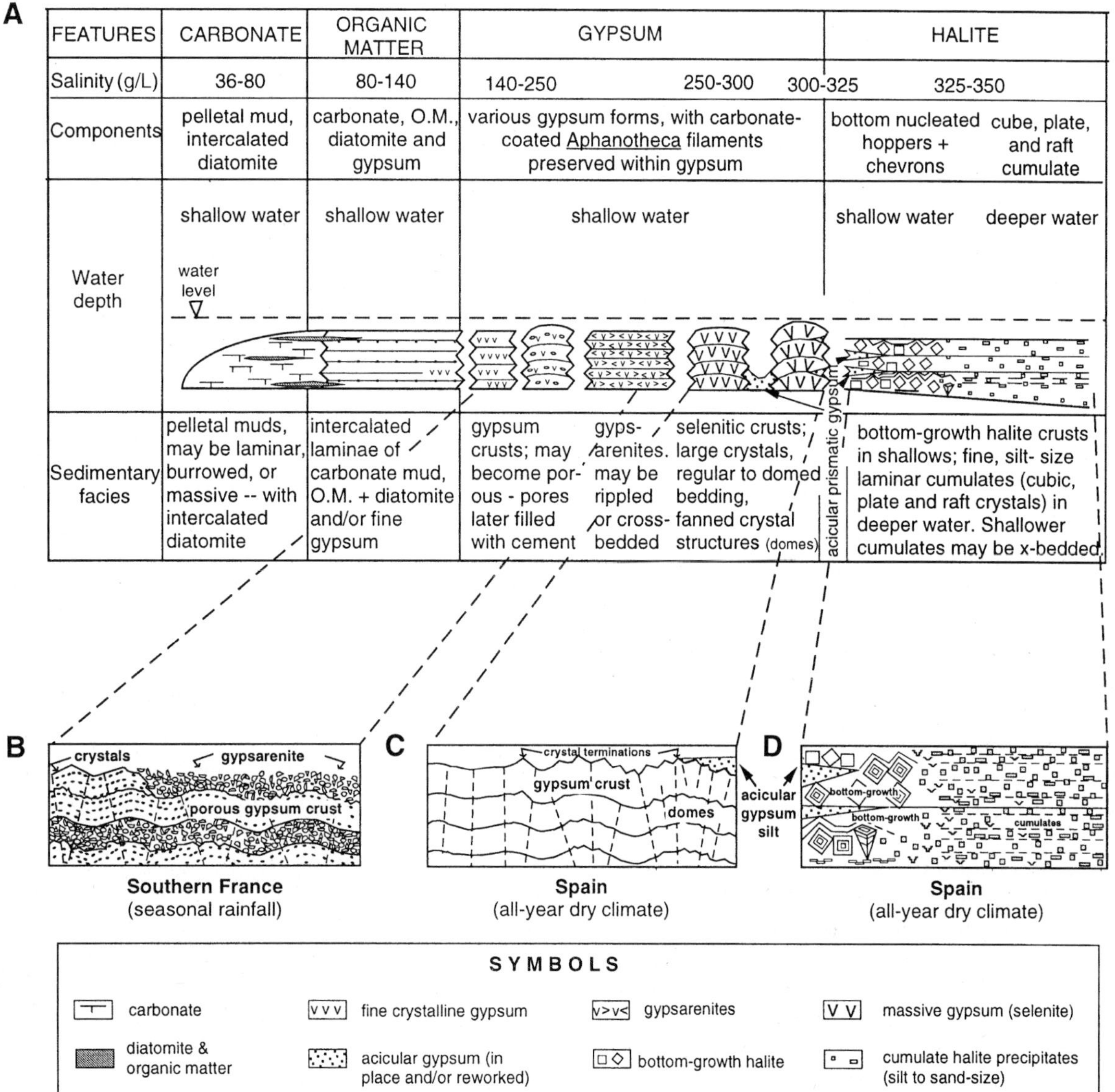

Fig. 4. Evaporative facies continuum, based on salinity. Taken from observations of Mediterranean salt works (Ortí *et al.*, 1984; Rosell *et al.*, 1998) and Lowenstein & Hardie (1985): (A) section modified to include facies apparently present in deeper water bodies; (B) gypsum facies – primary crystal crusts, porous and corroded, with gypsarenite accumulations; (C) gypsum facies – arid region morphologies; (D) halite facies – bottom-growth halite forms in the shallows and may interfinger with acicular gypsum silts. Halite raft accumulations, commonly with crystalline halite overgrowths present. Fine halite cubes and small rafted halite fragments present in deeper water (formed into regular, laminar beds).

has led numerous authors to declare that evaporative settings are barren. This is certainly not the case in modern evaporative settings, with high bacterial and brine shrimp productivity, although preserved biota are rare in bottom sediments. Geochemical studies of such carbonates may reveal the composition and richness of the original biota, many of which contain characteristic organic biomarkers, even where the original life forms left no actual shells or imprints. Evans and Kirkland (1988) first documented the high productivity of these regions, and numerous authors studied the biomarker signatures of the various fauna, flora and bacteria (Benalioulhaj *et al.*, 1993; Benali *et al.*, 1995; Rouchy *et al.*, 1998).

Gypsum

The gypsum 'crystallizer' ponds have a salinity of 150–320 g L^{-1} (Fig. 4). The crystals form as independent clusters, as even crusts, as undulating beds and, occasionally, as large domed, fanning clusters. The gypsum (within the crystals themselves) contains a surprising bacterial assemblage of its own. A fresh crust of gypsum is usually green near the top (photosynthetic filamentous forms), purple by 3–4 cm from the top (photosynthetic, coccoidal, sulphate-oxidizing bacteria) and black near the bottom of the crystals (nonphotosynthetic, sulphate-reducing bacteria). The green bacterial filaments living in the upper part of the crystals become coated with a thin film of micritic carbonate that is incorporated into the gypsum crystals (Cornée, 1982, 1984). Such layers of coated filaments 1–2 mm long are preserved as relics in many Neogene gypsums. Similarly, the characteristic organic biomarkers from such bacteria are found in many unaltered gypsums throughout the Neogene (Benalioulhaj *et al.*, 1993; Benali *et al.*, 1995; Rouchy *et al.*, 1998).

In the salinas of southern France (Busson *et al.*, 1982; Geisler-Cussey, 1997; Rosell *et al.*, 1998), gypsum deposits include primary *in situ* gypsum crystal crusts plus considerable amounts of residual, corroded gypsum sand (gypsarenite). However, the gypsum forming in the salinas of south-eastern Spain (both at Santa Pola, Alicante, and Cabo di Gata, Almeria) is largely in the form of primary gypsum clusters and layered crusts, and gypsarenites are rare. This difference is because of simple dissolution and also because of activity of bacteria during periods of temporary dilution in winter in the south of France. There, seasonal rains allow strong biological reactions. On dilution, sulphate-reducing bacteria grow rapidly, utilize gypsum in their biological processes, and actually break down much of the gypsum. This causes the generation of porous, corroded gypsum crusts together with layers of residual particulate gypsum. The breakdown to a gypsum sand also permits mechanical reworking, resulting in localized clastic bedding features in gypsum, such as graded bedding, ripples and cross-bedding.

In the salinas at the south-eastern corner of Spain (at Santa Pola and Cabo di Gata; Ortí *et al.*, 1984), one other facies is present in the gypsum ponds—very fine acicular gypsum (crystals are 0·5–2 mm long), precipitated within the water column and on the bottom at the most saline end of the gypsum crystallizers and in the first part of the halite crystallizers (300–320 g L^{-1}; Fig. 4). These fragile crystals are usually precipitated in the upper, less saline portions of temporarily stratified waters, sink to the bottom, and are interbedded and/or mixed with thin layers of halite. In the most saline ponds of Cabo di Gata, they accumulate as alternating gypsum and halite laminae, with some beds having low-angle cross-bedding and ripples.

Halite

The study by Shearman (1970) from Salina Ometapec was the first documented observation of sedimentation of primary halite (Fig. 5). Shearman noted that the first precipitate formed after evaporation from a major marine incursion was a thin layer of small prismatic gypsum crystals, overlain by vertically oriented, chevron-like growth of milky halite crystals. The crystals are commonly coarsely crystalline chevrons (cubes with corners systematically oriented upwards), in extensive, continuous beds. The more rapidly this halite grows, the greater is the volume of fluid inclusions within the crystals. These fluids are incorporated along the growing crystal faces in aligned rectangular voids, giving them a

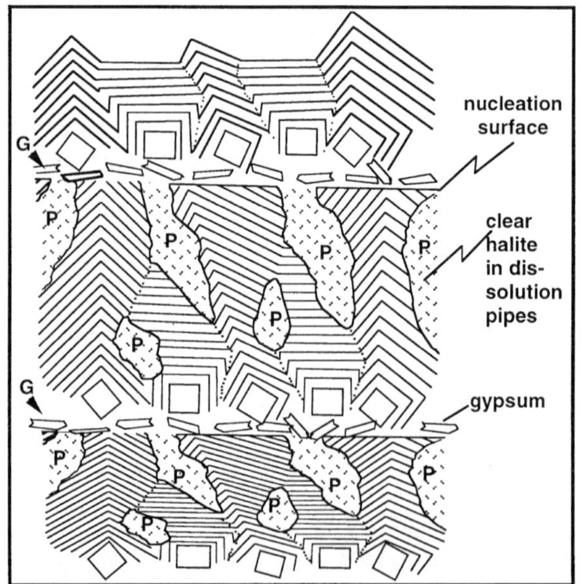

Fig. 5. Salt crust development. Vertical section through three successive layers, showing internal structures marked by fluid inclusions. Adapted from Shearman (1970).

milky appearance. The halite beds are composed of a series of depositional layers, separated from each other by flat dissolution surfaces. Each truncated surface is further marked by dissolution pits that extend down into the underlying layer (Fig. 5, 'P'). The solution pits are filled by clear halite associated with the overlying halite layer. The truncated surfaces are marked by a very thin gypsum crust (Fig. 5, arrow 'G').

Halite precipitation in artificial salinas may be divided into three morphological subgroupings based on the crystal forms. Geisler-Cussey (1997) noted that the first halite crystals form in the narrow range from 320 to 325 g L^{-1}, and usually are perfect cubes of very milky colour (very rich in fluid inclusions). The crystal forms change as the salinity rises. In the next area of the salt works (325–370 g L^{-1}), halite crystallizes at the surface in the form of floating, inverted halite pyramids (hollow shells) or as thin sheets of floating crusts (1–2 mm thick) that sink to the bottom almost as soon as they form. Once at the bottom, crystal overgrowth forms vertically elongated chevrons and/or skeletal halite cornets (or halite 'teeth'). The growth pattern of the halite is defined by aligned fluid inclusions. Each major influx of water causes partial dissolution of the previously deposited halite, and the development of vertical solution pits along the margins of the halite crystals (Fig. 5). In the final ponds (salinity greater than 370 g L^{-1}), skeletal halite crystals grow, with hollow depressed faces and pronounced raised corners and edges.

Halite rafts (nucleated at the surface of the water) may also develop into a slightly different morphology. On a windy day, rafts and thin millimetre-scale surface crusts are blown to one side of the salt ponds and form edge-wise conglomerates (see fig. 4.11 in Moretto & Curial, 1997). Another shallow-water feature, i.e. halite ooids (halolites), may form, apparently aided by bacterial action (Castanier *et al.*, 1999). Some halolites form during storms affecting the halite portions of a shallow saline water body, and are known to accumulate in cross-bedded and rippled structures (Weiler *et al.*, 1974; Nurmi & Friedman, 1977). All these features (chevrons, cubes, chip-like rafts and overgrown rafts) are preserved, even in Precambrian rocks, while still apparently retaining the original incorporated fluid inclusions.

In the marine-fed artificial salinas, the harvesting of salt ends at 370 g L^{-1} and the residual brines are shunted back into the sea. In nature, under very arid conditions, sylvite and/or carnallite may be formed (Lowenstein & Spencer, 1990). At present, no analogues are known for the marine environments of the potassic and magnesian salts, although nonmarine carnallite and sylvite are commonly formed in Andean salars (salt lakes), and carnallite presently forms in the Dead Sea (although both carnallite and sylvite formed in the past). Casas *et al.* (1992) showed that synsedimentary carnallite mineralization also takes place in the shallow subsurface (Dabuson Lake, Qaidam Basin, China), although all these examples are nonmarine.

Shallow-water summary

Documentation of the formation and variation in shallow subaqueous evaporite facies from modern salinas has illustrated the formation of most of the morphologies already known from unaltered and weakly altered evaporites. Gypsum and halite beds forming in modern salinas are of the same thicknesses, morphologies and chemistries as many unaltered ancient deposits. Many synsedimentary changes have also been observed—particularly the controls for and effects of early bacterial action. These biological variations, tied to slight changes in climate, are partial causes of clastic gypsum formation seen in some deposits. Interfingering facies, such as evaporative carbonates intercalated with gypsum and/or halite, are also now known from modern salina analogues. Most modern salinas are kept free of siliciclastic influx; however, numerous Neogene counterparts contain interfingering sands, silts and clays. Even so, many evaporites are free of substantial siliciclastic interfingering, because of the aridity of the climate as well as the rapidity of deposition.

Subaqueous deeper water evaporites

Gypsum

Gypsum precipitates as tiny prismatic needles in hypersaline water bodies, commonly at elevated salinity near or at halite saturation (Geisler-Cussey, 1997; Rosell *et al.*, 1998). These acicular gypsum crystals form either near the top of water columns or at boundaries between layers in a stratified water body, and accumulate at the bottom. The small crystals commonly form thin, regular, laminated beds in many gypsum deposits. However, in some cases, cross-bedded and rippled layers of the same fine acicular gypsum needles are present. The water depth for such cumulate gypsum deposits is either comparatively shallow (marked by current structures) or 'deep' (finely laminated and continuous). Sedimentary features such as oscillation ripples and bird footprints define shallow conditions, whereas sequences of gypsum turbidites with interbedded regular laminites define the deeper environments. Actual water depths are difficult to ascertain in ancient deposits, but the wavebase is shallower in hypersaline water than in normal marine water, so basinal stratification is more easily maintained for long periods.

There are no modern 'deep basin' evaporite deposits, but some Neogene deposits appear to have formed in deep water, based on their position in the basin and sedimentary features. The earliest basin from which deep-water gypsum was recognized is the tectonically active Messinian basin of Gibellina, Sicily (Schreiber *et al.*, 1976), where evaporative mass flows (with stromatolite-bearing gypsum clasts), turbidites and laminites are present. In that basin, based on basinal geometry and relative subsidence rates (due to loading), it has been estimated that the water depth was no greater than 250–275 m. Schlager and Bolz (1977) also recognized considerable volumes of down-slope reworked evaporites in Zechstein deposits. Resedimented evaporites contain all the usual depositional features known from nonevaporites (graded beds, turbidites with well-defined Bouma sequences), with bedding scour and prod marks. Resedimented sections (mass flows and turbidites) are commonly intercalated with beds of laminated carbonate and/or argillites. Reworked gypsum (and anhydrite) is common in many Cenozoic deposits, but is not readily recognized in most deposits after burial dehydration. At burial depths of only a few hundred metres, gypsum goes to anhydrite with about a 50% volume loss (Murray, 1964; Jowett *et al.*, 1993). In some cases, there are sufficient volumes of nonevaporite impurities to outline the original depositional structures but, in most cases, the final product is a nearly featureless nodular sulphate mosaic.

Halite

Deeper water salt basins become stratified, which produces a recognizable halite morphology, usually composed of thin beds or laminae made up of tiny cubic crystals together with swamped halite rafts (cumulates: Lowenstein & Hardie, 1985; Schubel & Lowenstein, 1997). Rafts and cubes from shallow, stratified basins contain abundant fluid inclusions, but deeper water cumulates, mostly tiny cubes and a few rafts, contain sparse fluid inclusions. The beds composed of such halite cumulate laminae are continuous over considerable distances (several kilometres) and each major bed (about 0·3–0·5 m) is made up of very thin layers, 1–3 cm thick (Decima & Wezel, 1973; Lugli *et al.*, 1999). The beds commonly begin and end with an anhydritic (and/or argillaceous) layer, and the first halite layers above the start of a new bed commonly contain finely disseminated organic material with a distinctly halophylic bacterial signature (Benalioulhaj *et al.*, 1994).

The sequences bearing these beds only rarely have dissolution/reprecipitation surfaces but, during periods of shallowing, they may be associated with weakly cross-bedded cumulates. Mass flows in halite sections are uncommon but have been observed in at least one mine in Messinan salt in Sicily, and includes clasts of carbonate and siliceous rocks. Deeper basins may become more or less saline through their evolution, as evidenced by increasing/decreasing bromine content. The chemical changes observed in many sections took place more slowly than in deposits formed from distinctly shallow-water environments (changes being buffered by a larger volume of water).

Deep-water summary

Large restricted marine basins are absent in the modern record (as compared to ancient basins) and this results in problems in interpreting palaeoenvironments of deposition of some ancient evaporite facies. However, evaporites that accumulated in deeper water have, as with other sediments, many components from reworked shallow-water components. Mass flows, in particular, contain large fragments that are distinctly shallow-water in origin. Turbidites composed of evaporites (gypsum, anhydrite, and rarely halite) contain many of the same sedimentary features as siliciclastic and calcareous deposits (graded beds, Bouma sequences, typical sole marks, etc.). Evaporative cumulates, which commonly precipitate in the uppermost portion of a water column (often in stratified water bodies), form laminated, regular carbonate, sulphate and halite deposits interbedded with the turbidites and mass flows.

Nonmarine deposits

In most nonmarine accumulations (saline pans and associated mud flats), the same morphologies develop as in marine settings, but they may have a different range of mineralogy (Fig. 6). Most of the modern analogues are found in comparatively shallow settings; however, thick, laterally extensive, correlatable deposits with deep-water depositional features are well known in the rock record (Green River: Eugster & Surdam, 1973; Fischer & Roberts, 1991; Lisan Formation, Neev & Emery, 1967; Begin *et al.*, 1980; 1985; Katz & Kolodny, 1989; Niemi, 1997; Frumkin, 1998). Some of the crystal forms in sulphate are modified as a result of the presence of trace amounts of continental organic matter (Cody, 1979; Cody & Cody, 1988). For example; in continental subaqueous gypsum, the crystals are commonly in large selenitic forms that are the Montmarte or 'Paris' twin-type (twinned on the *d*[010] composition plane), whereas swallow-tail twins (also called 'ferro-di-lancia') form in marine settings (twinned on the *a*[100] composition plane).

Nonmarine evaporative environments can become very concentrated (high TDS), but the chlorinity may be relatively low. For this reason, certain burrowing organisms can live (in large numbers) in some of these lake bottoms (Ortí & Salvany, 1997). Evaporated marine water would have a high chlorinity and almost no bottom-dwelling fauna. Intervals of refreshment (interbeds) in nonmarine evaporative settings produce remarkable fossil assemblages, including crocodiles and palm trees (as in the Paris Basin, Oligocene, France). Another unexpected aspect of at least one nonmarine setting is the interfingering of coal with evaporites. Nury and Schreiber (1997) document that the fresh end of an elongate graben-filling lake has a brackish-water marsh (with terrestrial marsh plants, now coal), which passes laterally into micritic carbonates and then into gypsum beds with no erosion surfaces or indicators of time breaks. Therefore, evaporites apparently can accumulate contemporaneously with considerable organic matter (that becomes coal) in a single water body, along a salinity gradient in a shallow lake.

Although carbonates, gypsum and/or anhydrite and halite are the most common evaporite minerals in continental settings, a host of other minerals may be present, including borates, fluorides and compounds of lithium, potassium, sodium, copper, iron, arsenic and selenium. These require substantial nonmarine input of

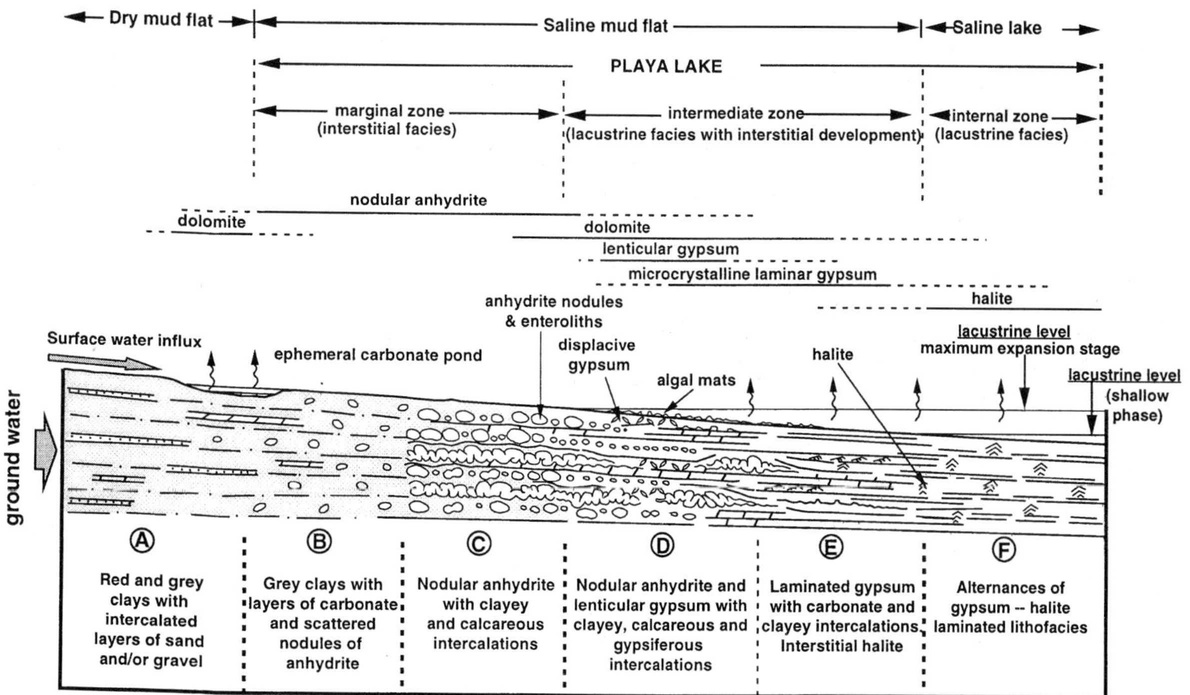

Fig. 6. Idealized section across a 'salar' (salt lake) with associated mudflat. Mineralogy of nodules and enterolithic layers may vary greatly (e.g. polyhalite, thenardite, glauberite, and various borates) but general morphology remains similar. Adapted from Salvany (1997).

Table 1. Listing of a few of the many informative papers that discuss continental evaporite deposits.

Author	Year	Area	Topic
Alonso *et al.*	1991	Andean volcanic arc	Playa deposits (borates, etc.)
Alonso and Viramonte	1990	Andes	Borates
de la Peña *et al.*	1982	La Mancha (Spain)	Modern Mg, Na salts
Ericksen, Chong and Vila	1976	Central Andes	Lithium salts
Ericksen, Vine and Ballon	1977	Bolivia	Lithium salts
Eugster	1969	General	General chemistry
Eugster	1984	Kenya	Magadi cherts
Fontes *et al.*	1991	Alsace, France	Recycled evaporites (isotopes)
Hardie, Smoot and Eugster	1978	General, saline lakes	General chemistry
Helvaci	1995	Turkey	Borate deposits
Helvaci and Orti	1998	Turkey	Borate deposits
Igarzábal	1991	Argentina	Quaternary evaporites
Keys and Williams	1981	Antarctica	Cold desert salts
Kunasz	1978	Chile	Brine compositions
Mees	1999	Mali	Holcene Na, Mg salts
Morris and Dickey	1957	Peru	General evaporites
Nakhla *et al.*	1985	Egypt	Thenardite formation
Niemi *et al.*	1997	Dead Sea	Multiple papers, chemistry, etc.
Ortí *et al.*	1998	Turkey	Sulphate/borate relationships
Pierre	1983	Ojo de Liebre (S. Calif.)	Brine mixing
Renault and Tiercelin	1994	Kenya	Saline lake overview of seds
Salvany and Ortí	1997	Spain (Ebro Basin)	Glauberite
Salvany *et al.*	1994	Spain (Ebro Basin)	General evaporites
Smith and Medranno	1996	S. America	Cenozoic borates
Stoertz and Ericksen	1974	Chile	Saline lakes and brines

dissolved solids. Concentrations vary from area to area, and unusual compounds are surprisingly common in some regions and may make up a large portion of a regional deposit—trona ($NaHCO_3 \cdot Na_2CO_3 \cdot 2H_2O$) and various borate compounds being the most common of these (Table 1).

Evaporite minerals found within nonevaporative sediments include cements and void fillings composed of anhydrite, gypsum and/or halite. These cements may be carried in by evaporatively enriched groundwaters but, in many cases, they are not directly precipitated as a result of evaporative concentration. For example, satin spar (fibrous gypsum), which may originate in association with evaporites, also precipitates as a result of the oxidation of sulphides and even from sulphate released by bacterial breakdown of organic matter, neither of which is related to evaporation (El Tabakh *et al.*, 1997).

Nonmarine summary

Nonmarine evaporites develop many of the same sedimentary features as do marine evaporites and sometimes are indistinguishable both mineralogically and sedimentologically. There are nonmarine sabkhas (often having a very broad lateral continuity), shallow- and deep-water bodies, as in marine settings. More often, however, there are trace element and isotopic differences, as well as some marked morphologic differences from marine deposits. Also common are broad fluctuations in both water depth and lateral extent, as a result of limited water sources (endorheic drainage systems).

DIAGENESIS

Synsedimentary changes

Desiccation of halite-floored salinas and salt lakes causes extensive early recrystallization, as a result of repeated dissolution and precipitation (Handford, 1982). The product of such recrystallization is an interlocking mosaic of halite crystals, with weak bedding and no discernible pattern of crystal orientation. Expansion and contraction of exposed beds (thicker than about 1 m) causes 'teepee' structures (Tucker, 1981). A drop in the groundwater level during desiccation results in the development of localized dissolution and widespread vertical piping features (Lowenstein & Hardie, 1985; Lowenstein *et al.*, 1998; Lugli *et al.*, 1999). These are all strong indicators of synsedimentary exposure and dissolution, even where exposure surfaces are not well defined.

In natural salinas and salt lakes, each yearly cycle of evaporation may produce alternations of gypsum crusts with well-defined tabular crystals and intercalated halite layers. Thin halite beds, formed in one season, may be dissolved by the succeeding influx of water, producing residual gypsum deposits with layering composed of horizontal rows of laterally continuous gypsum crystals that have little or no impurity along their boundaries and no crystal overgrowths passing into the overlying layers (Schreiber & Schreiber, 1977).

Exposure of primary halite beds is common in some basins and is evidenced by erosion/dissolution surfaces. In some instances, the exposure surfaces are related to recrystallization and dissolution of underlying halite (groundwater solution), but with primary beds overlying the recrystallized salt. When periods of exposure become ubiquitous (as in continental playas), the result is an entirely recrystallized, coarse, interlocking halite mosaic, hundreds or even thousands of feet thick (Faulds *et al.*, 1997). Because recrystallization (usually taking place by solution and reprecipitation) is a slower process, as compared with primary growth, the original incorporated impurities are forced out of the new crystal growth; hence, the muds are pushed to the crystal margins, even where they were originally enclosed within the primary crystals.

Numerous instances of 'vertical bedding' features have been reported in salt deposits, lying between horizontal, primary beds (Richter-Bernburg, 1955, 1980). Many of these anomalies are the product of expansion and contraction of halite beds, formed on exposure of halite either in shallow subaqueous or subaerial settings (Tucker, 1981; Lugli *et al.*, 1999). Secondary salt precipitation within the open fractures causes the net expansion of a salt bed; repeated heating causes further expansion and salt tepee structures develop with a typical polygonal outline (in plan view). Vertical piping structures caused by groundwater solution may also form on the planes created by expansion and contraction; sediments may also fill the cracks and pits, although clear halite from succeeding halite cycles is a common filling (Tucker, 1981; Lowenstein & Hardie, 1985; Casas & Lowenstein, 1989; Lugli *et al.*, 1999). When found in an otherwise deep-water salt sequence, such structures suggest that periods of desiccation and exposure may occur in any hypersaline basin, regardless of its initial depth.

Early replacement features

Under some circumstances, halite may replace gypsum (Hovorka, 1992; Schreiber & Walker, 1992) and vice versa. Both types of replacement are quite complete, preserving delicate internal crystal structures and associated impurities. Apparently, the replacements take place early in the burial history of the sediments – usually within a few tens of centimetres of burial (cycle by cycle). These replacements, when present, also may be associated with localized dissolution and possible collapse structures. Halite replacement of gypsum may be because of temperature variations, and associated changes in solubility of these two minerals (Schreiber & Walker, 1992). Additionally, chemical changes in the brine composition caused by the addition of new water may also drive replacement (Hovorka, 1992).

Regions that lie at basin margins or along structural highs commonly contain evaporative sediments with interfingering mineralogies. These sediments undergo early dissolution of soluble components, such as gypsum and halite, and develop into autobreccias. These fragments commonly are composed of the less soluble minerals (e.g. aragonite and calcite) and can be identified as solutional residues as a result of the presence of evaporite moulds, imprints and replacements variously formed by gypsum, calcite, dolomite and celestite. Ogniben (1957, 1963) first described some of these replacement features in a core in which halite was still present. Cubic halite moulds, filled by mud or sand, are ubiquitous in the rock record. Delicate replacement, in which the internal impurities trapped during crystal growth remain in the same pattern as when the original halite crystal formed, are also preserved in some places.

Shallow-water evaporative environments are associated with strong microbial activity, promoting dolomite formation and early dolomitization of other carbonate sediments at low temperature. Primary dolomite deposits may form in supratidal and upper intertidal environments. These conditions are identified in the modern coastal lagoon environment in Brazil (Vasconcelos & McKenzie, 1997). In this environment, the dolomite forms under anoxic hypersaline conditions with intense evaporation. Such environments are also well known as sites for syngenetic dolomitization of carbonates (e.g. McKenzie *et al.*, 1980; Chafetz *et al.*, 1999). As a result of elevated salinity in these environments, calcite cementation of the associated deposits may not be prominent, as it is in normal marine settings, and dolomite forms instead. Most of the dolomite is finely crystalline and these dolomite crystals may locally grade into coarser dolomite, suggesting early recrystallization of at least some of the initial dolomite as a result of prolonged periods of extreme salinity and high evaporation of marine water in associated supratidal flats. Diagenetic conditions generate early pervasive dolomitization similar to reflux dolomitization (Lucia, 1972; Chafetz *et al.*, 1999), and dolomite may develop penecontemporaneously from brine reflux related to the sinking of saline brine produced in shallow marine mudflats. In the marine sabkhas of the Arabian Gulf, dolomite commonly forms by the replacement of a $CaCO_3$ precursor during diagenesis and these dolomite crystals are small, ranging from 1 to 5 μm in size (McKenzie *et al.*, 1980; McKenzie, 1981).

Late diagenesis

Although this paper deals in detail with synsedimentary dissolution and diagenesis, and not with later burial changes, it is important to point out general types of diagenesis developed as a consequence of burial and exhumation. On burial below depths of a few hundred metres (Murray, 1964; Jowett *et al.*, 1993), geothermal heating and dehydration blur depositional features. Because gypsum converts to anhydrite with an attendant, major, volume loss (see also Shearman, 1985), it causes an apparent thinning of the section. The water that is released both as a result of compaction and as a result of changes in the flow of groundwater in the basin, may in turn affect associated halite and other soluble salts, such as potash. Exhumation and related groundwater effects have been addressed (in part) in Murray (1964), Johnson and Neal (1997), and in El Tabakh *et al.* (1998a).

Replacement, dissolution and resultant variations in evaporite thickness are features resulting from very early diagenetic processes that can be recognized once primary features are understood. Most of these alterations also suggest dissolution (10–50%) of the original sediment. Early dissolution can occur within days or weeks of initial deposition, and the thinning, collapse or alteration of the original deposits may be made out from sedimentary fabrics. An understanding of very early dissolution features is helpful, particularly when contemporaneous and interfingering sediments are also recognized. For example, influx of fresh water may carry muds into a salt basin, forming laterally equivalent and intercalated mudstones. This same water influx then percolates into the existing groundwater system, diluting it temporarily. Dissolution and collapse features may develop locally, and the resultant replacement features and/or solutional breccia are not the product of late processes but may instead form on a limited basis.

Late dissolution, however, is commonly found where uplift brings evaporites into the range of groundwater activity. Dissolution because of undersaturated fluids from underlying and interfingering aquifers must always be considered. Subformational dissolution may form a sort of 'upside down' caprock at the base of an evaporite section, as recently demonstrated in Thailand (El Tabakh *et al.*, 1998b) and in the area of Texas–New Mexico where the Salado overlies the Capitan (Hiss, 1975a,b, 1976, 1980; Hovorka, 1998). Karsting and attendant collapse is the least difficult of these solutional features to recognize.

Geochemical relationships

Geochemical information commonly is collected from evaporite deposits in order to help in the differentiation of marine from hydrothermal and/or nonmarine deposits, and to pinpoint water sources. Geochemical data may include the results of studies of fluid inclusions, trace elements, bromine/chlorine ratios, $^{87}Sr/^{86}Sr$ ratios, sulphur and oxygen, and oxygen and carbon isotope ratios in associated evaporative carbonates. Geochemical study is useful on a local basinal scale and also is an excellent tool for stratigraphic correlation in a worldwide framework.

Fluid inclusions

Primary fluid inclusions are present in most evaporative minerals (gypsum and halite) as well as in associated anhydrite, calcite, ankerite, dolomite, quartz and feldspar (Roedder, 1984; Goldstein & Reynolds, 1994). Initially, the most significant problem is to establish that the fluid inclusions in question are primary and were not compromised by heating or leakage (Hardie *et al.*, 1985). In halite, primary inclusions are usually plentiful and aligned parallel to growth faces. They are particularly concentrated on cube corners and reflect chevron and hopper morphologies (Goldstein & Reynolds, 1994). The more rapidly the crystal grows, the more it is able to incorporate fluids (and also solid impurities that are representative of the original environment).

Fluid inclusions that are the product of recrystallization are usually large and sporadic. Commonly the altered crystals have an equigranular mosaic texture. Where care is taken in the choice of samples, a great deal of detail can be obtained concerning the parent water from which evaporites formed (e.g. marine, nonmarine and mixed; Ayora *et al.*, 1994), and even its temperature, indicating regional climate markers and fluctuations (Roberts & Spencer, 1995; Lowenstein *et al.*, 1998; Benison & Goldstein, 1999). Fluid-inclusion studies tied to palaeontology and U-series chronology (Ku *et al.*, 1998) permit the establishment of detailed stratigraphic/climatic sections and the unravelling of regional history. Similar climatic studies have been carried out in other minerals, but none is as extensively studied and analysed as halite.

Bromine

The bromine content of seawater is not directly reflected in the solid precipitates that form from it, and it is also the case that bromine does not form its own minerals under normal sedimentary conditions. Because bromine ions are not the same size as the chlorine, they cannot fit into the growing halite lattice as readily, and part of the bromine accumulates in the residual water. Most bromine studies are carried out on halite and other chlorides that have been carefully prepared to analyse the solid salt and to eliminate any fluid inclusions (with excess bromine). The process of fractionation follows a predictable pattern and is not significantly affected by temperature (Holser, 1966; Braitsch, 1971). The degree of discrimination between such relatively similar ions is called the 'partition function' of that ion. For example, evaporated seawater, at the point of halite saturation, usually contains about 500–550 ppm bromine, but the first halite that forms from it only contains 65–75 ppm of bromine within the NaCl crystal lattice (fractionation function of about 0·12–0·14). As the seawater continues to evaporate and become more saturated and more halite is formed, increasingly more bromine is left behind in the residual solution. The last halite forming from the solution (about 2000 ppm Br) has much more bromine (270 ppm) than the earliest – and the residual brines now have a great deal of bromine compared with the remaining chlorine (after much of the sodium is used up).

Highly evaporated seawater, at the point of carnallite ($MgCl_2{\cdot}KCl{\cdot}6H_2O$) precipitation, may contain 2300 ppm bromine and, because the partition function of bromine (0·7) during formation of carnallite is much greater than for the precipitation of halite, much more of the bromine may be incorporated (>1600 ppm). Moreover, because carnallite readily alters to sylvite during burial (or any other heating), the alteration is apparently conservative of the bromine, and the bromine content of the newly formed sylvite contains nearly as much as the parent carnallite (Wardlaw, 1970). Sylvite may also form directly from concentrated water and has a partition function only slightly lower than that of carnallite. Zak (1997) has demonstrated that repeated solution and reprecipitation from brines, as in the Dead Sea, can give rise to extremely high bromine values both in the water and in any resulting salts.

Isotopic variations

Strontium

Isotope analyses of strontium in a large number of marine limestones of diverse ages have revealed that the $^{87}Sr/^{86}Sr$ ratio of seawater has varied systematically with time and the variation has been used to reconstruct the history of ancient seawater and to date sequences (Burke *et al.*, 1982; DePaolo & Ingram, 1985; McArthur, 1994; Ruppel *et al.*, 1996; Bralower *et al.*, 1997). These variations can be employed in dating marine carbonates, so may be used for stratigraphic correlations (Azmy *et al.*, 1999). Such variations may have been caused by flux in the isotopic compositions of strontium that entered the oceans from diverse sources and by changes in the relative proportions of these inputs. Most important of these changing processes are continental run-off, oceanic crust and seawater reactions, and hydrothermal activity (Stanley & Hardie, 1999). ^{87}Rb breakdown to ^{87}Sr is the controlling factor in the final $^{87}Sr/^{86}Sr$ ratio in any water body and, if there is little or no ^{87}Rb, then the ratio remains unchanged.

The isotopic composition of evaporites (evaporitic carbonates, sulphates, and halite) is a direct reflection of the isotopic composition of their waters of formation, i.e. marine, nonmarine or mixed. The major factors governing the available isotopic composition are as follows: (1) initial isotopic composition of the formative water reservoir; (2) the amount of local dissolution, reprecipitation and recrystallization of older evaporites in the basin; and (3) the presence of clastic intercalations.

If incoming waters are purely marine in origin and merely evaporated, then the strontium isotopic composition will reflect that value (about 0·7092 for a modern seawater value); however, the marine value has changed through time. Where there is a significant influx of nonmarine water (as near the mouth of a major river), waters with values of 0·7431 up to 0·9217 may be mixed with the original marine water (e.g. from a granitic terrain: Palmer & Edmond, 1992), providing a local reservoir that has very different values from the ocean itself. Therefore, stratigraphic variations of the $^{87}Sr/^{86}Sr$ ratios of some carbonate rocks can be caused by local variation in the formative waters (Neat *et al.*, 1979).

Evaporative carbonate rocks formed from marine sources include trace amounts of strontium that reflect the general changes in the oceans and may be utilized in detailed stratigraphic studies in evaporative basins, particularly where the other evaporites are recrystallized or otherwise altered. Faure (1986) suggests that strontium isotopes in carbonates are not further fractionated, even during moderate diagenesis by exchange reactions or by kinetic effects, although the total amount of strontium may change considerably as a result of recrystallization. Faure's (1986) argument is that there is a very low Rb/Sr ratio in pure marine-sourced calcite or aragonite, so that its $^{87}Sr/^{86}Sr$ ratio cannot be significantly altered by later radioactive decay of ^{87}Rb to ^{87}Sr after deposition. Therefore, the isotope composition of strontium in any form of marine $CaCO_3$ reflects its initial isotopic composition. Marked recrystallization in regions of significant water movement, however, may affect the original $^{87}Sr/^{86}Sr$ ratios and diagenetic fluids can impose radiogenic strontium ratios relative to the original composition (Machel *et al.*, 1996). Such highly altered sections should be considered with great care as their isotopic values may be reset.

Strontium isotopes are not appreciably fractionated by sulphate crystallization and the strontium incorporated into the sulphate minerals has the same $^{87}Sr/^{86}Sr$ ratio as that of the original brines. Influx of clays and other siliciclastics by riverine input adds radiogenic strontium into the evaporating pans, and may introduce more radiogenic strontium into the forming evaporative sediments. In more open marine basins, such isotopic variations may be used to document climatic changes during formation of these deposits (e.g. Azmy *et al.*, 1999). In nonmarine basins, unaffected by the introduction of hydrothermal or marine waters, the $^{87}Sr/^{86}Sr$ ratio of sulphate dissolved in water is directly dependent on the rocks to which the water was exposed and the ratio may be entirely localized (Neat *et al.*, 1979).

Generally, in an evaporite sequence, the presence of clastic intercalations and/or the evaporite mineral carnallite ($KMgCl_3{\cdot}6H_2O$) gives rise to anomalous strontium isotopic compositions (Baadsgaard, 1987). In the case of the clastics, radiogenic minerals in the sands or clays will alter the $^{87}Sr/^{86}Sr$ *in situ* without visible diagenetic changes in the rock. In the presence of carnallite, because (radiogenic) rubidium readily fits into the crystal lattice, it can also alter the $^{87}Sr/^{86}Sr$. Furthermore, carnallite is metastable on burial, giving rise to sylvite (KCl), and any new ^{87}Sr produced from the contained ^{87}Rb that was in the carnallite lattice is then incorporated in the daughter minerals and also in the waters released by the recrystallization (Baadsgaard, 1987). Not only does this process occur during and after burial, but it also takes place in the groundwaters of isolated basins where older evaporites are in outcrop or subcrop, resulting in $^{87}Sr/^{86}Sr$ ratios far in excess of the original values.

Sulphur isotopes

The study of sulphur isotopes from sedimentary rock usually involves the most common isotopes, ^{32}S (95%) and ^{34}S (4·22%; see Faure, 1986; Strauss, 1997). The $\delta^{34}S$ value of sulphur in modern marine water is fairly consistent, at about +20‰, but it has apparently varied systematically through geologic time (Claypool *et al.*, 1980; Holser *et al.*, 1986; Longinelli, 1989; Paytan *et al.*, 1998). This variation has been used to compare stratigraphic sections on a worldwide basis.

Compounds containing sulphur are ubiquitous in sedimentary rocks and are most commonly in the form of pyrite (or one of its variants). However, sulphates such as gypsum and anhydrite and sulphur in organic complexes also contain considerable amounts. Thode and Monster (1965) and Holser and Kaplan (1966) have shown that sulphur isotopes are not greatly fractionated by the precipitation of gypsum (or anhydrite) from sulphate-bearing brines (from 0 to +2·4‰). Therefore, a deposit composed of marine sulphate is representative of the brine from which it precipitated. Late-stage sulphates, however, formed within halite and potash–magnesian salts, are apparently depleted in ^{34}S by as much as 4‰ (Raab & Spiro, 1991). The isotopic values of sulphur in organic compounds and in pyrite, even when they originate from marine waters, cross a very broad isotopic range.

Most pyrite in sediments originates by bacterial reduction of sulphate, and substantial fractionation accompanies this process (Chambers & Trudinger, 1979). The resulting depletion in sulphur isotopic composition causes ^{34}S changes, ranging from 4‰ to 70‰ (CDT), and arises from the multistep reduction processes that take place in many environments (Strauss, 1997). Even in nonpyritic settings, primary (marine) gypsum may be attacked by bacteria in several stages, producing native sulphur (Feeley & Kulp, 1957; Jensen, 1962). Within the same layer, native sulphur may be partly preserved and partly oxidized by migrating groundwater, forming secondary gypsum in void spaces. The second-cycle gypsum has a different isotopic composition from the parent rock and from the intermediate native sulphur. It is also common to find fracture-filling selenitic gypsum with a significantly depleted ^{34}S ($\delta^{34}S$ of −10‰) within a fracture in a gypsum rock that has a normal marine signature ($\delta^{34}S$ of +20‰), reflecting localized biologically driven fractionation.

Nonevaporitic gypsum is present in many sedimentary rocks. The most striking occurrences are gypsums in pyritiferous shales (marine origin; Berner, 1984) that have been exhumed and exposed. Here, white, fibrous, gypsum, satin-spar veins, having a $\delta^{34}S$ value from +2‰ to +14‰, are found along bedding planes in black shales. This gypsum originates as a result of oxidation of finely disseminated pyrite in the shale, in the presence of meteoric water. Other examples of nonevaporitic fibrous gypsum veins, commonly associated with faults, nearly always have the isotopic composition of older evaporitic rock in the region (Utrilla *et al.*, 1992; El Tabakh *et al.*, 1998a).

Isotopic signatures of oxygen and carbon in evaporative carbonates and evaporites

Carbon

The isotopic composition of carbon in true evaporative carbonates formed from marine water is often employed as a partial proxy for ocean compositional changes through time. Isotopic changes in carbon in the ocean waters have apparently varied markedly through the past 545 million years (Holland, 1984), suggesting that significant geochemical changes in the ocean took place throughout the Phanerozoic (Broecker, 1970; Holser, 1984). The average isotopic composition of dissolved inorganic carbon in seawater is maintained by isotopic compositions of carbon fluxes into and out of the world ocean (Schidlowski *et al.*, 1983). Carbon entering the oceanic reservoir from weathering of continental rocks or from the mantle has a long-term average isotopic composition of approximately −5‰. Carbon leaves the oceans as either carbonate sediments or as organic carbon. However, isotopic compositions of primary marine carbonates ($\delta^{13}C_{carb}$) are slightly enriched (by 0–2‰) relative to dissolved inorganic carbon of the ambient seawater, and carbonate sedimentation has little net effect on $\delta^{13}C$ of the dissolved inorganic carbon in ocean water. Low $\delta^{13}C$ values in carbonates commonly indicate a biogenic source that is produced during (1) aerobic decomposition of the organic matter; and (2) respiration of plants. During sulphate reducing reactions, however, where organic matter is used as a source of energy by bacteria, the values of $\delta^{13}C$ are close to +20‰. Positive values of $\delta^{13}C$ are indications of strict anaerobic conditions during methanogenic reactions and are common under evaporative conditions (Nissenbaum *et al.*, 1972).

Differences in $\delta^{13}C$ of the dissolved inorganic carbon between surface and bottom waters result from poor mixing and stratification of ocean waters (Rouchy *et al.*, 1998). In the early stages of many evaporative basins, such stratification is pronounced and bottom sediments are sporadically euxenic with variable isotopic profiles, even while upper water was well oxygenated and highly productive. The result of this extreme variability may be seen clearly in the study of the Messinian Lorca Basin (south-eastern Spain) by Rouchy *et al.* (1998) and in Sicily (Decima *et al.*, 1988). On modern supratidal algal flats, magnesian-rich carbonates are common and include a mixture of magnesian calcite, finely crystalline dolomite and minor aragonite. The $\delta^{13}C$ values in calcium carbonates are low, reflecting a contribution of biogenic CO_2.

Oxygen

The ^{18}O values in carbonates are strongly temperature-dependent and, in carbonate rocks or sediments, ^{18}O is enriched by 30% compared with water at isotopic equilibrium (Craig, 1965). As water evaporates, and carbonates and evaporites form, the light isotopes are selectively removed by evaporation from the water body, so that the residual evaporated water contains proportionally higher ^{18}O values. Therefore, primary carbonates that form from evaporated waters (as in a salina or other hypersaline basin) may have elevated $\delta^{18}O$ values, despite their somewhat higher temperatures. Oxygen present within sulphates (gypsum) formed from evaporated waters may have significantly elevated values relative to simple thermal fractionation (Pierre, 1985). During crystallization of gypsum, the ^{18}O in the water of the crystal lattice is enriched by 4‰ relative to the evaporating brines (Sofer & Gat, 1975). Although Sofer and Gat do not specify the temperatures at which these experiments were carried out, because they were meant to approximate evaporative conditions in nature, they were probably run at relatively low temperatures. The ^{18}O isotope in the SO_4 of the sulphate is enriched up to 6‰ relative to the dissolved sulphate in the parent water (Lloyd, 1968).

In their study of Messinian evaporites from ODP Leg 42, Pierre and Fontes (1978) demonstrated the complexity of the gypsum studied (despite its apparent uniformity of appearance), and mixing of recycled evaporites was supported by the diverse oxygen isotope values present in the associated water of crystallization. Such variations can also be related to the following: (1) backreactions of different mineral phases as a result of diagenesis and mixing of waters of different sources; (2) influx of waters of different sources, during different periods of deposition; and (3) synsedimentary alteration of minerals during deposition, as a result of bacterial action as well as simple chemical reaction of new brines with existing minerals.

While isotopic data can define local depositional conditions and specific chemical and biological processes in the environment of sedimentation, isotopic information can help with large-scale geological issues. Holland *et al.* (1996) concluded that long-term changes in weathering and the magnitude of run-off into evaporite basins are controlled by the rate of sea floor spreading, as well as being the product of seawater recycling along spreading centres (Hardie, 1996; Stanley & Hardie, 1998, 1999). In addition, the isotopic record of evaporites in thick sedimentary successions can aid in defining 'stratigraphic markers' that can be tied to other stratigraphic information, such as the fossil record, sequence boundaries and the palaeomagnetic record.

CONCLUSIONS

Traditionally, evaporites were treated as no more than chemical compounds, studied by chemists and engineers working on commercial salt deposits or in the laboratory, and were never dealt with as true sediments. Schaller and Henderson (1932) were the first to treat evaporite deposits as true geological entities and address physical features that they saw as sedimentary deposits, rather than a series of experimental precipitates in a beaker. When geologists began working in the Middle East, a naturally occurring depositional laboratory was opened for all to consider. Those first studies were electrifying and changed the view geologists had of evaporative environments. Study of sabkhas, marine-fed salinas, continental salt lakes and hydrothermal springs have since given a valid sedimentary framework to theory. Present analyses, cognisant of depositional environments and their lithologies, have come full circle and are again employing extensive geochemical study—but these studies are based on ground truth, i.e. the rocks themselves.

Evaporites form in many environments under a relatively arid climate, and their depositional features reflect not only their chemistry (sulphates, chlorides, etc.) but also the depth and energy of the depositional environment. Because of the differences in chemistry, bedding and preserved morphology, it is possible to recognize the original facies in many cases, despite overprinting diagenesis. Shallow-water features, such as cyanobacterial structures within gypsum, require sunlight (photic zone) and oscillation ripples, and festooned cross-beds signal shallow, high-energy reworking of both sulphates and chlorides. Similarly, thick turbidite sequences only form from deposits mechanically reworked into deeper basins. Based on such recognizable features, the depositional environments of particular evaporite sequences can be assigned to formative facies as with other sediments. However, because the inflow and climatic restrictions of evaporites are so well defined, the overall water depths and basinal chemistry may have changed repeatedly through a given section. Basin depth does not imply a like and constant water depth, and the mineralogy and thickness of a bed do not signal the environment of deposition. Only the rock record, not borehole logs, can tell the story.

After analysing an evaporative section and fitting the observations into one or more hypothetical depositional frameworks, the unknown factors, such as water sources and timing of deposition intervals, must be addressed by geochemical techniques. Because evaporites have a very limited faunal and floral assemblage, every clue must be utilized. In this paper, we have introduced a few specialized methods (fluid inclusions, bromine content, and the isotopic ratios of strontium, sulphur, oxygen, and carbon), but there are many others. The characteristic biomarkers of organic-rich beds, trace elements such as lithium, boron and fluorine, are all first steps in tracing evaporative ancestry and diagenesis—and all await analysis. All observed lithologies must be tied to real sediments, not models and theories.

ACKNOWLEDGEMENTS

The authors wish to thank Steve Burns for his help in crafting our short presentation of the necessary isotope geochemistry for this paper. Thanks are also due to Susan Hovorka and Peter Mozley for their careful review of the first version of the manuscript.

REFERENCES

Alonso, R.N. and **Viramonte, J.G.** (1990) Borate deposits in the Andes. In: *Stratabound Ore Deposits in the Andes* (Ed. by L. Fontboté, G.C. Amstutz, M. Cardozo, E. Cedillo and J. Frutos), pp. 721–732, Springer, Heidelberg.

Alonso, R.N., Jordan, T.E., Tabbutt, K.T. and **Vandervoort, D.S.** (1991) Giant evaporite belts of the Neogene central Andes. *Geology*, **19**, 401–404.

Aref, M. (1998) Holocene stromatolites and microbial laminites associated with lenticular gypsum in a marine-dominated environment, Ras El Shetan area, Gulf of Aqaba, Egypt. *Sedimentology*, **45**, 245–262.

Ayora, C., Garcia-Veigas, J. and **Pueyo, J.J.** (1994) X-ray microanalysis of fluid inclusions and its application to the geochemical modeling of evaporite basins. *Geochim. Cosmochim. Acta*, **58**, 43–55.

Azmy, K., Veizer, J., Wenzer, B., Bassett, B.G. and **Copper, P.** (1999) Silurian strontium isotope stratigraphy. *Bull. Geol. Soc. Am.* **111**, 475–483.

Baadsgaard, H. (1987) Rb–Sr and K–Ca isotope systematics in minerals from potassium horizons in the Prairie Evaporite Formation, Saskatchewan, Canada. *Chem. Geol.*, **66**, 1–15.

Begin, Z.B., Ehrlich, A. and **Nathan, Y.** (1980) Stratigraphy and facies distribution in the Lisan Formation—New evidence from the area south of the Dead Sea. *Israel J. Earth Sci.*, **29**, 182–189.

Begin, Z.B., Broecker, W., Buchbinder, B., Druckman, Y., Kaufman, A., Magaritz, M. and **Neev, D.** (1985) Dead Sea and Lisan Lake levels in the last 30, 000 years. *Geol. Survey Israel, Prelim. Rep.*, GSI/29/85.

Benalioulhaj, S., Schreiber, B.C., Philp, R.P. and **Landais, P.** (1993) Assessment of biomarkers characteristic of hypersalinity using artificial maturation on sediments from diverse evaporitic environments. *Spec. Publ. Euro. Meet., Stavenger. Ext. Abs., Poster Sess., Organ. Geochem.*, **6**, 532–535.

Benalioulhaj, S., Schreiber, B.C. and **Philp, R.P.** (1994) Contribution to the reconstruction of depositional conditions during restricted and evaporitic sedimentation through the study of biomarkers: Preliminary results. In: *Sedimentology and Paleolimnological Record of Saline Lakes* (Ed. by R. Renaut and W. Last), *Soc. Econ. Paleont. Mineral. Spec. Publ.*, **50**, 315–324.

Benali, S., Schreiber, B.C., Philp, P.R. and **Helman, M.L.** (1995) Characterization of organic matter from a restricted/evaporative environment: Late Miocene of Lorca Basin, Southeastern Spain. *Bull. Am. Assoc. Petrol. Geol.*, **79**, 816–830.

Benison, K.C. and **Goldstein, R.H.** (1999) Permian paleoclimate data from fluid inclusions in halite. *Chem. Geol.*, **154**, 113–132.

Berner, R.A. (1984) Sedimentary pyrite formation: an update. *Geochim. Cosmochim. Acta*, **48**, 605–615.

Bowser, C.J., Rafter, T.A. and **Black, R.F.** (1970) Geochemical evidence for the origin of mirabilite deposits near Hobbs Glacier, Victoria Land, Antarctica. *Mineral. Soc. Am. Spec. Pap.*, **3**, 261–272.

Braitsch, O. (1964) The temperature of evaporite formation. In: *Problems in Paleoclimatology* (Ed. by A.E.M. Nairn), pp. 479–490. Wiley, New York.

Braitsch, O. (1971) *Salt Deposits, Their Origin, and Composition*. Springer, Berlin.

Bralower, T.J., Fullagar, P.D., Paull, C.K., Dwyer, G.S. and **Leckie, R.M.** (1997) Mid-Cretaceous strontium-isotope stratigraphy of deep-sea sections. *Bull. Geol. Soc. Am.*, **109**, 1421–1442.

Broecker, W.S. (1970) A boundary condition on the evolution of atmospheric oxygen. *J. Geophys. Res.*, **75**, 3553–3557.

Broecker, W.S. and **Peng, T.-S.** (1982) *Tracers in the Sea*. Eldigio Press, New York.

Burke, W.H., Denison, R.E., Hetherington, E.A., Koepnick, R.B., Nelson, H.F. and **Otto, J.B.** (1982) Variation of seawater $^{87}Sr/^{86}Sr$ throughout Phanerozoic time. *Geology*, **1**, 516–519.

Busson, G. and **Schreiber, B.C.** (Eds) (1997) *Sedimentary Deposition in Rift and Foreland Basins in France and Spain (Paleogene and Lower Neogene)*. Columbia University Press, New York.

Busson, G., Cornée, A., Dulau, N., Fontes, J.Ch., Geisler, D., Gouleau, D., Jaccard, J., Landry, J.C., Noël, D., Perthuisot, J.P., Pierre, C., Poumot, C., Tetard, J., Thomas, J.C., Thomas, M., Trauth, N. and **Zaninett, L.** (1982) Données hydrochimiques, biologiques, isotopiques, sédimentologiques et diagénétiques sur les marais salants de Salin-de-Giraud (Sud de la France). *Géol. Médit. IX*, **4**, 303–391.

Butler, G.P. (1970) Holocene gypsum and anhydrite of the Abu Dhabi sabkha, Trucial Coast: An alternative explanation of origin. In: *Third Symposium on Salt* (Ed. by J.L. Rau and

L.F. Dellwig), Vol. 1, pp. 120–152. Northern Ohio Geological Society, Cleveland, OH.

Butler, G.P., **Harris, P.M.**, **Kendall, C.G.** and **St. C.** (1982) Recent evaporites from the Abu Dhabi coastal flats. In: *Depositional and Diagenetic Spectra of Evaporites—A Core Workshop* (Ed. by C.R. Handford, R.G. Loucks and G.R. Davies), *SEPM Core Workshop*, **3**, 33–64.

Casas, S.E. and **Lowenstein, T.K.** (1989) Diagenesis of salt pan halite: comparison of petrographic features of modern, Quaternary and Permian halites. *J. Sedim. Petrol.*, **59**, 724–739.

Casas, S.E., **Lowenstein, T.K.**, **Spencer, R.J.** and **Zhang, P.** (1992) Carnallite mineralization in the nonmarine Quaidam Basin, China. *J. Sedim. Petrol.*, **62**, 881–898.

Castanier, S., **Perthuisot, J.-P.**, **Matrat, M.** and **Morvan, J.-Y.** (1999) The salt ooids of Berre salt works (Bouches du Rhone, France); the role of bacteria in salt crystallization. *Sedim. Geol.*, **125**, 9–21.

Chambers, L.A. and **Trudinger, P.A.** (1979) Microbiological fractionation of stable sulfur isotopes: a review and critique. *J. Geomicrobiol.*, **1**, 249–293.

Chafetz, H.S., **Alicia, A.**, **Imerito-Tetzlaff, A.** and **Zhang, J.** (1999) Stable-isotope and elemental trends in Pleistocene sabkha dolomites: descending meteoric water vs. sulfate reduction. *J. Sedim. Res.*, **69**, 256–266.

Claypool, G.E., **Holser, W.T.**, **Kaplan, I.R.**, **Sakai, H.** and **Zak, I.** (1980) The age curves of sulfur and oxygen isotopes in marine sulfate and their mutual interpretation. *Chem. Geol.*, **28**, 199–260.

Cody, R.D. (1979) Lenticular gypsum: occurrences in nature, and experimental determinations of soluble green plant material on its formation. *J. Sedim. Petrol.*, **49**, 1015–1028.

Cody, R.D. and **Cody, A.M.** (1988) Gypsum nucleation on crystal morphology in analog saline terrestrial environments. *J. Sedim. Petrol.*, **58**, 247–255.

Colwell, R.R., **Litchfield, C.D.**, **Vreeland, R.H.**, **Kiefer, L.A.** and **Gibbons, N.E.** (1979) Taxonomic studies of red halophillic bacteria. *Int. J. Sys. Bacter.*, **29**, 379–399.

Cornée, A. (1982) Bactéries des saumures et des sédiments des marais salants de Salin-de-Giraud (Sud de la France). *Géol. Médit IX*, **4**, 369–389.

Cornée, A. (1984) Etude préliminaire des bactéries des saumures des sédiments des salins de Santa Pola (Espagne). Comparaison avec les marais salants de Salin-de-Giraud (Sud de la France). *Rev. Invest. Geol., Barcelona*, **38–39**, 109–122.

Craig, H. (1965) The measurement of oxygen isotope palaeotemperatures. In: *Stable Isotopes in Oceanographic Studies and Palaeotemperatures* (Ed. by E. Tongiorgi), pp. 3–24. Consiglio Nazionale delle Richerche, Pisa.

d'Ans, J. (1947) Über die Bildung und Umbildung der Kalisalzlagerstätten. *Naturwissenschaften*, **34**, 295–301.

Decima, A. and **Wezel, F.C.** (1973) Late Miocene evaporites of the Central Sicilian Basin. In: *Initial Reports of the Deep Sea Drilling Project, Leg 42(1)* (Ed. by W.B.F. Ryan and K.J. Hsü), pp. 1234–1240. US Government Printing Office, Washington, DC.

Decima, A., **McKenzie, J.** and **Schreiber, B.C.** (1988) The origin of evaporative carbonates. *J. Sedim. Petrol.*, **58**, 256–272.

de la Peña, J.A., **Garcia Ruiz, J.M.**, **Marfil, R.** and **Prieto, M.** (1982) Growth features of magnesium and sodium salts in a recent playa lake of La Mancha (Spain). *Estud. Geol.*, **38**, 245–257.

DePaolo, D.J. and **Ingram, B.L.** (1985) High resolution stratigraphy with strontium isotopes. *Science*, **227**, 938–941.

Eckardt, F.D. and **Spiro, B.** (1999) Origin of sulphur in gypsum and dissolved sulphate in the Central Namib Desert, Namibia. *Sedim. Geol.*, **123**, 255–273.

El Tabakh, M., **Riccioni, R.M.** and **Schreiber, B.C.** (1997) Deposition and diagenesis of non-marine rift-basin evaporites: The Passaic Formation (Late Triassic). *Sedimentology*, **44**, 767–791.

El Tabakh, M., **Schreiber, B.C.** and **Warren, J.K.** (1998a) Late fibrous fracture-fill within non-marine strata of the Newark Rift Basin, Eastern North America. *J. Sedim. Res.*, **68**, 88–99.

El Tabakh, M., **Schreiber, B.C.**, **Utha-Aroon, C.**, **Warren, J.K.** and **Coshell, L.** (1998b) Diagenetic origin of Basal Anhydrite in the Cretaceous Maha Sarakham salt: Khorat Plateau, NE Thailand. *Sedimentology*, **45**, 579–594.

Ericksen, G.E., **Chong, G.** and **Vila, T.** (1976) Lithium resources of Salars in the central Andes. In: *Lithium Resources and Requirements by the Year 2000* (Ed. by J.D. Vine), *US Geol. Survey Prof. Paper*, **1005**, 66–74.

Ericksen, G.E., **Vine, J.D.** and **Ballon, A.R.** (1977) Chemical composition and distribution of lithium—rich brines in Salar de Uyuni and nearby salars in southwestern Bolivia. *Energy*, **3**, 355–363.

Eugster, H.P. (1969) Interbedded inorganic cherts from the Magadi area, Kenya. *Contrib. Mineral. Petrol.*, **22**, 1–31.

Eugster, H.P. (1984) Geochemistry and sedimentology of marine and non marine evaporites. *Ecol. Geol. Helv.*, **77**, 237–248.

Eugster, H.P. and **Surdam, R.C.** (1973) Depositional environment of the Green River Formation of Wyoming: a preliminary report. *Bull. Geol. Soc. Am.*, **84**, 1115–1120.

Evans, R. and **Kirkland, D.W.** (1988) Evaporitic environments as a source for petroleum. In: *Evaporites and Hydrocarbons* (Ed. by B.C. Schreiber), pp. 256–299. Columbia University Press, New York.

Faulds, J.E., **Schreiber, B.C.**, **Reynolds, S.J.** and **Okaya, D.** (1997) An immense, nonmarine Miocene salt deposit in the Basin and Range. *J. Geol.*, **105**, 19–36.

Faure, G. (1986) *Principles of Isotope Geology*. Wiley, New York.

Feeley, G.W. and **Kulp, J.L.** (1957) The origin of Gulf Coast salt dome sulfur deposits. *Bull. Am. Assoc. Petrol. Geol.*, **41**, 1802–1853.

Fischer, G.A. and **Roberts, L.T.** (1991) Cyclicity in the Green River Formation (Lacustrine Eocene) of Wyoming. *J. Sedim. Petrol.* **61**, 1146–1154.

Fontes, J.C., **Filly, A.**, **Gaudant, J.** and **Duringer, P.** (1991) Origine continentale des évaporites paléogènes de Haute Alsace: arguments paléoécologiques, sédimentologiques et isotopiques. *Bull. Soc. Géol. France*, **162**, 725–737.

Frumkin, A. (1998) The Holcene history of Dead Sea levels. In: *The Dead Sea: The Lake and its Setting* (Ed. by T.M. Niemi, Z. Ben-Abraham and J.R. Gat.), pp. 237–248. Oxford University Press, Oxford.

Geisler-Cussey, D. (1997) Modern depositional facies developed in evaporative environments (marine, mixed, and nonmarine). In: *Sedimentary Deposition in Rift and Foreland Basins in France and Spain (Paleogene and Lower Neogene)* (Ed. by G. Busson and B.C. Schreiber), pp. 3–42. Columbia University Press, New York.

Goldstein, R.H. and **Reynolds, T.J.** (1994) Systematics of fluid inclusions in diagenetic minerals. *SEPM Short Course*, **31**.

Gornitz, M.V. and **Schreiber, B.C.** (1981) Displacive halite hoppers from the Dead Sea: some implications for ancient evaporite depositions. *J. Sedim. Petrol.*, **51**, 787–794.

Handford, C.R. (1982) Sedimentology and evaporite genesis in a Holocene continental sabkha playa basin—Bristol Dry Lake, California. *Sedimentology*, **29**, 239–253.

Hardie, L.A. (1996) Secular variations in seawater chemistry: an explanation for the coupled secular variation in the mineralogies of marine limestones and potash evaporites over the past 600 m.y. *Geology*, **24**, 279–283.

Hardie, L.A., Smoot, J.P. and **Eugster, H.P.** (1978) Saline lakes and their deposits: A sedimentological approach. In: *Modern and Ancient Lake Sediments* (Ed. by A. Matter and M.E. Tucker), *Int. Assoc. Sedim. Spec. Publ.*, **2**, 7–41.

Hardie, L.A., Lowenstein, T.K. and **Spencer, R.J.** (1985) The problem of distinguishing between primary and secondary features in evaporites. In: *Sixth International Symposium on Salt* (Ed. by B.C. Schreiber and H.L. Harner), Vol. 1, pp. 11–39. Salt Institute, Alexandria, VA.

Harvie, C.E., Weare, J.M., Hardie, L.A. and **Eugster, H.P.** (1980) Evaporation of seawater: calculated mineral sequences. *Science*, **204**, 498–500.

Helvaci, C. (1995) Stratigraphy, mineralogy and genesis of the Bigadic borate deposits, western Turkey. *Econ. Geol.*, **90**, 1237–1260.

Helvaci, C. and **Ortí, F.** (1998) Sedimentology and diagenesis of Miocene colemanite–ulexite deposits (Western Anatolia, Turkey). *J. Sedim Petrol.*, **68**, 1021–1033.

Hiss, W.L. (1975a) Thickness of the Permian Guadalupian Capitan aquifer, southeast New Mexico and West Texas. *Texas Bureau of Mines and Mineral Resources Resource Map*, One sheet, Scale 1:510 000.

Hiss, W.L. (1975b) Chloride-ion concentration in ground water in Permian Guadalupian rocks, southeast New Mexico and West Texas, New Mexico. *Texas Bureau of Mines and Mineral Resources Resource Map*, One sheet, Scale 1:510 000.

Hiss, W.L. (1976) Structure of the Permian Ochoan Rustler Formation, southeast New Mexico and West Texas. *Texas Bureau of Mines and Mineral Resources Resource Map*, One sheet, Scale 1:510 000.

Hiss, W.L. (1980) Movement of ground water in Permian Guadalupian aquifer systems, southeastern New Mexico and western Texas. In: *Trans-Pecos Region; Southeastern New Mexico and West Texas* (Ed. by P.W. Dickerson, J.M. Hoffer and J.F. Callender), *New Mexico Geological Society Guidebook*, **31**, pp. 289–294.

Holland, H.D. (1984) *The Chemical Evolution of the Atmosphere and Oceans*. Princeton University Press, Princeton, NJ.

Holland, H.D., Horita, J. and **Seyfried, W.E. Jr** (1996) On the secular variations in the composition of Phaneozoic marine potash evaporites. *Geology*, **24**, 993–996.

Holser, W.T. (1966) Bromide geochemistry of salt rocks. In: *Second Symposium on Salt* (Ed. by J.L. Rau) 2, pp. 248–275. Northern Ohio Geological Society, Cleveland, OH.

Holser, W.T. (1984) Gradual and abrupt shifts in ocean chemistry during Phanerozoic time. In: *Patterns of Change in Earth Evolution* (Ed. by H.D. Holland and A.F. Trendall), pp. 123–144. Springer, Berlin.

Holser, W.T. and **Kaplan, I.R.** (1966) Isotope geochemistry of sedimentary sulfates. *Chem. Geol.*, **1**, 93–135.

Holser, W.T., Magaritz, M. and **Wright, J.** (1986) Chemical and isotopic variations in the world ocean during the Phanerozoic. In: *Global Bioevents* (Ed. by O. Walliser), pp. 63–74. Springer, Berlin.

Hovorka, S.D. (1992) Halite pseudomorphs after gypsum in bedded anhydrite – clue to gypsum-anhydrite relationships. *J. Sedim. Petrol.*, **62**, 1098–1111.

Hovorka, S.D. (1998) Characterization of bedded salt for storage caverns: case study from the Midland Basin. *Open File Report*, Bureau of Economic Geology, University of Texas, Austin, TX. http://www.utexas.edu/research/beg/salt/index.html.

Igarzábal, A. (1991) Evaporitas cuaternarias de la Puna Argentina. In: *Génesis de Formaciones Evaporíticas. Modelos Andinos E Ibéricos* (Ed. by J.J. Pueyo), *Estud. General*, **2**, pp. 333–374. Universitat de Barcelona, Barcelona.

Javor, B. (1989) *Hypersaline Environments*. Elsevier, Amsterdam.

Jensen, M.L. (1962) Biogenic sulfur and sulfide deposits. In: *Biochemistry of Sulfur Deposits: International Symposium on the Biochemistry of Sulfur Isotopes* (Ed. by M.L. Jensen), pp. 1–15. National Science Foundation Symposium.

Johnson, K.S. and **Neal, J.T.** (1997) Symposium on evaporite karst: origins, processes, landforms, examples and impacts. *Carbon. Evap.*, **12**, 1–116.

Jowett, E.C., Cathles, L.M., III and **Davis, B.W.** (1993) Predicting depths of gypsum dehydration in evaporitic sedimentary basins. *Bull. Am. Ass. Petrol. Geol.*, **77**, 402–413.

Katz, A. and **Kolodny, N.** (1989) Hypersaline brine diagenesis and evolution of the Dead Sea—Late Lisan system. *Geochim. Cosmochim. Acta*, **41**, 1609–1626.

Keys, J.R. and **Williams, K.** (1981) Origins of crystalline cold desert salts in the McMurdo region, Antarctica. *Geochim. Cosmochim. Acta*, **45**, 2299–2309.

Kinsman, D.J.J. (1969) Modes of formation, sedimentary associations and diagnostic features of shallow-water and supratidal evaporites. *Bull. Am. Assoc. Petrol. Geol.*, **53**, 830–840.

Kinsman, D.J.J. (1976) Evaporites: relative humidity control of primary mineral facies. *J. Sedim. Petrol.*, **46**, 273–279.

Kirkham, A. (1997) Shorline evolution, aeolian deflation and anhydrite distribution of the Holocene, Abu Dhabi. *Geoarabia*, **2**, 403–416.

Kirkland, D.W., Denison, R.E. and **Dean, W.E.** (2000) Parent brine of the Castile Evaporites (Upper Permian), Texas and New Mexico. *J. Sedim. Res.*, **70**, in press.

Ku, T.-L., Luo, B., Lowenstein, T.K., Li, J. and **Spencer, R.J.** (1998) U-Series chronology of lacustrine deposits in Death Valley, California. *Quat. Res.* **50**, 261–275.

Kunasz, I.A. (1978) Lithium in brines. In: *Fifth International Symposium on Salt* (Ed. by A.H. Coogan and L. Hauber), Vol. 1, pp. 115–117. Northern Ohio Geological Society, Cleveland, OH.

Longinelli, A. (1989) Oxygen-18 and sulphur-34 in dissolved oceanic sulphate and phosphate. In: *Handbook of Environmental Isotope Geochemistry* (Ed. by P. Fritz and J.Ch. Fontes), Vol. 3, pp. 219–255. Elsevier, Amsterdam.

Lloyd, R.M. (1968) Oxygen isotope behavior in the sulfate-water system. *J. Geophys. Res.*, **73**, 6099–6110.

Lowenstein, T.K. and **Hardie, L.A.** (1985) Criteria for recognition of salt-pan evaporites. *Sedimentology*, **32**, 627–644.

Lowenstein, T.K. and **Spencer, R.J.** (1990) Syndepositional origin of potash evaporites: petrographic and fluid inclusions evidence. *Am. J. Sci.*, **290**, 1–42.

Lowenstein, T.K., Li, J. and **Brown, C.B.** (1998) Paleotemperatures from fluid inclusions in halite: method verification and a 100,000 year paleotemperature record, Death Valley, CA. *Chem. Geol.*, **150**, 223–245.

Lucia, F.J. (1972) Recognition of evaporite-carbonate shoreline deposition. In: *Recognition of Ancient Sedimentary Environments* (Ed. by J.K. Rigby and W.K. Hamblin), *Soc. Econ. Palaeont. Mineral. Spec. Publ.*, **16**, 160–191.

Lugli, S., Schreiber, B.C. and **Triberti, B.** (1999) Giant polygons in the Messinian salt of the Realmonte Mine (Agrigento, Sicily): evidence for the desiccation of a Messinian halite basin. *J. Sedim. Res.*, **69**, 764–771.

Machel, H.G., Cavell, P.A. and **Patey, K.S.** (1996) Isotopic evidence for carbonate cementation and recrystallization, and for tectonic expulsion of fluids into the Western Canada sedimentary basin. *Bull. Geol. Soc. Am.*, **108**, 1108–1119.

McArthur, J.M. (1994) Recent trends in strontium isotope stratigraphy. *Terra Nova*, **6**, 331–358.

McKenzie, J.A., Hsü, K.J. and **Schneider, J.F.** (1980) Movement of subsurface waters under the sabkha, Abu Dhabi, U.A.E. & its relation to evaporative dolomite genesis. In: *Concepts and Models of Dolomitization* (Ed. by D.H. Zenger, J.B. Dunham and R.L. Ethington), *Soc. Econ. Palaeont. Mineral. Spec. Publ.*, **28**, 11–30.

McKenzie, J.A. (1981) Holocene dolomitization of calcium carbonate sediments from the coastal sabkhas of Abu Dhabi (U.A.E.): a stable isotope study. *J. Geol.*, **89**, 185–198.

Mees, F. (1999) Textural features of Holocene perennial saline lake deposits of Taoudenni—Agorgott basin, northern Mali. *Sedim. Geol.*, **127**, 65–84.

Momenzadeh, M. (1990) Saline deposits and alkaline magmatism: a genetic model. *J. Petrol. Geol.*, **13**, 341–356.

Moretto, R. and **Curial, A.** (1997) The salt basin of Bresse: Southern Saone Graben. In: *Sedimentary Deposition in Rift and Foreland Basins in France and Spain (Paleogene and Lower Neogene)* (Ed. by G. Busson and B.C. Schreiber), pp. 136–239. Columbia University Press, New York.

Morris, R.C. and **Dickey, P.A.** (1957) Modern evaporite deposition in Peru. *Bull. Am. Assoc. Petrol. Geol.*, **41**, 2467–2474.

Murray, R.C. (1964) Origin and diagenesis of gypsum and anhydrite. *J. Sedim. Petrol.*, **34**, 512–523.

Nakhla, F.M., Saleh, S.A. and **Gad, N.L.** (1985) Mineralogy, chemistry, and paragenesis of the thenardite deposits in Beida Lake, Wadi Natrun, Egypt. In: *Applied Mineralogy*, pp. 1001–1013. The Metallurgical Society of AIME, New York.

Neat, P., Faure, G. and **Pergram, W.** (1979) The isotopic composition of strontium in non-marine carbonate rocks: the Flagstaff Formation of Utah. *Sedimentology*, **26**, 271–282.

Neev, D. and **Emery, K.O.** (1967). The Dead Sea: depositional processes and environments of evaporites. *Geol. Survey Israel Bull.*, **41**.

Niemi, T.M. (1997) Fluctuations of Late Pleistocene Lake Lisan in the Dead Sea Rift. In: *The Dead Sea: The Lake and Its Setting* (Ed. by T.M. Niemi, Z. Ben-Abraham and J.R. Gat), pp. 226–236. Oxford University Press, Oxford.

Niemi, T.M., Ben-Abraham, Z. and **Gat, J.R. (Eds)** (1997) *The Dead Sea: The Lake and Its Setting.* Oxford University Press, Oxford.

Nissenbaum, A., Presley, B.J. and **Kaplan, I.R.** (1972) Early diagenesis in a reducing fjord, Saanich Inlet, British Columbia. *Geochim Cosmochim. Acta*, **36**, 1007–1027.

Noel, D. (1984) Les diatomées des saumures et des sédiments de surface du Salin de Bras del Port (Santa Pola province d'Alicante, Espagne). *Rev. Inst. Invest. Geol. Barcelona*, **38–39**, 79–107.

Nury, D. and **Schreiber, B.C.** (1997) Paleogene basins of Southern Provence. In: *Sedimentary Deposition in Rift and Foreland Basins in France and Spain (Paleogene and Lower Neogene* (Ed. by G. Busson and B.C. Schreiber), pp. 240–300. Columbia University Press, New York.

Nurmi, R. and **Friedman, G.M.** (1977) Sedimentology and depositional environments of basin-center evaporites, Lower Salina Group (Upper Silurian), Michigan Basin. In: *Reefs and Evaporites* (Ed. by J.H. Fisher), *Am. Ass. Petrol. Geol. Spec. Publ.*, **5**, 23–52.

Ogniben, L. (1957) Petrografia della Serie Solfifera Siciliana e considerazioni geologiche relative. *Mem. Descr. Carta Geol. d'Ital.*, **33**.

Ogniben, L. (1963) Sedimenti halitico-calcitici a struttura grumosa nel Calcare de Base Messiniano Sicilia. *Giornale Geol. Ser. 2*, **31**, 509–542.

Ortí, F. and **Salvany, J.M.** (1997) Continental evaporitic sedimentation in the Ebro Basin during the Miocene. In: *Sedimentary Deposition in Rift and Foreland Basins in France and Spain (Paleogene and Lower Neogene)* (Ed. by G. Busson and B.C. Schreiber), pp. 420–429. Columbia University Press, New York.

Ortí, F. and **Alonso, R.N.** (1999) Gypsum–hydroboracite association in the Sijes Formation (Miocene, NW Argentina). *J. Sedim. Petrol.*, **70**, in press.

Ortí, F., Busson, G., Cornée, A., Dulau, N., Geisler-Cussey, D., Gouleau, D., Jaccard, J., Landry, J.C., Mosser, C., Noël, D., Pierre, C., Poumot, C., Pueyo, J.J., Thomas, J.C., Thomas, M., Truc, G., Utrilla, R., Casal, R. and **Zaninetti, L.** (1984) Introduction to the sedimentology of the coastal salinas of Santa Pola (Alicante, Spain). *Rev. Inst. Invest Geol. Barcelona*, **38–39**, 9–235.

Ortí, F., Helvaci, C., Rosell, L. and **Gündogan, I.** (1998) Sulphate–borate relations in an evaporitic lacustrine environment: the Sultancayir Gypsum (Miocene, western Anatolia). *Sedimentology*, **45**, 697–710.

Palmer, M.R. and **Edmond, J.M.** (1992) Controls over the strontium isotope budget of river water. *Geochim. Cosmochim. Acta*, **56**, 2099–2111.

Paytan, A., Kastner, M., Campbell, D. and **Thiemens, M.H.** (1998) Sulfur isotopic composition of Cenozoic seawater. *Science*, **282**, 1259–1262.

Perthuisot, J.-P. (1975) La Sebkah el Melah de Zarzis. Genèse et évolution de un basin paralique. *Tra. Lab. Géol. Ecole Norm. Sup., Paris*, **9**.

Perthuisot, J.-P. (1980) Sites et processus de la formation d'évaporites dans la nature actuelle. *Bull. Cent. Rech. Explor. Prod. Elf Aquitaine*, **4**, 207–233.

Perthuisot, J.-P., Castanier, S., Rouchy, J.-P., Maurin, A. and **Guelorget, O.** (1993) Le role des bactéries dans le précipitation du sel. Exemple du Lac Asal (Djibouti). In: *Jurade du Sel*, pp. 119–144. Actes Coll. Int. du Sel, Salies de Béarn.

Peryt, T.-M. and **Pierre, C.** (1994) Origin of polyhalite deposits in the Zechstein (Upper Permian) Zdrada sulphate platform (northern Poland). *Int. Sedim. Congr.*, **14**, C1-10.

Pierre, C. (1985) Isotopic evidence for the dynamic redox cycle of dissolved sulphur compounds between free and interstitial solutions in marine salt pans. *Chem. Geol.* **53**, 191–196.

Pierre, C. (1989) Sedimentation and diagenesis in restricted marine basins. In: *Handbook of Environmental Isotope Geochemistry* (Ed. by P. Fritz and J.Ch. Fontes), Vol. 3A, pp. 257–315, Elsevier, Amsterdam.

Pierre, C. (1983) Polyhalite replacement after gypsum at Ojo de Liebre Lagoon (Baja California, Mexico): an early diagenesis by mixing of marine brines and continental waters. In: *Sixth International Symposium on Salt* (Ed. by B.C. Schreiber and L. Harner), Vol. 1, pp. 257–265. Salt Institute, Alexandria, VA.

Pierre, C. and **Fontes, J.Ch.** (1978) Isotope composition of Messinian sediments from the Mediterranean Sea as indicators of paleoenvironments and diagenesis. In: *Initial Reports of the Deep Sea Drilling Project, Leg XLIIA* (Ed. by K.J. Hsü and L. Montadert), Vol. 42A, pp. 635–650. US Government Printing Office, Washington, DC.

Purser, B.H. (1973) *The Persian Gulf. Holocene Carbonate Sedimentation and Diagenesis on a Shallow Epicontinental Sea.* Springer, Berlin.

Purser, B.H. (1985) Coastal evaporite systems. In: *Hypersaline Ecosystems: the Gavish Sabkha* (Ed. by G.M. Friedman and W.E. Krumbein), pp. 72–102. Springer, Berlin.

Raab, M. and **Spiro, B.** (1991) Sulfur isotopic variations during seawater evaporation with fractional crystallization. *Chem. Geol.*, **86**, 323–333.

Renaut, R.W. and **Tiercelin, J.-J.** (1994) Lake Begoria, Kenya Rift Valley—a sedimentological overview. In: *Sedimentology and Geochemistry of Modern and Ancient Saline Lakes* (Ed. by R.W. Renaut and W.M. Last), *Soc. Econ. Paleont. Mineral. Spec. Publ.*, **50**, 101–123.

Richter-Bernburg, G. (1955) Über salinare Sedimentation. *Zeitschr. Deutsch. Geol. Gesellsch.*, **105**, 593–645.

Richter-Bernburg, G. (1980) Aberrant vertical structures in well bedded halite deposits. In: *Fifth International Symposium on Salt* (Ed. by A.H. Coogan and L. Hauber), Vol. 1, pp. 159–166. Northern Ohio Geological Society, Cleveland, OH.

Roberts, S.M. and **Spencer, R.J.** (1995) Paleotemperatures preserved in fluid inclusions in halite. *Geochim. Cosmochim. Acta*, **59**, 3929–3942.

Roedder, E. (1984) The fluids in salt. *Am. Mineral.*, **69**, 413–439.

Rosell, L., Ortí, F., Kasprzyk, A., Playà, E. and **Peryt, T.M.** (1998) Strontium geochemistry of primary gypsum: Messinian of Southeastern Spain and Sicily and Badenian of Poland. *J. Sedim. Res.*, **68**, 63–79.

Rouchy, J.M., Taberner, C., Blanc-Valleron, M.-M., Sprovieri, R., Russell, M., Pierre, C., Di Stefano, E., Pueyo, J.J., Caruso, A., Dinarès-Turrell, J., Gomis-Coll, E., Wolff, G.A., Cespuglio, G., Ditchfield, P., Pestrea, S., Combourieu-Nebout, N., Santisteban, C. and **Grimalt, J.O.** (1998) Sedimentary and diagenetic markers of the restriction in a marine basin: the Lorca Basin (SE Spain) during the Messinian. *Sedim. Geol.*, **121**, 23–55.

Ruppel, S.C., James, E.W., Barrick, J.E., Nowlan, G. and **Uyeno, T.T.** (1996) High resolution $^{87}Sr/^{86}Sr$ chemostratigraphy of the Silurian: implications for event correlation and strontium flux. *Geology*, **24**, 831–834.

Salvany, J.M. (1997) Continental evaporitic sedimentation in Navarra during the Oligocene to Lower Miocene: Falces and Lerín Formations. In: *Sedimentary Deposition in Rift and Foreland Basins in France and Spain (Paleogene and Lower Neogene)* (Ed. by G. Busson and B.C. Schreiber), pp. 397–411. Columbia University Press, New York.

Salvany, J.M. and **Ortí, F.** (1997) Glauberite deposits of the Lerín formation (Lower Miocene: Alcanadre Zone, La Rioja-Navarra). In: *Sedimentary Deposition in Rift and Foreland Basins in France and Spain (Paleogene and Lower Neogene)* (Ed. by G. Busson and B.C. Schreiber), pp. 412–419. Columbia University Press, New York.

Salvany, J.M., Muñoz, A. and **Pérez, A.** (1994) Nonmarine evaporitic sedimentation and associated diagenetic processes of the southwestern margin of the Ebro Basin (Lower Miocene), Spain. *J. Sedim. Res.*, **A64**, 190–203.

Sammy, N. (1985) Biological systems in North-Western Australian Solar Salt Fields. In: *Sixth International Symposium on Salt* (Ed. by B.C. Schreiber and H.L. Harner), Vol. 1, pp. 207–215. Salt Institute, Alexandria, VA.

Schaller, W.T. and **Henderson, E.P.** (1932) Mineralogy of drill cores from the potash field of New Mexico and Texas. *Bull. U.S. Geol. Surv.*, **833.**

Schidlowski, M., Hayes, J.M. and **Kaplan, I.R.** (1983) Isotopic inferences of ancient biochemistries: carbon, sulfur, hydrogen, and nitrogen. In: *Earth's Earliest Biosphere: Its Origin and Evolution* (Ed. by J.W. Schopf), pp. 149–186. Princeton University Press, Princeton, N.J.

Schlager, W. and **Bolz, H.** (1977) Clastic accumulation of sulphate evaporites in deep water. *J. Sedim. Petrol.*, **47**, 600–609.

Schreiber, B.C., Friedman, G.M., Decima, A. and **Schreiber, E.** (1976) The depositional environments of the Upper Miocene (Messinian) evaporite deposits of the Sicilian Basin. *Sedimentology*, **23**, 729–760.

Schreiber, B.C. and **Schreiber, E.** (1977) The salt that was. *Geology*, **5**, 527–528.

Schreiber, B.C., Catalano, R. and **Schreiber, E.** (1977) Evaporitic facies observed in the Upper Miocene (Messinian) deposits of the Salemi Basin (Sicily) and a modern analog. In: *Reefs and Evaporites* (Ed. by J.H. Fisher), pp. 169–180, *AAPG Spec. Publ.*, **5.**

Schreiber, B.C. and **Hsü, K.J.** (1980) Evaporites. In: *Developments in Petroleum Geology* (Ed. by G.D. Hobson), Vol. 2, pp. 87–138. Elsevier, Barking.

Schreiber, B.C. and **Walker, D.** (1992) Halite pseudomorphs after gypsum: a suggested mechanism. *J. Sedim. Petrol.*, **62**, 61–70.

Schubel, K.A. and **Lowenstein, T.K.** (1997) Criteria for the recognition of shallow-perennial-saline-lake halites based on recent sediments from the Qaidam Basin, Western China. *J. Sedim. Res.*, **67**, 74–87.

Shearman, D.J. (1963) Recent anhydrite, dolomite and halite from the coastal flats of the Arabian shore of the Persian Gulf. *Proc. Geol. Soc. London*, **1607**, 63–65.

Shearman, D.J. (1970) Recent halite rock, Baja California, Mexico. *Trans. Inst. Min. Metall.*, **79B**, 155–162.

Shearman, D.J. (1978) Halite in sabkha environments. In: *Marine Evaporites* (Ed. by W.E. Dean and B.C. Schreiber), *SEPM Short Course*, **4**, 30–42.

Shearman, D.J. (1985) Syndepositional and late diagenetic alteration of primary gypsum to anhydrite. In: *Sixth International Symposium on Salt* (Ed. by B.C. Schreiber and L. Harner), Vol. 1, pp. 41–50. Salt Institute, Alexandria, VA.

Shearman, D.J. and **Fuller, J.G.** (1969) Anhydrite diagenesis, calcitization, and organic laminites, Winnipegois Formation, Middle Devonian, Saskatchewan. *Bull. Can. Petrol. Geol.*, **17**, 496–525.

Smith, G.I. and **Medranno, M.D.** (1996) Continental borate deposits of Cenozoic age. In: *Boron: Mineralogy, Petrology and Geochemistry* (Ed. by E.S. Grew and L.M. Anovitz). *Rev. Mineral.*, **33**, 263–298.

Smoot, J.P. and **Castens-Seidell, B.** (1994) Sedimentary features produced by efflorescent salt crusts, Saline Valley and Death Valley, California. In: *Sedimentology and Geochemistry of Modern and Ancient Saline Lakes* (Ed. by R.W. Renaut and W.M. Last), *Soc. Econ. Paleont. Mineral. Spec. Publ.*, **50**, 73–90.

Sofer, Z. and **Gat, J.R.** (1975) Activities and concentrations of oxygen 18 in concentrated aqueous solutions: analytical and geophysical implications. *Earth Planet. Sci. Lett.*, **15**, 232–238.

Southgate, P.N. (1982) Cambrian skeletal halite crystals and experimental analogues. *Sedimentology*, **29**, 391–407.

Stanley, S.M. and **Hardie, L.A.** (1998) Secular oscillations in the carbonate mineralogy of reef-building and sediment-producing organisms driven by tectonically forced shifts in seawater chemistry. *Palaeogeog. Palaeoclimat. Palaeoecol.*, **144**, 3–19.

Stanley, S.M. and **Hardie, L.A.** (1999) Hyper-calcification: paleontology links plate tectonics and geochemistry to sedimentology. *GSA Today*, **9**, 1–7.

Steinhorn, I. (1997) Evaporation estimate for the Dead Sea: Essential considerations for saline lakes. In: *The Dead Sea: The Lake and Is Setting* (Ed. by T.M. Niemi, Z. Ben-Abraham and J.R. Gat), pp. 122–132. Oxford University Press, Oxford.

Steinhorn, I. and **Assaf, G.** (1977) The physical structure of the Dead Sea water column. In: *Hypersaline Brines and Evaporitic Environments* (Ed. by A. Nissenbaum), pp. 145–153. Elsevier, Amsterdam.

Stoertz, G.E. and **Ericksen, G.E.** (1974) Geology of salars in Northern Chile. *US Geol. Survey Prof. Paper*, **811**.

Strauss, H. (1997) The isotopic composition of sedimentary sulfur through time. *Palaeogeog. Palaeoclimat. Palaeoecol.*, **132**, 97–118.

Thode, H.G. and **Monster, J.** (1965) Sulfur-isotope geochemistry of petroleum, evaporites and ancient seas. In: *Fluids in Subsurface Environments* (Ed. by A. Young and J.E. Galley), *Am. Ass. Petrol. Geol. Mem.*, **4**, 367–377.

Tucker, R.M. (1981) Giant polygons in the Triassic salt of Cheshire, England: a thermal contraction model for their origin. *J. Sedim. Petrol.*, **51**, 779–786.

Utrilla, R., **Pierre. C.**, **Ortí, F.** and **Pueyo, J.J.** (1992) Oxygen and sulphur isotope compositions as indicators of the origin of Mesozoic and Cenozoic evaporites from Spain. *Chem. Geol. (Isotope Geosci.)*, **102**, 229–244.

Vasconcelos, C. and **McKenzie, J.A.** (1997) Microbial mediation of modern dolomite precipitation and diagenesis under anoxic conditions (Lagoa Vermelha, Rio de Janeiro, Brazil). *J. Sedim. Res.*, **67**, 378–390.

Wardlaw, N. (1970) Effects of fusion, rates of crystallization and leaching on bromide and rubidium solid solutions in halite, sylvite and carnallite. In: *Third Symposium on Salt* (Ed. by J.L. Rau and F.L. Dellwig), Vol. 1, pp. 223–231. Northern Ohio Geological Society, Cleveland, OH.

Warren, J.K. (1982) The hydrological setting, occurrence and significance of gypsum in late Quaternary salt lakes in South Australia. *Sedimentology*, **29**, 609–627.

Warren, J.K. and **Kendall, C.G.St.C.** (1985) Comparison of marine sabkhas (subaerial) and salina (subaqueous) evaporites: modern and ancient. *Bull. Am. Ass. Petrol. Geol.*, **69**, 1013–1023.

Weiler, Y., **Sass, E.** and **Zak, I.** (1974) Halite oolites and ripples in the Dead Sea, Israel. *Sedimentology*, **21**, 623–632.

West, I. (1979) Primary gypsum nodules in a modern sabkha on the Mediterranean coast of Egypt. *Geology*, **7**, 354–358.

Zak, I. (1997) Evolution of the Dead Sea brines. In: *The Dead Sea: The Lake and Its Setting* (Ed. by T.M. Niemi, Z. Ben-Abraham and J.R. Gat), pp. 133–144. Oxford University Press, Oxford.

Sedimentology (2000) **47** (Suppl. 1), 239–256

Continental margin sedimentation, with special reference to the north-east Atlantic margin

PHILIP P. E. WEAVER, RUSSELL B. WYNN, NEIL H. KENYON and JEREMY EVANS
Southampton Oceanography Centre, Empress Dock, Southampton SO14 3ZH, UK
(E-mail: p.weaver@soc.soton.ac.uk)

ABSTRACT

The north-east Atlantic continental margin displays a wide range of sediment transport systems with both along-slope and down-slope processes. Off most of the north-west African margin, south of 26°N, upwelling produces elevated accumulation rates, although there is little fluvial input. This area is subject to infrequent but large-scale mass movements, giving rise to debris flows and turbidity currents. The turbidity currents traverse the slope and deposit thick layers on the abyssal plains, while debris flows deposit on the continental slope and rise. From the Atlas Mountains northwards to 56°N, the margin is less prone to mass movements, but is cut by a large number of canyons, which also funnel turbidity currents to the abyssal plains. The presence of a lithospheric plate boundary off SW Iberia is believed to have led to high rates of sediment transport to the deep sea. Even larger quantities of coarse sediments have fed the canyons and abyssal plains in the Bay of Biscay as a result of drainage from melting icecaps. Bottom currents have built sediment waves off the African and Iberian margins, and created erosional furrows south of the Canaries. The Mediterranean outflow is a particularly strong bottom current near the Straits of Gibraltar, depositing sand waves and mud waves in the Gulf of Cadiz. North of 56°N, the margin is heavily influenced by glacial and glaciomarine processes active during glacial times, which built glacial trough-mouth fans, such as the North Sea Fan, and left iceberg scour marks on the upper slope and shelf. Over a long period, especially during interglacials, this part of the margin has been greatly affected by along-slope currents, with less effect by turbidity currents than on the lower latitude margins. Large-scale mass movements are again a prominent feature, particularly off Norway and the Faeroes. Some of these mass movements have occurred during the Holocene, although high glacial sedimentation rates may have contributed to the instability.

Keywords: Continental margin, sediments, slope processes, debris flow, turbidity current, bottom current

INTRODUCTION

In this paper, we describe the sedimentary processes, and map the major sedimentary features, on the north-east Atlantic continental margin from north-west Africa to northern Norway (Fig. 1). This region includes areas of passive margin with low sediment input, areas with high glacial sedimentation, areas incised by canyons, areas of fan sedimentation, areas with low gradient continental slopes, and areas with steep continental slopes.

Continental margins are important areas for investigation, since they form a 'new frontier' for exploitation by the hydrocarbons industry, and increasing numbers of telecommunication cables are laid across them. Nevertheless, many of the sedimentary processes displayed here remain poorly understood. Major advances in our knowledge of continental margins have followed the development and deployment of long-range side-scan sonars, such as GLORIA, which can image large areas in relatively short time intervals. Further advances have been achieved following the development of much higher-resolution deep-towed side-scan, and of swath bathymetry. There has been some ocean drilling of passive continental margins by Deep Sea Drilling Project (DSDP) and the Ocean Drilling Program (ODP), especially of modern fans (Legs 96, 116, 155), sediment drifts (e.g. Legs 94, 172, 181), and abyssal plains (Legs 149, 157). Down-slope processes have not been specifically targetted except in Legs 150 and 174 A, both on the New Jersey margin, where a down-slope transect of cores was drilled.

In this paper, we have limited our discussion to sedimentary processes operating from the shelf edge to the abyssal plain, so including the continental slope, continental rise and abyssal plain. In all cases, it is important to include the sediments of the abyssal plains as part of the continental margin, because the plains are formed by sediments derived from the margin. Within this area, the dominant sedimentary processes are along-slope sediment transport, which builds sediment drifts and waves, and down-slope processes involving mass wasting (sediment slides, debris flows and turbidity currents). In addition, canyons can be incised into the slope, in some cases cutting across the shelf, and large submarine fans may be built off major points of sediment supply. The scale of some of the processes can be very large; for example, the Storegga slide off Norway has an estimated volume of 5700 km^3 (Bugge *et al.*, 1988), while a single turbidite in the Madeira Abyssal Plain has a volume in excess of 300 km^3 (Weaver *et al.*, 1992, 1995).

MAIN SEDIMENTARY PROCESSES

Submarine slides and debris flows

Sediment instability resulting in submarine slides and debris flows is very common on some continental margins, and single events are frequently very large (of the order of hundreds of square kilometres). Should failure occur, then the ocean floor offers unimpeded slopes and flat-floored basins hundreds of kilometres in length, allowing flow over enormous distances. Slides are defined as the downward and outward movement of slope-forming materials, wherein shear failure occurs along one or several surfaces (see Hampton *et al.*, 1996, for a thorough review). Many submarine slides disintegrate down-slope into debris flows which are described as the movement of granular solids, sometimes mixed with minor amounts of entrained water on a low slope. The typical end product is a mass of displaced blocks of sediment or rock embedded in a more highly disrupted matrix. Debris avalanches are used here to describe submarine slides where little or no water entrainment occurs, and the resulting run-out distances are less. High-latitude slopes commonly consist of stacked debris flows (Vorren *et al.*, 1998). Such flows have been studied in the field and experimentally (e.g. Mohrig *et al.*, 1999), and are believed to flow on a lubricating layer of water between the flow and the underlying material—a process referred to as hydroplaning.

Turbidity currents

Turbidity currents occur on a variety of scales, and are a common feature of many continental margins. The largest turbidity currents form deposits in the deep ocean, and are termed megaturbidites; they are often linked to mass movements occurring higher up the slope. These currents transport tens to hundreds of cubic kilometres of sediment, and appear to be able to travel immense distances over extremely low-gradient slopes. They dump their sediment load only when they become ponded on the flat abyssal plains at the base of the slope. Individual mud-dominated turbidites commonly form layers up to a few metres in thickness over the whole abyssal plain (Pilkey, 1987; Rothwell *et al.*, 1992).

On the continental slope and rise, turbidity currents are usually funnelled down-slope by canyons and channels. Canyons are generally V-shaped in profile and are often eroded into bedrock. In plan view, they display tributary canyon systems with a number of orders of tributary, including steep, gullied slopes. They may originate offshore of a sediment point source, such as a river, and generally occur on, or just seaward of the shelf-break. Canyons commonly pass down-slope into channels, which are shallower than canyons, more U-shaped in profile, and may have depositional elements such as levees and aggradational floors. Deep-water channels occur in a variety of forms, from deep, narrow erosive channels, to broad, shallow distributary channels. Meandering channels tend to occur on fans with gentle gradients and a

dominantly fine-grained sediment supply (e.g. Clark *et al.*, 1992). Both canyon and channel floors can have bedforms such as gravel and sand waves and erosional scours.

Turbidity currents generally deposit the majority of their load on turbidite fans. The rivers running into the north-east Atlantic are relatively small and the resultant fans are usually built near the foot of the slope, beyond the canyon mouths.

Bottom currents

Bottom currents often play a major role in continental margin sedimentation. They can erode, mould, transport and redistribute sediments supplied to the slope and rise by down-slope flows and vertical settling. Bottom current features include erosional furrows and moats, and depositional bedforms, such as contourite drifts and sediment waves.

Contourite drifts are the primary deposits of bottom currents, and are widespread throughout the World Ocean. Contourite drifts occur on a variety of scales, from small patch drifts (<100 km^2) to giant elongate drifts (>100 000 km^2). Controls on contourite accumulation include the following: (1) active geostrophic circulation—this varies through time, for example, bottom currents are generally more active during interglacials, so contourite deposition is increased during these periods; (2) sea-floor topography—this can influence bottom current flow velocity, leading to sediment erosion or deposition in certain areas; (3) sediment supply—many contourite deposits actually derive the majority of their sediment from turbidite systems, because turbidity currents are one of the main mechanisms for sediment to be transported to the slope, rise and basins; (4) nepheloid layer turbidity—this also affects the ability of bottom currents to deposit or erode sediment (Faugeres *et al.*, 1993).

PROCESS INTERACTION ON THE NORTH-EAST ATLANTIC MARGIN

The nonglaciated margin south of 26°N (mass-wasting processes dominant)

Down-slope processes

The passive continental margin off north-west Africa, between 15°N and 26°N, is dominated by mass-wasting processes (submarine slides, debris flows and megaturbidites) (Fig. 2c). This is because of a number of factors, including low terrigenous input and local upwelling. This part of the margin receives very little fluvial sediment, particularly the margin off the western Sahara. However, a number of upwelling cells (Sarnthein *et al.*, 1982) give rise to high accumulation rates along the upper slope and shelf edge. The megaturbidites derived from this margin and deposited on the Madeira Abyssal Plain are distinguished by their high organic contents (0·5–2%; Rothwell *et al.*, 1992), suggesting that it is the sediments resulting from the upwelling which are unstable, presumably as a result of slope oversteepening.

The upper parts of submarine slides often show spectacular detail when imaged with high-resolution side-scan sonar (Fig. 3), or swath bathymetry. The Saharan Slide off west Africa (Simm & Kidd, 1984; Masson *et al.*, 1992) included about 600 km^3 of sediment, which was transported up to 700 km down-slope, across gradients ranging from 1·5° to 0·1°. Although not strictly part of the continental margin, the Canary Islands are also subject to large-scale mass wasting. The head and side-walls of the El Golfo debris avalanche in the Canary Islands extend from about 3000 m below sea-level to almost 1000 m above sea-level, forming a distinct arcuate cliff in the side of the island of El Hierro. This debris avalanche contains 150–180 km^3 of rock and debris which extends some 60 km down-slope, to a water depth of about 3500–4000 m (Masson, 1996). Some of the blocks in the debris avalanche are 1 km across.

In contrast to the margin to the south and to the north, well-developed turbidite fans are not present on the north-west African margin. This is because of the lack of major sediment point sources and because terrigenous input overall is limited and intermittent (Wynn *et al.*, 2000a). The limited input means that turbidity currents on the margin are infrequent, but on a large scale. They flow along linear turbidity current pathways that transport material up to 1200 km offshore to the deep abyssal plains. In addition, many parts of the margin have complex seafloor topography, which further inhibits fan development. Sandy lobes may occur at the terminations of channels on the proximal margins of abyssal plains. For example, although the fill of the Madeira Abyssal Plain is dominated by a thick sequence of ponded turbidite muds (Fig. 4), a sandy lobe has developed on the eastern margin of the plain near the termination of the Madeira Distributary Channel System (MDCS).

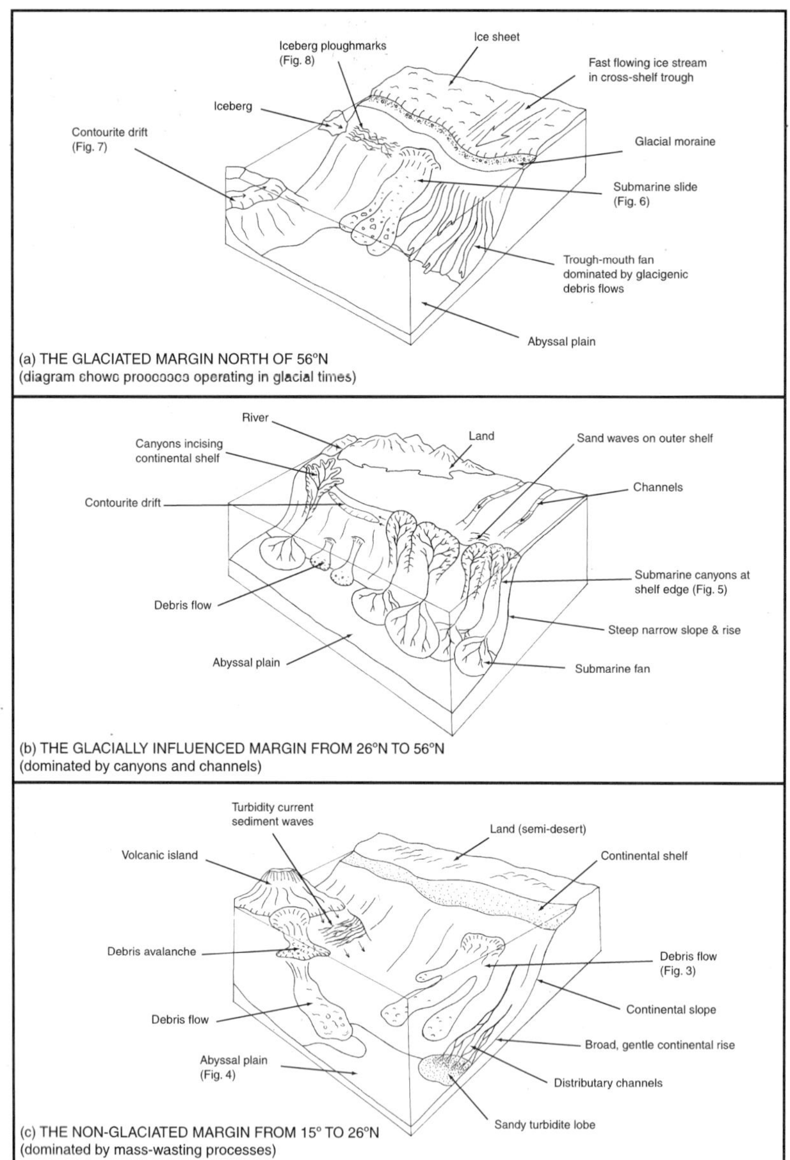

Fig. 2. Schematic diagrams, showing the idealized sedimentary features on (a) the glaciated margin north of 56°N, (b) the margin from 26°N to 56°N (dominated by canyons and channels, and (c) the margin from 15°N to 26°N (dominated by mass-wasting processes).

The MDCS is the best-developed distributary channel network on the north-west African margin and is some 500 km long. It extends from the western edge of the intraslope Agadir Basin to the eastern edge of the Madeira Abyssal Plain, and occurs on gradients of 0·3° to 0·05°. The shallower (<20 m deep), braided channels within the system occur on slopes of about 0·07°, whereas the narrower, more incised (up to 50 m deep) channels occur on slightly steeper slopes of about 0·3°. Unlike most distributary channels, levees are not developed in the MDCS, because there is no significant turbidity current overspill. This is because the sandy bedload is wholly confined within the channel, and the fine-grained suspension cloud travels as an unchannelized sheet flow across the rise (Masson, 1994). It is likely that other well-developed distributary channel systems supply material to the Cape Verde Basin (Jacobi & Hayes, 1992) and the southern margin of the Madeira Abyssal Plain; however, data for these areas are lacking.

A major canyon system occurs on the western Saharan margin at 26°N (von Rad & Wissmann, 1982). Twenty-five closely spaced canyons occur at a water depth of 1000–3000 m. They have an average spacing of just 10 km, and are different from most other canyons on the north-east Atlantic margin, in that they do not generally reach back to the shelf edge, and rarely extend down-slope beyond the slope–rise boundary. There is currently no fluvial input on this part

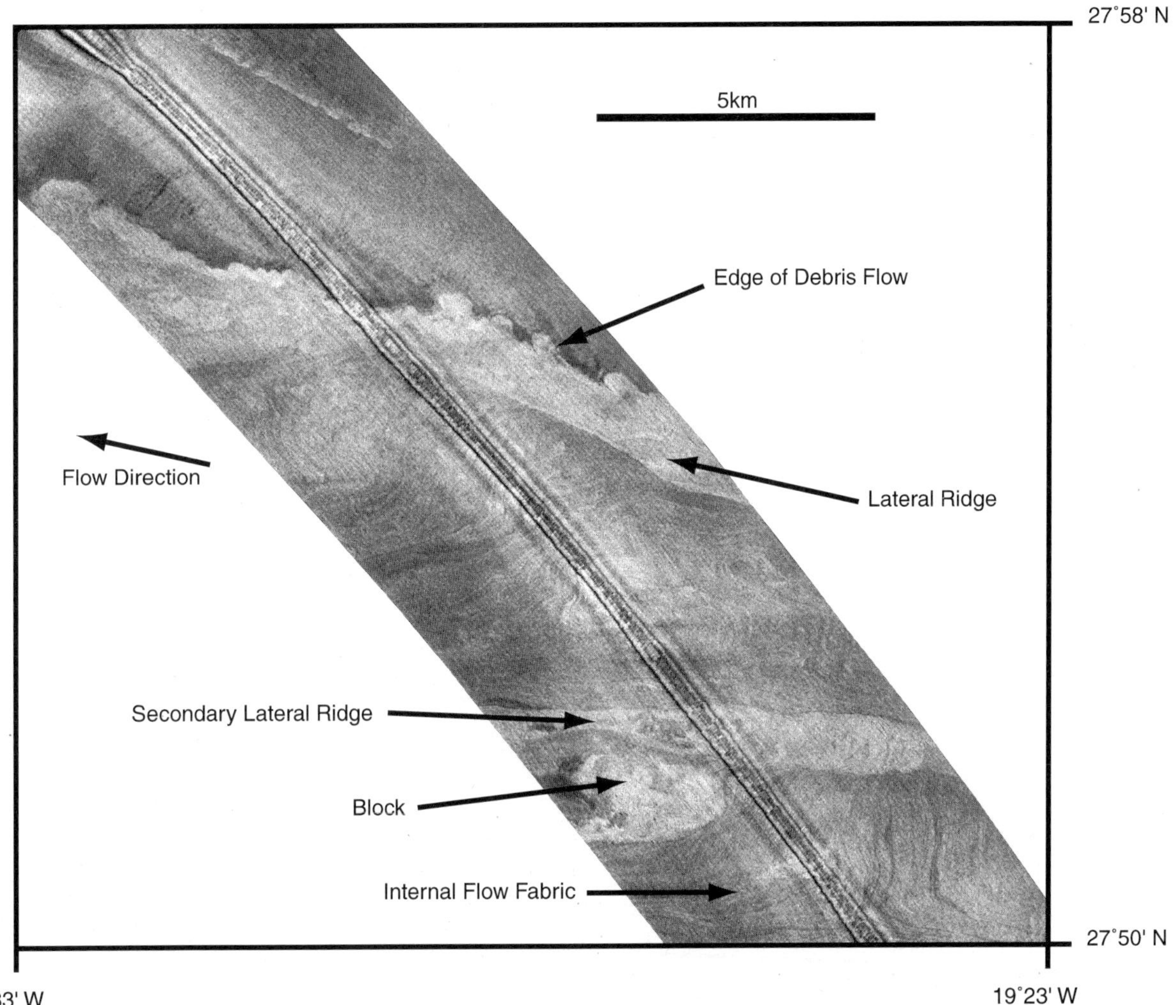

Fig. 3. TOBI (30 kHz deeptowed, side-scan sonar) image, showing the edge of the Saharan debris flow, west of the relatively starved north-west African margin. High levels of backscatter are shown as light tones. For location, see Fig. 1.

of the margin, and most aeolian input onto the shelf is swept southwards by strong currents (Sarnthein *et al.*, 1982). Therefore, the canyons are believed to have formed during previous lowstands of sea-level, when there was increased terrigenous input (von Rad & Wissmann, 1982). The close spacing of the canyons appears to be controlled by the steeper slope angles in this area. The north-west African continental slope generally has a slope angle of 0·5°–2° but, in the area of the canyon system, the slope angle is increased to 3°, leading to closely spaced canyon incision (von Rad & Wissmann, 1982). This highlights how subtle changes in seafloor gradients can have a dramatic control on slope processes. A similar example of control on canyon formation by slope gradient is found in the Porcupine Seabight (Kenyon, 1987).

Recent investigations on the lower slope and rise north-west of the Canary Islands have revealed that fields of sediment waves on open slopes may be generated by unconfined turbidity currents (Wynn *et al.*, 2000b). Sediment waves can also be generated by bottom currents, but a number of criteria can be used to distinguish the two types.

These criteria are as follows. (1) Turbidity current waves typically show a regular downslope decrease in dimensions, and bottom current waves are generally more irregular. (2) Turbidity current wave crestlines are aligned parallel to the slope, and are often sinuous and bifurcating. Bottom current waves on slopes are generally developed at an angle to the slope and are more uniform. (3) Core studies of different wave types typically reveal evidence of turbidity current deposition *or* bottom current deposition.

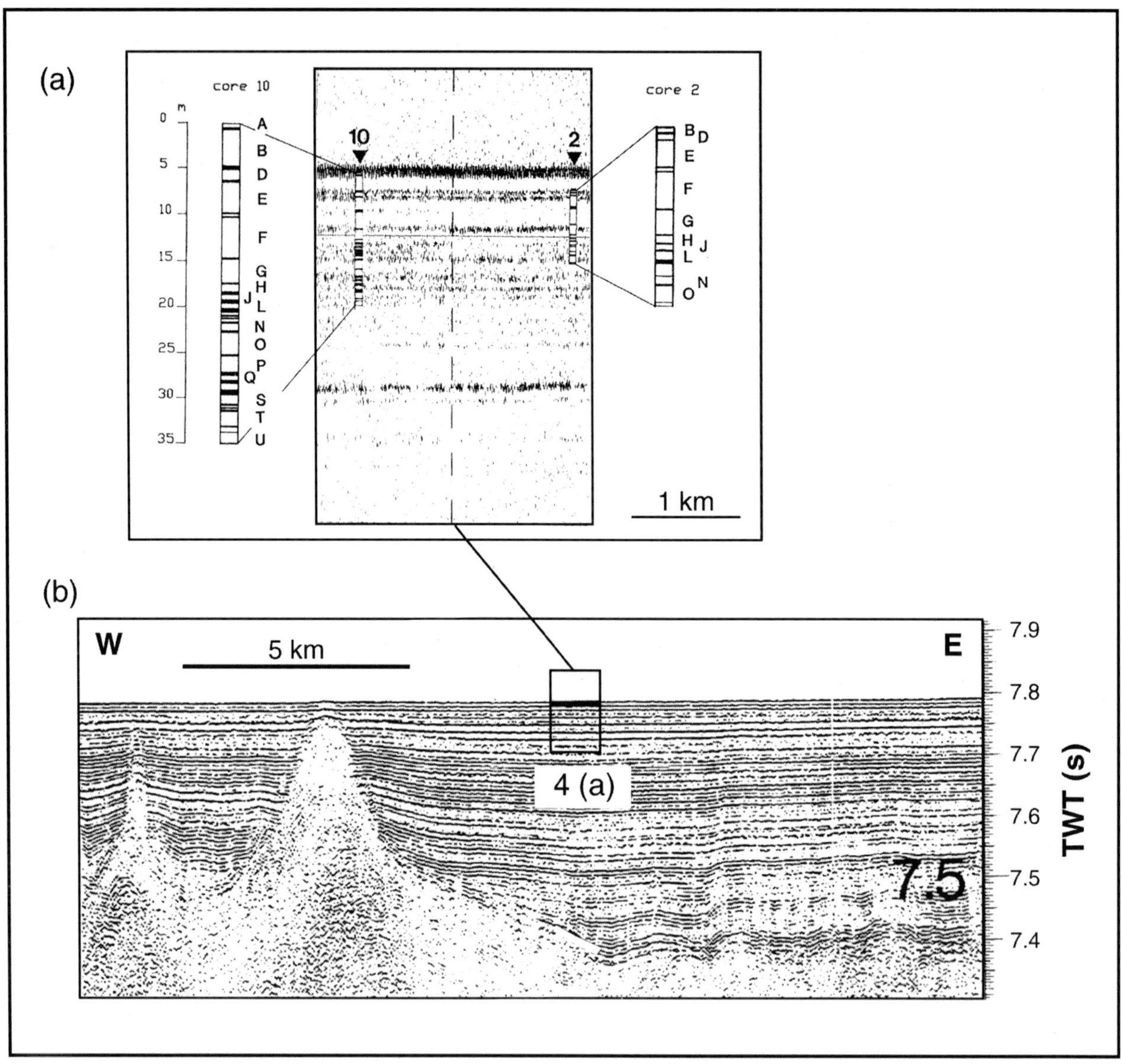

Fig. 4. (a) 3·5 kHz profile across a section of the Madeira Abyssal Plain (31°N, 25°W), with two cores showing how the megaturbidite sequence can be roughly correlated with seismic reflectors. The lettering scheme on the two cores represents individual megaturbidites as defined by Weaver *et al.* (1992). (b) Airgun seismic profile of ponded megaturbidites from the Madeira Abyssal Plain. Boxed area represents the high-resolution profile shown in (a). For location, see Fig. 1.

(4) Bottom current sediment waves only form beneath bottom currents of 9–50 cm s^{-1} (e.g. Flood, 1988). They do not occur in areas with weak or absent bottom currents (although the bottom current regime in the past must be taken into account). (5) A thorough examination of the oceanography, seafloor topography, and sedimentary regime in the area of a wave field will often give conclusive evidence of its origin. Wave heights of turbidity current sediment waves around the Canary Islands are up to 70 m, and wavelengths are up to 2·4 km. The waves are aligned roughly parallel to the slope and migrate upslope. Cores taken through the waves show a sequence of interbedded turbidites and pelagic/hemipelagic sediments. Overall, the dimensions, morphology and mode of formation of the waves are very similar to those described for sediment waves on channel-levee backslopes (e.g. Nakajima *et al.*, 1998). Sediment waves in turbidity current channels have also been described from the flanks of the western and northern Canary Islands (Wynn *et al.*, 2000b). These waves are smaller and coarser-grained than those formed

by unconfined flows. They have wave heights up to 6 m and wavelengths up to 1·2 km.

Along-slope processes

Antarctic bottom water (AABW) is responsible for forming erosional furrows and sediment waves on the continental rise off north-west Africa (Jacobi & Hayes, 1984, 1992). In the area of the Saharan Seamounts, AABW is funnelled between the seamounts and strengthened by tidal currents, leading to the formation of erosional furrows.

The glacially influenced margin from 26°N to 56°N (canyon and channel processes dominant)

Down-slope processes

This margin has a series of large submarine canyons and little or no evidence of submarine slides (Fig. 2b). North-east of the Canary Islands is an area of canyoned margin beginning with the Agadir Canyon. These canyons funnel sediment derived from the Atlas Mountains across a continental slope which is considerably narrower than that to the south. The turbidity currents deliver sediments directly to the Seine Abyssal Plain (Davies *et al.*, 1997) and the Agadir Basin. The Agadir Basin is an intraslope basin from which sediment flows now spill over and flow down-slope to the Madeira Abyssal Plain (Masson, 1994; Wynn *et al.*, 2000a).

TOBI deep-towed side-scan sonar surveys at the mouth of the Agadir and Lisbon Canyons have revealed extensive zones of erosional scours and bedforms associated with the transition zone between canyon mouth and depositional lobe. These transition zones are commonly developed at the mouths of submarine canyons and channels (see Normark & Piper, 1991, for a review), and are created by the increased turbulence and erosive power of turbidity currents undergoing flow expansion. The flow expansion may be linked to a hydraulic jump (Komar, 1971) in areas where there is a significant break-of-slope at the canyon/channel mouth. At the mouth of the Agadir and Lisbon Canyons, individual erosional scours up to 21 m deep and 500 m wide have been imaged. In addition, fields of chevron-shaped scours, erosional scarps and sediment waves have all been imaged in this zone. The Agadir Canyon system feeds infrequent (one every 30 000 years), large-volume (up to 250 km^3) turbidity currents to the Agadir Basin (Weaver *et al.*, 1992). These turbidity currents spread out and flow as sheet flows across the basin floor, which is topographically constrained by volcanic islands and seamounts. Many of the flows continue westwards to the Madeira Abyssal Plain but, in doing so, they often deposit much of their sandy bedload in the Agadir Basin as a sheet sand (Ercilla *et al.*, 1998).

The margin from Iberia to the southern Rockall Trough is dominated by canyons which are often deeply incised. The canyons on the eastern side of the Rockall Trough are, as yet, poorly mapped but are known to exist even on the Porcupine Bank margin, which is relatively isolated from areas of fluvial input (Roberts, 1975). It has been noted before (e.g. Emery & Uchupi, 1984) that this concentration of canyons is near the southern limits of the Quaternary ice cover, and coarse sediment carried by meltwater may have fed them. Canyons were most active during lowstands of sea-level, because more material was transported to the canyon head at these times. Many canyons are largely inactive during interglacials, and can become partly infilled with pelagic and hemipelagic sediments. However, infrequent events, such as earthquakes, may trigger large turbidity currents at any time. For example, the 1755 Lisbon earthquake triggered a massive turbidity current that is believed to have travelled along the Setubal Canyon and out onto the Tagus Abyssal Plain (Thomson & Weaver, 1994). Location, morphology and spacing of canyons is controlled by a combination of sediment supply, slope angle, seafloor topography and structure (e.g. faults). On the west Portuguese margin, the canyons are dominantly controlled by the positions of river mouths which would have been closer to the canyon heads during lowstands of sea-level. Several canyons on this margin (e.g. Setubal Canyon) even cut back to the inner shelf, and feed directly from the lower reaches of major river valleys. The west Portuguese continental margin is steep and narrow, and the turbidity currents have built large abyssal plains at the foot of the slope (Milkert & Weaver, 1996; Lebreiro *et al.*, 1997). On the extensively canyoned north Biscay margin, the canyons are fed by cross-shelf glacial drainage channels, and by tidal shelf sand transport (Kenyon *et al.*, 1978; Reynaud *et al.*, 1999). The canyons are restricted to areas of steeper gradients with an angle of 4°–9°. Many of the canyons are aligned

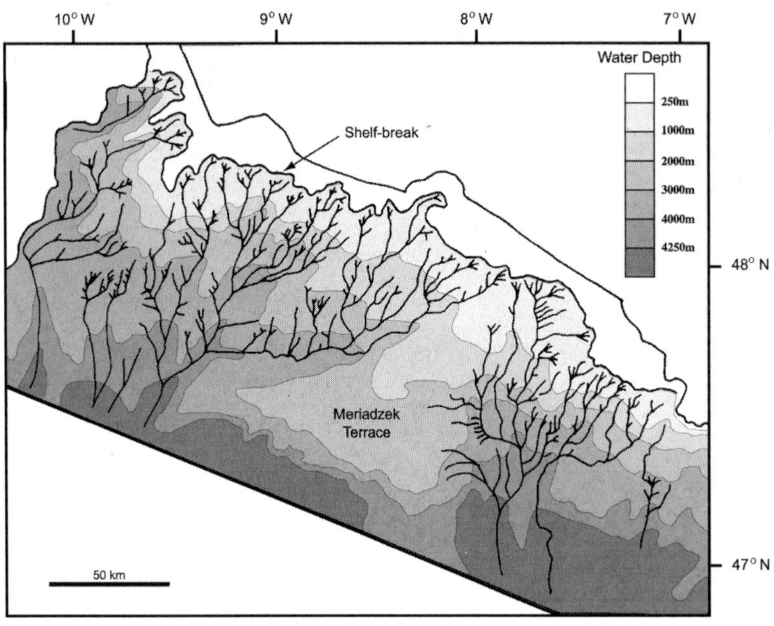

Fig. 5. Line diagram based on swath bathymetry, showing the detailed and complex morphology of canyons on the continental slope bordering the Celtic Sea Shelf (Biscay margin). Modified from Bourillet and Loubrieu (1995). For location, see Fig. 1.

oblique to the slope, and their morphology is clearly related to underlying basement topography and faulting (Kenyon *et al.*, 1978; Lallemand & Sibuet, 1986).

Most canyons in the study area have gullied walls and a number of minor tributary canyons feeding sediment from the continental shelf (Fig. 5). They merge down-slope into one major canyon before terminating directly onto a basin floor. Small distributary channels are known in a few cases to have developed beyond the canyon mouths, e.g. on the Celtic Fan (Droz *et al.*, 1999), and beyond the Gollum Canyon system in the Porcupine Seabight. The Gollum Canyon system has canyons up to 280 m deep and 1·5 km wide that develop down-slope into a dendritic pattern of slightly sinuous channels running across the floor of the Seabight (Kenyon *et al.*, 1978). Levees are absent from this channel system, probably for the same reasons as for the MDCS. Beyond the Seabight, distributary channels are only up to 40 m deep. The largest abyssal plains are found on this sector of the margin as demonstrated in Fig. 1 (which is an equal area projection). This is presumably because of higher turbidity current activity in this area, as a result of increased glaciomarine sediment supply at the edge of the Quaternary ice sheet.

Along-slope processes

The Mediterranean outflow water (MOW) is responsible for the creation of a wide range of bottom current features on the south and west Iberian margin (Kenyon & Belderson, 1973; Nelson *et al.*, 1999). Immediately west of the Straits of Gibraltar, the MOW flows at up to 250 cm s^{-1}, and erosional scours and sand ribbons cover the seafloor. Further west, the current velocity decreases to 40–75 cm s^{-1} and sand waves are developed. A progressive decrease in velocity to around 20 cm s^{-1} leads to the development of mud waves up to 40 m high. Further west, on the south and west Portuguese margin, the MOW is responsible for shaping the seafloor sediments into a series of contourite drifts down to depths of about 1500 m. These drifts are generally small and isolated, for example, the Faro Drift on the southern Portuguese margin is about 50 km long and 10–25 km wide (Faugeres *et al.*, 1985).

The glaciated margin north of 56°N (glacial and along-slope processes dominant)

Down-slope processes

There is a marked change in shelf and slope morphology where the Quaternary icecaps were in proximity to the margin (Fig. 2a). The shelves as far south as 56°N (west of Scotland) are crossed by troughs that were cut near to the shelf edge by fast-flowing ice streams, and the maximum southerly extent of icesheets during the glaciations was to southern Ireland. South of 56°N, the slope is dominated by canyons, probably formed by erosive turbidity currents fed by a plentiful

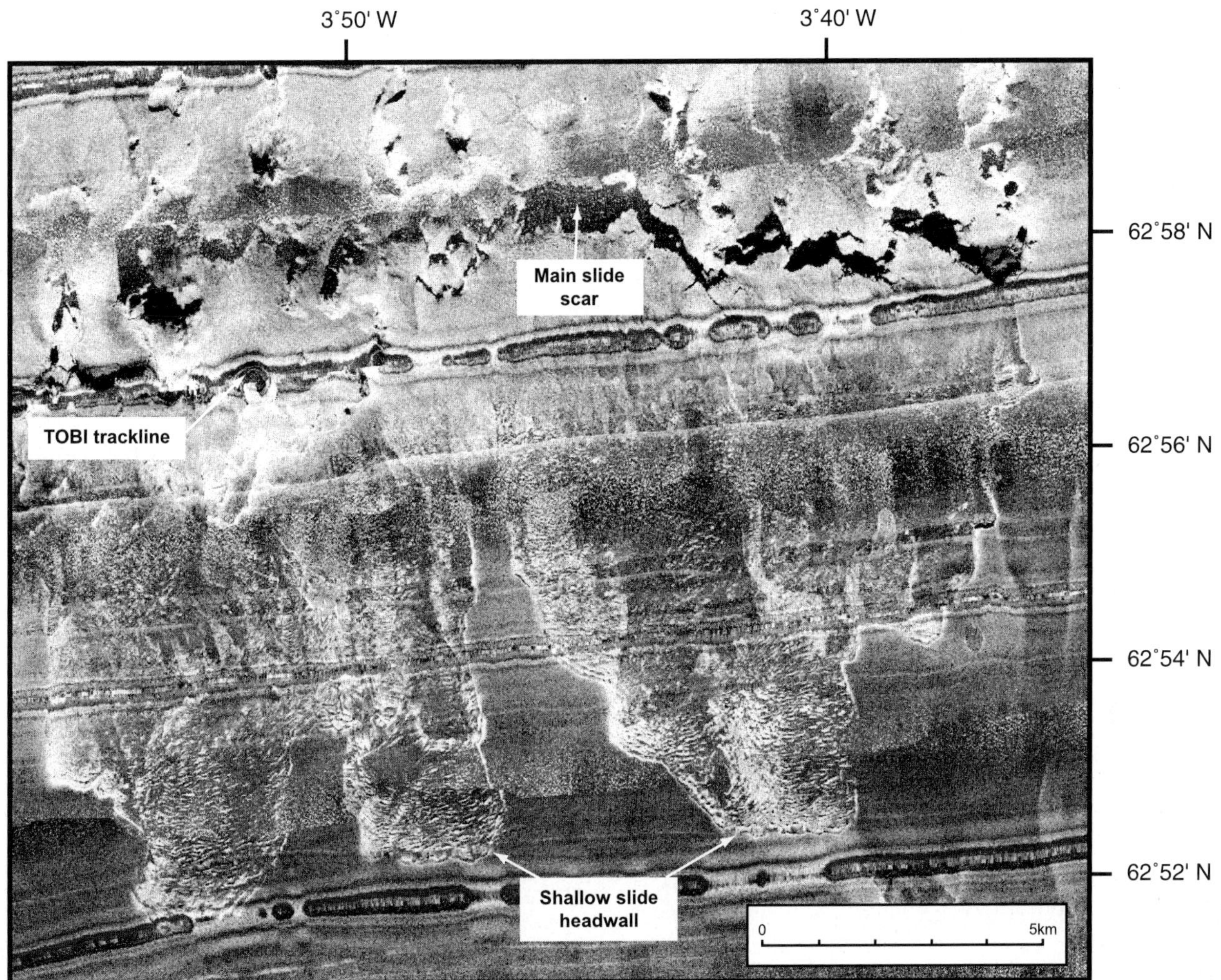

Fig. 6. TOBI side-scan sonar mosaic, showing an area of mass movements north of the Faeroe Islands. Shallow slides occur upslope of a major slide scar (seen as black shadows), representing a deeper failure. Down-slope is to the north. High levels of backscatter are shown as light tones. Modified from van Weering *et al.* (1998b). For location, see Fig. 1.

supply of coarse sediment. North of 56°N, there are few large canyons. The reasons for this are not established. Possibly, the glacial meltwater that must have carried large amounts of material across the shelf was largely diverted into the cross-shelf troughs and fed to the slope at these few point sources, causing sliding rather than turbidity currents and canyon cutting. It has also been suggested that the strong currents along the upper slope have prevented canyon formation, as is the case on the north-western Atlantic margin to the south of Cape Hatteras (Kenyon, 1987).

The geometry and processes of sedimentation along the margin in the Norwegian Sea have been reviewed by Vorren *et al.* (1998), and similar processes act along the margin west of Scotland. At the mouth of the cross-shelf troughs, there are usually trough-mouth fans, accumulations of sediment on the slope, which have a characteristic internal geometry very different from typical low-latitude turbidite fans. Of the two main small-scale geometries, the most common are stacks of glacial debris flows: long, narrow tongues of a fairly consistent size and composition that extend down-slope. Less frequent are classic debris flows which are usually much more extensive and varied in thickness (King *et al.*, 1996; Baltzer *et al.*, 1998).

Modelling of the flows within ice sheets by Dowdeswell and Siegert (1999) shows the predicted pattern of ice flux to the shelf edge along the Eurasian margin. Sediment flux transported by the icestreams has also been modelled and agrees well with the known distribution and size of trough mouth fans. The southernmost glacial

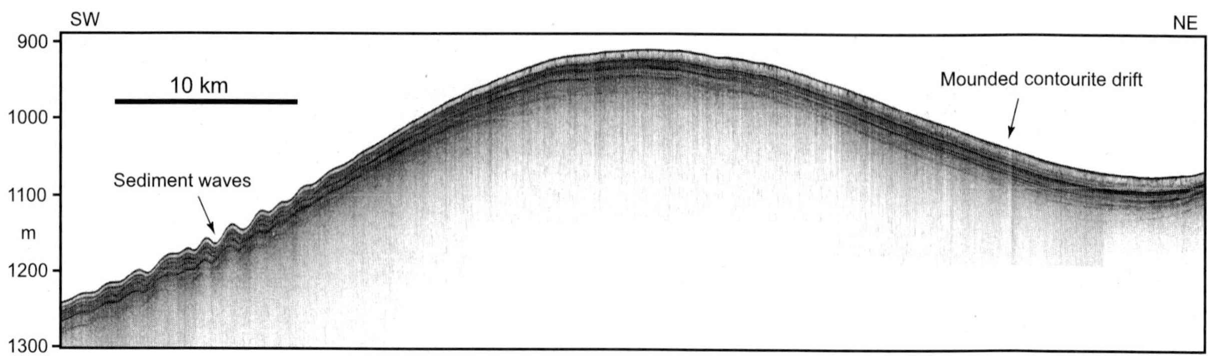

Fig. 7. 3·5 kHz profile, showing sediment waves on the flank of a sediment drift in the north-east Rockall Trough (59°45′N, 07°40′W). For location, see Fig. 1.

fan is the Barra Fan. The presence of extensive turbidite deposition at the foot of the Barra Fan (Roberts, 1975; Faugeres *et al.*, 1981) and the network of tributary gullies at the top of the fan (Armishaw *et al.*, 1998) are unusual glacial fan features, so this is possibly a transitional example of a glacial fan. A further example of a glacial fan west of Scotland is the Sula Sgeir Fan (Baltzer *et al.*, 1998). West of the Shetland Islands, there are no well-developed cross-shelf troughs and no major depocentres. The extensive areas of both classic and glacial debris flows form what could be called a slope apron.

The two major areas of glacial drainage result in the enormous North Sea Fan (King *et al.*, 1996) and the Bear Island Fan (Dowdeswell *et al.*, 1997; Vorren *et al.*, 1998). Both consist mainly of large numbers of stacked glacial debris flows. The Bear Island Fan is comparable in size with the largest low-latitude turbidite fans, such as the Mississippi Fan, which consists mainly of large numbers of stacked channel-levee complexes. Both the largest fans are modified by major slides. The slides either bound the fans, such as the Storegga and North Faeroes Slides (Fig. 6) on the northern and southern margins of the North Sea Fan (van Weering *et al.*, 1998a,b), or are on top of them, such as the Bjornoyrenna Slide (Laberg & Vorren, 1993) on top of the Bear Island Fan. The best-studied slide is the Storegga Slide (Bugge *et al.*, 1987, 1988; Evans *et al.*, 1996). This is a site of repeated failure. Less well-studied slides include that east of Rockall Bank (Roberts, 1972; Flood *et al.*, 1979), the Traenadjupet Slide (e.g. Dowdeswell *et al.*, 1996) and the Andoya Slide (Laberg *et al.*, 1999). It seems likely that seismic investigations will reveal a complex history of repeated failure for most of these slides. It has been suggested that these slides could be caused by detachment of slabs of sediment charged with solid gas hydrate, released as the seawater warms up (Bugge *et al.*, 1988; Mienert *et al.*, 1998). Bottom-simulating reflectors have been detected near the Storegga Slide and modelling predicts that the Norwegian Sea margin should be a likely place for gas hydrate formation (Miles, 1995).

Along-slope processes

Bottom currents have developed an extensive suite of contourite drifts and mud waves in the north-east Rockall Trough area (Fig. 7). In this region, there is a complex circulation of bottom water, and mixing between three or four different water masses may occur. Generally, bottom currents have a greater influence on sedimentation during interglacials, because much of the slope was affected by large-scale mass movements and turbidity currents during glacial periods. North-east of the Faeroe Islands, the Norwegian Sea deep water (NSDW) has formed contourite deposits on the upper slope (van Weering *et al.*, 1998a). Part of the NSDW then flows south through the Faeroe–Shetland Channel as Norwegian Sea overflow water (NSOW), before moving west over the Wyville–Thomson Ridge. It partially mixes with Labrador Sea water (LSW) and AABW to form North Atlantic deep water (NADW), which circulates as an anticlockwise gyre in the Rockall Trough. In this region, a number of contourite drifts have formed, and three distinct types can be recognized (Stoker *et al.*, 1994). These are as follows: distinctly mounded elongate drifts; broad sheet drifts; and isolated patch drifts (including moat-related drifts). Sediment waves occur on the flanks of these drifts, and have wave heights up to 20 m

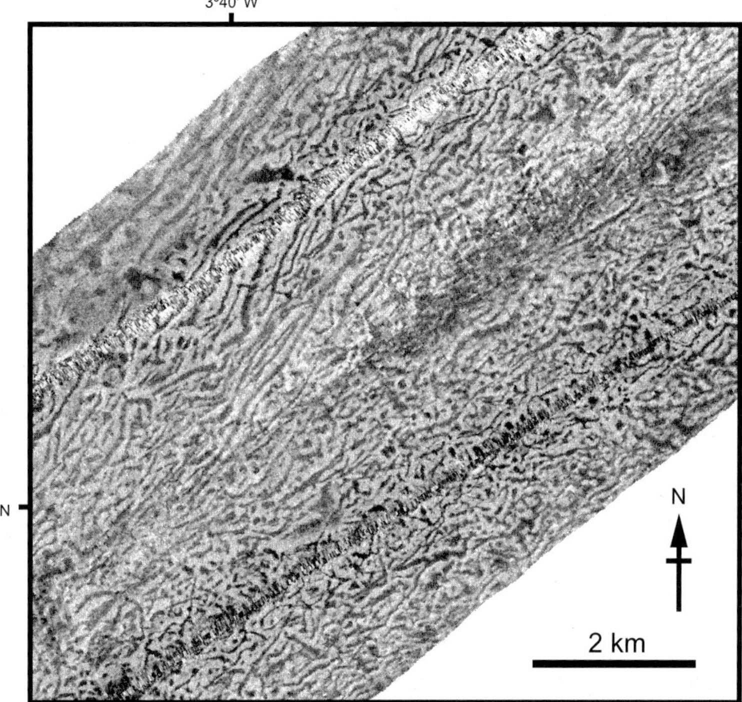

Fig. 8. TOBI side-scan sonar mosaic, showing iceberg plough marks on the upper slope west of Scotland (60°N, 03°W; Masson *et al.*, 1997). The pattern of plough marks changes from criss-crossing to parallel as the slope steepens to the north-west. High levels of backscatter are shown as light tones. For location, see Fig. 1.

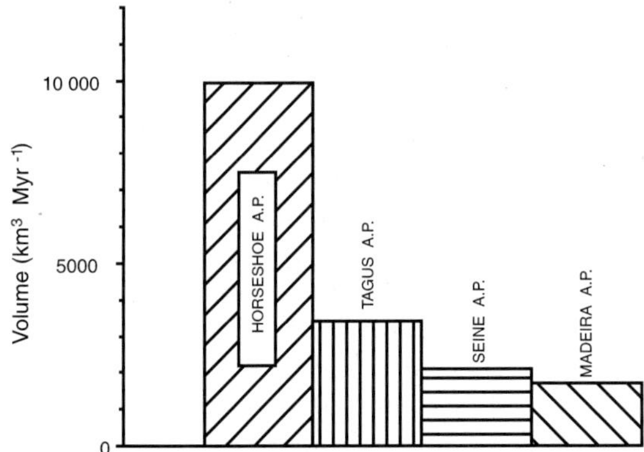

Fig. 9. Graph showing how the volume of turbidites per million years varies on four different abyssal plains. The volume of material contributed to the abyssal plains increases as one goes northwards, because the margin is both more tectonically active and has a greater sediment supply in this area.

and wavelengths up to 2 km (Fig. 6). Further south and west, adjacent to the Rockall Plateau, are the giant Feni and Hatton Drifts (McCave & Tucholke, 1986).

During glacial periods, the currents moved large icebergs around the north-east Atlantic. Where they impinge on the continental slope and outer shelf, they form patterns of iceberg plough marks (Fig. 8; Belderson *et al.*, 1973). They generally occur down to about 700 m, except where buried by later sedimentation. It is usual to find a dense network of criss-crossing iceberg plough marks on the shelf and upper slope, and isolated plough marks running along-slope in deeper water. The southernmost limit of known plough marks is west of Ireland.

Relatively strong currents are found both bounding the drifts and in other areas where topography enhances flow speeds, such as the gateway of NSDW across the Iceland–Faeroe–Scotland Ridge. These can winnow away the finer sediments and prevent deposition. Sandy bed-forms and thin sheet sands are found, such as those along the upper slope from Scotland to northern Norway (Kenyon, 1986), and in the Faeroe–Shetland Channel (Masson *et al.*, 1997). The exit from the Faeroe Bank Channel resembles the northern Gulf of Cadiz, in having small drifts bounded by coarse floored channels, some of which appear to be aggradational (Kenyon *et al.*, 1998).

DISCUSSION

Variations in sedimentation pattern along the margin

The map (Fig. 1) and the preceding description show that sedimentary processes vary along the length of the continental margin. There are three primary margin types: nonglaciated, glacially influenced and glaciated. Mass movements are common on the nonglaciated margin south of about 26°N. From here to 56°N, on the glacially influenced margin, there are few mass movements but canyons are common, cutting into a steep continental slope. North of 56°, on the glaciated margin, most of the prominent seafloor features were generated during the last ice age and are now being reshaped by along-slope currents. An exception to this simple model is the submarine sliding which has occurred off Norway during the Holocene (e.g. the Storegga Slide with a date of emplacement of 7100 years; Bondevik *et al.*, 1997), although these features may also be related to climate change, such as via earthquakes associated with isostatic rebound initiating the sliding.

Off north-west Africa (the nonglaciated margin), there is little sediment input from the continent, but high productivity along the upper slope produces accumulation rates of up to 147 m Myr^{-1}, in contrast to mid- and lower slope accumulation rates of 20 m Myr^{-1} (Ruddiman *et al.*, 1988). This leads to oversteepening of the slope and potential instability. Therefore, sediment sequences along the upper part of this margin show a series of hiatuses, each representing tens to hundreds of metres of missing sediment, and with durations of hundreds of thousands of years (Weaver, 1994). Further downslope, the sediment sequence contains a number of debris flow deposits and, at the base of the slope, the abyssal plain fills gradually with a series of megaturbidites. The large sediment slides off Norway operate in a similar fashion but, in this area, there are also substantial deposits derived from ice sheet transport. The rate of accumulation of these ice-derived sediments is very high and they rapidly build large fan-like bodies. At the present day, they are largely inactive and Holocene sedimentation has added only a thin veneer to the surface of the glacial fans.

The glacially influenced margin is characterized by a steep continental slope and relatively large abyssal plains which almost coalesce. Off the Iberian margin, frequent earthquakes are correlated with frequent turbidite deposition, although the frequency of turbidite emplacement seems to have been higher during glacial periods (Lebreiro *et al.*, 1997). Although there is limited information on the canyons and abyssal plains in the Bay of Biscay, the density of canyons is striking. This may be because of the large amounts of sediment which would have been transported across this margin during ice ages, and particularly during ice retreat. There was no ice sheet near the shelf edge in this area during the ice ages, but the inland drainage basins of the UK, France and Iberia would have been much more active in transporting sediment eroded by ice at these times. South of the Atlas Mountains, there would have been no influence of ice and, north of southern Ireland, grounded ice processes would have operated near the shelf edge, especially in front of cross-shelf troughs, producing stacked debris flow deposits. The high sediment inputs and steep, narrow continental slopes have led to most of the deposition being on the abyssal plains, such that these have grown to a very large size. Therefore, to summarize, the large size of these abyssal plains is a result of a combination of (1) high sediment supply from glacial outwash; (2) a narrow, steep slope and rise which leads to sediment bypassing; and (3) the high seismicity of the margin off south-west Iberia triggering frequent turbidity currents (Fig. 9).

One consequence of the differing margin physiographies is the distribution of sand from the numerous turbidity currents. For basins at the base of steep continental slopes, e.g. the Tagus and Horseshoe Basins off Portugal (Lebreiro *et al.*, 1997) and the Bay of Biscay (Droz *et al.*, 1999), turbidity currents have enough velocity to carry large volumes of sand to the abyssal plain. The sand is then dumped across a large area centred around the point of entry. On the Tagus Abyssal Plain, this sandy area forms a very low-relief fan with shallow channels, and covers approximately 20% of the total plain area. At the other extreme, the Madeira Abyssal Plain lies at the base of a very low-angle continental slope (1:300) and, in this case, the ratio of mud to sand is much higher (of the order of just a few per cent of sand). The sand component in these turbidity currents may be deposited in channels and intraslope basins on the continental slope and rise. Sands deposited on the plain are concentrated around distributary channel terminations (Wynn *et al.*, 2000a). This concentration of sand in one part of the plain produces laterally as well as vertically graded sediment units.

Turbidity currents can be generated by hyperpycnal discharge at the mouths of 'dirty' rivers. The flows will follow the basin floor if the plume density is greater than that of the surrounding salty water. Such flows rarely occur off large rivers but they may occur off small rivers with a high suspended-sediment load (Mulder & Syvitski, 1995). Such rivers may have been common on the steep, narrow land surrounding the Iberian Peninsula and, during periods of high glacial meltwater drainage, on the shelf to the north.

It is interesting to note that the spectacular meandering channels that are so common on the west African margin to the south of the study area are absent from the north-west African and west European margin. This is probably because of a combination of gradient and sediment composition. On the north-west African and west European margin, the outer shelf is dominated by coarse-grained sediments, so most canyons and channels are formed by sand-rich turbidity currents.

Long-term history of the margin

The abyssal plain turbidites provide an excellent record of large-scale mass-wasting events in the Canary Basin, because each major event here seems to produce both a debris flow deposited on the continental slope and rise, and a turbidity current that deposits on the abyssal plain. Therefore, drill-holes through the abyssal plain sequence can provide a record of the largest mass-wasting events that have occurred at a variety of locations on the basin margins. In this way ODP Legs 157 (Schmincke *et al.*, 1995) and 149 (Whitmarsh *et al.*, 1996), which drilled the Madeira and Iberia Abyssal Plains, respectively, each provided a history of their basin. In the case of the Iberia Abyssal Plain, the turbidite input and, hence, basin filling began at about 2·4 Ma, coincident with the onset of Northern Hemisphere glaciations (Milkert & Weaver, 1996). In the Madeira Abyssal Plain, turbidite sedimentation began abruptly in the middle Miocene, at a time when ice began to build up rapidly on Antarctica (Weaver *et al.*, 1998). This external climatic change could have caused significant reorganisation of sedimentation along the world's continental margins. It would have initiated a time of lowered sea-level, which may have allowed sediments to be deposited beyond the shelf edge on the upper slope for the first time. In addition, the build up of ice would have led to the production of cold bottom waters in Antarctica, which would, in turn, have caused a reorganization of ocean circulation. This may have led to increased upwelling and associated sedimentation along the West African margin (see Sarnthein *et al.*, 1982). Both these processes would have caused an oversteepening of the upper continental slope, possibly leading to mass wasting and initiation of organic-rich turbidites which reached the abyssal plains.

CONCLUSIONS

A range of sedimentary processes affect the north-east Atlantic continental margin from north-west Africa to northern Norway. Processes vary considerably along the margin, with some areas being dominated by down-slope and others by along-slope transport and deposition. In the more northern areas, there are also significant differences between the processes that operated during glacial periods and those that operate today. The main sectors of the margin and their respective dominant processes are as follows.

1 *The nonglaciated margin south of 26°N.* High rates of upwelling here produce elevated accumulation rates along the upper slope, although there is little fluvial input and canyons are rare. This sediment is subject to mass movements involving dislocation of hundreds to thousands of square kilometres of sediment, on timescales of a few tens of thousands of years. The resultant debris flows are deposited across the continental slope and rise, and large-scale turbidity currents transport hundreds of square kilometres of sediment to the abyssal plains.

2 *The glacially influenced margin from 30°N to 56°N.* In this area, there is more fluvial input and the margin is less prone to large-scale mass movements. A number of canyons funnel turbidity currents down the steep slopes to the extensive abyssal plains. Some canyons, such as the Agadir Canyon, have built high levees. The presence of a lithospheric plate boundary off south-west Iberia is associated with frequent earthquakes and has led to high rates of sediment transport to the deep sea via the canyon systems. In addition, there is a high sediment supply to the shelf edge, because a lot of sediment was transported by glacial outwash during and at the end of glacial periods. The Mediterranean outflow has built a series of sediment drifts in the Gulf of Cadiz and off south-west Portugal, although the along-slope transport is interrupted by a number of large canyons, some of which cut back into river mouths.

3 *The glaciated margin north of 56°N.* Here, the margin is heavily influenced by glaciomarine processes which build fans composed largely of

stacked debris flows. These occur in front of the fast-flowing ice streams that have cut cross-shelf troughs. There are iceberg plough marks on most of the upper slope and outer shelf. During interglacials, this part of the margin has been affected by along-slope currents. Giant submarine slides are again a prominent feature, particularly off Norway and the Faeroe Islands. Some of these slides have occurred during the Holocene, although high glacial sedimentation rates and failure at melting gas hydrate layers may have contributed to the instability.

ACKNOWLEDGEMENTS

This work was carried out as part of the European Union-supported STEAM programme (Contract MAS2-CT94-0083) and ENAM 2 programme (Contract MAS3-CT95-0003). RBW acknowledges PhD funding from the University of Southampton and the Southampton Oceanography Centre (Challenger Division). Tim le Bas is thanked for providing technical assistance. Finally, the manuscript benefited greatly from the detailed reviews of David Piper, William Morris and Ian Jarvis.

REFERENCES

Armishaw, J.E., Holmes, R.W. and **Stow, D.A.V.** (1998) Morphology and sedimentation on the Hebrides Slope and Barra Fan, NW UK continental margin. In: *Geological Processes on Continental Margins: Sedimentation, Mass-wasting and Stability* (Eds M.S. Stoker, D. Evans and A. Cramp), *Geol. Soc. London Spec. Publ.*, **129**, 81–104.

Baltzer, A., Holmes, R. and **Evans, D.** (1998) Debris flow on the Sula Sgeir Fan, NW of Scotland. In: *Geological Processes on Continental Margins: Sedimentation, Mass-wasting and Stability* (Eds M.S. Stoker, D. Evans and A. Cramp), *Geol. Soc. London Spec. Publ.*, **129**, 105–115.

Belderson, R.H., Kenyon, N.H. and **Wilson, J.B.** (1973) Iceberg plough marks in the Northeast Atlantic. *Palaeogeog., Palaeoclimatol., Palaeoecol.*, **13**, 215–224.

Bondevik, S., Svendsen, J.L. and **Mangerud, J.** (1997) Tsunami sedimentary facies deposited by the Storegga tsunami in shallow marine basins and coastal lakes, western Norway. *Sedimentology*, **44**, 1115–1131.

Bourillet, J.-F. and **Loubrieu, B.** (1995) *Atlantique nord-est, bathy-morphologie de la marge des entrees de la Manche. Echelle 1: 250 000*. IFREMER, Brest.

Bugge, T., Befring, S., Belderson, R.H., Eidvin, T., Jansen, E., Kenyon, N.H., Holtedahl, H. and **Sejrup, H.P.** (1987) A giant three-stage submarine slide off Norway. *Geo-Mar. Lett.*, **7**, 191–198.

Bugge, T., Belderson, R.H. and **Kenyon, N.H.** (1988) The Storegga Slide. *Phil. Trans. R. Soc. London Ser. A*, **325**, 357–388.

Clark, J.D., Kenyon, N.H. and **Pickering, K.T.** (1992) Quantitative analysis of the geometry of submarine channels: Implications for the classification of submarine fans. *Geology*, **20**, 633–636.

Coumes, F., Delteil, J., Gairaud, H., Ravenne, C. and **Cremer, M.** (1982) Cap Ferret Deep Sea Fan (Bay of Biscay). *AAPG Bull.*, **42**, 583–591.

Davies, T.L., van Neil, B., Kidd, R.B. and **Weaver, P.P.E.** (1997) High resolution stratigraphy and turbidite processes in the Seine Abyssal Plain, north-west Africa. *Geo-Mar. Lett.*, **17**, 147–153.

Dowdeswell, J.A. and **Kenyon, N.H.** (1997) Long-range side-scan sonar (GLORIA) imagery of the eastern continental margin of the glaciated polar North Atlantic. In: *Glaciated Continental Margins: an Atlas of Acoustic Images* (Eds T.A. Davies, T. Bell, A.K. Cooper, H. Josenhans, L. Ployak, A. Solheim, M.S. Stoker and J.A. Stravers), pp. 260–263. Chapman & Hall, London.

Dowdeswell, J.A. and **Siegert, M.J.** (1999) Ice-sheet numerical modeling and marine geophysical measurements of glacier-derived sedimentation on the Eurasian Arctic continental margins. *Geol. Soc. Am. Bull.*, **111**, 1080–1097.

Dowdeswell, J.A., Kenyon, N.H., Elverhoi, A., Laberg, J.S., Hollender, F.-J., Mienert, J. and **Siegert, M.J.** (1996) Large-scale sedimentation on the glacier-influenced Polar North Atlantic margins: Long-range side-scan sonar evidence. *Geophys. Res. Lett.*, **23**, 3535–3538.

Dowdeswell, J.A., Kenyon, N.H., Laberg, J.S. and **Elverhoi, A.** (1997) Submarine debris flows on glacier-influenced margins: GLORIA imagery of the Bear Island fan. In: *Glaciated Continental Margins: an Atlas of Acoustic Images* (Eds T.A. Davies, T. Bell, A.K. Cooper, H. Josenhans, L. Ployak, A. Solheim, M.S. Stoker and J.A. Stravers), pp. 118–119. Chapman & Hall, London.

Droz, L., Auffret, G.A., Savoye, B. and **Bourillet, J.-F.** (1999) The Celtic Deep-sea fan: stratigraphy and sedimentary evolution. *CR Acad. Sci. Paris. Earth Planet. Sci.*, **328**, 173–180.

Embley, R.W. (1982) Anatomy of some Atlantic margin sediment slides and some comments on ages and mechanisms. In: *Marine Slides and Other Mass Movements* (Eds S. Saxov and J.K. Nieuwenhuis), pp. 189–213. Plenum, New York.

Embley, R.W. and **Jacobi, R.D.** (1977) Distribution and morphology of large submarine sediment slides and slumps on Atlantic continental margins. *Mar. Geotechnol.*, **2**, 205–228.

Emery, K.O. and **Uchupi, E.** (1984) *The Geology of the Atlantic Ocean*. Springer, New York.

Ercilla, G., Alonso, B., Perez-Belzuz, F., Estrada, F., Baraza, G., Farran, M., Canals, M. and **Masson, D.** (1998) Origin, sedimentary processes and depositional evolution of the Agadir turbidite system (Central Eastern Atlantic). *J. Geol. Soc. London*, **155**, 929–939.

Evans, D., King, E.L., Kenyon, N.H., Brett, C. and **Wallis, D.** (1996) Evidence for long-term instability in the Storegga Slide region off Western Norway. *Mar. Geol.*, **130**, 281–292.

Faugeres, J.C., Gonthier, E., Grousset, F. and **Poutiers, J.** (1981) The Feni Drift: The importance and meaning of slump deposits on the eastern slope of the Rockall Bank. *Mar. Geol.*, **40**, 49–57.

Faugeres, J.C., Cremer, M., Monteiro, H. and **Gaspar, L.** (1985) Essai de reconstitution des prosessus d'edification de la ride sedimentaire de Faro (Marge sud-Portugaise). *Bull. Inst. Géol. Bassin Aquit.*, **37,** 229–258.

Faugeres, J.-C., Mezerais, M.-L. and **Stow, D.A.V.** (1993) Contourite drift types and their distribution in the North and South Atlantic Ocean basins. *Sedim. Geol.*, **82,** 189–203.

Flood, R.D. (1988) A lee wave model for deep-sea mudwave activity. *Deep-Sea Res.*, **35,** 973–983.

Flood, R.D., Hollister, C.D. and **Lonsdale, P.** (1979) Disruption of the Feni sediment drift by debris flows from Rockall Bank. *Mar. Geol.*, **32,** 311–334.

Hampton, M.A., Locat, J. and **Lee, H.J.** (1996) Submarine landslides. *Rev. Geophys.*, **34,** 33–59.

Holmes, R., Long, D. and **Dodd, L.R.** (1998) Large-scale debrites and submarine landslides on the Barra Fan, west of Britain. In: *Geological Processes on Continental Margins: Sedimentation, Mass-wasting and Stability* (Eds M.S. Stoker, D. Evans and A. Cramp), *Geol. Soc. London Spec. Publ.*, **129**, 67–79.

Jacobi, R.D. and **Hayes, D.E.** (1984) Echo-character, microphysiography and geologic hazards. In: *Northwest African Continental Margin and Adjacent Ocean Floor Off Morocco* (Eds D.E. Hayes, P.D. Rabinowitz and K. Hinz), *Ocean Drilling Program Regional Atlas Series,* Atlas 12, 14. Marine Sci. Internat., Woods Hole, MA.

Jacobi, R.D. and **Hayes, D.E.** (1992) Northwest African continental rise: Effects of near-bottom processes inferred from high-resolution seismic data. In: *Geologic Evolution of Atlantic Continental Rises* (Eds C.W. Poag and P.C. de Graciansky), pp. 293–325. Van Nostrand Reinhold, New York.

Kenyon, N.H. (1986) Evidence from bedforms for a strong poleward current along the upper continental slope of north-west Europe. *Mar. Geol.*, **72,** 187–198.

Kenyon, N.H. (1987) Mass wasting features on the continental slope of north-west Europe. *Mar. Geol.*, **74,** 57–77.

Kenyon, N.H. and **Belderson, R.H.** (1973) Bed forms of the Mediterranean undercurrent observed with side-scan sonar. *Sedim. Geol.*, **9,** 77–99.

Kenyon, N.H., Belderson, R.H. and **Stride, A.H.** (1978) Channels, canyons and slump folds on the continental slope between south-west Ireland and Spain. *Oceanol. Acta*, **1,** 369–380.

Kenyon, N.H., Ivanov, M.K. and **Akhmetzhanov, A.M.** (Eds) (1998) Cold water carbonate mounds and sediment transport on the north-east Atlantic margin: Preliminary results of geological and geophysical investigations during the TTR-7 cruise of R/V Professor Logachev in co-operation with the CORSAIRES and ENAM 2 programmes, July–August 1997. *IOC Tech. Ser.*, **52**, 178pp.

King, E.L., Serjup, H.P., Haflidason, H., Elverhoi, A. and **Aarseth, I.** (1996) Quaternary seismic stratigraphy of the North Sea Fan: glacially-fed gravity flow aprons, hemipelagic sediments, and large submarine slides. *Mar. Geol.*, **130,** 293–315.

Komar, P.D. (1971) Hydraulic jumps in turbidity currents. *Geol. Soc. Am. Bull.*, **82,** 1477–1488.

Kuijpers, A., Andersen, M.S., Kenyon, N.H., Kunzendorf, H. and **van Weering, T.C.E.** (1998) Quaternary sedimentation and Norwegian Sea overflow pathways around Bill Bailey Bank, northeastern Atlantic. *Mar. Geol.*, **152,** 101–127.

Laberg, J.S. and **Vorren, T.O.** (1993) Late Pleistocene submarine slide on the Bear Island trough mouth fan. *Geo-Mar. Lett.*, **13,** 227–234.

Laberg, J.S., Vorren, T.O. and **Knutsen, S.-M.** (1999) The Lofoten contourite drift off Norway. *Mar. Geol.*, **159,** 1–6.

Lallemand, S. and **Sibuet, J.-C.** (1986) Tectonic implications of canyon directions over the northeast Atlantic continental margin. *Tectonics*, **5,** 1125–1143.

Lebreiro, S.M., McCave, I.N. and **Weaver, P.P.E.** (1997) Late quaternary emplacement of turbidites on the Horseshoe Abyssal Plain. *J. Sedim. Res.*, **67,** 856–870.

Masson, D.G. (1994) Late quaternary turbidity current pathways to the Madeira Abyssal Plain and some constraints on turbidity current mechanisms. *Basin Res.*, **6,** 17–33.

Masson, D.G. (1996) Catastrophic collapse of the volcanic island of Hierro 15 ka ago and the history of landslides in the Canary Islands. *Geology*, **24,** 231–234.

Masson, D.G., Kidd, R.B., Gardner, J.V., Huggett, Q.J. and **Weaver, P.P.E.** (1992) Saharan continental rise: Facies distribution and sediment slides. In: *Geologic Evolution of Atlantic Continental Rises* (Eds C.W. Poag and P.C. de Graciansky), pp. 327–343. Van Nostrand Reinhold, New York.

Masson, D.G., Bett, B.J. and **Birch, K.G.** (1997) Atlantic margin environmental survey. *Sea Technol.*, **38,** 52–59.

McCave, I.N. and **Tucholke, B.E.** (1986) Deep current-controlled sedimentation in the western North Atlantic. In: *Geology of North America*, Vol. M. *The Western North Atlantic Region* (Eds P.R. Vogt and B.E. Tucholke), pp. 1117–1126. Geol. Soc. Am., Boulder, CO.

Mienert, J., Posewand, J. and **Baumann, M.** (1998) Gas hydrates along the northeastern Atlantic margin: possible hydrate bound margin instabilities and possible release of methane. In: *Gas Hydrates: Relevance to World Margin Stability and Climate Change* (Eds J.P. Henriet and J. Mienert), *Geol. Soc. London Spec. Publ.*, **137**, 275–291.

Miles, P.R. (1995) Potential distribution of methane hydrate beneath the European continental margins. *Geophys. Res. Lett.*, **22,** 3179–3182.

Milkert, D. and **Weaver, P.P.E.** (1996) Pleistocene and Pliocene turbidites from the Iberia Abyssal Plain drilled during ODP Leg 149. *Proc. ODP Init. Reports*, **149,** 281–294. Ocean Drilling Program, College Station, TX.

Mohrig, D., Elverhoi, A. and **Parker, G.** (1999) Experiments on the relative mobility of muddy subaqueous and subaerial debris flows, and their capacity to remobilize antecedent deposits. *Mar. Geol.*, **154,** 117–129.

Mougenot, D. (1985) Progradation on the Portuguese continental margin: interpretation of seismic facies. *Mar. Geol.*, **69,** 113–130.

Mulder, T. and **Syvitski, P.M.** (1995) Turbidity currents generated at river mouths during exceptional discharges to the World Ocean. *J. Geol.*, **103,** 285–299.

Nakajima, T., Satoh, M. and **Okamura, Y.** (1998) Channel-levee complexes, terminal deep sea fan and sediment wave fields associated with the Toyama deep-sea channel system in the Sea of Japan. *Mar. Geol.*, **147,** 25–41.

Nelson, H.C., Baraza, J., Maldonado, A., Rodero, J., Escutia, C. and **Barber, J.H. Jr** (1999) Influence of the Atlantic inflow and Mediterranean outflow currents on Late Quaternary sediment facies of the Gulf of Cadiz continental margin. *Mar. Geol.*, **155,** 99–129.

Normark, W.R. and **Piper, D.J.W.** (1991) Initiation processes and flow evolution of turbidity currents. *Implications for the Depositional Record. SEPM Spec. Publ.*, **46,** 207–230.

Pilkey, O.H. (1987) Sedimentology of basin plains. In: *Geology and Geochemistry of Abyssal Plains* (Eds P.P.E. Weaver and J. Thomson), *Geol. Soc. London Spec. Publ.*, **31,** 1–12.

von Rad, U. and **Wissmann, G.** (1982) Cretaceous-Cenozoic history of the West Saharan continental margin (NW Africa): Development, destruction and gravitational sedimentation. In: *Geology of the Northwest African Continental Margin* (Eds U. von Rad, K. Hinz, M. Sarnthein and E. Seibold), pp. 106–131. Springer, Berlin.

Reynaud, J.-C., Tessier, B., Proust, J.-N., Proust, D., Bourillet, J.-F., de Batist, M., Lericolais., B.e.r.n.e., S. and Marsset, T. (1999) Architecture and sequence stratigraphy of a late Neogene incised valley at the shelf margin, Southern Celtic Sea. *J. Sedim. Res.*, **69**, 351–364.

Roberts, D.G. (1975) Marine geology of the Rockall Plateau and Trough. *Phil. Trans. R. Soc. London. Ser. A.*, **278**, 447–509.

Roberts, D.G. (1972) Slumping on the eastern margin of the Rockall Bank, North Atlantic Ocean. *Mar. Geol.*, **13**, 225–237.

Roberts, D.G. and **Kidd, R.** (1979) Abyssal sediment wave fields on Feni Ridge, Rockall Trough: Long range sonar studies. *Mar. Geol.*, **33**, 175–191.

Rothwell, R.G., Pearce, T.J. and **Weaver, P.P.E.** (1992) Late Quaternary evolution of the Madeira Abyssal Plain, Canary Basin, NE Atlantic. *Basin Res.*, **4**, 103–131.

Ruddiman, W.F., Sarnthein, M. and **Baldauf, J.** (1988) *Proc. ODP, Init. Rep. Leg 108.* Ocean Drilling Program, College Station, TX.

Sarnthein, M., Thiede, J. and **Pflaumann, U.** (1982) Atmospheric and ocean circulation patterns off Northwest Africa during the past 25 million years. In: *Geology of the Northwest African Continental Margin* (Eds U. von Rad, K. Hinz, M. Sarnthein and E. Seibold), pp. 545–604. Springer, Berlin.

Schmincke, H.-U., Weaver, P.P.E., Firth, J.V. *et al.* (1995). *Proc. ODP Init. Rep. Leg 157.* Ocean Drilling Program, College Station, TX.

Simm, R.W. and **Kidd, R.B.** (1984) Submarine debris flow deposits detected by long-range side-scan sonar 1,000-kilometers from source. *Geo-Mar. Lett.*, **3**, 13–16.

Stoker, M.S. (1995) The influence of glacigenic sedimentation on slope-apron development on the continental margin off northwest Britain. In: *The Tectonics, Sedimentation and Palaeoceanography of the North Atlantic Region* (Eds R.A. Scrutton, M.S. Stoker, G.B. Shimmield and A.W. Tudhope), *Geol. Soc. London Spec. Publ.*, **90**.

Stoker, M.S., Leslie, A.B., Scott, W.D., Briden, J.C., Hine, N.M., Harland, R., Wilkinson, I.P., Evans, D. and **Ardus, D.A.** (1994) A record of Late Cenozoic stratigraphy, sedimentation and climate change from the Hebrides Slope, NE Atlantic Ocean. *J. Geol. Soc. London.*, **151**, 235–249.

Thomson, J. and **Weaver, P.P.E.** (1994) An AMS radiocarbon method to determine the emplacement time of recent deep-sea turbidites. *Sedim. Geol.*, **89**, 1–7.

Vorren, T.O., Laberg, J.S., Blaume, F., Dowdeswell, J.A., Kenyon, N.H., Mienert, J., Rumohr, J. and **Werner, F.** (1998) The Norwegian–Greenland Sea continental margins: morphology and late Quaternary sedimentary processes and environment. *Quat. Sci. Rev.*, **17**, 273–302.

Weaver, P.P.E. (1994) Determination of turbidity current erosional characteristics from reworked coccolith assemblages, Canary Basin, N. E. Atlantic. *Sedimentology*, **41**, 1025–1038.

Weaver, P.P.E., Rothwell, R.G., Ebbing, J., Gunn, D.E. and **Hunter, P.M.** (1992) Correlation, frequency of emplacement and source directions of megaturbidites on the Madeira Abyssal Plain. *Mar. Geol.*, **109**, 1–20.

Weaver, P.P.E., Masson, D.G., Gunn, D.E., Kidd, R.B., Rothwell, R.G. and **Maddison, D.A.** (1995) Sediment mass wasting in the Canary Basin. In: *Atlas of Deep Water Environments: Architectural Style in Turbidite Systems* (Eds K.T. Pickering, R.N. Hiscott, N.H. Kenyon, F. Ricci Lucchi and R.D.A. Smith), pp. 287–229. Chapman & Hall, London.

Weaver, P.P.E., Jarvis, I., Lebreiro, S.M., Alibes, B., Baraza, J., Howe, R. and Rothwell, R.G. (1998) Neogene turbidite sequence on the Madeira Abyssal Plain: basin filling and early diagenesis in the deep ocean. *Proc. ODP Sci. Res.*, **157**, 619–634. Ocean Drilling Program, College Station, TX.

van Weering, T.C.E., Nielsen, T., Kenyon, N.H., Akentieva, K. and **Kuijpers, A.H.** (1998a) Large submarine slides on the NE Faeroe continental margin. In: *Geological Processes on Continental Margins: Sedimentation, Mass-Wasting and Stability* (Eds M.S. Stoker, D. Evans and A. Cramp), *Geol. Soc. London Spec. Publ.*, **129**, 5–17.

van Weering, T.C.E., Nielsen, T., Kenyon, N.H., Akentieva, K. and **Kuijpers, A.H.** (1998b) Sediments and sedimentation at the NE Faeroe continental margin; contourites and large-scale sliding. *Mar. Geol.*, **152**, 159–176.

Wessel, P. and **Smith, W.H.F.** (1991) Free software helps map and display data, EOS Trans. *Am. Geophys. Union*, **72**, 445–446.

Whitmarsh, R.B., Sawyer, S., Klaus, A. and **Masson, D.G.** (1996) Iberia Abyssal Plain: covering Leg 149 of the cruises of the drilling vessel 'Joides Resolution', Balboa Harbor, Panama, to Lisbon, Portugal, Sites 897–901. *Proc. ODP Sci. Res.*, **149**, 785pp. Ocean Drilling Program, College Station, TX.

Wynn, R.B., Masson, D.G., Stow, D.A.V. and **Weaver, P.P.E.** (2000a) The northwest African slope apron: a modern analogue for deep water systems with complex seafloor topography. *Mar. Petrol. Geol.*, **17**, 253–265.

Wynn, R.B., Masson, D.G., Stow, D.A.V. and **Weaver, P.P.E.** (2000b) Turbidity current sediment waves on the submarine slopes off the western Canary Islands. *Mar. Geol.*, **163**, 185–198.